1 MONTH OF
FREE
READING

at

www.ForgottenBooks.com

By purchasing this book you are eligible for one month membership to ForgottenBooks.com, giving you unlimited access to our entire collection of over 700,000 titles via our web site and mobile apps.

To claim your free month visit:
www.forgottenbooks.com/free59568

ISBN 978-1-5279-6469-3
PIBN 10059568

BOTANICON SINICUM

Part III

BOTANICAL INVESTIGATIONS INTO THE MATERIA
MEDICA OF THE ANCIENT CHINESE

BOTANICON SINICUM

NOTES ON CHINESE BOTANY

FROM NATIVE AND WESTERN SOURCES

BY

E. BRETSCHNEIDER, M.D.

Late Physician to the Russian Legation at Peking

Membre corresp. de l'Institut de France

(Académie des Inscriptions et Belles Lettres)

Part III

ANICAL INVESTIGATIONS INTO THE MATERI MEDICA OF THE ANCIENT CHINESE.

SHANGHAI, HONGKONG, YOKOHAMA & SINGAPORE

KELLY & WALSH, LIMITED

1895

6728
5 3

CONTENTS.

		Pages.
Introduction		1
Abbreviated References to Chinese, Japanese and European Books ...		*v*
Medicinal Plants of the *Shen nung Pen ts'ao king* and the *Pie lu* ...		13
Appendix :—		
Chinese Geographical Names		547
Alphabetical Index of Chinese Names of Plants		606
Alphabetical Index of Genus Names of Plants		616

BOTANICON SINICUM.

BOTANICAL INVESTIGATIONS INTO THE MATERIA MEDICA OF THE ANCIENT CHINESE.

———◆———

INTRODUCTION.

———

In connection with a former paper dealing with the economic plants known to the Chinese in the classical period, and forming the second part of the *Botanicon sinicum*,[1] the author of those notes now attempts to examine and identify the drugs of vegetable origin noticed in the earliest Chinese works on Materia Medica,—the *Shen nung Pen ts'ao king* and the *Ming i pie lu*.

The first of these works, the 神農本草經 *Shen nung Pen ts'ao king*, or Herbal of the Emperor SHEN NUNG, of which a detailed notice has been given in Part I of the *Botanicon sinicum* [p. 27 seqq.] has, as the word *king* (classic) in the title indicates, always been and is still considered by Chinese practitioners a book of the highest authority and a model of pharmacological wisdom. Therefore most of the drugs mentioned in this ancient pharmacopœia are still kept in store and sold for medical use, and are still known by the same names as they appear in that ancient book.

[1] See *Journ. China Br. Asiat. Soc.*, Vol. XXV.

Although the authorship of this work has always been ascribed to the legendary Emperor SHEN NUNG [B.C. in the 28th century], there is internal evidence in it, at least in that which was current with the above title in the 5th century, that it had been compiled in the Han period [B.C. 202–A.D. 221], but presumably from earlier traditions on the subject.

The *Shen nung Pen ts'ao king* or *Pen ts'ao king* (Herbal Classic), also simply termed *Pen king*, was originally a book treating of 365 different drugs, in accordance with the number of days in the year, arranged in three classes according to their medical virtues. LI SHI-CHEN in his *Pen ts'ao kang mu* [Chap. IV] gives the Index of the original work, in which appear 252 names of vegetable drugs. These are nearly all spoken of and commented upon in the *Pen ts'ao kang mu*, and all that is known regarding the drugs of the *Pen ts'ao king* is from the quotations found in LI SHI-CHEN's Materia Medica, from which it appears that the *Pen ts'ao king* gives only particulars regarding the mode of preparing the officinal parts of the plants for medical use, their specific virtues and their therapeutic use. It is quite exceptional to find in this ancient book any descriptive details with respect to the plants from which the drugs are derived.

The 名醫別錄 *Ming i pie lu*, called also simply *Pie lu*, is a supplement to the *Shen nung Pen ts'ao*, adding to the original Materia Medica 365 more drugs, employed by eminent physicians in the Han and Wei periods. [The Wei dynasty reigned A.D. 221–264.] In the first part of the *Botanicon sinicum* [p. 42] I have said that this work, as is indeed stated by LI SHI-CHEN in his account of it, was compiled by 陶宏景 T'AO HUNG-KING, who lived A.D. 452–536. But the frequent quotations from it in the *Pen ts'ao kang mu*, together with T'AO HUNG-KING's commentaries thereupon, prove that the *Pie lu* was an independent treatise which

existed before his day, not a work of his compilation.   He as
well as other ancient authors when referring to the *Pie lu*
call it sometimes *Pen ts'ao king*, and thus seem to comprise
both the *Shen nung Pen ts'ao* and the *Pie lu* under this
general appellation of Herbal Classic.   In the History of the
Sui dynasty [A.D. 589-618], Chap. 36, section on Literature,
we find the title of a work 陶宏景本草經集注, *i.e.* the
Pen ts'ao king, collected and explained by T'AO HUNG-KING.
This work contained probably the *Shen nung Pen ts'ao* and
the *Pie lu* with the commentary of T'AO HUNG-KING.

The *Pie lu* is an enlargement of the Herbal of SHEN NUNG.
We meet in it with notices of all the plants mentioned in
the earlier work to which an account of the drugs used in
the Han and Wei periods is added.   These accounts are very
short, giving only in a few cases descriptive details of the
respective drugs (plants).   But the provinces or districts
where the drug in question is produced are generally
indicated.   Nearly all these geographical names refer to the
Ts'in [3rd cent. B.C.] or Han periods, although some of
them can be traced to the Chou dynasty [B.C. 1122-249].
In a few cases they cannot be ascertained.   The part of the
plant which is used in medicine and the time of gathering
it are also noticed.   The *Pie lu* uses generally four phrases
to distinguish the localities in which the plants grow, *viz.* :—

生山谷 it grows in mountain valleys (in the mountains).

生川谷 it grows in river valleys (the plain is probably
　　　　 meant, meadows).

生田野 it grows in fields.

生平澤 it grows in level marshes (low marshy land).

As detailed accounts of the *Pen ts'ao kang mu*, the great
repertory of Chinese Materia Medica, published by LI SHI-
CHEN in the second half of the 16th century, and of the authors
and books quoted in this important work have been given

in the first part of the *Botanicon sinicum*, I shall therefore confine myself here to an enumeration of such books and authors' names as appear more frequently in the following pages.

There are first two ancient treatises on Materia Medica 1. (2)[2] the 采藥錄 *Ts'ai yao lu* and 2. (3) the 雷公 藥對 *Lei kung Yao tui*, the compilation of which tradition refers to the time of the legendary Emperor HUANG TI [B.C. 27th cent.]. The *Ts'ai yao lu*, or directions for gathering drugs, is ascribed to 桐君 T'UNG KÜN, one of the ministers of HUANG TI, the other, the Materia Medica of LEI KUNG, to one of the sages who assisted the emperor in his investigations into the Art of Healing.

The next in order are two works on the same subject :— 3. (4) the 李氏藥錄 *Li shi Yao lu*, by 李當之 LI TANG-CHI, and 4. (5) the 吳氏本草 *Wu shi Pen ts'ao*, by 吳普 WU P'U, both written in the first half of the third century.

5. (6).—The 炮炙論 *P'ao chi lun*, a work explaining the medical virtues of drugs, written about the middle of the 5th century, by 雷公 LEI KUNG or properly 雷斅 LEI HIAO.

6. (9).—The 千金食治 *Ts'ien kin shi chi*, by 孫思邈 SUN SZ'-MO [beginning of 7th cent.].

7. (10).—The 藥性本草 *Yao sing pen ts'ao* and the 藥性論 *Yao sing lun*, two works, both by 甄權 CHEN KUAN [6th and 7th cent.].

8. (11).—The 唐本草 *T'ang Pen ts'ao*, by 蘇恭 SU KUNG [7th cent.].

9. (12).—The 食療本草 *Shi liao pen ts'ao*, by 孟詵 MENG SHEN [second half of 7th cent.].

[2] The figures in parentheses refer to *Botanicon sinicum*, I, p. 40-54.

10. (13).—The 本草拾遺 *Pen ts'ao shi i*, by 陳藏器 Ch'en Ts'ang-k'i [first half of 8th cent.].

11. (14).—The 海藥本草 *Hai yao pen ts'ao*, by 李珣 Li Sün [second half of 8th cent.].

12. (15).—The 四聲本草 *Sz' sheng pen ts'ao*, by 蕭炳 Siao Ping. [T'ang period, 7th to 9th cent.].

13. (18).—The 本草性事類 *Pen ts'ao sing shi lei*, by 杜善方 Tu Shan-fang [T'ang period].

14. (19).—The 食性本草 *Shi sing pen ts'ao*, by 陳士瓨 Ch'en Shi-liang [10th cent.].

15. (20).—The 蜀本草 *Shu pen ts'ao*, by 韓保昇 Han Pao-sheng [10th cent.].

16. (21).—The 日華諸家本草 *Ji hua Chu kia pen ts'ao*, by 大明 Ta Ming [A.D. 970].

17. (22).—The 開寶本草 *K'ai pao Pen ts'ao*, by 馬志 Ma Chi [second half of 10th cent.].

18. (23).—The 嘉祐補註本草 *Kia yu Pu chu Pen ts'ao*, by 掌禹錫 Chang Yü-si and 林億 Lin I [A.D. 1057].

19. (24).—The 圖經本草 *T'u king pen ts'ao*, by 蘇頌 Su Sung [end of 11th cent.].

20. (26).—The 證類本草 *Cheng lei pen ts'ao*, called also 大觀本草 *Ta kuan Pen ts'ao*, by 唐慎微 T'ang Shen-wei [A.D. 1108].

21. (27).—The 本草衍義 *Pen ts'ao yen i*, by 寇宗奭 K'ou Tsung-shi [A.D. 1115].

22. (29).—The 用藥法象 *Yung yao fa siang*, by 李杲 Li Kao, called also 明之自 Ming chi-tsz' and 東垣 Tung yüan [12th and 13th cent.].

23. (30).—The 湯液本草 *T'ang i pen ts'ao*, by 王好古 Wang Hao-ku, called also 海藏 Hai Ts'ang and 進之 Tsin chi [first half of 13th cent.].

24. (31).—The 日用本草 *Ji yung pen ts'ao*, by 吳瑞 Wu Shui. [Mongol period, 13th and 14th cent.].

25. (33).—The 本草衍義補遺 *Pen ts'ao yen i pu i*, by 朱震享 Chu Chen-heng [second half of 14th cent.].

26. (40).—The 本草會編 *Pen ts'ao hui pien*, by 汪機 Wang Ki [16th cent.].

In the subsequent account of the vegetable drugs mentioned in the *Shen nung Pen ts'ao* and the *Pie lu* the reader will find them treated of in the same order as in the text of the *Pen ts'ao kang mu*, where the names of drugs first given in the *Shen nung Pen ts'ao* are always placed at the head of the respective articles. The principal object kept in view by the author in extracting the following notes from Li Shi-chen's work, is the botanical identification of the drugs of vegetable origin mentioned in the ancient Chinese works on Materia Medica. Notice is therefore taken only of such details in the ancient descriptions of drugs and plants as may be serviceable to this end. Statements of no interest for European readers have generally been omitted.

The style used by the ancient Chinese authors in describing plants is generally very simple, but owing to the vagueness of the expressions and terms, the translator often meets with great difficulty, and is constrained to guess should the plant described be unknown to him. Thus the character 子 *tsz'*, which means "seed," is frequently used in ancient books for 實 *shi*, fruit, and the latter character again often occurs there with the meaning "solid," opposed to hollow. 莖 *heng* is the stem of herbaceous plants, but it is also used for petiole and for 幹 *kan*, the trunk of a tree. 苗 *miao*, which

originally means " tender blade of herbs and grass, sprouts,"
is more generally used in the sense of herb (stem and leaves
together). 穟 *sui* is properly a spike of flowers, an ear of
corn, but the ancient authors use this term also to designate
a panicle, raceme, etc.

Great confusion and vagueness prevail in these ancient
descriptions of plants with respect to colours. 青 *ts'ing*
originally means " blue." The dictionaries say it is the
colour of indigo. But when applied to plants it always
means " green," the character 綠 *lü*, now the common term
for green, being but rarely used in the *Pen ts'ao kang mu*.
The character 碧 *pi* [WILLIAMS' *Dict.*, 691] means green or
blue jade. It is occasionally used in the *Pen ts'ao* to
indicate the colour of flowers, and I think blue is meant.
紫 *ts'* is originally a purple colour, but frequently it must
be translated by violet or brown. 赤 *ch'i* and 紅 *hung* are
used for red in the *Pen ts'ao*, the first being the older term.
The term 五色 *wu se*, the five primary colours, occurring
in the classics [see *Shu king*, p. 80], is defined by the commentators by 青 blue, 黃 yellow, 赤 red, 白 white, 黑 black.
When meeting in the *Pen ts'ao* with a term like 紅白花
we are, if the plant described be unknown to us, left in doubt
whether we have to translate red and [or] white flowers or
reddish white flowers.

Chinese pharmacy and therapeutics with complicated prescriptions, which fill up the greater part of the text of the
Pen ts'ao kang mu, do not lie within the compass of our
investigations, and the medical uses of the drugs are only
occasionally noticed. In our opinion European science can
learn nothing in this department of Chinese knowledge. We
do not mean to deny that there are in China vegetable
drugs possessed of powerful medical virtues, but the Chinese
Faculty in employing them in their practice of medicine

are seldom guided by experience, but rather by fanciful
suppositions regarding the virtues of drugs. The Chinese
are much addicted to the doctrine of "signatures," which
prevailed also in Europe centuries ago and which is based
upon a belief that an external mark or character on a plant
indicates its suitableness to cure particular diseases. Thus
they employ internally a decoction of thorns of Gleditschia
or Zizyphus to accelerate the bursting of abscesses. The
pods of a Gleditschia, which resemble in shape the tusk of a
boar, are administered in toothache; the yellow bark, or
wood, of Berberry in jaundice; emmenagogue properties are
ascribed to the red coloured root of Rubia cordifolia. The
reader interested in the Chinese views with respect to the
medical virtues of drugs may find information on the subject
in Dr. F. PORTER SMITH's *Contribution towards the Materia
Medica, etc. of China*, 1871.

We constantly meet in the ancient Chinese accounts of
plants and drugs with names of provinces, prefectures, districts, etc. where the drugs were produced. The author has
bestowed a peculiar attention upon the correct identification
of the geographical names of various periods appearing in
the *Pen ts'ao kang mu*. This is by no means an easy task.
As has already been pointed out in the first part of my
Botanicon sinicum [p. 67–69] the same names at different
times were applied to quite different localities of the empire.[3]
T'AO HUNG-KING, in speaking of the localities where medicinal plants are produced, frequently employs the terms
近道 *kin tao* and 邊道 *pien tao*, adjoining and border
provinces. The first we understand to mean the provinces
not far distant from the capital and translate it by Central
China.

---

[3] Comp. *eg.* in the Appendix: 160, *Kuang chou*: 234, *Ning chou*; 187,
Liang chou; 228, *Nan hai*; 389. *Wu*: 146. *King chou*; 124. *Kiang nan*; 229.
Nan k'ang.

To avoid complicated and frequently repeated explanations in the text of our translations from the *Pen ts'ao kang mu*, and to save space, it has been considered advisable to omit all Chinese characters referring to geographical names and to consign the geographical identifications to the Appendix.

TITLES OF SOME CHINESE, JAPANESE AND EUROPEAN BOOKS QUOTED IN MY RESEARCHES BY ABBREVIATED REFERENCES.

P. = *Pen ts'ao kang mu*, the great Chinese Materia Medica and Natural History by LI SHI-CHEN [second half of the 16th century. See *Botanicon sinicum*, I, p. 54].

T. = *T'u shu tsi ch'eng*, the great Chinese Cyclopædia, published in 1726. [See *Botanicon sinicum*, I, p. 71.] My quotations refer to the chapters of the Botanical Section.

Ch. = *Chi wu ming shi t'u k'ao*, a Chinese Botany illustrated by woodcuts, published in 1848. [See *Botanicon sinicum*, I, p. 73.] My quotations refer to the drawings.

The quotations from the *Rh ya* and the *Classics* refer to *Botanicon sinicum*, Part II.

Kiu huang = *Kiu huang Pen ts'ao*, a treatise on plants which can be used for food; accompanied with woodcuts dating from the end of the 14th century. [See *Botanicon sinicum*, I, p. 49–53.]

App. = Appendix to the present volume on ancient Chinese geographical names.

2

K. D. = K'ANG HI's *Dictionary*, the *K'ang hi Tsz' t'ien* or Chinese Dictionary, published in 1716 by order of the Emperor K'ANG HI.

W. D. = WILLIAMS' *Syllabic Dictionary of the Chinese Language*, 1874.

Amœn. exot. = E. KÆMPFER's *Amœnitates Exoticæ*, 1712, in which a great number of Japanese plants are described, sometimes also figured. The Chinese names in Chinese characters are generally added. [See *Botanicon sinicum*, I, p. 126.]

Kwa wi, a Japanese Botany, illustrated by woodcuts, published in the middle of the last century and translated into French by Dr. L. SAVATIER in 1873. [See *Botanicon sinicum*, I, 99].

Phon zo = *Phon zo dzu fu*, a large Japanese work on Botany with nearly 1,800 coloured drawings illustrating the Chinese *Pen ts'ao kang mu*, and published in 1828. [See *Botanicon sinicum*, I, p. 100, *Hon zo dzu fu.*]

So moku = *So moku dzu setsu*, another Japanese Botany with 1,215 excellent drawings representing herbaceous plants. [See *Botanicon sinicum*, p. 101.] I quote these drawings in preference to those of the *Phon zo*. To the latter I generally refer only in the cases in which the plant in question is not mentioned in the *So moku* as *e.g.* trees.

THBG. *Fl. jap.* = C. P. THUNBERG's *Flora Japonica*, 1784. [See *Botanicon sinicum*, I, p. 126.]

SIEB. *Icon.* = Dr. PH. FR. SIEBOLD's Coloured drawings representing Japanese plants, about 600, unpublished. [See *Botanicon sinicum*, I. p. 127.]

SIEB. ZUCC. *Fl. Jap.* = Dr. PH. FR. SIEBOLD et Dr. ZUCCARINI, *Flora Japonica*, 1835–1870; described and pictured 150 plants. [See *Botanicon sinicum*, p. 127.]

SIEB. *œcon.* = SIEBOLD's *Synopsis Plantarum Œcono-micarum Universi Regni Japonici*, 1827.

HOFFM. SCHLT. = J. HOFFMANN et H. SCHULTES, *Noms indigènes d'un choix de Plantes du Japon et de la Chine.* 2nd edition 1864. [See *Botanicon sinicum*, I, 127.] The botanical identifications of Japanese and Chinese names of plants are based upon SIEBOLD's statements.

MIQ. *Prol. Fl. Jap.* = F. A. G. MIQUEL, *Prolusio Floræ Japonicæ*, 1866.

FRANCH. SAV. *Pl. Jap.* = A. FRANCHET et L. SAVATIER, *Enumeratio plantarum in Japonia sponte crescentium*, 1874-1876.

J. MATSUMURA, *Nomenclature of Japanese Plants, in Latin, Japanese and Chinese*, 1884.

GAUGER = G. GAUGER, *Chinesische Roharzneiwaaren*, 1848. Descriptions of Peking drugs, with drawings. [See *Botanicon sinicum*, I, p. 122.]

TATAR. *Cat.* = A. TATARINOV, *Catalogus Medicamentorum Sinensium*, 1856. [See *Botanicon sinicum*, I, p. 122.]

HAN. *Sc. pap.* = D. HANBURY's *Science Papers*, 1875. P. 209–277 his Notes on Chinese Materia Medica are reprinted, which originally appeared in 1860, 1861. [See *Botanicon sinicum*, I, p. 128.]

P. SMITH = Dr. FR. PORTER SMITH, *Contribution towards the Materia Medica and Natural History of China*, 1871. [See *Botanicon sinicum*, I, p. 128.]

Cust. Med. = *List of Chinese Medicines passing through the Chinese Maritime Customs*, 1889.

Hank. Med. = R. BRAUN, *List of Medicines exported from Hankow and the other Yangtze Ports*, 1888.[4]

[4] Both the last-named books were published by order of the Inspector-General of the Chinese Maritime Customs. The first comprises the Chinese names of all the drugs appearing in the Chinese Customs tariff and which from the first of November 1884 to the 31st October 1885 passed inwards and outwards through the Customs of the 19 principal Chinese ports opened to European trade. Quantity and value and the places of production of

PARKER, *Sz ch'uan plants* = *Chinese Names of Plants collected by E. H. Parker in Sz ch'uan 1880 and 1881, and determined by Dr. Hance.* [See *China Review*, XI, 1883, p. 339.]

PARKER, *Canton plants* = *Chinese Names of Canton Plants*, by E. H. PARKER. [See *China Review*, 1886, p. 104-119.]

GARDNER, *Ichang plants* = *Vegetable Products of the Consular District of Ichang, in the Province of the Hupeh*, by C. T. GARDNER. [See *Journ. Ch. Br. Asiat. Soc.*, XIX, (1884), p. 6-26.]

HENRY, *Chin. plants* = *Chinese Names of Plants*, by AUGUSTINE HENRY, M.A.-L.R.C.P. [See *Journ. Ch. Br. Asiat. Soc.*, XXII, 1887, p 233 seqq.] These names refer to plants of the province of Hupeh.

Ind. Fl. sin. = *Index Florœ sinensis*, by F. B. FORBES, F.L.S., and W. B. HEMSLEY, A.L.S., 1888. Now in course of publication.

the drugs are noticed. These two books would have been very useful compilations had the compilers confined themselves to giving only such particulars as could be derived from the official documents of the Customs, adding occasionally a short popular description of the drug from their own observation. But unhappily they attempted to identify the drugs from the Chinese names only, without examining them. They satisfy themselves with relying almost entirely upon the identifications of Chinese drugs as put forth in Dr. F. PORTER SMITH's *Contributions towards the Materia Medica of China*, etc., a work full of erroneous notions, as has been elsewhere detailed. It is a matter of regret that the valuable collections of Chinese drugs in possession of the Chinese Maritime Customs were not sent to Europe to be examined by competent specialists before the costly publication of the List of Chinese Medicines was undertaken. Dr. A. HENRY wrote me, before the book was published, that he had revised it. His corrections (referring evidently to plants well-known to him and determined at Kew) seem to be contained in Part II, Alphab. Index. But unhappily his name does not appear in the book, and the reader is unable to decide for what identifications he is responsible.

[With regard to my share in the publication of "List of Chinese Medicines" I may state :—Part II was sent to me for revision : but I am not responsible for the Index as it stands, as additions and "corrections" were made after it left my hands. My revision consisted in (1) omitting all the grave wrong identifications, (2) in grouping under one heading the many different Chinese names which often exist for a single drug. This was a very laborious piece of work. Part I of this work is entirely free from any editing and is full of errors. The scientific names are not to be depended on. The classification has not been carried out, as *e.g.* bulbs are entered under "seeds," etc. etc.,—A. HENRY.]

THE MEDICINAL PLANTS OF THE *SHEN NUNG PEN TS'AO KING* AND THE *PIE LU*.

1.—甘草 *Kan ts'ao.* P., XII *a*, 1. T. CLX.

Comp. also *Rh ya*, 199, *Classics*, 425.

Pen king :—*Kan ts'ao* (sweet herb). The root is sweet of a uniform nature, non-poisonous.

Pie lu :—*Kan ts'ao*, also 蜜甘 *mi kan* (honey sweet), ｜草 *mi ts'ao* (honey herb), 美草 *mei ts'ao* (excellent herb), 蕗草 *lu ts'ao* [this name occurs in the ancient dictionary *Ki tsiu pien* and in *Tung fang so*, second century B.C.], 國老 *kuo lao*. The latter name, old man of the Empire, which properly is a title given to meritorious statesmen, according to T'AO HUNG-KING is applied to the plant on account of its eminent virtues as a remedy. The *Pie lu* says :—The *kan ts'ao* grows in the river valleys (plains) of Ho si [west of the Yellow River, *v.* App. 79] and on sand-hills. It is also found in Shang kün [north-eastern part of Shen si, *v.* App. 273]. The people gather the root towards the end of the 8th month and dry it in the sun during 10 days.

T'AO HUNG-KING considers the *kan ts'ao* to be one of the most important of medicines, which takes the same place among drugs as the *Ch'en hiang* [Aloewood, *see* 307] among fragrances. It enters into almost all prescriptions. Besides this it has the property of neutralizing the effect of poison. He says that in his time the drug was not brought from Ho si and Shang kün, but the places of production were Shu [Sz' ch'uan, *v.* App. 292], Han chung [Southern Shen si, *v.* App. 54] and especially it was supplied by the barbarian tribes who dwelt in or about the district of **Wen shan** [in N. Sz' ch'uan, *v.* App. 388]. The root has a red rind, is of a hard solid structure. The best sort is called 抱罕草 *Pao han ts'ao*, from a place in the country of the Si K'iang barbarians [Kukonor, N.E. Tibet, App. 300— Ancient Pao han is now Ho chou in Kan su, *v.* App. 242]. It is not advisable to dry the root by means of artificial heat, for it then becomes fissured. Another kind of *kan ts'ao* resembles fish-bowels. It is not advisable to cut it with a knife. A drug of an inferior quality is produced in Ts'ing chou [East Shan tung, *v.* App. 363]. There is also the *tsz'* (purplish or violet) *kan ts'ao*. It is slender, but for fault of a better drug it may also be used.

Su Sung [11th century]:—The *kan ts'ao* grows in all the prefectures of Shen si [modern Shensi and E. Kansu, *v.* App. 284] and Ho tung [present Shan si, *v.* App. 80]. It is a plant from one to two feet high. The leaves resemble those of the *huai* (*Sophora*). In the 7th month it produces violet flowers resembling those of the *nai tung* [unknown to me] which are followed by pods like pea-pods. The root has a red rind, is from 3 to 4 feet long, coarse or slender. In its upper part the principal root emits horizontal branches (runners) which are beset with rootlets. After the crown and the red rind have been removed, the root is dried in the shade.

Kan ts'ao is still the common Chinese name for *Liquorice* or *Glycyrrhiza*. A good drawing of a *Glycyrrhiza* sub *kan ts'ao*, roots, flowers and echinate legumes, is found in the *Ch.* [VII, 6].

Comp. *Phon zo*, V, 1, 甘草 *Glycyrrhiza*.

SIEB. *œcon.*, 305, *Glycyrrhiza, kan soo*. E. China introducta, rarius et quidem in provinciis insulæ Sikok culta.

TATAR. *Cat.* 25,—P. SMITH, 136. Liquorice root,

Cust. Med., p. 4 (22). New chwang exported in 1885 Liquorice to other Chinese ports 1,767 piculs,—p. 30 (100), Tientsin exported 4,576 piculs,—p. 46 (18), Chefoo exported 8,690,—p. 68 (34), Hankow exported 1,148,—p. 455 (587). Liquorice, places of production :—Chili, Shan tung, Shen si, Kan su.

In 1882 I sent some specimens of Chinese Liquorice root from Shan si to Prof. Dr. FLÜCKIGER, who in the 2nd edition of his *Pharmacognosie* [p. 355] writes that he is not able to distinguish it from Spanish Liquorice of the first quality.

The Liquorice root used in medicine in Europe is derived from *Glycyrrhiza glabra*, L., indigenous in Southern Europe. The typical form of this supplies the Spanish Liquorice, which is considered to be the best. The variety *glandulifera*, which grows in Hungary, South Russia, yields the Russian Liquorice, which is likewise derived from *Gl. echinata*, L.

LOUREIRO [*Fl. cochin.*, 543] states that Chinese Liquorice root is yielded by *Gl. echinata* and *glabra* of the northern provinces of China. [*See* my *Early Europ. Res. Fl. China*, p. 145.]

BUNGE [*Enum. pl. Chinæ bor.*, 97] records *Gl. glandulifera* from the neighbourhood of Peking and the Great Wall.

PRZEVALSKY [*Mongolia, Tangut, etc.*, Engl. edition, I, 191] states that the root of *Gl. uralensis*, Fischer, one of the characteristic plants of the Ordos, is dug up there by the Mongols, hired by the Chinese, who despatch the drug down the Huang ho to supply the Chinese markets. The same plant is recorded by Father DAVID [FRANCHET, *Plantæ David. Mongol.*, 93] from the Peking plain and Southern Mongolia. It grows also in the Altai and Ural mountains.

2.—黃耆 *huang k'i.*   P., XII *a*, 6.   T. CLII.

Pen king:—Huang (yellow) *k'i*, 戴糁 *tai sun.*   Root used in medicine.   Taste sweet, nature slightly warm.   Non-poisonous.

Pie lu:—Other names: 戴椹 *tai shen*, 菱草 *ki ts'ao*, 百本 *po pen* (a hundred roots), 獨 | *tu shen*, 蜀脂 *Shu chi.* The *huang k'i* grows in the mountain valleys of Shu [W. Sz' ch'uan, *v.* App. 292], in Pai shui [in Mid Shen si, *v.* App. 239], in Han chung [S. Shen si, *v.* App. 54].   It is gathered in the second and tenth months, and dried in the shade.

T'AO HUNG KING :—Now the drug brought from Lung si and T'ao yang [both in Mid Kansu, *v.* App. 216, 336], which is of a yellowish white colour and sweet, is considered the best, but it is scarce, and more generally an inferior sort of a white colour and coarse-grained, which comes from Hei shui [*v.* App. 60] and Tang ch'ang [in Mid Kan su, App. 330] is used.   There is also a red kind which is employed for making plasters.

The *Yao sing lun* [7th century] calls this plant 王孫 *Wang sun.*

SU KUNG [7th century]:—Now the best sort comes from Yüan chou [in E. Kan su, App. 414] and Hua yüan [in E. Shensi, App. 88].   That from Shu and Han chung is not much employed.   That from I chou [in Kuang si, App. 103] and Ning chou [in N.E. Kan su, App. 234] is also of a superior quality.

SU SUNG [11th century]:—This drug is abundantly produced in all the prefectures of Ho tung [Shan si, App. 80] and Shen si (modern Shen si and E. Kan su).   There are several sorts of it.   The rind of one of them furnishes textile fibres.

LI SHI-CHEN :—The common name of the plant nowadays is 黃芪 *huang k'i.*   The original name (*c.s.*) is sometimes erroneously written 黄耆 *huang shi*. The leaves of the *huang k'i* resemble those of the *huai* (*Sophora japonica*) but are smaller and pointed. They resemble also the leaves of the *tsi li* (*Tribulus terrestris*) but are broader and larger, and of a whitish green colour. The flowers are of a yellowish purple colour, as large as those of the *huai*. The fruit is a pointed pod, one inch and more long. The root is from 2 to 3 feet long. That which is tight and solid, like the shaft of an arrow, is the best. The young leaves are edible, and therefore the plant is also cultivated as a vegetable.

The *Kiu huang* [XLVI, 13] and the *Ch.* [VII, 3] represent sub *huang k'i* a leguminous plant.

TATAR. *Cat.*, 10: *huang k'i*, Radix *Sophoræ flavescentis*. The latter is a common plant in North China.

GAUGER [8], who describes and depicts this Chinese drug, is of opinion that it is the root of an *Astragalus*. This view is confirmed by Father DAVID, who in the account of his journey in S. Mongolia speaks of a large herbaceous plant *hoang tchy*, of the order *Leguminosæ*, the root of which is dug up there and sent to China as a medicine. FRANCHET [*Plantæ David. Mongol.*, 86] described it as *Astragalus hoang tchy*. In Hupeh the drug *huang k'i* (*ch'i*) is derived from *Astragalus Henryi*, Oliv. [*See* HENRY's memorandum in HOOKER's *Icones. Plant.*, tab. 1959.]

Comp. also my *Early Europ. Res. Bot. Chin.* [p. 147], LOUREIRO's *Robinia flava* and *infra* 7 sub. *huang tsing*.

P. SMITH, 202: *huang k'i*, *Sophora tomentosa*. But [p. 180] he identifies erroneously the same Chinese name with *Ptarmica sibirica*.

Cust. Med. p. 24 (25), *huang k'i* exported from Tien tsin 3,545 piculs,—p. 58 (10) I chang 224 piculs,—p. 68 (26) Han kow 1,450.—P. 451 (510): Places of production:

2

Manchuria, Chi li, Shan tung, Sz ch'uan, Shen si. Several
sorts are distinguished on the Port lists and are probably
yielded by different plants. There are the *hung* (red) *k'i*,
451 (510) exported from I chang, the *pai* (white) *k'i* and
other kinds.

So moku, XIV, 3:—黃芪 *Hedysarum esculentum,* Ledeb.
(edible root).

Kwa wi, 30:—Same Chinese name, *Astragalus lotoides,*
Lam. (same as *Astr. sinicus,* L.).

So moku, XIV, 4:—木黃芪 (the first character means
woody). *Astragalus reflexistipulus,* Miq. SIEBOLD states
that it was introduced into Japan from China.

3.—人蔘 *jen shen,* the famous Ginseng root. *P.,* XIIa,
11. *T.,* CXXV.

Pen king:—人蔘, also written 人薓 *jen shen,* 人銜
jen hien, 鬼蓋 *kui kai.* The root used officinally.[5] Taste
sweet, nature slightly cold. Non-poisonous.

Pie lu:—Other names : 血蔘 *hue shen,* 神草 *shen ts'ao*
(divine herb), 土精 *t'u tsing* (terrestrial essence). The *jen
shen* grows in the mountain valleys of Shang tang [S.E.
Shan si, App. 275] and Liao tung [S. Manchuria, App. 191].
The root is dug up during the first decade of the 2nd, 4th
and 8th months. It is scraped with a bamboo knife and
dried in the sun, protected from the wind. This root in its
shape resembles a man (人 *jen,* whence the name) and is
possessed of Divine power.

Wu P'u [3rd cent.]:—The plant is also called 黃蔘
huang (yellow) *shen.* It grows in Han tan [S. Chi li, App.
56]. In the 3rd month it shoots forth leaves which are

---

[5] The leaves of the Ginseng, 蔘蘆 *shen lu,* are also employed in
medicine.

dentated. The branches (petioles?) are black, the stem is covered with hair. The root is dug up in the 3rd and 9th months. It is hairy, has hands, feet, a face and eyes like a man possessed of a god.

T'AO HUNG-KING:—Shang tang (the locality mentioned in the *Pie lu*) is south-west of Ki chou [Northern part of present Chi li, App. 119]. The drug which now comes from that locality is a long root of a yellow colour resembling the *fang feng* [an umbelliferous plant. *See below*, 31]. It is succulent and sweet, and highly valued. The drug brought from Po tsi [in the present Corea, App. 261] is slender and hard, of a white colour. In taste it is weaker than the Shang tang drug. There is a third sort produced in Kao li [Corea, App. 116]. This is the same as the Liao tung drug [mentioned in the *Pie lu*]. It is of large size but devoid of juice, soft and inferior to that from Po tsi. But the best of all is the Shang tang drug.[6] The plant sends up only one straight stalk. Its leaves are four or five together (*i.e.* four or five leaflets at the top of a common petiole). The flowers are of a purple colour. There is a Corean song in praise of the Ginseng (高麗人作人參讚) saying: the branches (petioles) which grow from my stalk are three in number, and my leaves are five by five. The back part of the leaves is turned to the sky, the upper side downwards. Whoever would find me must look for the 椵 *kia* tree. T'AO HUNG-KING explains that the *kia* tree resembles the *t'ung* (*Paulownia*), growing very high and casting a large shade.[7] In this kind of place the Ginseng is found in great abundance. The gathering and preparing of the drug require a great deal of experience. There is some Ginseng found at present

---

[6] Nowadays, on the contrary, the Ginseng from Liao tung, Manchuria and the northern part of Corea is considered the best. No Ginseng is now produced in S.E. Shan si or ancient Shang tang.

[7] Regarding the *kia* tree, see *Bot. sin.*, II, 226.

in the mountains not far distant [the author lived, it seems, in Kien k'ang, the present Nan king], but it is not good.

Su Kong [7th cent.]:—The Ginseng which is now used comes chiefly from Kao li and Po tsi [*v. supra*]. That which grows on the Tsz' t'uan shen mountain of the 太行 T'ai hang range [App. 323] in Lu chou [in S.E. Shan si, App. 204] is called 紫團參 *Tsz' t'uan shen.*

Han Pao-sheng [10th cent.]:—Now Ginseng is produced in Ts'in chou [in S.E. Shan si, App. 362], in Liao chou [in S.E. Shan si, App. 190], in Tse chou [in S.E. Shan si, App. 345], in P'ing chou [in Chi li, App. 255], in I chou [in Chi li, App. 101], in T'an chou [in Chi li, App. 329], in Yu chou [present Peking, App. 411], in Kui chou [in N. Chi li, App. 168], in Ping chou [App. 253]. All these prefectures are situated near the T'ai hang range.

Li Sün [8th cent.]:—The Ginseng with which the kingdom of Sin lo [in S. Corea, App. 311] pays tribute has hands and feet and resembles a man. It is above a foot long. It is kept pressed between boards of *shan mu* [see *Bot. sin.,* II, 228], bound and wrapped up with red silk. The Ginseng of Sha chou [App. 266] has a small short root and is not of any practical value.

Su Sung [11th cent.]:—All the prefectures of Ho tung [Shan si, App. 80] and also the mountain T'ai shan [in Shan tung, App. 322], produce Ginseng. That which is imported under the name of Ginseng from Sin lo (Corea) through the provinces of Ho pei [Chi li, App. 78] and Min [Fu kien, App. 222] is not so valuable as that of Shang tang [S.E. Shan si. *V. supra*]. The plant begins to shoot in the spring. It is found in the depths of the mountains in shady, moist places, growing beneath the *kia* tree [*v. supra*] and the *ts'i* (varnish) tree. When the plant is young and not above 3 or 4 inches high it shoots forth a branch with five leaves (five leaflets at the top of a common petiole) and after four or five years it sends out a second with the same number of leaves; it has, however, neither stalk nor flowers as yet. At the end of ten years it shoots forth a third branch, and many years after a fourth, each of them having five leaves. It then begins to produce from the middle of the heart a stalk which is commonly called 百尺杵 *po chi chu* (pestle of a hundred feet). In the third and fourth months it bears small pale violet flowers about the size of a grain of millet, the filaments of which are like untwisted silk. The fruits (berries) which ripen at the end of autumn are of the size of a *ta tou* (Soy bean) and from 7 to 8 together. They are at first green but become red as they ripen. When they are quite ripe they fall off of themselves. The root is in figure like a man. The Ginseng which is found on the T'ai shan mountain [*v. supra*] has a green stalk and green leaves and a white root. Another kind of Ginseng grows in Kiang Huai [An hui, Kiang su, App. 124, 89]. It is called *t'u jen shen* (native Ginseng), grows two feet high. Its leaves are opposite, resemble a small spoon, like the leaves of the *kie keng* [*Platycodon. See* 6]. The root, which shows five joints, bears likewise a resemblance to the *kie keng* root but is more tender and of a sweeter, more pleasant taste. Its flowers appear in the autumn. They are of a purple colour tinged with green. The root is dug up in spring and in autumn.—It is said that in order to know the true Ginseng of Shang tang, two persons walk together, one going with Ginseng in his mouth and the other with his mouth empty. At the end of three or five *li* he who has the Ginseng in his mouth does not find himself at all out of breath, whilst the other on the contrary is tired and breathless. This is a mark of the goodness of the drug.

K'OU TSUNG-SHI [12th cent.]:—The Ginseng of Shang tang has a long thin root which sometimes reaches above a foot deep in the earth and often divides itself into ten branches. It is sold for near its weight in silver and is obtainable with difficulty.

LI SHI-CHEN :—Ancient Shang tang is what is now called Lu chou [in S.E. Shan si, App. 204]. The people look upon the Ginseng as a calamity for the country where it grows (for the drug must be delivered to the emperor without compensation). That which is made use of at present comes from Liao tung (Manchuria). The three kingdoms Kao li, Po tsi and Sin lo (mentioned by the ancient authors as producing Ginseng) now constitute the kingdom of Chao sien [Corea, App. 9]. Corean Ginseng is much carried to China. The people there cultivate it also like a vegetable. The seeds are sown in the 10th month. That (root) which is dug up in autumn and in winter is firm and full of juice ; on the contrary that taken in the spring and summer seasons is soft and devoid of juice, which difference does not proceed from the good or bad quality of the ground where it grows. The Ginseng root of Liao tung when it has its rind on is of a smooth yellow colour like the *fang feng* [*v. supra*], but when the rind is taken off it is firm and white like starch. Other roots are frequently substituted for the true Ginseng, namely the *sha shen* [*Adenophora*, see the next], the *tsi ni* [*Adenophora*, see 5] and the *kie keng* [*Platycodon*, see 6]. The *sha shen* and the *tsi ni* are entirely devoid of juice, have no heart (無心 without energy ?) and are of an insipid taste. The root of the *kie keng* is hard, has a heart (有心 has active properties ?) and is of a bitter taste. But the root of the true Ginseng is of a juicy substance, has a heart and an agreeable sweet taste with a spice of bitterness. It is commonly called 金并玉蘭 *kin tsing yü lan*. That which is in the shape of a man is called 孩兒參 *hai rh shen* (infant's Ginseng). It is frequently adulterated. SU SUNG [in the *T'u king Pen ts'ao*

—11th cent.] figures the Ginseng from Lu chou [v. supra] with three branches and five leaves [i.e. five leaflets at the top of a common petiole, v. supra]. This is the true Ginseng. But his figure of the Ginseng from Ch'u chou [in An hui, App. 25] is, judging from the form of the leaves, the sha shen, and his Ginseng from Ts'in chou [in S.E. Shan si, App. 362] and Yen chou [in Shan tung, App. 404] and likewise his native Ginseng from Kiang Huai [v. supra] must all be referred to the plant tsi ni. These drugs are frequently confounded with the true Ginseng. At present [LI SHI-CHEN says] the true Ginseng is no more met with in the prefecture of Lu chou [in S.E. Shan si]. Compare also infra 4, at the end.

T'AO HUNG-KING and other ancient authors say that the Ginseng root is very apt to breed worms, especially when exposed to the sun or the wind.

LI YEN-WEN (an author of the Ming period, who wrote a treatise on Ginseng) says that Ginseng grows in such a manner that the back part of its leaves is turned towards the sky, and therefore it does not like either the wind or the sun. Taken as a medicine it is generally chewed crude without any other preparation, or it is dried before the fire on a sheet of paper for medical use. Sometimes it is also steeped in a kind of wine called 醇酒 shun tsiu. Ginseng must neither be kept in an iron vessel nor prepared with any instrument made of that metal.

The drawing given in the Ch. [VII, i] of the jen shen plant is bad and incorrect.

TATAR. Catal., 64: jen shen, Radix Panacis Ginseng.— P. SMITH, 103.

Cust. Med., p. 4 (21):—Ginseng exported from New chwang in 1885 about 180,000 piculs to other ports of China. The list enumerates several sorts. The wild Ginseng from Manchuria is the highest in price = 6,400 Taels per picul ; next in order stands the first quality Corean Ginseng = 1,000 Taels. The best cultivated Manchurian Ginseng = 200 Taels per picul.

It is well known that the most highly valued specimens of Ginseng are the property of the Chinese Emperor and come only occasionally to the market. Superior sorts are sold from 20 to 250 times their weight in silver. The Chinese consider this drug the most powerful and even life-prolonging medicine. The wild Manchurian root is the most prized. But the experiments made repeatedly by European physicians with genuine Ginseng proved that it does not possess any important medicinal properties.

From the ancient Chinese accounts of the Ginseng plant it would appear that in ancient times it grew in the mountains of Shan si and Chi li. The Shan si drug was considered the best. At present it is met with in a wild state only in Manchuria and Corea. As the wild plant even in these countries is very rare, Ginseng is much cultivated in Manchuria, Corea and Japan.

About 50 years ago Dr. P. KIRILLOV, then physician to the Russian Eccles. Mission at Peking, sent a complete herbarium specimen of the wild-growing Manchurian Ginseng to St. Petersburg, where it was described and depicted in GAUGER's *Repert. f. Pharmacie, etc.* [I (1842), p. 516] by C. A. MEYER under the name of *Panax Ginseng*. The Manchurian species is closely allied to the N. American *Panax quinquefolium*. Both have palmate leaves with five-toothed leaflets, minute flowers, arranged in an umbellate manner, and red berry-like fruits. The difference between them is principally in the shape, and according to the Chinese also in the medical properties, of the roots.

In Japan *Panax Ginseng* occurs only in a cultivated state. *Amœn. exot.*, 818 : 人蔘 *sju sjin*, vulgo *nisji, nindsin,* sinice *som.* *Sisarum montanum Coræense.* On the annexed plate [fig. 4] KÆMPFER correctly depicts the Ginseng root, but the plant he figures on the same plate as the true Ginseng is *Sium Ninsi*, L., an umbelliferous plant, in Japanese *mukago ninzin.*

Phon zo [V, 4, 5] and *So moku* [IV, 46]:—人蔘 *Panax Ginseng,* japonice *ninzin.*

According to the ancient Chinese authors the best sort of the true Ginseng was considered the 上黨蔘 *Shang Tang shen* or Ginseng from Shang Tang in S.E. Shan si. But nowadays this Chinese name is applied to the root of *Codonopsis tang shen,* Oliv., a *Campanulacea. See* Dr. HENRY's memorandum in HOOKER's *Icon. Plant.* [tab. 1966]. The *Tang shen* is figured in *Ch.* [VII, 49] as a climbing *Campanulacea.*

4.—沙蔘 *sha shen* (Sand Ginseng). *P.,* XIIa, 23.— *T.,* CXL.

Pen king :—Sha shen. Root officinal. Taste bitter. Nature slightly cold. Non-poisonous.

Pie lu :—Synonyms: 鈴兒草 *ling rh ts'ao* (bellwort, campanula), 知母 *chi mu* [this name is properly applied to another plant. *V.* 9], 羊乳 *yang ju* (goat's teat), 虎鬚 *hu sü* (tiger's beard), 苦心 *k'u sin* (bitter heart). The *sha shen* grows in the river valleys (meadows) of Ho nei [S.E. Shan si, N. Honan, App. 77], in Yüan kü [in S.W. Shan tung, App. 415] and 殷陽 Pan yang [in N.W. Shan tung, App. 241], in the mountains. The root is dug up in the 2nd and 8th months and dried in the sun.

WU P'U [3rd cent.]:—The *sha shen* is also called 白蔘 *pai* (white) *shen.* In the second month, when the plant first begins to grow, it resembles the *k'ui* (Malva). The root is white, juicy, like the root of the mustard plant and as large as the *wu tsing* (turnip).

4

T‘AO HUNG-KING :—There are five drugs to which the
name *shen* (Ginseng) is applied,[8] *viz.* the *jen shen* (true
Ginseng), the *sha shen* (the plant under review), the 玄參
huan (black) *shen* [*v. infra* 18], the 丹 | *tan* (red) *shen*
[*v.* 20], the 苦 | *k‘u* (bitter) *shen* [*v.* 34]. These are
termed the "five *shen*." There is also a drug called 紫 |
tsz‘ shen [purple *shen*, *v.* 21]. The *sha shen* grows in the
central provinces. Its leaves resemble those of the *kou k‘i*
(*Lycium*). The root is white, juicy.

SU KUNG [7th cent.]:—The best drug comes from the
Hua shan mountain [in S.E. Shen si, App. 86].

SU SUNG [11th cent.]:—The *sha shen* is common in the
central provinces. The plant grows in the mountains, in a
bushy manner, two feet high. The leaves resemble the *kou
k‘i* [*v. supra*], vary in size, are pronged (or lobed). In the
7th month it opens its violet flowers. The root resembles
the mallow root and is about the size of a finger, of a reddish
yellow colour outside, white within, juicy.

LI SHI-CHEN :—The name *sha shen* (Sand Ginseng)
refers to its growing in a sandy soil. The name *pai shen*
(white Ginseng) is applied to it on account of the white
juice contained in the root. The *sha shen* is a common
mountain plant. When it first begins to grow, in the second
month, the young leaves resemble those of the *shui k‘ui*
(water mallow, *Limnanthemum*), but are thinner, not shining.
In the 8th or 9th month it is from 1 to 2 feet high.
The leaves are collected around the stem ; they are long,
pointed, resemble those of the *kou k‘i* but are smaller in size
and toothed. In autumn small violet flowers appear between
the leaves ; they resemble a bell in shape, the corolla is five-
lobed. The filaments are white. Sometimes the corolla is
also white. The fruit is as large as that of the *tung ts‘ing*

---

[8] Comp. *infra*, 20.

(*Ilex*); it contains small seeds. When the plant grows in a sandy soil its root is large and becomes more than a foot long, but when produced in a loamy soil it has a small short root. The root as well as the stem contain a white juice. The root is more juicy when dug up in autumn.

The *sha jen* plant with root, leaves and flowers is figured in the *Kiu huang* [LI, i] and *Ch.* [VII, ii]. The drawings represent a *Campanulacea*, probably an *Adenophora*.

GAUGER, 31:—Description of the *sha shen* root.— TATAR., *Cat.*, 50: *sha shen*, Radix *Adenophorœ* seu *Campanulæ*.—P. SMITH, 4.

Cust. Med., p. 46 (25):—*Sha shen* exported in 1885 from Chefoo to other Chinese ports, 2,894 piculs.—*P.* 474 (1078). Place of production Shan tung, An hui.

HENRY, *Chin. pl.*, 405:—*Sha shen* in Hu pei is *Adenophora polymorpha*, Ledeb., and other species.

Kiu huang, LI, 17, and *Ch.*, VIII, 68:—細葉 | | *si ye sha shen* (*sha shen* with small leaves), bad drawings, roots and leaves.

Amœn. exot., 822:—沙蔘 *sadsin*. Lychnis sylvestris, foliis Leucoji lanuginosis, flosculis albis pentapetalis . . . . radice Pastinacæ, ab impostoribus pro radice *ninsin* (Ginseng) supponi solita.—KÆMPFER seems to be mistaken. The description of the flowers does not agree with a *Campanulacea*. In the *So moku* [III, 6, 7] 沙蔘 is *Adenophora verticillata*, Fischer. *Ibid.*, 5:—細葉 | | *Wahlenbergia marginata*, A. DC. (*Campanulacea*). Both these species are known also from China.

There is a plant 薫蔘 *Tang shen* figured and described in the *Ch.* [VII, 49]. Large root, bluish white bell-shaped flowers. Creeping plant, frequent in Shan si. It is said

there that the full name should be 上黨參 *Shang Tang shen*
(Ginseng from Shang Tang in S.E. Shan si), which name
in ancient times was applied to the best sort of the true
Ginseng. But as this latter has long disappeared in that
locality [*v. supra*, 3] the people have applied the above name
to the plant substituted for the genuine drug.

GAUGER [14] figures and describes the drug *Tang shen*,
cylindrical roots. The plant has a square stem. He conjec-
tures that it may be a *Rubiacea*.—TATAR. *Cat.* [19]:—*Tang
shen. Radix Convolvuli?*

Comp. also P. SMITH, 104, 48.

Cust. Med., No. 1251 :—*Tang shen, Campanumœa pilosula*,
Franch.

Cust. Med., p. 70 (64):—*Tang shen* exported in 1885 from
Hankow to other Chinese ports, 8,330 piculs. *Ibid.*, 60 (24),
from I chang 197 piculs.—*Ibid.*, p. 481 (1251), places of
production : Shan si, Shen si, Sz ch'uan, Hu peh.—See also
Hank. Med., p. 43.

5.—薺苨 *tsi ni. P.*, XIIa, 25.—*T.*, CXLIII.

Comp. *Rh ya*, 45.

In the *Pen king* this name is given as a synonym for
kie keng [see the next], but the *Pie lu*, which is followed by
LI SHI-CHEN, keeps these drugs apart.

According to the *Pie lu* the root of the *tsi ni* is used in
medicine. Its taste is sweet, its nature cold. Non-poisonous.
It counteracts the effects of poison.

T'AO HUNG-KING :—In its root and stem the *tsi ni*
much resembles Ginseng, but the leaves are different in shape
and smaller in the *tsi ni*. The root is of a sweet taste and
has the property of neutralizing poison.

SU SUNG [11th cent.]:—The *tsi ni* is a common plant
in Mid China, especially in Jun chou [in Kiang su,

App. 111], Shen chou [in Ho nan, App. 283]. The root is dug up in the 2nd and 8th months and dried in the sun. It is of a pleasant sweet taste. The people use it also for food. According to LI SHI-CHEN the 杏參 *hing shen* (apricot [leaved] Ginseng) mentioned by SU SUNG is the same as *tsi ni*.

LI SHI-CHEN says :—The *tsi ni* in its leaves resembles the *kie keng* (*Platycodon*), in its roots the Ginseng, for which it is fraudulently substituted. The *Kiu huang pen ts'ao* calls it 杏葉沙參 *hing ye sha shen* (*sha shen* with apricot leaves), also 白麵根 *pai mien ken* (white flour root). Another name is 甜桔梗 *t'ien kie keng* (sweet *kie keng*, see the next). Its leaves resemble apricot leaves, but they are smaller, slightly pointed, toothed and white underneath. The corolla of the flower is bowl shaped, 5 cleft, white, sometimes blue. The root is like a wild carrot, gray outside, and with white hairs (filaments) within. The leaves as well as the root are used for food. The leaves are also known under the name 隱荵 *yin yen* and employed to destroy intestinal worms. This name is found in the *Rh ya* [84].

The *Kiu huang* [LI, 6] and *Ch.* [VIII, 69] figure sub *hing ye sha shen* a *Campanula* or *Adenophora*. Blue flowers.

So moku [III, 10]:—薺苨 *Adenophora remotiflora*, Miq.

Ibid. [III, 9]:—杏葉沙參 *Adenophora latifolia*, Fischer.

6.—桔梗 *kie keng*. P., XIIa, 28.—*T.*, CLI.

The *Pen king* makes the *kie keng* and the *tsi ni* [see the preceding] to be the same, but the *Pie lu* and all subsequent writers agree in keeping them apart. Of both of these plants the root is officinal. That of the *kie keng* is of a pungent taste; nature somewhat warm and slightly poisonous. The stem and the leaves (薺苨) are also used in medicine.

Pie lu :—The *kie keng*, which is also called 白藥 *pai yao* (white drug), 梗草 *keng ts'ao*, grows in the mountain valleys of Sung kao [in Ho nan, App. 317] and in Yüan kü [in S.W. Shan tung, App. 415]. The root is dug up in the second month and dried in the sun.

Wu P'u [3rd cent.]:—The *kie keng* is also called 利如 *li ju*, 符扈 *fu hu*, 房圖 *fang t'u*. Its leaves resemble those of the *tsi ni* [*v.* 5]. The stem is like a pencil of a purple colour. It begins to grow in the second month.

T'ao Hung-king :—The *kie keng* is a common plant in Mid China. The young plant can be eaten boiled. It is also used as a vermifuge. The root is fraudulently substituted for the true Ginseng.

Su Sung [11th cent.]:—Its root is as thick as a finger, of a yellowish white colour. The plant grows one foot and more high. Its leaves resemble apricot leaves but are longer, and stand opposite, four together. Can be eaten boiled. In summer it opens its small blue flowers resembling those of the *k'ien niu* (*Pharbitis*). The root is dug up in the 8th month. It has a heart [*comp. above, sub.* 3]. The root of the *tsi ni* has no heart.

Li Shi-chen :—The *kie keng* and the *tsi ni* are plants of the same order. The difference is that the *tsi ni* is sweet and the *kie keng* is bitter. Therefore the *tsi ni* is also called 甜桔梗 *t'ien* (sweet) *kie keng.*

The plant *kie keng* represented in the *Ch.* [VIII, 11] is *Platycodon grandiflorum*, A. DC.—See also *Kiu huang* [XLVII, 1]. Bad drawing.

The drug *kie keng* is described and depicted in Gauger [49].

Tatar. *Cat.*, 58, *kie keng.* Radix *Platycodonis grandiflori.*—P. Smith, 173.

HENRY, *Chin. pl.*, 52 :—*Kie keng, Platycodon grandiflorum.* Common in Hu peh.

Cust. Med., p. 120 (7) :—*Kie keng* in 1885 exported from Chin kiang 2,162 piculs,—p. 44 (5) from Che foo 138,—p. 96 (5), from Wu hu 60,—p. 22 (7), from Tien tsin 10.—*Hank. Med.*, 3 : Exported from Han kow. Places of production : An hui, Chi li, Hu nan, Hu peh, Sz ch'uan.

Amœn. exot., 822 :—桔梗 *kekko*, vulgo *kikjo* and *kirakoo*. Rapunculus medicamentosus, foliis oblongis denticulatis. Radice palmari, pingui, lactescente, multiplicis virtutis, & secundum in usu medico locum obtinente a radice *nindsin* (Ginseng) ; flore campanulæ, cœruleo.—THBG. [*Fl. jap.*, 88] identifies KÆMPFER's plant with *Campanula glauca*, which is the same as *Platycodon grandiflorum.* [See *China Review*, XV, p. 346.]

So moku, III, 4 :—Same Chinese name, *Platycodon grandiflorum.*

7.—黃精 *huang tsing.* *P.*, XIIa, 32.—*T.*, CLII.

The above name appears first in the *Pie lu.* Synonyms given in the same work : 兔竹 *t'u chu*, 鹿竹 *lu chu* (deer bamboo), 救窮草 *kiu k'iung ts'ao* (poor man's relief), 重樓 *chung lou*, 雞格 *ki ko.* The *huang tsing* is a mountain plant. The root is dug up in the second month and dried in the shade. Its taste is sweet, its nature is uniform. Non-poisonous.

T'AO HUNG-KING :—It is a common plant luxurious in foliage. The leaves resemble bamboo-leaves, but are shorter. The root resembles that of the *wei jui* (*Polygonatum*, see the next) and also the root of the 荻 *ti* (a rush) and the *ch'ang p'u* [*Acorus*, see 194]. It has large joints, is succulent. It is not commonly used in medicine, but the root is highly valued by Taoists. Root, leaves, flowers, fruit, all are eaten. For medical use the root is steeped in wine or administered in powder. The leaves much resemble the leaves of the poisonous plant *kou wen* [*see* 162], and people frequently confound this plant with the *huang tsing*. In Taoist books the *huang tsing* is also called 仙人餘糧 *sien jen yü liang* (extra ration of the immortals).

The Taoists consider the *huang tsing* to be a kind of 芝 *chi* (plant of immortality) and therefore call it also 黃芝 *huang* (yellow) *chi*, 戊巳芝 *wu ki chi*.

LEI HIAO [5th cent.]:—The *kou wen* [*v. supra*], which is injurious to life, resembles the *huang tsing*.

SU KUNG [7th cent.]:—When growing in a fat soil, the root of the *huang tsing* attains the size of a fist, but in poor soil it is not larger than the thumb. It is akin to the *wei jui*. The *kou wen* is quite a different plant.

CH'EN TS'ANG K'I [8th cent.]:—The true *huang tsing* has opposite leaves. There is one kind of it in which the leaves are all inclined on one side. This is called 偏精 *p'ien tsing* (*p'ien* = inclined on one side). T'AO HUNG-KING is incorrect in stating that the *kou wen* resembles the *huang tsing*.

SU SUNG [11th cent.]:—The *huang tsing* is as common in North China as in the South. The best drug comes from Sung shan [in Honan, App. 317] and Mao shan [in Kiang su, App. 218]. The plant grows from one to two feet high. The leaves resemble bamboo-leaves, but are shorter. They stand opposite, two and two together. The root is soft, of a yellow colour, its lower part is red. In the fourth month the plant opens its greenish white flowers, which resemble the flowers of small beans. The seed is white, resembles millet. The yellow root, which has some resemblance to young ginger-root, is very sweet and of a pleasant taste. It is dug up in the second month, boiled, and dried in the sun. The mountain people make a preserve of it for sale which is very sweet and pleasant. They gather also the young plants to use them for food. It is a very palatable dish ; they call it 蕨菜 *pi ts'ai.*

Lɪ Sʜɪ-cʜɛɴ :—The *huang tsing* grows wild in the mountains. It is also cultivated. Its leaves resemble bamboo-leaves but are not pointed. They are arranged in a whorl from three to five around the joints of the stem. The root grows in a horizontal direction, resembles that of the *wei jui.* The people gather the plant (stem and leaves) to use it for food. The bitter taste disappears after macerating and cooking. This is the 筆管菜 *pi kuan ts'ai* (pencil-holder vegetable). The *Po wu-chi* [3rd cent.] relates the following legend :—Emperor Hᴜᴀɴɢ ᴛɪ once asked one of his sages whether he knew a plant which when eaten gives immortality. The sage replied : There is the plant of the great male (or bright) principle (太陽) which is called 黄精 *huang tsing* (yellow ethereal essence). When eaten it prolongs life. There is also the plant of the great female (or dark) principle (太陰) which is called 鉤吻 *kou wen.* When it enters the mouth, it kills man.

The *huang tsing* is also called 垂珠 *ch'ui chu* (beads hanging down), which name refers to its drooping flowers and berries.

An author of the 16th century states that its root resembles young ginger, whence the popular name 野生薑 *ye sheng kiang* (wild ginger). By partly steaming and drying it is prepared for food and used as a substitute for corn. Another name is 米餔 *mi pu.*

Kiu huang [LIII, 2] and *Ch.* [VIII, 18–21, also X, 43] sub *huang tsing*, representations of various species of *Polygonatum.* The above Chinese descriptions of the *huang tsing* agree in a general way.

At Peking the name *huang tsing* is applied to *Polygonatum macropodum*, Turcz., and *P. chinense*, Kth., both wild and cultivated. The roots are eaten. In the Peking mountains these plants are more generally known by the name 甜草根 *t'ien ts'ao ken* (sweet root) and distinguished as large-leaved and small-leaved.

LOUREIRO [*Fl. cochin.*, 99] applies the Chinese name *hoam cim* (*huang tsing*) to *Galium tuberosum* (a dubious plant) and states that the root of this plant is eaten boiled.

TATAR. [*Cat.*, 10] identifies erroneously the drug *huang tsing* with Radix *Caraganæ flavæ*. He refers it evidently to LOUREIRO's *Robinia flava*, sinice *hoam khin* [*Fl. cochin.*, 556]. P. SMITH, 51, has the same erroneous identification, but the sweet mucilaginous drug *huang tsing* which he describes is *Polygonatum*.

The drug *huang tsing* I obtained from a Peking apothecary shop, was the root of a *Polygonatum*.

Cust. Med., p. 342 (52):—*Huang tsing* exported 1885 from Canton to other ports of China 64 piculs,—p. 210 (22) from Wen chou 64,—p. 186 (39) from Ningpo 12.—Small quantities also exported from Amoy and Swatow.

SIEB., *Œcon.*, 76:—*Convallaria multiflora*, var. *odora* (*Polygonatum*). Japonice *narukojuri;* sinice 黃精 Radices rarius eduntur.

So moku, VI, 6:—黃精 *Polygonatum multiflorum*, All.—*Ibid.*, VI, 7, 大葉黃精 (large-leaved *huang tsing*) *P. canaliculatum*, Pursh.

8.—萎蕤 *wei jui*. *P.*, XIIa, 35. *T.*, CXLII, CXXX.

Pen king:—*Wei jui* (the second character means pendent twigs or leaves), 女萎 *nü wei* [comp. *Rh ya*, 52]. The root is officinal. Taste sweet. Nature uniform. Non-poisonous.

Pie lu :—The *wei jui* is also called 玉竹 *yü chu* (jade bamboo) and 地節 *ti tsie* (earth nodes). It grows in the valleys of the T'ai shan mountain [in Shan tung, App. 322]. The root is dug up in the beginning of spring and dried in the sun.

In the dictionary *Shuo wen* [A.D. 100] the plant is called 萎蕤 *wei i*, in the *Wu P'u* [3rd cent.] it is 萎蕤 *wei jui*.

T'AO HUNG-KING :—The *wei jui* is a common plant. Its root resembles that of the *huang tsing* [r. 7] but is smaller. The people eat it.

SU SUNG [11th cent.]:—The plant is common in Ch'u chou [in An hui, App. 25], in Shu chou [in An hui, App. 294], in Han chung [S. Shen si], in Kün chou [in Hu pei, App. 172]. The stem is straight like a bamboo arrow-shaft ; the leaves are narrow and long, white on the upper side, green below. It is a kind of *huang tsing*. The root is of the thickness of a finger, covered with radical fibres, one or two feet long, edible. The flowers appear in the third month, and are of a green colour. The fruit is globular (berry).

LI SHI-CHEN :—It is a common plant in the mountains. Its root grows in a horizontal direction like that of the *huang tsing*, but is smaller, of a yellowish white colour, soft, covered with many radical fibres. It is very difficult to dry. The leaves grow two and two together, resemble bamboo-leaves. The plant is very easily propagated from the roots. The leaves and the root both are eaten boiled.

Kiu huang [LI, 3] and *Ch.* [VII, 14] sub *wei jui*, representations of a *Polygonatum*.

HAN., *Sc. pap.*, 255, examined and described the drug received from Shanghai under the name of *yü chu* (jade bamboo). It seems what HANBURY had before him were the rhizomes of a *Polygonatum*, but, having been misled by the Chinese name, he believed they belonged to a bamboo.

P. SMITH identifies *yü chu* and *wei jui* erroneously [p. 31] with bamboo rhizomes and [p. 175] with *Polygonatum aviculare*.

Cust., Med., p. 48 (36) :— *Yü chu*, exported 1885 from Chefoo to other Chinese ports 367 piculs,—p. 352 (169) from Canton 160 piculs,—p. 26 (71) Tientsin 86,—p. 8 (68) New chwang 66,—p. 102 (73) Wu hu 41.—Smaller quantities exported from Kiu kiang, Ning po.—P. 492 (1547). Places of production : Manchuria, Chi li, Shan tung, An hui, Che kiang, Sz ch'uan, Kuang si.

SIEB., *Œcon.*, 75 :— *Convallaria latifolia*. Japonice *hanemumasu*. Sinice 萎 蕤.

So moku [VI, 3] same Chin. name, *Polygonatum vulgare*, All.

Polygonatum vulgare is a common plant in the mountains of Northern China. The starchy mucilaginous root of it is eaten by the natives. The Chinese drug *wei jui* or *yü chu* is probably derived from this species.

9.—知 母 *chi mu.* P., XIIa, 39. T., CXXIX.

Pen king :—*Chi mu*, also written 蚔 母 *ch'i mu.* Other names 貨 母 *huo mu*, 地 参 *ti shen*, 連 母 *lien mu.* According to LI SHI-CHEN the second name (*ch'i mu*) means, mother of the eggs of ants, and is derived from the resemblance of the seeds, when they first begin to germinate, to ant's eggs. The root is officinal. Taste bitter. Nature cold. Non-poisonous.

For other ancient names see *Rh ya*, 94.

Pie lu:—The *chi mu* is also called 苦心 *k'u sin* (bitter heart), 兒草 *rh ts'ao.* It grows in the river valleys (plains) of Ho nei [S.E. Shan si, App. 77]. The root is dug up in the 2nd and 8th months and dried in the sun.

T'AO HUNG-KING:—Now the *chi mu* is met with in P'eng ch'eng [in Kiang su, App. 247]. It (the root) resembles the *ch'ang p'u* [*Acorus*, see 195]. The leaves are succulent and have a great vitality, and even when dried by fire the plant survives.

SU SUNG [11th cent.]:—The plant is found in the prefectures of Pin chou [in Shan tung, App. 251], Ho chou [in Kan su, App. 73], in Huai chou [in Ho nan, App. 93], in Wei chou [in Ho nan, App. 381], Chang te [in Ho nan, App. 5], likewise in Kie chou [in Shan si, App. 135], Ch'u chou [in An hui, App. 25]. In the 4th month it opens its green flowers resembling those of *Allium.* In the 8th month its fruit is formed.

Under the name of *chi mu* the *Ch.* [VII, 41] figures three different plants, all bad drawings. One of them represents a plant with lanceolate leaves and may perhaps be intended for *Anemarrhena asphodeloides*, Bge., which plant in the Peking mountains, where it is common, is known as *chi mu.*

TATAR., *Cat.*, 16:—*Chi mu*, Radix *Anemarrhenæ asphodeloides* et *Ophiopogon.*—GAUGER [42] describes and figures the *chi mu.* He says: Rhizomes of a monocotyl plant, having the appearance of the root of *Acorus Calamus.* Comp. also HAN., *Sc. pap.*, 259.

P. SMITH [57] identifies the *chi mu* erroneously with *Chelidonium.*

Cust. Med., p. 22 (9):—*Chi mu* exported 1885 from Tientsin to other Chinese ports 3,400 piculs.—A small quantity is also exported from Chefoo, p. 44 (6).—P. 436 (186):—Place of production : Chi li.

The plant figured in the *So moku* [II, 14] sub 知母
is, according to FRANCHET, *Aletris japonica*, Lamb. I should
rather think that the Japanese drawing represents *Anemarrhena.* The plant represented under the same Chinese
name in the *Phon zo* [V, 20, 21] has been identified by
FRANCHET with *Anemarrhena asphodeloides.*

10.—肉蓯蓉 *jou ts'ung yung.*   *P.*, XIIa, 41.—*T.*, CLXXI.

Pen king:—*Jou ts'ung yung.* The root and the stem
are officinal. Taste sweet. Nature slightly warm. Non-
poisonous.

Pie lu:—The *jou ts'ung yung* grows in Ho si [west of
the Yellow River, App. 79], in mountain valleys, also in
Tai [in N. Shan si, App. 321] and Yen men [in N. Shan si].
It is gathered on the 5th day of the 5th month, and dried
in the shade.

WU P'U [3rd cent.]:—This plant, which is also called
肉松容 *jou sung yung* and 黑司命 *hei sz' ling*, grows
in the mountains of Ho si, in shady places. It is gathered
from the 2nd to the 8th month.

T'AO HUNG-KING :—The localities Tai and Yen men
(mentioned in the *Pie lu*) are in the province of Ping chou
[N. Shan si, App. 253] and are rich in horses. People
say that the *jou ts'ung yung* plant is produced from the semen
of the wild horses dropping on the ground. The growing
plant has the appearance of flesh. It is cooked with mutton
broth and is very restorative. It may also be eaten raw.
It is also common in Ho nan. The best drug comes from
Lung si [Mid Kan su, App. 216]. It is flat [having been
pressed], of a yellow colour, succulent, has many flowers.
Its taste is sweet. That brought from the northern countries
is considered of a second quality. It is short and has but
few flowers. That from Pa tung [E. Sz' ch'uan, App. 237]

and Kien p'ing [E. Sz' ch'uan and W. Hu pei, App. 139] is of an inferior quality.

Su Kung [7th cent.]:—The drug spoken of by T'ao Hung-king is the 草 | | ts'ao (herbaceous) ts'ung yung, he had not seen the fleshy sort or jou ts'ung yung. The drug now commonly used in China is the first, of which the flowers have been scraped off. It is less potent than the fleshy kind.

Han Pao-sheng [10th cent.]:—The jou ts'ung yung grows in the Fu lu hien district in Su chou [in Kan su, App. 47, 313], in a sandy soil. In the 3rd and 4th months the people dig up the root, which is more than a foot long, cut out from the centre three or four inches, pass a string through it and dry it in the sun. In the 8th month it is ready for use.

The skin (of the plant) is scaly like the cone of a fir. As to the ts'ao (herbaceous) ts'ung yung it is gathered in the middle of the 4th month. It is from five or six inches to one foot long, has a round stem of a purple (violet) colour.

Su Sung [11th cent.]:—The plant is found in all prefectures of the province of Shen si [modern Shen si and E. Kan su, App. 284] but this is inferior to the drug brought from the border of Si k'iang [N.E. Tibet, App. 300] which is fleshy, thick and more potent. Ancient writers say that it is produced from the semen of the wild horse.

Ch. [VII, 17]:—Jou ts'ung yung. A rude drawing. No inference can be drawn from it.

Tatar., Cat., 13:—Ts'ung yung and [64] jou ts'ung yung, Orobanche praeparata.—The same drug figured and described in Gauger [51]. He says it is a large tongue-shaped fleshy root covered with scales, in a salted condition. In 1879 I procured the same drug at Peking. It was said to be brought from Mongolia. It proved to be Phelipæa salsa,

C. A. Meyer, order *Orobanchaceæ*. It consisted of the whole
plant, salted, the stem about three inches thick and scaly.
This plant is common in S. Siberia, Dsungaria, Mongolia.
It has been gathered in 1874 by Dr. PIASSETSKY in Kan su
[*Ind. Fl. sin.*, II, 222]. The name *ts'ung yung* in China is
probably applied to several plants of the order *Orobanchaceæ*.

Cust. Med., p. 70 (71) :—*Ts'ung yung* exported 1885
from Hankow to other Chinese ports 78 piculs. The drug
is referred there, it is unknown to me on what authority, to
Ruta angustiflora [sic !], which is most probably a mistake.
BRAUN [*Hankow Med.*, 46] states that *ts'ung yung* in Hankow
is *Æginetia japonica*, and gives 大芸 *ta yün* as a synonym.
According to the *Cust. Med.*, p. 26 (54) of the drug
ta yün, in 1885, 562 piculs were exported. *Ibid.*, p. 485
(1359) :—*Ts'ung yung*, *Æginetia*, sp. Places of production :
Chi li, Shan si, Sz ch'uan, Hu peh. The drug *ta yün* is
unknown to me.

HOFFM. & SCHLT., 26 :—草蓯蓉 *Æginetia japonica*,
Sieb. Zucc. [on the authority of SIEBOLD].

Phon zo, V, 21, 22 :—肉蓯蓉. The plant figured
there under this Chinese name is an *Orobanchacea*.

The *Ts'ao ts'ung yung* noticed in the above account has
a separate notice given of it in the *P.* [XIIa, 43] under
the name of 列當 *lie tang*, which plant is said there to be
the same as the *ts'ao* (herbaceous) *ts'ung yung*.

The *lie tang*, also called 栗當 *li tang* is first spoken of in
the *K'ai pao Pen ts'ao* [10th cent.] as a plant growing on
rocks in the province of Shan nan [S. Shen si, App. 268].
Its root resembles the root of *Nelumbium speciosum*. It is
used in medicine.

HAN PAO-SHENG [10th cent.]:—It grows in Yüan chou,
Ts'in chou, Wei chou, Ling chou [all in the eastern part

of present Kan su, App. 414, 358, 383, 195]. The root is dug up in the middle of the 4th month. It is from five or six inches to one foot long. The stem is round and of a white colour. It is also gathered for use, and pressed and dried in the sun.

SU SUNG [11th cent.]:—The root of the *ts'ao ts'ung yung* is very like the *jou ts'ung yung*, and after the flowers have been scraped off and the drug has been pressed it is substituted for the latter ; but it is less potent than the *jou ts'ung yung*. It is also called *lie tang*.

The *Ji hua Pen ts'ao* [10th cent.] terms the herbaceous *ts'ung yung* = 花 | | *hua* (flowering) *ts'ung yung*.

The *Ch.* [XVI, 58] represents sub *lie tang* a cylindrical stem covered with scales.

The plant figured in the *Phon zo* [V, 22] sub 列當 is a small *Orobanchacea*.

Under the name of 鎖陽 *so yang* the *P.* [XIIa, 43] mentions yet another plant which the Chinese believe to be a kind of *ts'ung yung*.

LI SHI-CHEN says :—The *so yang* is produced in Su chou (in Kan su). According to the CHO KENG-LU [14th cent.] it grows in the steppes of the Ta ta (Tatars, Mongols) in such places where the wild horse and the scaly dragon have happened to copulate. From the semen dropping upon the ground, sprouts like those of the bamboo shoot forth. The upper part is more succulent than the lower. It is covered with scales, resembles the penis and is a kind of *ts'ao ts'ung yung*. It is reported that lecherous women of the Ta ta use the *so yang* for purpose of masturbation. It is said that from the contact with the female organ it assumes the characteristics of the natural organ. The natives dig it up, wash it, take off the skin, dry it in the sun, and then use it as a medicine.

6

This is probably the plant *so yen* mentioned by the Chinese mediæval traveller Ch'ang chun [A.D. 1221] in North Mongolia. [See my book *Chin. Mediæv. Travellers*, p. 52.]

Ch., VIII, 63:—*So yang*. Rude drawing representing a cylindrical stem covered with scales.

The drug *so yang* which I procured in an apothecary's shop at Peking, and which has been examined by Professor Flückiger, seemed to be *Cynomorium coccineum*, L., order Balanophoreæ. This is a singular, fleshy, red, herbaceous plant, which, as the genus name indicates, resembles a canine penis. According to Potanin and Przewalski, it is a common plant in the Mongolian desert.

Braun, *Hankow Med.*, p. 40:—*So yang*, reddish brown roots with wrinkled surface. Sz ch'uan, Shan si, Ho nan.

Cust. Med., p. 70 (60):—*So yang* exported 1885 from Hankow to other Chinese ports 24 piculs.—*Ibid.*, 478 (1189): *so yang*, Balanophora. Places of production: Sz ch'uan, Hu peh.

11.—赤箭 *ch'i tsien.*—*P.*, XII*b*, 1.—*T.*, CLVI.

Pen king:—*Ch'i tsien* (red arrow), also 離母 *li mu*, 鬼督郵 *kui tu yu*. Root officinal. Taste pungent. Nature warm. Non-poisonous.

Pie lu:—The *ch'i tsien* grows in the river valleys (plains) of Ch'en ts'ang [in Shen si, App. 14], in Yung chou [Mid Shen si, App. 424], also in T'ai shan [in Shan tung, App. 322] and Shao shi [in Ho nan, App. 281]. The root is dug up in the 3rd, 4th and 8th months, and dried in the sun.

Pao P'o-tsz' [4th cent.] calls this plant 合離草 *ho li ts'ao* and 獨搖芝 *tu yao chi*, or self-moving *chi*, for it is said to move even when the air is still. It grows in the depths of the high mountains; no other plants are found near it. The root is very large and has twelve smaller tubers of the size of a hen's egg on the sides. The people use the tubers for food.

T‘AO HUNG-KING :—The *ch‘i tsien* is a kind of 芝 *chi* [plant of immortality, properly a Fungus, *v. infra*, 266]. It has a stem like the *tsien kan* [a reed used for arrow-shafts. *Bot. sin.*, II, 564] of a red colour. The leaves grow at the top. The root is very large, resembles that of the *yü* (*Colocasia antiquorum*), has twelve secondary tubers on the sides. The plant is not moved by wind, it moves only in still air.

In the *Yao sing Pen ts‘ao* [7th cent] this plant is termed 赤箭芝 *ch‘i tsien chi* (red arrow *chi*).

SU KUNG [7th cent.]:—The *ch‘i tsien* is a kind of *chi* [*c. supra*]. The stem resembles the *tsien kan* [*v. supra*] is of a red colour. The flowers and the leaves, which are likewise red, are at the top of the stem. It has wings like an arrow. The flowers open in the 4th month. The fruit resembles a decayed *k‘u lien* [*Melia. V.* 321]. The kernels are five or six angled and contain a mealy matter. When dried in the sun it is spoiled. The root is juicy, from five to six inches long, has ten and more smaller tubers on the sides. It is eaten raw, for in a dry state it is uneatable.

In the *K‘ai pao Pen ts‘ao* [10th cent.] the plant in question is called 天麻 *t‘ien ma* (heavenly hemp) and said to grow in Yün chou [in Shan tung, App. 421], Li chou [in Shan tung, App. 184], T‘ai shan [in Shan tung, App. 322], Lao shan [in Shan tung, App. 180]. The root is dug up in the 5th month and dried in the sun. Its leaves resemble those of the *shao yao* (*Pœonia albiflora*) but are smaller. From the midst of them rises a straight stem resembling the *tsien kan* [*v. supra*]. The fruit is produced at the top and resembles the fruit of the *su sui tsz.*[9] It becomes yellow and ripe at the time of the withering of the leaves. The principal root is connected with 12 secondary tubers of various sizes like the *t'ien men tung* [*Asparagus lucidus*. See 176]. Some of them are in shape like cucumbers, others like radishes. They are much used for food, both raw and steamed. The best sort is produced in Yün chou [*v. supra*].

I omit the details given regarding this plant by other authors quoted in the *P.*, for these accounts are confused and contradictory. Su Sung [11th cent.] seems to take the *ch'i tsien* and the *t'ien ma* to be distinct plants, both common in Middle China.

The figure of the *ch'i tsien* or *t'ien ma* in the *Ch.* [VII, 8] is a fancy drawing.

TATAR., *Cat.*, 56:—*T'ien ma*, Radix *Urticœ tuberosœ ?*—GAUGER [52] describes and figures the drug, a fleshy root, egg-shaped, three inches long.

HENRY, *Chin. plants*, 464:—*T'ien ma* or *ch'i tsien*, a *Gastrodia*, order *Orchideœ*.

Cust. Med., p. 60 (25):—*T'ien ma* exported 1885 from I chang to other Chinese ports 76 piculs.—*Ibid.*, 70 (67): from Han kow 500 piculs.—P. 493 (1296) Places of production: Hu peh, Sz ch'uan.

Phon zo, V, 23, 24:—赤箭 or 天麻. Figured a plant with a large root. It has the appearance of an *Orobanchacea*. But according to FRANCHET this drawing refers to *Gastrodia elata*, Bl. I may observe that *G. sesamoides*, R. Br., in Australia has a root which is full of starch and which is much eaten by the natives.

⁹ 續隨子 *Euphorbia Lathyris* in Japan. *So moku* [IX, 23].

12.—朮 *shu* (*chu*). *P.*, XII*b*, 4.—*T.*, CII.

Comp. *Rh ya*, 7, 8, 159.

Pen king:—Shu, also 山薊 *shan ki*. The root is officinal. Taste sweet. Nature warm. Non-poisonous.

Pie lu:—The *shu* is also called 山薑 *shan kiang* (mountain ginger) and 山連 *shan lien*. It grows in the mountain valleys of Cheng shan and Nan cheng in Han chung [in Southern Shen si, App. 16, 226, 54]. The root is dug up in the 2nd, 3rd, 8th and 9th months and dried in the sun.

Wu P'u [3rd cent.]:—The *shu* is also called 山芥 *shan kie* (mountain mustard) and 天薊 *t'ien ki* (heavenly thistle).

T'ao Hung-king:—Cheng shan and Nan cheng (the localities mentioned in the *Pie lu*) are identical. The *shu* is a common plant. The best drug comes from the mountains Tsiang shan [unknown to me. App. 351], Pai shan and Mao shan [both in Kiang su, App. 238, 218]. The root is dug up in the 11th and 12th months. At this time of the year it is very fat (resinous), and sweet. From the leaves a pleasant fragrant beverage is made. There are two kinds of *shu*. One of them is the 白朮 *pai shu* (white *shu* or *shu* simply so called). Its leaves are large, covered with hair and lobed. Its root is sweet, contains little resin, is used in the form of pills and powder. The other kind is the 赤朮 *ch'i shu* or red *shu*. It has small leaves, not lobed. The root is small and of a bitter taste, contains much resin. It is used in a fried state. That brought from the eastern frontier is large, not strong, and is not much used.

Su Sung [11th cent.]:—The *shu* is a common plant. The best drug comes from Mao shan [*v. supra*] and Sung shan [in Ho nan, App. 317]. The plant grows from two to three feet high. In summer it opens its violet flowers

resembling those of the *ts‘z‘ ki* [spiny thistle. *See* 83].
Sometimes the flowers are yellowish white. After summer
it produces seeds and in autumn the plant withers. The root
resembles ginger, is beset with radical fibres. Its skin is
black, the heart is yellowish white, contains a resinous juice
of a purplish (brown) colour. T‘AO HUNG-KING distinguishes
two kinds of *shu*. His *pai* (white) *shu* is the same as the
yang fu of the *Rh ya* [see *Rh ya*, 8]. At present it is
found on the high mountains in the prefectures of Hang chou
[in Che kiang, App. 58], Yüe chou [in Che kiang, App.
418], in Shu chou [in An hui, App. 294], in Süan chou
[in An hui, App. 315]. Its leaves stand opposite each other,
are covered with hair. The stem is square, on its top are
the flowers. They are of a pale purple colour, or blue, or
red. The root is branched. That from the violet flowered
kind yields the best drug. By the drug *shu* mentioned in
ancient prescriptions always the *pai shu* is to be understood.

LI SHI-CHEN says that the *pai shu* plant resembles
the *ki* (thistle). The taste of the root is like ginger and
mustard. It is much cultivated in Yang chou [Che kiang,
Kiang su, App. 400] and also known under the name of
吳尤 or *shu* from Wu [Che kiang, Kiang su, App. 389].
The ancient prescriptions do not discriminate between the
white and the red *shu*. It was only in later times that the
latter was distinguished and termed *ts‘ang shu*.

The 蒼尤 *ts‘ang shu* is first mentioned by K‘ou TSUNG-
SHI [12th cent.] as a succulent root of the size of a finger,
with a gray skin and of a pungent, bitter taste. It is used
in the same way as the *pai shu*, which is sweet.

LI SHI-CHEN identifies the *ts‘ang shu* with the 赤尤
ch‘i (red) *shu* which is mentioned in the *Pie lu*, in the *Rh ya*
[159] and by T‘AO HUNG-KING [*c. supra*]. It is also known

under the names of 山 薊 *shan ki* (mountain thistle), 山 精 *shan tsing*, 仙 术 *sien shu*. It is a common mountain plant with leaves clasping the stem. The leaves are from three to four lobed, serrated on the margin and provided with small spines. The root resembles ginger.

Kiu huang, LI, 8, and *Ch.*, VII, 9, sub *ts'ang shu*, representation of an *Atractylis*.

TATAR., *Cat.*, 5:—*Ts'ang shu (chu)*. Radix?

GAUGER describes and depicts [5] the drug *pai shu* and [46] the *ts'ang shu*.

HANB., *Sc. pap.*, 255:—*Ts'ang shu*. Rhizome of *Atracty-loides*, sp.

P. SMITH, 28:—*Peh shuh*, *Atractylodes alba* and *tsang shuh*, *Atractylodes rubra* [species names invented by P. SMITH].

Cust. Med., p. 138 (68-70):—*Pai shu* 1885 exported from Ning po to other Chinese ports about 28,000 piculs. In small quantities also exported from Han kow, Wen chow, Kiu kiang.—*Ibid.*, p. 469 (961) Places of production: Che kiang, Kiang si, An hui, Yün nan.

Ibid., p. 8 (59):—*Ts'ang shu* exported from New chwang 1,708 piculs,—p. 26 (62), from Tientsin 1,620 piculs,—p. 48 (34), from Che foo 74 piculs,—p. 70 (69), from Han kow 2,254 piculs,—p. 102 (64), from Wu hu 7 piculs.—*Ibid.*, 484 (1330). Places of production: Manchuria, Chi li, Shan tung, Sz ch'uan, Hu peh, An hui, Che kiang.

See also *Hank. Med.*, p. 39.

Phon zo, V, 24, 25:—白 术 *Atractylis ovata*, Thbg.,—26, 27 蒼 术 *A. lancea*.

So moku, XV, 52:—蒼 术 *Atractylis orata*,—51, same Chinese name, *A. lancea*, said to be from China,—50, 白 术 *A. ovata*, said to be from China.

SIEB., *Icon. ined.*, V:—蒼 术 *Atractylis chinensis*, DC.

The latter is a common plant in the Peking mountains. It is very spiny. The leaves very variable in shape. Flower-heads small, of a pale violet colour. All the Eastern Asiatic species of *Atractylis* are now generally reduced to one species— *A. ovata*. [See *Ind. Fl. sin.*, I, 459.]

13.—狗脊 *kou tsi*.   *P.*, XII*b*, 16.—*T.*, CLXVI.

Pen king :—Kou tsi (dog's backbone) or 百枝 *po chi* (hundred branches). The root is officinal. Taste bitter. Nature uniform. Non-poisonous.

Pie lu :—The *kou tsi*, also called 强脊 *k'iang lü* (worm's back), 扶筋 *fu kin*, 扶蓋 *fu kai*, grows in Ch'ang shan [in Chi li, App. 8] in river valleys. The root is dug up in the 2nd and 8th months and dried in the sun.

Wu P'u [3rd cent.]:—It is also called 狗青 *kou ts'ing*. The root of the *kou tsi* resembles the *pi hiai* [*Smilax*. See 178]. The stem has joints like the bamboo and is prickly. The leaves are round, of a red colour. The root is yellowish white, resembles the bamboo-root, is covered with hair.

T'ao Hung-king :—The *kou tsi* is a common plant in the mountains. It resembles the *pa k'ai* [*Smilax China*. See 179]. The stem is prickly, the leaves are round and have red veins. The root presents protuberances and excavations, resembles a ram's horn.

Su Kung [7th cent.]:—The *kou tsi* is a medicinal plant, which resembles the *kuan chung* [a Fern, *see the next*]. It has a long root with many protuberances and resembles the backbone of a dog. The flesh of the root is of a greenish colour, whence the [above] ancient name *kou ts'ing* (dog green).

Su Sung [11th cent.]:—The *kou tsi* is found in the T'ai Hang mountain range (between Chi li and Shan si,

App. 323] and in the prefectures of Tsz' chou [in Shan tung, App. 367], Wen chou [in Che kiang, App. 385] and Mei chou [in Sz ch'uan, App. 219]. The plant grows one foot high, has no flowers. In its stem and leaves it resembles the *kuan chung* [*v. supra*]. Its leaves are sharp, finely divided. The root is of a black colour three or four inches long, has many protuberances, resembles the backbone of a dog. Large specimens are two fingers' thick. The flesh of the root is of a greenish colour. It is dug up in spring and in autumn and dried in the sun. The drug now used in medicine is that beset with gold-coloured hairs (radical fibres). The plant spoken of by T'AO HUNG-KING as having prickles is the *pi hiai* (*Smilax*) not the *kou tsi*.

LI SHI-CHEN :—Of the *kou tsi* there are two kinds. One has a black root resembling the backbone of a dog, the other is covered with gold-coloured hairs and resembles a dog in shape. Both are used in medicine. The plant described as the *kou tsi* by WU P'U and T'AO HUNG-KING is not the true *kou tsi* but the *pa kia*, which according to the *Po wu-chi* [3rd cent.] is akin to the *pi hiai* [*v. supra*] and sometimes is also called *kou tsi*. The plant described by SU KUNG and SU SUNG is the true *kou tsi*.

Ch. [VIII, 2] sub *kou tsi*, representation of a fern with bipinnate fronds.

LOUREIRO [*Fl. cochin.*, 829] *Polypodium Baromez*, L., Agnus scythicus, sinice *keu tsie*. Radix oblonga, crassa, carnosa, multiformis, supra terram exerta : radiculis crassis, pilis densissimis, tenuibus rufis tota vestita.—This is the *Cibotium glaucum* in *Flora hongk.* [466], a fern.

HANB., *Sc. pap.*, 121.—P. SMITH, 194, *Tartarian. Lamb.*

Cust. Med., p. 186 (47):—*Kou tsi* exported 1885 from Ning po to other Chinese ports 52 piculs,—p. 210 (25), from

Wen chow 45 piculs,—p. 344 (66), from Canton 42 piculs.—
Ibid., p. 455 (606) Places of production: Che kiang, Kuang
tung, Kuang si.

The *Phon zo* [VI, i] figures sub 狗脊 a fern with
bipinnate fronds. FRANCHET, *Enum. Pl. Jap.* [II, 204]
refers this drawing with a ? to *Onoclea germanica*, Willd.

14.—貫眾 *kuan chung*. P., XIIb, 18.—T., CXXXIII.
Comp. *Rh ya*, 110.

Pen king :—*Kuan chung.* Other names : 貫節 *kuan
tsie,* 貫渠 *kuan k'ü,* 百頭 *po t'ou* (hundred heads), 虎卷
hu küan, 扁府 *pien fu.* The root is officinal. Taste bitter.
Nature slightly cold. Poisonous.

Pie lu :—The *kuan chung* is also called 草鴟頭 *ts'ao
ch'i t'ou* (herbaceous owl's head). It grows in the mountain
valleys of Yüan shan [unknown to me, App. 416], in Yüan
kü [in Shan tung, App. 415] and on the Shao shi mountain
[in Ho nan, App. 281]. The root is dug up in the 2nd and
8th months and dried in the shade.

T'AO HUNG-KING :—The *kuan chung* is common in Middle
China. Its leaves resemble those of the great *küe* [a fern.
See *Rh ya*, 185]. The root in its shape, colour and hairy
appearance recalls the head of an owl [*v. supra*].

SU SUNG [11th cent.] :—·The *kuan chung*, otherwise
called 鳳尾草 *feng wei ts'ao* (phœnix-tail plant) is common
in the provinces of Shen si [now Shen si and E. Kan su,
App. 284] and Ho tung [Shan si, App. 80] and also in
King and Siang [Hu pei, App. 145, 305].

LI SHI-CHEN :—The *kuan chung* is a common plant in
the mountains, in shady places. Several stalks issue from
the same root. They are as thick as a quill, slippery. The
leaves are in opposite pairs, resemble those of the *kou tsi*

[*see* 13] but are not serrated or dentated. They are of a yellowish green colour, the under side is paler. The root is crooked and covered with erect dense black hair, like that of the *kou tsi*, but it is larger and resembles an owl.

Ch. [VIII, 18]:—*Kuan chung*. Representation of a fern with large pinnate fronds, covered with spores. Root covered with dense hair.

According to M. FAUVEL ["Trip of a Naturalist to the Chin. Far East," 11], in Shan tung the name *kuan chung* is applied to *Aspidium falcatum*, Sw., a fern, known also from Fu kien and S. Shen si.

HENRY [*Chin. pl.*, 199, 200]:—*Kuan chung*. This name in Hu peh is applied to several ferns, viz. *Woodwardia radicans*, Sm., *Onoclea orientalis*, Hook., *Nephrodium filix mas*, Rich.

Cust. Med., p. 344 (69) :—*Kuang chung* exported from Canton to other Chinese ports 15 piculs,—p. 278 (61), from Amoy 0.43 piculs.—*Ibid.*, p. 457 (647) Places of production: Fu kien, Kuang tung.

Phon zo [VI, 3, 4]:—貫 眾 Fern. According to FRANCHET : *Lomaria nipponica*.

15.—巴 戟 天 *pa ki t'ien*. P., XII*b*, 20. T., CLXXIV.

Pen king :—*Pa ki t'ien*. The root is officinal. Taste bitter and sweet. Nature slightly warm. Non-poisonous.

Pie lu :—The *pa ki t'ien* grows in Pa [E. Sz ch'uan, App. 235] and in Hia p'ei [in Kiang su, App. 63] in mountain valleys. The root is dug up in the 2nd and 8th months and dried in the shade.

T'AO HUNG-KING :—At present the people use also the drug which comes from Kien p'ing [in Sz ch'uan and

Hu pei, App. 139] and I tu [in Hu pei, App. 104]. The root
resembles that of the *Pæonia moutan*, but is smaller, of a
red colour outside, black within. When prepared for use,
the heart is taken out.

Su Kong [7th cent.]:—A popular name for this plant
is 三蔓草 *san man ts'ao*. The leaves resemble tea-leaves;
they do not fall off in winter. The root consists of fleshy
tubers.

Su Sung [11th cent.]:—The plant grows in Kiang
Huai [An hui, Kiang su, App. 124, 89], in Ho tung [Shan si,
App. 80], but the best kind comes from Shu chou [part of
Sz ch'uan, App. 292]. Description of the plant not quite
clear.

The *Ji hua Pen ts'ao* [10th cent.] calls it 不凋草
pu tiao ts'ao (plant which does not fade).—The plant *pa ki
t'ien* seems to have been unknown to Li Shi-chen.

The *Ch.* [VII, 16] gives under the name of *pa ki t'ien*
two rude drawings representing two different plants, one
from Ch'u chou [in An hui, App. 25], the other from Kui
chou [in Hu pei, App. 169].

Tatar., *Cat.*, 1:—*Pa ki t'ien*. Not identified.—Gauger
[3] describes and figures the root.

This is perhaps the *pa tsi hien* in Lour., *Fl. cochin.*, 477,
the Chinese name for *Septas repens*, now called *Herpestis
Monnieria*, Benth. Order *Scrophularineæ*.

Cust. Med., p. 346 (103):—*Pa ki t'ien* exported 1885
from Canton to other Chinese ports c. 600 piculs,—p. 318
(52), from Swatow 38 piculs,—p. 280 (84), from Amoy
2 piculs.—*Ibid.*, p. 467 (926) Places of production: Kuang
tung, Kuang si, Che kiang.

The drawing sub 巴戟天 in Sieb., *Icon. ined.* [II]
seems to represent a *Polygala*. The plant described and

figured in the *Kwa wi* [11] under the same Chinese name is, according to FRANCHET, *Polygala Reinii*. Comp. also *Phon zo*, VI, 6, under the same Chinese name. The figure of the plant shows only leaves.

16.—遠志 *yüan chi*. *P.*, XIIb, 21.—*T.*, CXXXIII.

Comp. *Rh ya* and *Classics*, 194, 443.

Pen king:—*Yüan chi*, 小草 *siao ts'ao* (small herb). name applied to the leaves. Other names 細草 *si ts'ao* (same meaning), 棘菀 *ki yüen*, 葽繞 *yao yao*. Root and leaves officinal. Root bitter. Nature warm. Non-poisonous.

Pie lu:—The *yüan chi* grows in T'ai shan [in Shan tung, App. 322] and in Yüan kü [in Shan tung, App. 415]. The root is gathered in the 4th month and dried in the shade.

T'AO HUNG-KING :—Yüan kü is the prefecture of Tsi yin in Yen chou [in Shan tung, App. 347 and 404]. At present the drug commonly used comes from P'eng ch'eng [in Kiang su, App. 247] and Lan ling [Shan tung, App. 174]. In the 仙方 *sien fang* (Taoist prescriptions for procuring long life) the *siao ts'ao* [*v. supra*, the leaves of the plant] is used. It resembles the *ma huang* [*Ephedra*, see 97] but is green.

MA CHI [10th cent.]:—In its stem and its leaves it resembles the *ta ts'ing* [*Justicia*. See 89] but is smaller.

CHANG YÜ-HI [11th cent.]:—The plant is mentioned in the *Rh ya*. It resembles the *ma huang*, has red flowers, yellow, sharp leaves. The upper part of the plant is called *siao ts'ao*.

SU SUNG [11th cent.]:—This plant grows in North China. Its root resembles that of the *hao* (*Artemisia*). It is of a yellow colour. The leaves resemble those of the *ma huang, ta ts'ing* and other plants. In the 3rd month it produces white flowers. The root is nearly one foot long.

There is one kind of this plant in Sz chou [in An hui, App. 319] with red flowers and a large root. Another kind is produced in Shang chou [in Shen si, App. 278]; it has a black root. Now the *siao ts'ao*, used in medicine in ancient times, is seldom employed.

Li Shi-chen:—There are two kinds of *yüan chi*, one has larger, the other smaller leaves. The *siao ts'ao* spoken of by T'ao Hung-king belongs to the latter, the plant mentioned by Ma Chi is the large-leaved. It has red flowers.

Ch. [VII, 13] *yüan chi* and *Kiu huang* [LIII, 5] representations of *Polygala*.

Tatar. [*Cat.*, 31] *yüan chi*, Rad. *Polygalæ tenuifoliæ.*— The roots of the *yüan chi* described and figured in Gauger, 24.—P. Smith, 175.

Cust. Med., p. 72 (84):—*Yüan chi* exported 1885 from Han kow to other Chinese ports and Japan, 350 piculs,— p. 28 (73), from Tien tsin 150 piculs,—p. 48 (49), from Che foo 119 piculs,—small quantities from 1 chang, Chinkiang.—*Ibid.*, p. 493 (1557) Places of production: North and Mid China.

The drug *yüan chi* in China and Japan is yielded by *Polygala sibirica* (large-leaved), *P. tenuifolia* (small-leaved) and *P. japonica*. For further particulars see *Bot. sin.*, II, 194, 443.

17.—淫羊藿 *yin yang huo.* P., XIIb, 24.—T., CLXXIX.

Pen king:—*Ying yang huo*, also 剛前 *kang ts'ien.* The root and the leaves are officinal. Taste pungent. Nature cold. Non-poisonous.

Pie lu:—The *yin yang huo* grows in Shang kün [in N.E. Shen si, App. 273] and in Yang shan [in N. China, App. 399], in mountain valleys,

Su Kung [7th cent.]:—The plant is commonly called
仙靈脾 *sien ling p'i*. The leaves resemble pea-leaves, but
are round and thin. Stem slender but hard.

Su Sung:—It is a common plant in North and Mid
China. The stem like that of millet. The leaves like apricot-
leaves, provided with prickles (on the margin). The root
is of a purple colour, beset with radical fibres. In the
4th month it bears small white or purple flowers. In the
5th month the leaves are gathered and dried in the sun. One
kind, which is produced in Hu and Siang [Hu nan, App. 83,
307] has leaves like the pea on slender petioles, which do
not fade in winter. The root resembles that of the *huang lien*
[*Coptis*. See 26]. In Kuan chung [Shen si, App. 158]
the plant (another kind) is called 三枝九葉 *san chi kiu ye*
(three twigs—common petioles—nine leaves). It grows
from one to two feet high. The root and the leaves much
used in medicine. The *Shu pen ts'ao* says, the best for use
is that plant which grows where the sound of water is not
heard.

Li Shi-chen:—The plant grows in high mountainous
regions. Several coarse stems like thread issue from the
same root and grow one or two feet high. Each stem
divides into two branches, each branch (common petiole)
bears three leaves. The leaves are from two to three inches
long, resemble apricot-leaves or pea-leaves. They are very
thin, dentated, with small prickles, glabrous on the upper
side, glaucous beneath.

Ch. [VIII, i]:— *Yin yang huo*. The drawing seems to
represent *Epimedium sagittatum*, Bak. [*Aceranthus sagittatus*,
S. & Z.], which agrees with the above Chinese description.

Tatar. [*Cat.* 30]:—*Yin yang huo*, *Folia Populi* (an
erroneous identification).—P. Smith [176] identifies the same
Chinese name with *Populus spinosa*. This name is purely

imaginary on his part.—The *yin yang huo* which I obtained from an apothecary's shop in Peking were the leaves of *Aceranthus sagittatus*, a plant recorded by our botanists from Hu pei.—Comp. HENRY, *Chin. pl.*, 556.

Cust. Med., p. 362 (286):—*Yin yang huo* exported 1885 from Canton to other ports of China, 30 piculs,—p. 74 (111), from Hankow 25 piculs,—p. 288 (227) a small quantity from Amoy.—*Ibid.*, p. 492 (1,536) Places of production: Shen si, Hu peh, Fu kien, Kuang tung, Kuang si.

Ind. Fl. sin. [I, 32]. Three species of *Epimedium* reported from Mid and North China.

So moku [II, 45]:—淫羊藿, *Aceranthus sagittatus*. According to SIEBOLD this plant has been introduced from China into Japan.

18.—玄參 *hüan shen* (the first character is frequently substituted by 元 *yüan*). *P.*, XIIb, 28.—*T.*, CLIX.

Pen king:—*Hüan shen* (dark ginseng), also 重臺 *chung t'ai*. The root is officinal. Taste bitter. Nature slightly cold. Non-poisonous.

Pie lu:—The plant is also called 端參 *tuan shen*, 咸｜ *hien shen*, 正馬 *cheng ma*. It grows in the river valleys of Ho kien [in Chi li, App. 75] and in Yüan kü [in Shan tung, App. 415]. The root is dug up in the 3rd and 4th months and dried in the sun.

Wu P'u [3rd cent.]:—The plant is also called 玄臺 *hüan t'ai*, 鹿腸 *lu ch'ang*, 鬼藏 *kui ts'ang*. It grows in Yüan kü [*v. supra*] and Shan yang [App. 270]. The leaves are covered with hair. They stand four opposite, resemble the leaves of the *shao yo* [*Pœonia.* See 52]. The stem is black, square, from four to five feet high. The fruit is black.

T'AO HUNG-KING:—The *hüan shen* is a common plant in Mid China. Its stem is like that of the true ginseng but higher. The root is very black, slightly fragrant. The Taoists use it as a perfume.

SU KUNG [7th cent.]:—The *yüan shen* has a disagreeable smell. Its stem does not resemble that of the ginseng. It is unknown to the author that the drug is used as a perfume.

MA CHI [10th cent.]:—It has a square stem, from four to five feet high, of a brown colour, covered with fine hair. The leaves are large, like the palm of the hand, and sharp. The fresh root is greenish white, succulent; when dried it becomes purplish black.

SU SUNG [11th cent.]:—The leaves stand opposite, resemble those of the *chi ma* [*Sesamum*. See 216]. They are long, serrated. The stem is slender, of a purplish green colour. In the 7th month it opens its greenish flowers, in the 8th month it produces black fruits. Another kind has white flowers, a large square stem covered with fine hair. The root consists of from five to seven different pieces.

LI SHI-CHEN:—The *yüan shen* which is now used, is that described by SU SUNG. SU KUNG is right in stating that the root has an unpleasant smell. It is perennial and frequently worm-eaten, and therefore full of holes. The flowers are purple or white. It is also called 黑參 *hei shen* (black ginseng) or 野脂麻 *ye chi ma* (wild *Sesamum*).

Ch., VII, 43:—*Yüan shen.* The drawing is too indistinct to permit of identification.

TATAR., *Cat.*, 48:—*Huan shen*, Radix.—GAUGER, 40, the same root described and figured.—HANB. [*Sc. pap.*, 255] also describes this root. It is internally fleshy and black. P. SMITH, 104:—Black Ginseng.

8

Cust. Med., p. 192 (115, 116):—*Yüan shen* exported in 1885 from Ning po 4,700 piculs.—*Ibid.*, 493 (1563) Places of production : Che kiang.—According to BRAUN, [*Hank. Med.*] exported also from Hankow.

So moku, XI, 57:—玄參 *Scrophularia Oldhami*, Oliv. The above Chinese description might perhaps agree.

19.—地 榆 *ti yü*. *P.*, XII*b*, 30.—*T.*, CXXXVII.

Pen king:—*Ti yü* (ground elm). The root and the leaves used in medicine. Taste of the root bitter. Nature slightly cold. Non-poisonous.

Pie lu:—The *ti yü* grows in T'ung po [in Ho nan, App. 379] and Yüan kü [in Shan tung, App. 415] in mountain valleys. The root is dug up in the 2nd and 8th months and dried in the sun.

The same work says the 酸赭 *suan che* grows in the mountains of Ch'ang yang [in Shan tung, App. 7]. Comp. *infra* LI SHI-CHEN.

T'AO HUNG-KING :—The leaves of *ti yü* resemble elm leaves, but are longer and, as they cover the ground when the plant first begins to grow, the latter is called *ti yü* or ground elm. Its flowers and seeds (or fruits) are of a brown colour like the *shi* [soy. See *Bot. sin.*, II, 355] wherefore it is also called 玉 豉 *yü shi*. The root is used for fermenting liquors. The Taoists burn it and use it for alchemistic purposes (?). The mountain people substitute the leaves for tea. They may also be eaten fried.

SU SUNG [11th cent.]:—The *ti yü* is a common plant in the plain, on meadows and marshes. From the perennial root, in the 3rd month, the leaves issue. They cover the ground. After this a single stem shoots forth, from three to four feet high. It divides and produces leaves which stand opposite. The leaves resemble elm-leaves but are narrower and longer, serrated on the margin. In the 7th month the flowers appear. They are of a dark red colour and resemble (the flower-head) a mulberry. The root is black outside, red internally, resembles the willow root.

LI SHI-CHEN:—The *Wai tan fang* (alchemistic prescriptions) says, the *ti yü* is also called *suan che* [sour *che*, *v. supra*] for it is of a sour taste and of a brown colour. The same name is still applied to the *ti yü* by the people in K'i chou [in Hu pei, App. 121, LI SHI-CHEN's native place]. It is sometimes erroneously written *suan tsao* (sour jujube).

The rude drawings of the *ti yü* as given in the *Kiu huang* [XLVI, 30] and in the *Ch.* [VIII, 4] seem to be intended for *Sanguisorba officinalis*, L. (*Poterium*), the Burnet, with which the above Chinese description agrees.

TATAR., *Cat.*, 21:—*Ti yü*, *Rad. Hedysari.*—P. SMITH, 110.

Cust. Med., p. 350 (150):—*Ti yü* exported 1885 from Canton to other Chinese ports 47 piculs,—p. 146 (116), from from Shanghai, 8 piculs.—*Ibid.*, p. 482 (1,273) Places of production : Che kiang, Kuang tung, Kuang si.

Our common burnet is a common plant in North and Mid China. In the Peking mountains it is known by the name of *ti yü*. The Canton drug may be yielded by another plant, for in the *Ind. Fl. sin.* [I, 246] no *Poterium* is reported from South China.

Kwa wi, 29 :—地榆 or 玉豉 *Sanguisorba officinalis.* *So moku*, II, 24 :—地榆 *Poterium officinals*, which is the same.

20.—丹參 *tan shen.* P., XIIb, 32.—T., CLVI.

Pen king:—*Tan shen* (cinnabar coloured ginseng), 郄蟬草 *hi ch'an ts'ao.* The root is officinal. Taste bitter. Nature slightly cold, Non-poisonous.

Pie lu:—The *tan shen*, also called 赤蔘 *ch'i shen* (red ginseng) grows in T'ung po [in Ho nan, App. 379], in river valleys, and in T'ai shan [in Shan tung, App. 322]. The root is dug up in the 5th month and dried in the sun.

T'AO HUNG-KING :—The T'ung po [mentioned in the *Pie lu*] is a mountain situated in I yang [a district in Ho nan, App. 107]. The Huai River takes rise on it. It is not to be confounded with another mountain of the same name in Lin hai in the province Kiang tung [Che kiang, App. 192 and 124]. The *tan shen* is a common plant in Mid China. It has a square stem covered with hair, purple (or violet) flowers. The people call it sometimes 逐馬 *chu ma.*

WU P'U [3rd cent.]:—In its stem, leaves and seed capsules it resembles the *jen* [*Perilla.* See 67]. The plant is covered with hair. The root is of a red colour. In the 4th month it opens its purple (violet) flowers.

SU SUNG [11th cent.]:—It grows in all the prefectures of the provinces of Shen si [modern Shen si and E. Kan su] and Ho tung [Shan si, App. 80], also in Sui chou [in Hu pei, App. 314]. The leaves resemble those of the *po ho* (*Mentha*), are covered with hair. Purple flowers in spikes. The root is of a red colour, of the thickness of a finger. One plant has many roots.

LI SHI-CHEN :—There are five kinds of *shen* [ginseng. Comp. above, 4], distinguished according to their colour and their effect upon the five viscera :—

1.—The 人蔘 *jen shen* [true ginseng. *See* 3]. It operates upon the spleen and is called the *huang* (yellow) *shen.*

2.—The 沙| *sha shen* [see 4]. It operates upon the lungs and is called the *pai* (white) *shen.*

3.—The 玄 | *hüan shen* [see 18]. It operates upon the kidneys and is called the *hei* (black) *shen*.

4.—The 牡蒙 *mou meng* [see 21]. It operates upon the liver and is called *tsz'* (purple) *shen*.

5.—The 丹 | *tan shen*. It operates upon the heart and is called *ch'i* (red) *shen*.

The latter is a common plant in the mountains. Its leaves (leaflets) are five together on a common petiole, resemble those of the wild *su* (*Perilla*). The root is red externally and has a purple flesh.

Ch., VII, 20:— *Tan shen*. Rude drawing, but it seems a *Salvia* is intended. [Comp. also X, 37] *siao* (small) *tan shen*, likewise a *Salvia*.

TATAR., *Cat.*, 20 :— *Tan shen*. Rad. *Salviæ miltiorhizæ.*— P. SMITH, 194.

Salvia miltiorhiza, Bge., is a common plant in the Peking mountains. It has been recorded also from Shan tung, Hu pei, etc. It has a cinnabar red root, from three to seven foliate leaves, large violet flowers.

Cust. Med., p. 70 (62):— *Tan shen* exported 1885 from Han kow to other Chinese ports 405 piculs,—p. 124 (55) from Chen kiang 257 piculs,—p. 46 (30) from Che foo 233,—p. 26 (56) from Tien tsin 17 piculs.— *Ibid.*, p. 480 (1246) Places of production : Chi li, Shan tung, Shan si, Shen si, Sz ch'uan.

Phon zo, VI, 18 :—丹参 *Salvia nipponica*, Miq.

21.—紫参 *tsz' shen*. *P.*, XIIb, 34.— *T.*, CLIX.

Pen king :— *Tsz' shen* (purple ginseng), 牡蒙 *mou meng*. The root is officinal. Taste bitter. Nature cold. Non-poisonous.

Pie lu :—The *tsz'* *shen* is also called 童腸 *t'ung ch'ang*, 馬行 *ma hing*, 聚戎 *chung jung*. It grows in Ho si [West of the Yellow River, App. 79] and in the mountain valleys of Yüan kü [in Shan tung, App. 415]. The root is dug up in the 3rd month. After drying by fire it becomes purple.

T'AO HUNG-KING says that it is not much used in medicine.

The ancient authors give confused contradictory accounts of the plant. Some say it resembles in its leaves the Sorrel (*Rumex*), others liken the flowers to a *Polygonum*. The root when dried is of a dark purplish colour, the flesh within is pale red. It resembles the root of the *tsz' ts'ao* [*Lithospermum erythrorhizon*. See 23] but is smaller.

Ch. VII, 44 :—*Tsz' shen*. Rude drawing. A quite different plant is represented under the same Chinese name in *Ch.* [XXIII, 31].

In the Peking mountains the name *tsz' shen* is applied to the root of *Polygonum bistorta*, L.

So moku, VII, 54 :—紫參 *Polygonum bistorta*, var. foliis ovatis and other varieties.—*Phon zo*, VI, 20 *r.*—Same identification.

The typical form of *Polygonum bistorta* is called 拳參 *k'üan shen* (fist ginseng in Chinese) in the *So moku* [VII, 53].—Same identification in the *Kwa wi* [57].

The *k'üan shen* is mentioned in the *P.* [XIII, 63]. This name appears first in the *T'u king Pen ts'ao* [11th cent.]. The plant is said there to grow wild in Tsz' chou [in Shan tung, App. 367]. Its leaves resemble those of the *yang ti* [*Rumex*. See 193], the root is like a lobster, and of a black colour. It is dug up in the 5th month.—The *k'üan shen* is not mentioned in the *Ch.*

P. SMITH, 39 :—Bistort root.

22.—王孫 *wang sun*.　*P.*, XIІ*b*, 35.　*T.*, CXXVI.

Pen king:—*Wang sun*. The root is officinal. Taste bitter. Nature uniform. Non-poisonous.

Pie lu:—The *wang sun*, otherwise called 黃孫 *huang* (yellow) *sun*, 黃昏 *huang hun*, grows in Hai si [in Kiang su, App. 50] in river valleys, also near the wall of the city of Ju nan [in Ho nan, App. 110].

Wᴜ Pʻᴜ [3rd cent.] says, in Chʻu [Hu kuang, App. 24] it is called *wang sun*, in Tsʻi [N.E. Shan tung, App. 348] it is 長孫 *chʻang sun* or 海孫 *hai sun*, in Wu [Kiang su, An hui, App. 389] it is 白功草 *pai kung tsʻao* or 蔓延 *man yen*.

Tʻᴀᴏ Hᴜɴɢ-ᴋɪɴɢ says that in prescriptions the *wang sun* is also termed 黃昏 *huang hun* and 牡蒙 *mou meng*. But later authors prove that this is a mistake, for *huang hun* is the same as the *ho huan* [*Acacia*. See 324] and *mou meng* is the *tszʻ shen* [*v.* 21].

Cʜʻᴇɴ Tsʻᴀɴɢ-ᴋʻɪ [8th cent.] calls it 旱藕 *han ou* (*Nelumbium* root in a dry soil). The root resembles that of *Nelumbium speciosum*. It grows in the Tʻai Hang mountain range [in N. China, App. 323].

Lɪ Sʜɪ-ᴄʜᴇɴ:—The leaves of the *wang sun* are crowded towards the top of the stem as in the *tszʻ ho chʻe* [*Paris*. See 151]. These leaves resemble the leaves of the *ki ki* [*Chloranthus*. See 42]. The drawing of the *wang sun* in the *Pen tsʻao kang mu* seems to be intended for a *Paris*, the leaves of this genus, as is known, being whorled at the apex of the stem.

Ch., VIII, 3 :—*Wang sun*. Drawing indistinct.

Comp. also Hᴇɴʀʏ, *Chin. pl.*, 320.

So moku, VII, 81, 82 :—王孫 *Paris quadrifolia*, L., and *P. tetraphylla*, A. Gray.

23.—紫草 *tsz' ts'ao.* *P.,* XII*b*, 36. *T.,* CLIX.

Comp. *Rh ya,* 142, for other ancient names.

Pen king:—*Tsz' ts'ao* (purple herb). The root is officinal. Taste bitter. Nature cold. Non-poisonous.

Pie lu:—The *tsz' ts'ao* is also called 紫丹 *tsz' tan* (purple cinnabar red). It grows in the mountain valleys of Tang shan [in Kiang su, App. 334] and in the country of Ch'u [Hu kuang, App. 24]. The root is dug up in the 3rd month and dried in the shade.

In the WU P'U [3rd cent.] it is called 地血 *ti hüe* (earth blood). This name is properly applied to *Rubia.* [See 182.]

T'AO HUNG-KING:—This plant is produced in Siang yang [in Hu pei, App. 306]. Much of the drug is also brought from the district of Sin ye in the Nan yang prefecture [in S.W. Ho nan, App. 312, 231]. The people there cultivate it and employ it for dyeing a purple colour. It is not much used in medicine.

LI SHI-CHEN:—This plant has purple flowers and a purple root, whence the name *tsz' ts'ao.* It is cultivated for the colour yielded by its root. This root must be dug up in spring before the plant has flowered. Then the colouring matter will be found to be very bright. But if gathered after flowering the colour has·become deeper and is consequently inferior in quality. The top of the root [1 should rather think the plant above the ground] is covered with white hair. By certain processes a yellow colour can be produced from the root. The Yao chuang people [*v.* App. 402] call this plant 鴉衛草 *ya hien ts'ao.*

Ch., VII, 46:—*Tsz' ts'ao.* Representation of *Lithospermum erythrorhizon,* S. & Z.

LOUR., *Fl. cochin.*, 127 :—*Anchusa officinalis*, radice longa, rubra. Sinice *tsu tsao.*—TATAR., *Cat.*, 61 :—*Tsz' ts'ao.* Rad. *Anchusæ.*—GAUGER [48] describes and figures the root.

HENRY, *Chin. pl.*, 508 :—In Hu pei *tsz' ts'ao* is *Lithospermum erythrorhizon.*—*Ind. Fl. sin.*, II, 154.

Cust. Med., p. 60 (40) :—*Tsz' ts'ao* exported 1885 from I chang to other Chinese ports 73 piculs,[10]—p. 8 (65) from New chwang 27 piculs,—p. 354 (203) from Canton 2 piculs.— *Hank. Med.*, 48 :—Exported from Han kow.

Regarding *L. erythrorhizon* in Japan see *Bot. sin.*, II, 142.

24.—白頭翁 pai t'ou weng. P., XIIb, 38. T., CLXXII.

Pen king :—*Pai t'ou weng* (the gray head), 野丈人 *ye chang jen* (wild old man), 胡王使者 *hu wang shi che* (barbarian prince's envoy). Root and flowers used in medicine. Taste of the root bitter. Nature warm. Non-poisonous.

Pie lu :—The *pai t'ou weng*, also called 柰何草 *nai ho ts'ao*, grows in Kao shan [in Kiang su, App. 118], in mountain valleys and in waste places. It is dug up in the 4th month.

T'AO HUNG-KING :—It is a common plant. Near the root it is covered with a soft down and thus resembles the gray head of an old man, whence the above names.

SU KUNG [7th cent.]:—Its leaves resemble those of the *shao yo* [*Pænia*. See 52], but are larger. It produces one stem, at the top of which is a purple flower resembling that of the *mu kin* [*Hibiscus syriacus*. See *Rh ya*, 6]. The

---

[10] The export from I chang of 73 piculs is 紫 皮 *tzŭ p'i*, which seems to be a different substance from *tzŭ ts'ao.* The point at any rate requires elucidation.—A. HENRY.

fruit is as large as a hen's egg and covered with white hairs more than an inch long, which hang down like tassels, thus resembling the head of an old man.

Other authors quoted in the *P.* describe the *pai t'ou weng* differently, not quite distinct. The plant intended by Sʋ Kuɴɢ may perhaps be a *Pulsatilla* (*Anemone*), the fruit of which, with the long feathery tails of the seeds, may be compared to an old man's head.

The *Ch.* figures under the name *pai t'ou weng* two different plants,—IX, 52, a plant unknown to me, and VIII, 14, it seems, an *Eupatorium*. At Peking *pai t'ou weng* is *Eupatorium Kirillowii*, Turcz.

So moku, X, 36 :—白頭翁 *Anemone cernua*, Thbg.

Cust. Med., p. 68 (48) :—*Pai t'ou weng* exported 1885 from Han kow to other ports of China 69 piculs.

25.—白及 *pai ki.* *P.*, XII*b*, 39. *T.*, CLVII.

Pen king :—*Pai ki*, also 連及草 *lien ki ts'ao*, 甘根 *kan ken* (sweet root). The root is officinal. Taste bitter. Nature uniform. Non-poisonous.

Pie lu :—The 白及 *pai ki* grows in Pei shan (northern mountains) in river valleys, also in Yüan kü [in Shan tung, App. 415] and in Yüe shan [mountains of Yüe? App. 420]. The same work says, the 白給 *pai ki* grows in mountain valleys. Its leaves resemble those of the *li lu* [*Veratrum.* See 142], the root is like a mortar. It is dug up in the 9th month. [Lɪ Sʜɪ-ᴄʜᴇɴ says that the two drugs *pai ki*, differently written, are the same.]

Wʋ Pʻʋ [3rd cent.] :—The *pai ki* in its stem and leaves resembles the ginger plant and the *li lu*. In the 10th month purplish red flowers appear on the top of the plant. The root resembles a mortar, wherefore it is also called 臼根 *kiu ken* (mortar root).

T‘AO HUNG-KING :—It is a common plant in Mid China. The leaves resemble those of the *tu jo* [*Alpinia*. See 55], the root resembles that of the *ling* [*Trapa*. See 296]. It has hairs between the joints. It is not much used in medicine, but it is good for making paste.

HAN PAO-SHENG [10th cent.]:—The *pai ki* is produced in Shen chou [in Ho nan, App. 282]. The leaves resemble the young (unexpanded) leaves of the *tsung* (*Chamærops*), also the leaves of the *li lu* (*Veratrum*). One stem shoots forth in the 3rd or 4th month and produces purple flowers. In the 7th month the fruit ripens and assumes a dirty colour. The root is white and resembles that of the *Trapa* and has three horns. From the top of the root the sprouts issue. The root is dug up in the 8th month.

SU SUNG [11th cent.]:—The plant is common in North and Mid China. It grows on rocks, one foot and more high. The leaves resemble palm-leaves, are as long as two fingers. It bears purple flowers in summer.

LI SHI-CHEN :—HAN PAO-SHENG'S account of the plant is correct. It produces only one stem. The flowers are an inch long, of a reddish purple colour. The heart of the flower is like a tongue. The root resembles the *Trapa*. It has a navel like the *fu tsz‘* [*Scirpus tuberosus*. See *Rh ya*, 59] and is difficult to dry.

Ch., VIII, 12 :—*Pai ki*. Representation of an *Orchidea*, probably *Bletia hyacinthina*. It is stated there that the viscid juice of the root is used in manufacturing porcelain.

TATAR., *Cat.*, 4:—*Pai ki*. Rad. *Amomaceæ*.—GAUGER, who describes and figures the root [6], means that it belongs to an Orchid.—P. SMITH, 13.

The drug which I procured, under the name of *pai ki* from a Peking apothecary's shop, agreed in shape with the above Chinese descriptions and seemed to be the bulb of

Bletia hyacinthina, R. Br., an Orchid with purplish violet
flowers, much cultivated at Peking under the popular name of
lan hua. This bulb, when put in water, forms a mucilage which
at Peking is used by the manufacturers of "cloisonnés."

HENRY, *Chin. pl.*, 361 :—*Pai ki* in Hu pei is *Bletia
hyacinthina*.

Cust. Med., p. 68 (43):—*Pai ki* exported 1885 from
Han kow to other Chinese ports 385 piculs. See also an
interesting note regarding the *pai ki* in the Report on Trade,
Chin. Mar. Cust., 1869, p. 68, Ning po, from which port
the drug is likewise exported.

So moku, XVIII, 34 :—白及 *Bletia hyacinthina*.—Same
identification in SIEB., *Icon. ined.*, VIII.

26.—黃連 *huang lien.   P.*, XIII, i.   *T.*, CLIII.

Pen king :—*Huang* (yellow) *lien*, 王連 *wang lien.*—
The root is officinal. Taste bitter. Nature cold. Non-
poisonous.

Pie lu :—The *huang lien* grows in Wu yang [in E.
Sz ch'uan, App. 396], in river valleys, and in Shu [W. Sz
ch'uan, App. 292], also on the southern slope of the T'ai shan
[in Shan tung, App. 322]. The root is dug up in the
2nd and 8th months.

T'AO HUNG-KING :—Wu yang is in Kien p'ing [in
Sz ch'uan and Hu pei, App. 139]. Now the drug brought
from Western China is of a paler colour and less juicy than
that from Tung yang [in Che kiang, App. 376] and Sin an
[in Che kiang, App. 310] which are considered the best.
That from Lin hai [in Che kiang, App. 192] is of an inferior
quality. Before use the smaller fibres of the root are
removed, and then it looks like a string of beads.

In the *Yao sing lun* [7th cent.] the drug is called
支連 *chi lien.*

HAN PAO-SHENG [10th cent.]:—The plant grows in a bushy manner, about one foot high. One stem (petiole) bears three leaves, which do not fall off in winter. The flowers are of a yellow colour. The drug produced in Kiang tso [S. An hui, Kiang su, App. 124] is like a string of beads, that from Shu (Sz ch'uan) does not show this peculiarity. Now the best sort is considered to come from the country of Ts'in [Shen si, App. 358], also from Hang chou [in Che kiang, App. 58], Liu chou [in Kuang si, App. 199].

SU KUNG [7th cent.]:—This plant is produced in Shu [Sz ch'uan, App. 292], in Kiang tung [Kiang su, Che kiang, App. 124], and in Li chou [in Hu nan, App. 185]. The drug from the latter place is the best.

SU SUNG [11th cent.]:—The plant grows in Kiang and Hu [Mid China, App. 124, 83], in King chou [in Hu pei, App. 146], K'ui chou [in Sz ch'uan, App. 170] and Süan ch'eng [in An hui, App. 315]. The drug from the latter place is of a superior quality. Inferior sorts come from Shi [in Hu pei, App. 288], K'ien [N. Kui chou, App. 141], Tung yang [in Che kiang, App. 376], Hi chou [in An hui, App. 62], Ch'u chou [in Che kiang, App. 23]. The plant is one foot high, the leaves resemble those of the *kan kü* (sweet *Chrysanthemum*, unknown to me). In the 4th month it bears yellow flowers, in the 6th it produces fruits like those of the *k'in* (celery) and likewise yellow.

LI SHI-CHEN:—At the time when LI TANG-CHI wrote his herbal, at the end of the Han dynasty, the *huang lien* from Shu [W. Sz ch'uan] was considered the best. In the T'ang period the drug from Li chou [*v. supra*] was preferred. Now the drug used in medicine comes from Wu [Kiang su, App. 389] and Shu [Sz ch'uan]. The best is that produced in Ya chou and Mei chou [both in Sz ch'uan, App. 398, 219]. There are two sorts of *huang lien*. One has a coarse root destitute of hair (radical fibres), and forming a series of

united tubers. The root (with its branches) resembles the
claw of a bird. It is firm and juicy, of a deep yellow colour.
The other sort has no tubers, is densely covered with hair.
(radical fibres), is not juicy and is of a pale yellow colour.

Ch., VII, 32 :—*Huang lien.* Only leaves and root
figured. It seems a *Ranunculacea* is intended.

TATAR., *Cat.*, 10 :—*Huang lien.* Radix *Leontice.*—
GAUGER [9], *huang lien* and [10], *Ch'uan huang lien (huang
lien* from Sz ch'uan), figured and described. Jointed, yellow
rhizomes, very bitter.,

P. SMITH, 126 :—*Huang lien* identified with *Justicia.*[11]

According to PARKER [*China Review*, X, 28] the *huang
lien* plant is much cultivated in the mountains of Sz ch'uan.
It is *Coptis teeta*, Wall. (Order *Ranunculaceæ*). See HENRY,
Chin. pl., 137.

The yellow bitter root *huang lien*, which I obtained
from a Peking apothecary's shop, and which was examined by
Prof. FLÜCKIGER, seemed to belong to *Coptis teeta*. The
root has sometimes the appearance of a bird's claw.

DYMOCK, in his *Veget. Mat. Med. of W. India*, p. 13,
states that the root of *Coptis teeta* is much exported from
China to India.

Cust. Med., p. 70 (59) :—Exported 1885 under the
name of *shui lien* from Hankow to other Chinese ports 856
piculs,—p. 58 (21) from I chang 300 piculs.—*Ibid.*, p. 452
(516) Places of production : Sz ch'uan, Hu peh, Shen si,
Yün nan.

So moku, X, 38 :—黃連 *Coptis anemonæfolia*, Sieb. &
Zucc.—*Ibid.*, 39, 五加藜黃連 (five-leaved *huang lien*) =
C. quinquefolia, Miq. *Ibid.*, 40 :—三葉黃連 (three-leaved
huang lien) = *C. trifolia*, Salisb.

---

[11] " I have visited a *huang lien* plantation in the mountains of Hu peh,
and the plant is undoubtly *Coptis teeta*, Wall."—A. HENRY.

27.—黃芩 *huang k'in.* P., XIII, ii. T., CLIII.

Pen king:—Huang (yellow) *k'in*, 腐腸 *fu ch'ang* (putrid bowels). The root is officinal. Taste bitter. Nature uniform. Non-poisonous.

Pie lu:—Other names: 空腸 *k'ung ch'ang* (hollow bowels), 內虛 *nei hü* (empty inside), 經芩 *king k'in*, 黃文 *huang wen*. The *huang k'in* grows in Tsz' kui [in Hu pei, App. 368] in river valleys, and in Yüan kü [in Shan tung, App. 415]. The root is dug up on the 3rd day of the 3rd month and dried in the shade.

Wu P'u :—Other names: 妒婦 *tu fu*, 邱頭 *yin t'ou*.

T'ao Hung-king :—The fresh juicy and solid drug is called 子芩 *tsz' k'in*. The old broken drug with holes within is 宿芩 *su k'in*. As the drug is frequently rotten it is also called *fu ch'ang* [v. s.]. Now the best comes from P'eng ch'eng [in Kiang su, App. 247]. It is also produced in Yü chou [in Kuang si? App. 412]. That of a superior quality is of a deep yellow colour, firm, without holes. It is much used in medicine, but not in Taoist prescriptions.

Su Kung [7th cent.]:—Now the best comes from I chou [in Kuang si, App. 103], Fu chou [in Shen si, App. 43], King chou [in Kan su, App. 153]. That from Yen chou [in Shan tung, App. 404], which is also of a good quality, is known by the name of 狆尾芩 *t'un wei k'in* (sucking-pig's tail).

Su Sung [11th cent.]:—The *huang k'in* is common in Mid and North China. The plant grows about one foot high. Leaves small and long, purple flowers. The root resembles that of the *chi mu* [see 8].

Li Shi-chen :—The *huang k'in* is a bitter root of a yellow colour. The old root is yellow outside, black within, with many holes. The ancient dictionary *Shuo wen* writes the name 黃芩 *huang k'in*. The fresh root is also called 條芩 *t'iao k'in*. The drug produced in the north is of a deep yellow colour.

Ch., VII, 36 :—*Huang k'in*. The figure seems to intend a *Scutellaria.*

TATAR., *Cat.*, 10 :—*Huang k'in*, Radix *Scutellariæ viscidulæ*, Bge. (which has yellow flowers). But in his drawings of Peking plants, TATARINOV applies the name *huang k'in* to *S. macrantha*, Fischer, with blue flowers. This latter is a Siberian plant, common also in North China. It is the *Cassida montana* in AMMAN's *Stirp. rar. ruth.* (1739), p. 42, tab. 6. Radix carnosa extus et intus flava saporis subamari, etc.

P. SMITH, 194 :—*Sc. viscidula.*

The drug *huang k'in* obtained from an apothecary's shop at Peking were thin transversally cut slices of a yellow, bitter root.

HENRY, *Chin. pl.*, 141 :—*Huang k'in*. In Hu pei a name for *Berberis nepalensis*, Spr.[12]

Cust. Med., p. 24 (26) :—*Huang k'in* exported 1885 from Tien tsin 5,530 piculs,—p. 44 (16), from Che foo, 517 piculs,—p. 68 (27), from Han kow 162 piculs.

So moku, XI, 48 :—黃芩 *Scutellaria macrantha.*—See also *Kwa wi*, 14.

28.—秦芃 *Ts'in kiao.* P., XIII, 16. T., CXXXIX.

Pen king :—*Ts'in kiao.* The root is officinal. Taste bitter. Nature uniform. Non-poisonous.

Pie lu :—The *ts'in kiao* grows in Fei wu [in Sz ch'uan, App. 37], in mountain valleys. The root is dug up in the 2nd and 8th months and dried in the sun.

---

[12] "This name I have since found to be erroneous."—A. HENRY.

T'AO HUNG-KING:—Now this drug is produced in Kan sung [in Sz ch'uan, App. 114] and Lung tung [App. 214]. Worms eat the root and produce reticulate figures on it. It is twisted and contorted. The large root of a yellowish white colour is the best.

SU KUNG [7th cent.]:—The drug is commonly called 秦膠 Ts'in kiao. The original name was 秦札 Ts'in kiu. The best is produced in King chou [in Kan su, App. 153], Fu chou [in Shen si, App. 43], K'i chou [in Shen si, App. 120].

According to SIAO PING [T'ang period] it is also called 秦爪 Ts'in chao.

SU SUNG [11th cent.]:—It grows plentifully in North China (Shan si and Shen si). The root is of a dark yellow colour, twisted and contorted, one foot long. The leaves resemble Lettuce-leaves. In the 6th month it produces purple flowers resembling those of the ko [Pachyrrhizus. See 174].

LI SHI-CHEN does not seem to know this plant. He says only that the name is derived from the country of 秦 Ts'in [Kan su, Shen si, App. 358], where it is produced, and 札 twisted.

Ch., VII, 48:—Ts'in kiao. The drawing shows only the root and leaves. The text says that the plant grows on the Wu t'ai shan mountain, in Shan si.

P. SMITH [102] thinks that the Ts'in kiao is the Justicia gendarussa, but he does not say whereupon this identification is based.

Cust. Med., p. 66 (8):—Ts'in kiao exported 1885 from Han kow 1,280 piculs, and p. 72 (86), the rind of the same drug 308 piculs,—p. 22 (11):—Ts'in kiao exported from Tien tsin 786 piculs. It is said there to be the bark of Xanthorylon alatum.[13]—P., 58 (2) from I chang exported 217 piculs.

---

[13] Evidently confounded with 288 Ts'in tsiao.

The drug *ts'in kiao* is unknown to me. I suppose the above identifications are both wrong.

29.—芷胡 *ts'z' hu*.   *P.*, XIII, 17.   *T.*, CXLIX.

Pen king :—Ts'z' hu, 地薰 *ti hūn*. The root is officinal. Taste bitter. Nature uniform. Non-poisonous.

*Pie lu :—*The *ts'z' hu* leaves are called 芸蒿 *yün hao*. They are of a pungent taste, fragrant, and can be eaten. The plant grows in Hung nung [in Ho nan, App. 99], in river valleys, and in Yüan kü [in Shan tung, App. 415]. The root is dug up in the 2nd and 8th months and dried in the sun.

WU P'U [3rd cent.] calls it 山菜 *shan ts'ai* (mountain vegetable) and 茹草 *ju ts'ao* (edible herb).

T'AO HUNG-KING :—It grows in Mid China, resembles the *ts'ien hu* [*Angelica.* See 30]. The *Po wu chi* [3rd cent.] says : The leaves of the *yün hao* resemble those of the *sie hao*[14] (an umbelliferous plant). The young white shoots, which appear in spring and in autumn, and are from four to five inches long and fragrant, may be eaten. It is found in Ch'ang an [present Si an fu in Shen si, App. 6] and Ho nei [in N. Ho nan, App. 77].

SU KUNG [7th cent.]:—芷 is the ancient form for 柴 *ch'ai* (firewood), and the name of the drug is more commonly written 柴胡 *ch'ai hu*.

SU SUNG [11th cent.]:—This plant is common in North and Mid China. The best drug comes from Yin chou [in Shen si, App. 407]. The plant is very fragrant. Its stem is purplish green, rigid, shows fine lines (is channelled). The leaves resemble bamboo-leaves but are narrower and

----

[14] 邪蒿 *P.*, XXVI, 55.

smaller. In the 7th month it bears yellow flowers. The root is of a pale red colour, resembles that of the *ts'ien hu* [see 30]. A peculiar sort of the *ch'ai hu* grows in Tan chou [in Shen si, App. 327].

Li Shi-chen observes that the character 茈 in ancient times was also used for 紫 *ts:'* "purple" and refers to *Rh ya*, 142. He states that the *ch'ai hu* plant when young may be eaten, but the old plant is used for firewood, whence the name. He further proves that the ancient authors confounded under the name of *ch'ai hu* several umbelliferous plants. The northern *ch'ai hu* is not the same as that produced in the south.

Kiu huang, XLVI, 25, and *Ch.*, VIII, 27 :—*Ch'ai hu*. Rude figures, probably intended for *Bupleurum*.

Tatar., *Cat.*, 14 :—*Ch'ai hu*, Radix *Bupleuri octoradiati.*—Gauger [41] describes and figures the root, which he believes to belong to an umbelliferous plant.—P. Smith, 45.

In the Peking mountains the name *ch'ai hu* is applied to *Bupleurum falcatum*, L., and *B. octoradiatum*, Bge. Both have yellow flowers. The above Chinese descriptions of the *ch'ai hu* agree.

Cust. Med., p. 66 (1):—*Ch'ai hu* exported 1885 from Han kow 3,340 piculs,—p. 120 (1), from Chin kiang 197 piculs,—p. 22 (1), from Tien tsin 157 piculs.

So moku, V, 41:—柴胡 *Bupleurum falcatum*, L.— Comp. also *Kwa wi*, 47.

30.—前胡 *ts'ien hu*. P., XIII, 21. T., CXLIX.

Pie lu :—The root of the *ts'ien hu* is dug up in the 2nd and 8th months and dried in the sun. It is of a bitter taste. Nature slightly cold. Non-poisonous.

T'AO HUNG-KING:—It is a common plant in the marshes of Mid China. The best drug comes from Wu hing [in Che kiang, App. 390]. The root resembles the *ch'ai hu* [*v.* 29] but is softer.

SU SUNG [11th cent.]:—It is a common plant in Mid and North China. The fragrant young sprouts are eaten. White flowers. The root is of a greenish purple colour. There are several sorts.

LI SHI-CHEN:—The plant grows 2 feet high. The young leaves are eaten. Flowers of a dirty white colour resembling those of the *she ch'uang* [*Cnidium.* See 49]. The root is black outside, white internally, fragrant. The best is produced in the northern provinces.

Kiu huang, XVI, 29, and *Ch.*, VIII, 24:—*Ts'ien hu*. Representations of an umbelliferous plant, probably an *Angelica*.

TATAR., *Cat.*, 11:—*Ts'ien hu*. Rad. *Angelicæ?*— P. SMITH, 17.

HENRY, *Chin. pl.*, 59:—*Ts'ien hu* in Hu pei is an *Angelica*.

Cust. Med., p. 184 (8):—*Ts'ien hu* exported 1885 from Ning po 1,084 piculs,—p. 66 (5), from Han kow 277 piculs.—Small quantities exported from Fu chow and Pak hoi, p. 220, 414.

So moku, V, 33:—前胡 *Angelica refracta*, Fr. Schm.

31.—防風 *fang feng*. P., XIII, 22. T., CXXXVI.

Pen king:—*Fang feng*, 銅芸 *t'ung yün*. The root is officinal. Taste sweet. Nature warm. Non-poisonous.

Leaves, flowers and seeds also used in medicine.

Pie lu:—Other names of the plant: 囘草 *hui ts'ao*, 屏風 *p'ing feng*, 蘭根 *kien ken*, 百枝 *po chi* [hundred branches. Comp. also 13]. The *fang feng* grows in Sha yüan [in Shen si, App. 267], in marshes, also in Han tan [in Chi li, App. 56], Lang ye [in Shan tung, App. 178] and Shang ts'ai [in Ho nan, App. 276]. The root is dug up in the 2nd and 10th months, and dried in the sun.

Wu P'u [3rd cent.]:—The plant is also called 回芸 hui yün and 百蜚 po fei. The leaves are slender, round, dark green and yellowish white. In the 5th month it bears yellow flowers, in the 6th black fruits.

T'AO HUNG-KING observes that a place Sha yüan does not exist [comp. App. 267]. The best drug comes from P'eng ch'eng [in Kiang su, App. 247] and Lan ling [in Shan tung, App. 174] which is not far from Lang ye. It is also exported from Yü chou [App. 412]. A drug of an inferior quality is produced in Siang chou at the frontier of I yang hien [in Ho nan, App. 305, 107].

SU KUNG [7th cent.]:—The drug produced in Ts'i chou [in Shan tung, App. 348] and Lung shan [in Chi li, App. 213] is considered the best, but that from Tsz' chou [in Shan tung, App. 367], Yen chou [in Shan tung, App. 404] and Ts'ing chou [in Shan tung, App. 363] is also good. The leaves resemble those of the *mou hao* [*Artemisia*. *Bot. sin.*, II, 432] and the *fu tsz'* [*Aconitum*. See 134]. T'AO HUNG-KING is wrong in stating that a place Sha yüan does not exist. Sha yüan lies south of T'ung chou. But the drug which comes from that locality is inferior to that from the eastern provinces.

SU SUNG [11th cent.]:—It is a common plant in Mid China. Its leaves resemble the *ts'ing hao* (*Artemisia*) but are shorter and smaller. When young they are of a purplish red colour. The people of Kiang tung, of Sung and Po[15] [in An hui, App. 124, 316, 259] eat the young

[15] Or " Po " in the country of Sung. [See App. 259.]

leaves as a vegetable. They have a pleasant taste. In the
5th month the plant opens its small white flowers. The
inflorescence is like that of the *shi lo* (an umbelliferous plant).
The seeds resemble those of the *hu sui* (*Coriander*) but are
larger. The root is large, of a dirty yellow colour, resembles
the root of the *shu k'ui* (*Althæa*). The best comes from
Ts'i chou [*v. supra*]. The sort called 石防鳳 *shi fang feng*
grows in Ho chung fu [in S.W. Shan si, App. 74].

Li Shi-chen :—The drug which is produced in Kiang
and in Huai [Kiang su and An hui, App. 124, 89] is the
shi (stone) *fang feng*. It grows on rocks, whence the name.
In the second month the people there gather the young
leaves for food. They are pungent, sweet and fragrant.
It is called 珊瑚菜 *shan hu ts'ai* (coral vegetable). The
root looks coarse and deformed. The plant can be raised
from seed.

Ch., VII, 23 :—*Fang feng*. Rude drawing. Umbelliferous plant.

Lour., *Fl. cochin.*, 622 :—*Coreopsis leucorhiza*, sinice
fam fum (*fang feng*). Ibid., 452 :—*Origanum Dictamnus*,
sinice : *Quam tum* (Canton) *fam fum*. Notandum in
provinciis borealibus Sinarum nasci aliam plantam eodem
nomine *fam fum* vocatam, radice carnosa alba, subfuseformi,
magni usus in medicina, sed prorsus alienam ab Origano.

D'Incarville, in his list of Peking plants, calls the
fang feng Persil des montagnes.

Tatar., *Cat.*, 23 :—*Fang feng*. Rad. *Libanotidis ?*—
Gauger [17] describes and figures the same drug, the root
of an umbelliferous plant.—P. Smith, 133.

[16] At Peking the name *fang feng* is applied to two
umbelliferous plants, viz. *Stenocœlium divaricatum*, Turcz. (*Siler
divaricatum*, Benth. & Hook.) and *Peucedanum rigidum*, Bge.

---

[16] *Fang fêng* in mountains of Hu peh is the name given to *Peucedanum
terebinthaceum*, F.—A. Henry.

Cust. Med., p. 22 (21):—*Fang feng* exported in 1885 from Tien tsin 2,319 piculs,—p. 44 (10), from Che foo 1,063 piculs,—p. 2 (11), from New chwang 746 piculs.

Amœn. exot., 825 :—防風 *boofu & fofu*. *Ligusticum vulgare*.—*Ibid.*, 山防風 *san bofu*, vulgo *jamma bofu*. Apium littorale folio Aquilegiæ pinguiore. According to THBG. [*Fl. jap.*, 117] this is *Peucedanum japonicum*. Comp. also *infra*, 133.

So moku, V, 10 :—防風 *Siler divaricatum*.

32.—獨活 *tu huo*. *P.*, XIII, 25. *T.*, CXXXIV.

Pen king :—*Tu huo*, 羌活 *k'iang huo*, 羌青 *k'iang ts'ing*, 護羌使者 *hu k'iang shi che*. The root is officinal. Taste bitter and sweet. Nature uniform. Non-poisonous.

Pie lu :—Other name 獨搖草 *tu yao ts'ao* (self-moving plant). The *tu huo* grows in the river valleys of Yung chou [Mid Shen si, App. 424], also in Lung si [in Kan su, App. 216] and Nan yao [unknown, App. 232]. The root is dug up in the 2nd and 8th months and dried in the sun. This plant is not moved by wind, it moves only in still air, whence the above names [*tu huo* means self-moving].

WU P'U [3rd cent.] calls it 胡王使者 *hu wang shi che.*

T'AO HUNG-KING :—The localities mentioned in the *Pie lu* all belonged in former times to the country of the K'iang [Tibetans, Kukonor, App. 131]. The drug *k'iang huo* which is produced in those localities is smaller, full of joints, succulent, and of a strong nature. That which comes from I chou [in Sz ch'uan, App. 102], Pei tu [in Shan si, App. 246], Si ch'uan [in Kan su, App. 296] is called *tu huo*. It is of a whitish colour, larger. Both are used in the same way.

Su Sung [11th cent.]:—The best sorts of the *tu huo* and the *k'iang huo* are now produced in Shu Han [Sz ch'uan, App. 293]. The *Pen king* takes the *tu huo* and the *k'iang huo* to be the same, but these names refer to two different although allied plants. That drug (root) which is of a purple colour and has the joints close together is the *k'iang huo*, that of a yellow colour and forming a large mass is the *tu huo*.

Li Shi-chen :—The *tu huo* and the *k'iang huo* are two different species of the same genus. That which grows in China is the *tu huo*, that produced in Si K'iang [Kukonor, Tibet, App. 300] is the *k'iang huo*. The *tu huo* is also called 長生草 *ch'ang sheng ts'ao* (high-growing plant).

Ch., VII, 24 :—*Tu huo*. Representation of an umbelliferous plant.

D'Incarville [*Peking Plants*] saw the plant *tu huo* in the mountains three or four days' journey from Peking. He thinks it was an *Angelica* [probably *A. grosseserrata*, Max.].— Tatar., *Cat.*, 21 :—*Tu huo*. Rad. *Angelicæ ?*—Tatar., *Cat.*, ii :—*K'iang huo*. Radix.—Gauger [50] describes and figures the same root, which he states has an unpleasant balsamic smell.—P. Smith [18] describes both these drugs as *Angelica*.

[17] Henry [*Chin. pl.*, 36, 475] identifies the *k'iang huo*, as well as the *tu huo*, with *Peucedanum decursivum*, Max.

Cust. Med., No. 1,364 :—*Tu huo*, *Angelica inæqualis*, Max.

Cust. Med., p. 70 (72):—*Tu huo* exported 1885 from Han kow 1,327 piculs,—p. 8 (63), from New chwang 203 piculs.

---

[17] *Peucedanum decursivum* is called by some of the natives in the I chang mountains *tu huo* and *ch'iang huo*; but I do not consider that the sources of the drugs occurring in commerce under these two names have yet been made out. Several umbelliferous plants are probably concerned.— A. Henry.

Ibid., p. 66 (11) Sz ch'uan:—*K'iang huo* exported from Han kow 1,821 piculs,—p. 58 (1), from I chang 383 piculs.

Amœn. exot., 826 :—獨活 *doku quats*, vulgo *dosjen*, *udo*. Frutex annuus, rad e eduli, etc.—SIEB. & ZUCC. [*Fl. jap.*, I, p. 57] ide :fy this plant, described by KÆMPFER, with *Aralia cordata*, Thbg., or *A. edulis*, S. & Z. KÆMPFER is mistaken with respect to the Chinese and the first Japanese name, which do not belong to the plant he describes and which is *Aralia edulis* and has another Chinese name. [*See* 46.]

So moku, V, 34, 35 :—獨活 *Angelica japonica*, A. Gray. —*Phon zo*, VII, 16 :—Same Chinese name, *Archangelica Gmelini*, D.C. Japonice *dokukwats* [comp. *supra*, KÆMPFER]. The umbelliferous plant figured in the *So moku* [V, 37] and the *Phon zo* [VII, 16, 17] sub 羌活 has not been identified by FRANCHET.

33.—升麻 *sheng ma*. P., XIII, 29. T., CXLIX.

Pen king :—*Sheng ma* (ascending hemp). The root is officinal. Taste sweet and bitter. Nature uniform, slightly cold. Non-poisonous.

Pie lu :—Other names : 周麻 *chou ma* or *chou sheng ma* [some believe (KUANG YA, WU P'U) that *Chou* here denotes the ancient state of Chou in Mid Shen si, on the Wei River]. The *sheng ma* grows in the mountain valleys of I chou [in Sz ch'uan, App. 102]. The root is dug up in the 2nd and 8th months and dried in the sun.

T'AO HUNG-KING :—In ancient times the best kind was reckoned to be produced in Ning chou [Yün nan, App. 234]. The drug is small, black and firm. Now that from I chou (Sz ch'uan) is of a good quality. It is small, and after the skin has been taken off shows a green colour. It is called

11

雞骨 | | *ki ku* (chicken - bones') *sheng ma*. Another sort is produced in Pei pu [*see* App. 244]. It is juiceless, large and of a yellow colour. A third sort, which comes from Kien p'ing [App. 139], is also large, of a feeble taste and not much used. It is known by the name of 落新婦 *lo sin fu* and reputed for neutralizing poison. A decoction of the leaves is used in infantile diseases.

CH'EN TS'ANG-KI [8th cent.]:—The *lo sin fu* is now more commonly called the 小升麻 *siao* (little) *sheng ma*. Medical virtues the same as those of the *sheng ma*.

MA CHI [10th cent.]:—The *sheng ma* which is brought from Sung Kao [in Ho nan, App. 317] is of a green colour. It has less power than the drug from Shu [Sz ch'uan, App. 292].

SU SUNG [11th cent.]:—The *sheng ma* is produced all over Mid China. The best sort comes from Shu ch'uan [Sz ch'uan, App. 292]. The plant grows three feet high. Its leaves resemble hemp-leaves. In the 4th or 5th month it produces white flowers arranged in a spike like that of the *su* (*Setaria*). The fruit is black. The root is like the root of the *hao* (*Artemisia*) of a purplish black colour, covered with hair (radical fibres).

LI SHI-CHEN:—The name *sheng ma* is derived from the resemblance of the leaves to hemp-leaves. Its common name nowadays is 川升麻 *ch'uan sheng ma* (*sheng ma* from Sz ch'uan).

Ch., VII, 18:—*Sheng ma*. Rude drawing.

The name *sheng ma* seems to be applied to various ranunculaceous and saxifragaceous plants.

The *sheng ma* is noticed in GROSIER'S *Chine* [III, 349, *chine ma*].

TATAR., *Cat.*, 53:—*Sheng ma*. Radix.

HANBURY, *Sc. pap.*, 261 :—*Shing ma*, rhizome of *Thalictrum rubellum*. H. relies only upon SIEBOLD's identification of the Japanese drug with the above Chinese name. He is wrong in referring to CLEYER'S *Med. simpl.*, 90, *sien mao*.[18] CLEYER's [16] *xim ma* is intended for *sheng ma*.—P. SMITH, 216 :—*Thalictrum*.

In the Peking mountains the people apply the name *sheng ma* to *Cimicifuga daurica*, Max. (*Ranunculaceæ*), also to *Astilbe chinensis*, Max. (*Saxifrageæ*). The two plants, indeed resemble each other in their outer appearance.

HENRY, *Chin. pl.*, 420 :—*Sheng ma*. In Hu pei this name is applied to several species of *Astilbe* and also to *Cimicifuga*.

Cust. Med., p. 348 (131):—*Sheng ma* 1885 exported from Canton 2,047 piculs,—p. 190 (81), from Ning po 145 piculs,—p. 6 (45), from New chwang 44 piculs.—*Hank. Med.* [37]. The drug is also exported from Han kow.

So moku, X, 12-14, 16 :—升麻 various species of *Cimicifuga*, and *ibid.* [19] same Chinese name, *Astilbe japonica*, Miq., *ibid.* [28] same Chinese name, *Anemonopsis macrophylla*, S. & Z.—See also *Phon zo* [VII, 19-24] under the same Chinese name, *Cimicifuga, Anemonopsis, Trautvetteria, Astilbe*.

HOFFM. & SCHLT, 578 :—*Thalictrum rubellum*, S. & Z., sinice (*vide* SIEBOLD), a kind of 升麻.

34.—苦參 *k'u shen*. *P.*, XIII, 32. *T.*, CLXI.

Pen king :—*K'u shen* (bitter ginseng), 苦薽 *k'u shi* [this name is also applied to another plant. *See* 106, *Physalis*, and *Bot. sin.*, II, 547], 水槐 *shui huai* (water *Sophora*). The root and the fruit are officinal. Taste of the root bitter. Nature cold. Non-poisonous.

Pie lu :—Other names : 地槐 *ti huai* (ground *Sophora*),
菟槐 *t'u huai*, 驕 | *kiao huai*, 白莖 *pai heng* (white
stem), 芩莖 *k'in heng*, 綠白 *lu pai* (green and white),
陵郎 *ling lang*, 虎麻 *hu ma* (tiger's hemp). The *k'u shen*
grows in Ju nan [in Ho nan, App. 110] in the mountains
and in fields. The root is dug up in the 3rd, 8th and 10th
months and dried in the sun.

T'AO HUNG-KING :—It is a common plant in Mid China.
The leaves have a strong resemblance to those of the *huai*
(*Sophora japonica*). Yellow flowers. The fruit is a pod.
The root is very bitter.

SU SUNG [11th cent.]:—The root is of a yellow colour,
from five to seven inches and more long. From three to five
stems issue from it, three to four feet high. The leaves are
very like those of the *huai* and deciduous. Flowers yellowish
white. Fruit (or seeds) small beans.

LI SHI-CHEN likens the pods of the plant to the siliqua
of the radish. Other names 野槐 *ye* (wild) *huai*, 苦骨
k'u ku (bitter bone).

Ch., VIII, 5 :—*K'u shen.* Rude drawing. Leguminous
plant.

LOUR., *Fl. cochin.,* 556 :—*Robinia amara,* Cochinchina,
China, sinice *khu sem.* Radix subcarnosa, multiplex luteo-
fusca, *amarissima. Ibid.,* 555 :—*R. mitis.* Same Chinese
name. Both these species are known only from LOUREIRO'S
description.

TATAR., *Cat.,* 33 :—*K'u shen.* Radix *Robiniæ amaræ*
[TATARINOV evidently relies upon LOUREIRO'S identification].
—P. SMITH, 186.

I have seen the drug *k'u shen,* obtained from Wen chou
fu, transversally cut slices of an exceedingly bitter root,
one inch in diameter,

HENRY, *Chin. pl.*, 190 :—*K'u shen* in Hu pei is *Sophora Kronei*, Hance.

Cust. Med., p. 344 (68):—*K'u shen* root exported 1885 from Canton 56 piculs,—p. 122 (33), from Chin kiang 4 piculs,—p. 294 (304), *k'u shen* seeds exported from Amoy 3 piculs.

So moku, XIV, 28 :—苦參 *Sophora angustifolia*, S. & Z. According to the *Ind. Fl. sin.* [I, 202] this is identical with *S. flavescens*, Ait., a common plant in North and Mid China. *S. Kronei* has also been reduced to this species, which at Peking is called 野槐 *ye* (wild) *huai*.

35.—白鮮 *pai sien.* P., XIII, 36. T., CLVIII.

Pen king :—*Pai* (white) *sien* (strong smell). The rind of the root is officinal. Taste bitter. Nature cold. Non-poisonous.

Pie lu :—The rind of the *pai sien* (root) is produced in Shang ku [in Chi li, App. 272] in river valleys, also in Yüan kü [in Shan tung, App. 415]. It is gathered in the 4th and 5th months and dried in the shade.

T'AO HUNG-KING :—It is also called 白羊鮮 *pai yang* (goat) *sien*. These names refer to the white colour of the root and its strong smell, like the odor of goats. It is therefore also termed 白羶 *pai shen* (the second character = odor of goats). It is a common plant in Mid China. The best sort is produced in Shu [Sz ch'uan, App. 292].

SU KUNG [7th cent.]:—The plant grows more than a foot high. Its leaves resemble those of the Chu yü [*Boymia rutaecarpa*. See 291]. The root has a white rind, is juicy. Purplish white (violet) flowers. The root should be dug up in the second month, for in the 4th or 5th it has already lost its power.

Su Sung [11th cent.]:—It is also called 地羊鮮
ti yang sien. It is common in Ho ch'un r [S.W. Shan si,
App. 74], Kiang ning fu [present Nan king, App. 129],
Ch'u chou [in An hui, App. 25], Jun chou [in Kiang su,
App. 111]. Flowers of a pale purple colour, resemble
small *shu k'ui* (*Althœa*) flowers. Root like a small turnip
with a yellowish white rind. The mountain people eat the
young leaves.

Li Shi-chen:—The fruit consists of several carpels like
that of the *tsiao* (*Zanthoxylon*). The plant is therefore
also called 金雀兒椒 *kin tsio rh tsiao* (golden bird's
Zanthoxylon).

Ch., VII, 40:—*Pai sien.* Rude figure. Plant with
pinnate leaves. *Ibid.,* X, 6:—*Pai sien p'i.* This figure may
be intended for *Dictamnus.* The drug *pai sien* obtained
from an apothecary's shop in Peking, a white root cut in
slices, seemed to belong, according to Prof. Flückiger,
to a *Dictamnus.* The above Chinese description agrees in a
general way. *Dictamnus albus,* L. (*D. Fraxinella,* Pers.)
is a plant which exhales a strong but not unpleasant odor.
It has a white root which in former times was used in
medicine in Europe. It is known from Manchuria, Corea,
Japan. [*Ind. Fl. sin.,* I, 104.]

Cust. Med., p. 10 (78):—*Pai sien p'i* exported 1885
from New chwang 170 piculs,—p. 104 (86), from Wu hu
4 piculs.—*Hank. Med.,* 31:—Exported also from Han kow.

Phon zo, VII, 26:—白鮮 *Dictamnus Fraxinella.*

36.—貝母 *pei mu.* P., XIII, 39. T., CXXIX.

Comp. for other ancient names *Bot. sin.,* II, 75, 423.

Pen king:—*Pei mu* (mother of cowry shell). In the
Index of the *Pen king* it is called 藥實根 *yao shi ken.*

The root (bulbs) officinal.   Taste pungent.   Nature uniform.
Non-poisonous.

Pie lu :—Other names : 勤母 *k'in mu,* 苦菜 *k'u ts'ai,*
苦花 *k'u hua,* 空草 *k'ung ts'ao.*   The *pei mu* grows in the
country of Tsin [Shan si, App. 353].   The root is dug up
in the 10th month and dried in the sun.

T'AO HUNG-KING :—The root resembles cowry shells
collected together, whence the name *pei mu.*

SU KUNG [7th cent.]:—Leaves like garlic-leaves.   The
root ought to be gathered in the 4th month, when garlic is
ready.   In later months it is not good.   The best drug is
brought from Jun chou [in Kiang su, App. 111], King chou
[in Hu pei, App. 146] and Siang chou [in Hu pei, App.
305].   It is also produced in Kiang nan [South of the
Yang tsz', App. 124].

SU SUNG [11th cent.]:—Localities enumerated where
the drug is produced, in present Kiang su, An hui, Ho nan,
Hu pei, S. Shan si.   Its leaves resemble those of buckwheat.
It bears greenish flowers resembling in shape the *ku tsz'*
[*Convolvulus.*   See 169].   The root is dug up in the 8th
month.   It consists of many (small) bulbs collected together
and resembling cowry shells.   There are many sorts of
pei mu.

Ch., VII, 42 :—*Pei mu.*   Representation of a tuberous
plant with hastate leaves.

LOUR., *Flor. cochin.,* 423 :—*Thalictrum sinense* (a plant
known only from LOUREIRO's description), sinice *poi mu.*
Radix tuber subrotundus, solidus, albissimus.   Root used in
medicine.

P. SMITH, 225 :—*Pei mu, Uvularia grandiflora,* and
[112] *pei mu, Hermodactyle* or corms of *Colchicum.*   Both
identifications are wrong.

Interesting notices regarding the *pei mu*, cultivated near Ning po, are found in the Reports on Trade [Chin. M. Customs] for 1869, p. 61, and for 1880, p. 142. It is there stated that these bulbs are much larger than the *pei mu* produced in Sz ch'uan, but do not possess medicinal properties to the same extent as the Sz ch'uan drug.—See also the same Reports for 1879, p. 32, Han kow, regarding the 川貝母 or *pei mu* from Sz ch'uan, where it is much cultivated and is in great repute for the treatment of several diseases.

Father DAVID states [*Journ. N. Ch. Br. As. Soc.*, VII, 212] that the *pei mu* which grows in the high mountains of Mu pin [Tibet, on the border of Sz ch'uan] and the corms of which are much used in medicine, is a *Fritillaria* with yellow flowers. This is, according to FRANCHET [*Plantæ David.*, II, 130], *Fritillaria Roylii*, Hook.— FORTUNE [*Res. am. the Chinese*, 261] speaks of a *Fritillaria* with grayish white flowers, cultivated near Ning po for its bulbs, used in medicine. This is, it seems, the *pei mu* mentioned in the Reports on Trade.—HENRY, *Chin. pl.*, 366:— The name *pei mu* in Hu pei is applied to a *Pleione* (*Orchid*), but this is not the Sz ch'uan drug of the same name.

Cust. Med., p. 76 (135):—川貝母 *pei mu* from Sz ch'uan exported 1885 from Han kow 356 piculs,—p. 62 (51), from I chang 281 piculs.—*Ibid.*, 190 (104), 土貝母 native *pei mu* exported from Ning po 2,474 piculs.

So moku, V, 81:—貝母 *Fritillaria Thunbergii*, Miq. (*Uvularia cirrhosa*, Thbg.). Yellow flowers.—See also *Kwa wi*, 10.

37.—白茅 *pai mao*.   *P.*, XIII, 45.   *T.*, CIII.

Comp. for other ancient names *Bot. sin.*, II, 183, 459.

Pen king:—*Pai mao* (white grass). The root is called 茹根 *ju ken*, 茅根 *mao ken*, 蘭根 *lan ken*. The root is officinal. Taste sweet. Nature cold. Non-poisonous. The young sprouts and the flowers are also used in medicine.

Pie lu :—The *mao ken* grows in the country of Ch'u [Hu kuang, App. 24], in mountain valleys and fields. The root is dug up in the 6th month.

T'AO HUNG-KING :—It is now called 白茅菅 *pai mao kien.* The root resembles the *cha k'in* (a kind of celery), is of a sweet, pleasant taste.

SU SUNG [11th cent.]:—It is a common plant. Its sprouts, which shoot forth in spring and cover the ground, are like needles. The people call them 茅針 *mao chen* (*mao needles*). These sprouts are edible and good for children. In summer the plant bears white, plushy flowers, and withers in autumn. The root is very white. It is dug up in the 6th month. The plant 菅 *kien* is a kind of *mao.*

LI SHI-CHEN :—There are several sorts of *mao,* viz. the *pai* (white) *mao,* the *kien mao,* the *huang* (yellow) *mao,* the *hiang* (fragrant) *mao,* the 芭 | *pa mao.* The leaves in all these plants are the same (for they are all grasses).

The *pai mao* plant is short and small. In the 3rd month it bears white flowers in panicles, followed by small fruit (seeds). The root is white, very long, flexible like a tendon, provided with joints, of a sweet taste. The people call it 絲 | *sz' mao* (floss silk *mao*). The plant can be used for thatching. It is likewise employed for wrapping up things offered in sacrifice. This is the drug *mao ken,* spoken of in the *Pen king.* The dry root, at night, gives out a light, and after decaying changes into glow-worms.

The *pai mao* is a grass, *Imperata.* For further particulars see *Bot. sin.,* II, 459.

The *Cust. Med.* [p. 278 (76)] notices the *mao ken* as exported in a small quantity from Amoy, and [p. 344 (89)] c. 290 piculs imported from Hong kong into Canton. It is not stated from what Chinese port it was brought to Hong kong.

38.—地筋 *ti kin* (earth tendon). *P.*, XIII, 49. *T.*, CXXXVIII.

Pie lu:—Other names : 菅根 *kien ken*, 土筋 *t'u kin* (earth tendon). The *ti kin* is produced in Han chung [S. Shen si, App. 54]. The root is covered with hair (radical fibres). It is dug up on the 3rd day of the 3rd month. It is used in the same way as the *pai mao*.

T'AO HUNG-KING :—It (the root) is smaller than the *pai mao*.

LI SHI-CHEN states (sub *pai mao*) that the *kien mao* resembles the *pai mao* but it is longer. It grows in the mountains. It flowers in autumn (the *pai mao* in summer). When in seed it bears sharp pointed bristles which stick to clothes. The root is short, hard, like a small bamboo-root, jointless. As a medicine it is less potent than the *pai mao* root.

For further particulars see *Bot. sin.*, II, 460.

39.—龍膽 *lung tan*. *P.*, XIII, 50. *T.*, CLXIV.

Pen king:—*Lung tan* (dragon's gall). The root is officinal. Taste bitter and harsh. Nature very cold. Non-poisonous.

In the *Kuang ya*, 陵游 *ling yu* is given as an old name for *lung tan*.

Pie lu:—The *lung tan* grows in Ts'i k'ü [unknown, App. 349] in mountain valleys, also in Yüan kü [in Shan tung, App. 415]. The root is dug up in the 2nd, 8th, 11th and 12th months and dried in the shade.

T'AO HUNG-KING :—It is a common plant in Mid China. The drug from Wu hing [in Che kiang, App. 390] is the best. The root resembles that of the *niu si* [*Achyranthes*. See 101], is exceedingly bitter.

Su Sung [11th cent.]:—Its root is perennial, of a yellowish white colour. It has about ten branches, resembles the *niu si* [*v. supra*] but is shorter The plant grows about one foot high. Leaves like young garlic-leaves. The stem is slender like a small bamboo-branch. The flowers, which appear in the 7th month, are bell-shaped, of a blue colour, resemble the *k'ien niu* [*Pharbitis.* See 168]. It is commonly called 草 | | *ts'ao* (herbaceous) *lung tan*. There is another kind which is called 山 | | *shan* (mountain) *lung tan*. It is of a bitter harsh taste. It does not lose its leaves by hoar frost. The mountain people employ it for the cure of various diseases.

Kiu huang, XLVI, 27, and *Ch.*, VIII, 6, and X, 40 :— *Lung tan ts'ao*. The figures represent *Gentiana*.

In Morrison's *Engl. Chin. Dictionary* (1822), *lung tan ts'ao* is given as the Chinese name for *Dictamnus albus*. See also P. Smith, 87. Probably a mistake.

Tatar., *Cat.*, 36 :—*Lung tan ts'ao*. Rad. *Gentianæ.*— P. Smith [102] suggests that the *lung tan ts'ao* may probably be the *Gentiana asclepiadea*, but this species (European) has not been recorded from China.

It seems that in China the name *lung tan* is applied in various provinces to different species of *Gentiana*. In the Peking mountains *lung tan* is *G. barbata*, Froel., and *G. Olivieri*, DC. The *Ind. Fl. sin.* [II, 123-138] enumerates 57 Chinese species of this genus.

Cust. Med., No. 791 :—*Lung tan, Gentiana scabra*, Bge. (Hu peh).

Cust. Med., p. 6 (26):—*Lung tan ts'ao* exported 1885 from New chwang 64 piculs,—p. 202 (277), from Ning po 28 piculs, —p. 132 (178), from Chin kiang 32 piculs,— p. 114 (210), from Wu hu 3 piculs,

So moku, IV, 48, 49:—龍膽草 *Gentiana Buergeri*, Miq.—*Phon zo* [VIII, 5, 6] same Chinese name. Several species of *Gentiana*.

40.—細辛 *si sin*. P., XIII, 51. T., CLXI.

Pen king:—*Si sin* (slender and pungent), 小辛 *siao sin*. The root is officinal. Taste pungent. Nature warm. Non-poisonous.

In the *Shan hai king* it is called 少辛 *shao sin*.

Pie lu:—The *si sin* grows in the mountain valleys of Hua yin [in Shen si, App. 87]. Root dug up in the 2nd and 8th months and dried in the shade.

T'AO HUNG-KING:—Now the drugs from Tung yang and Lin hai [both in Che kiang, App. 376, 192] are used, but they are inferior to the *si sin* from Hua yin.

SU SUNG [11th cent.]:—The true *si sin* from Hua chou [in Shen si, App. 85] is slender (fine rootlets) and of a very pungent taste, whence the name. It is frequently substituted for the *tu heng* [*v. infra*].

LI SHI-CHEN:—The ancient *Po wu chi* [3rd cent.] observes that the *si sin* is frequently confounded with the *tu heng*. The difference between these two plants is that the *si sin* has tender leaves resembling those of a small mallow. The stem is slender, the root is straight, of a purple colour and very pungent in taste. The *tu heng* has leaves resembling a horse's hoof, a coarser stem, a contorted root of a yellowish white colour and a pungent taste.

For identification see the next.

41.—杜衡 *tu heng*. P., XIII, 54. T., CXLVIII.

The *Pen king* gives this name as a synonym for *tu ju* [*v. infra*, 55], but the *Pie lu* applies it to another plant.

Comp. also *Bot. sin.*, II, 156, 414.

Pie lu :—The *tu heng* is a mountain plant. The root is dug up on the 3rd day of the 3rd month and dried in the sun.

T'AO HUNG-KING :—In its root and leaves it much resembles the *si sin*, but it is less potent. It is not much used in medicine. The Taoists employ it for scenting clothes.

The *T'ang Pen ts'ao* [7th cent.] calls it 馬蹄香 *ma t'i hiang* (horse's-hoof fragrance).

SU SUNG [11th cent.]:—It is a common plant in Mid China. Perennial root. A small branchless plant, two to three inches high. Leaves like a horse's hoof in shape. Purple flowers. Fruit of the size of a pea, contains small seeds.

The *si sin* and the *tu heng* are both species of *Asarum*. See *Bot. sin.*, II, *l.c.*

TATAR., *Cat.*, 44 :—*Si sin.* Folia *Heterotropœ asaroidcs* (= *Asarum Thunbergii*, Al. Br.).—P. SMITH, 112.

The drug *si sin* received from an apothecary's shop in Peking consisted of the tender, fibrous roots with some leaves of an *Asarum*.

Cust. Med., p. 2 (14):—*Si sin* exported 1885 from New chwang 2,044 piculs,—p. 44 (13), from Che foo 20 piculs,—p. 68 (37), *ma sin* (*ma t'i hiang*) from Han kow 132 piculs,—p. 302 (407), from Amoy the same exported in a small quantity.

42.—及已 *ki ki. P.*, XIII, 55. *T.*, CXXXIV.

The *Pie lu* has only the name (*ki ki*), no details.

SU KUNG [7th cent.]:—The *ki ki* grows in the mountains, in shady valleys. The plant has only one stem, at the top of which are four leaves. The flowers are white, issue between the leaves. The root resembles that of the *si sin* [*Asarum*. See 40] but is of a black colour, bitter and poisonous. It is dug up in the 2nd month and dried in the sun.

LI SHI-CHEN adds that the plant is also called 獐耳 細辛 *chang rh si sin*. *Chang rh* (= deer's ear) refers to the shape of the leaves.

Ch., VIII, 29 :—*Ki ki*. Representation of a plant with leaves in accordance with the above description.

So moku, II, 49 :—及已 *Chloranthus serratus*, Roem. & Sch.—*DC. Prodr.* [XVI, I, 475]. Caule simplici ultra-pedali herbaceo .... foliis 4 approximatis .... See also *Kwa wi*, 12.

SIEB., *Icon. ined.*, VII :—及已 *Tricercandra quadrifolia*, A. Gray. (Same as *Chloranthus japonicus*, Sieb.), known also from China.

43.—徐長卿 *sü ch'ang k'ing*. P., XIII, 57. T., CLXXI.

Pen king :—*Sü ch'ang k'ing*, 鬼督郵 *kui tu yu*. The root is officinal. Taste pungent and bitter. Nature uniform. Non-poisonous.

As LI SHI-CHEN explains, *Sü Ch'ang k'ing* is properly the name of a man, a physician in whose memory the plant was named. In the Index of the *Pen king* we find besides *sü ch'ang k'ing* also a name of a plant 石下長卿 *shi hia ch'ang k'ing*, which name seems to refer to its growing beneath (among) stones. Some ancient authors consider it identical with the *sü ch'ang k'ing*, others say it is a distinct plant.

Pie lu :—The *sü ch'ang k'ing* grows on the T'ai shan mountain [in Shan tung, App. 322], also in Lung si [in Kan su, App. 216]. It is dug up in the 3rd month. The *shi hia ch'ang k'ing* grows likewise in Lung si, in marshes.

The description given of the *sŭ ch'ang k'ing* by the ancient authors is not characteristic. A rude drawing of the plant is found in the *Ch.* [VII, 21]. Comp. HENRY, *Chin. pl.*, 154.

So moku, IV, 30 :—徐長卿 *Pycnostelma chinensis,* Bge.—See also *Kwa wi,* 33.

The 鬼督郵 *kui tu yu,* which name in the *Pen king* is given as a synonym for the *sŭ ch'ang k'ing* and also for the *ch'i tsien* [*vide* 11], is considered by the authors who commented upon the ancient Materia Medica, to be a distinct plant which resembles the former only in its root. See *P.*, XIII, 56.

SU KUNG [7th cent.] describes the *kui tu yu* as a plant with a single always moving stem, at the top of which the leaves are inserted like an umbrella.—Another author says that the flowers come out between the leaves and are of a yellowish white colour. It is figured in the *Ch.* [VIII, 30] in accordance with the above description.

So moku, XVI, 2 :—鬼都郵 *Macroclinidium verticillatum,* Franchet, a *Composita.* The leaves are arranged in a whorl at the top of the stem. White flowers.

44.—白薇 *pai wei.* *P.*, XIII, 58. *T.*, CLVII.

Pen king:—*Pai* (white) *wei,* 春草 *ch'un ts'ao* (spring herb). The root is official. Taste bitter, saltish. Nature uniform. Non-poisonous.

LI SHI-CHEN refers this plant to *Rh ya,* 147. It has been erroneously identified by the commentators of the *Rh ya* with the *mang ts'ao* [see 158].

Pie lu:—Other names : 薇草 *wei ts'ao*, 白幕 *pai mo*, 骨美 *ku mei*. The *pai wei* grows in the river valleys of P'ing yüan [in Shan tung, App. 258]. The root is dug up on the 3rd day of the 3rd month and dried in the shade.

T'AO HUNG-KING says it is a common plant in Mid China.

SU SUNG [11th cent.]:—It grows in Mid and North China. Its leaves resemble willow-leaves. Red flowers. Root yellowish white.

Kiu huang, LII, 26, and *Ch.*, VII, 39, sub *pai wei*, rude drawings showing large follicles. An *Asclepiadea* seems to be intended.

Cust. Med., p. 346 (113) :—*Pai wei* exported 1885 from Canton 13 piculs,—p. 188 (71), from Ning po 1.75 picul.

So moku, IV, 26 :—白薇 *Vincetoxicum acuminatum*, Morr. & Dcn. (Maxim.),—[27] same Chinese name, *V. atratum*, Morr. & Dcn. and [28] *V. macrophyllum*, S. & Z.

45.—白前 *pai ts'ien*.   P., XIII, 60.   T., CLVII.

This is mentioned in the *Pie lu* as a drug (root) useful in cough. Taste sweet. Nature slightly warm. Non-poisonous.

T'AO HUNG-KING :—The *pai ts'ien* grows in Mid China. The root resembles that of the *si sin* (*Asarum*) but is larger, of a white colour, not soft, breaks easily. It is much used in curing cough.

SU KUNG [7th cent.]:—The plant grows a foot and more high. Leaves like willow-leaves, also like the leaves of the *yüan hua* [*Daphne*. See 156]. The root is longer than the *si sin* root, white. It grows on islets and on sandy ground. It is commonly called 石藍 *shi lan* also 嗽藥 *sou yao* (cough medicine).

Ch., VIII, 26 :—*Pai ts'ien.*   The plant represented shows only leaves.

Cust. Med., p. 346 (105):—*Pai ts'ien* exported 1885 from Canton 9 piculs.

So moku, IV, 32 :—白前 *Vincetoxicum purpurascens*, Morr. & Den.

46.—當歸 *tang kui.   P.*, XIVa, 1.—*T.*, CXXXII.

Pen king:—Tang kui, 乾歸 *kan kui.*   The root is officinal.   Taste bitter.   Nature warm.   Non-poisonous.

For other ancient names see *Bot. sin.*, II, 5, 49.

Pie lu:—The *tang kui* grows in Lung si [in Kan su, App. 216] in river valleys.   The root is dug up in the 2nd and 8th months and dried in the shade.

In the *Ku kin chu* [4th cent.] the plant is called 文無 *wen wu.*

T'AO HUNG-KING:—The *tang kui* which comes from Lung si [in Kan su, App. 216], Si yang [in Hu pei, App. 302], Hei shui [in Kan su and Sz ch'uan, App. 60] is very fleshy, not much branched, and fragrant. It is called 馬尾 | | *ma wei* (horse's-tail) *tang kui*.   The drug from Si ch'uan [in Kan su, App. 296] and Pei pu [in Sz ch'uan or Kan su, App. 244] has many branches and is smaller.   That from Li yang [in An hui, App. 186] is of a white colour and has but little taste and smell. It is called 草 | | *ts'ao* (herbaceous) *tang kui*.   It is sometimes substituted for the true *tang kui*.

SU KUNG [7th cent.]:—The *tang kui* is produced in Tang chou [in Sz ch'uan, App. 332], in Tang chou [in Kan su or Sz ch'uan, App. 331], this is of a superior quality, in I chou [in Sz ch'uan, App. 102], in Sung chou [in Sz ch'uan, App. 318].   There are two kinds.   One resembles the large-leaved *kung k'iung* [*Angelica*. See 47] and is called *ma wei tang kui* [*v. supra*].   This is now much used.   The

13

other resembles the small-leaved *kung k'iung*. It is called
霠頭 | | *ts'an t'ou* (silk-worm's-head) *tang kui*. This is
the drug from Li yang spoken of by T'AO HUNG-KING. It is
not much used.

SU SUNG [11th cent.]:—The *tang kui* grows in Ch'uan
Shu [Sz ch'uan, App. 26], Shen si [App. 284], also in
Kiang ning fu [Nan king, App. 129], Ch'u chou [in An hui,
App. 284]. The best drug comes from Shu (Sz ch'uan).
The leaf is divided into three segments. It flowers in the
7th or 8th months. The flowers resemble those of the *shi lo*
(*Anethum ?*), and are of a pale purple colour. The drug
which is thick and fleshy, of a dark yellow colour and not
rotten, is the best.

LI SHI-CHEN :—The drug is now much cultivated for
sale by the people of Sz ch'uan, Shen si, Ts'in chou [in
Kan su, App. 358] and Wen chou [in Sz ch'uan, App. 387].
The *ma wei tang kui* from Ts'in chou is the best.

Ch., XXV, 14 :— *Tang kui*. Rude drawing representing,
it seems, an umbelliferous plant.

The aromatic root *tang kui* brought from Sz ch'uan, and
much valued by the Chinese, was sent to Paris, in 1723
by the Jesuit Father PARENNIN. [See my *Earl. Eur. Res.
Fl. Ch.*, p. 31]. D'INCARVILLE [*Peking Plants*] says it is a
kind of " Ache " (Celery).

TATAR., *Cat.*, 19 :—*Tang kui*. Rad. *Levistici chinensis ?*
—GAUGER [13] describes and figures the root. He thinks
that it belongs to an umbelliferous plant.—HANB., *Sc. pap.*,
260 : — *Tang kwei*, described as a fleshy branchy root . . . .
approaching in odour that of Celery or *Angelica*. HANBURY
identifies it erroneously with *Aralia edulis*, as does also
P. SMITH [20], but [p. 133] the latter refers the name *tang
kui* to *Levisticum*.

Cust. Med., p. 70 (63) :—*Tang kui* exported 1885 from
Han kow 11,700 piculs,—p. 60 (23), from I chang 650

piculs,—p. 26 (57), from Tien tsin 441 piculs,—p. 46 (31), from Che foo 80 piculs.

Cust. Med., No. 1,250 :—The Sz ch'uan *tang kui* is the root of an Umbellifer not yet determined.

So moku, V, 5 :—當歸 japonice *toki*, *Legusticum acutilobum*, S. & Z.

SIEB., *Œcon.*, 246 :—*Apium ternatum*, japonice *toki*, sinice 當歸 Herba edulis ac medici usus. SIEBOLD'S *Apium ternatum* is *Ligusticum acutilobum*.—This identification is not in contradiction with the Chinese description of the *tang kui*. *L. acutilobum* is known from Japan, Corea, Formosa. According to Dr. HENRY, *Angelica polymorpha*, var. *sinensis*, Oliv., is the source of the drug *tang kui* exported from I chang and Han kow. See HOOKER'S *Icones. Plant*, tab. 1999.

There is a plant 土當歸 *t'u* (native) *tang kui* noticed in the *P.* [XIII, 28], but the plant is not described. Drawings of it are given in the *Kiu huang*, XLIX, 27, and *Ch.*, XXV, 5. From these drawings and the descriptions there it may be concluded that it is an umbelliferous plant. It is said to grow wild in the mountains of Kiang si and Hu nan.

So moku, V, 53 :—土當歸 japonice *udo*, *Aralia cordata*, Thbg. This is the same as *A. edulis*, S. & Z., *Flora japon.* [I, p. 57, tab. 25] and SIEB. [*Œcon.*, 242]. According to SIEBOLD the *udo* is universally cultivated in Japan, in fields and gardens, and valued chiefly on account of its root, which is eaten like *Scorzonera*. The young stalks are also a delicious vegetable. S. thinks that the plant has been introduced into Japan from China. It is, however, not mentioned in the *Ind. Fl. sin.* Whether the Chinese *t'u tang kui* is *Aralia edulis* is doubtful.

The *udo* is described by KÆMPFER, *Amœn. exot.* [826] but the Chinese characters there and the names *doku quatz, do sjen* are erroneous, for they are applied to an *Angelica* [see 32].

47.—芎藭 *kung k'iung.* P., XIVa, 5. T., CXLV.

Pen king :—Kung k'iung. The root is officinal. Taste pungent. Nature warm. Non-poisonous.

The plant *kung k'iung* is repeatedly mentioned in the *Shan hai king.*

*Pie lu :—*Other names: 胡 | *hu k'iung,* 香果 *hiang kuo.* Name of the leaves 蘼蕪 *mi wu.* The *kung k'iung* grows in the river valleys of Wu kung, also in Sie ku and Si ling [all in Shen si, App. 393, 309, 301]. The root is dug up in the 3rd and 4th months and dried in the sun.

T'AO HUNG-KING :—The localities Wu kung, Sie ku Si ling are all near Ch'ang an [in Shen si, App. 393, 309, 301, 3]. The drug is now produced in Li yang [in An hui, App. 186]. The plant is also much cultivated. It has fragrant leaves which resemble those of the *she ch'uang* [*Cnidium.* See 49]. Large joints. The stem is slender, looks like a horse's bit, whence the name 馬銜 | | *ma hien* (horse's-bit) *kung k'iung.* That found in Shu (Sz ch'uan) is smaller.

SU KUNG [7th cent.]:—The drug *kung k'iung* is now produced in Ts'in chou [in Kan su, App. 358]. That from Li yang is no longer in use. The *kung k'iung* is also cultivated. This drug (the root) represents large fleshy masses which contain much resin. That brought from the mountains is smaller in size, poor, and of a bitter, pungent taste. The best time for digging it up are the 9th and 10th months.

Su Sung [11th cent.]:—The plant grows in Shen si [App. 284], Ch'uan Shu [Sz ch'uan. App. 26] and in the mountains of Kiang tung [An hui, Kiang su, etc., App. 124], the best is that from Shu Ch'uan (Sz ch'uan). Its leaves resemble those of the *shui k'in* [*Œnanthe*. See 250], the *hu sui* (*Coriandrum*), the *she ch'uang* (*Cnidium*). They grow in a bushy manner, but the stem is slender. The leaves are very fragrant. The people of Kiang tung and Shu gather them for preparing a beverage. It flowers in the 7th or 8th month. Small white flowers like those of the *she ch'uang*. The root is hard and poor, of a yellowish black colour. The drug which comes from Kuan chung [Shen si, App. 158] consists of compact masses resembling the brain of a bird, whence the name 雀腦芎 *tsio nao* (bird's-brain) *kung*. This is very potent.

Li Shi-chen :—The best sort comes from Hu Jung [N.E. Tibet, App. 82]. Ancient authors call it *ma hien kung k'iung*, from the resemblance of the root with its joints to a horse's bit. Another kind is called *tsio nao kung*. That from Kuan chung [Shen si, App. 158] is called 京 | *king kung*, also 西 | *si kung*, that produced in Shu (Sz ch'uan) is 川 | *ch'uan kung*, that brought from T'ien t'ai [in Che kiang, App. 340] is called 台 | *t'ai kung*, that from Kiang nan [App. 124] is 撫 | *fu kung*. In Shu (Sz ch'uan) the *kung k'iung* is much cultivated. The leaves continue without withering till late in autumn. The root is perennial.

The 川 | *ch'uan kung* is figured in the *Kiu huang* [XLVI, 31], only leaves and the root, a nodular roundish mass. Evidently an umbelliferous plant. See also *Ch.*, XXV, 4.

Tatar., *Cat.*, 18:—大川芎, the great *kung* from Sz ch'uan, Rad. tuberosa *Levistici?*.—Gauger [12]:—The same drug figured and described: the thick, globular, nodose rhizome of an umbelliferous plant resembling in taste and odour that of Parsley.

TATAR., *Cat.*, 24 :—茯芎 *fu kung* [I suspect the first character is a mistake for 撫 *fu* (*v. supra*)], Rad. tuberosa *Levistici.*—GAUGER [19]:—The drug *fu kung* figured and described as resembling the *ch'uan kung*, but smaller.

HANB., *Sc. pap.*, 260 :—*Ch'uan kung*. Nodular masses consisting apparently of the root stock of some umbelliferous plant allied to *Angelica*, etc. The odour of the drug resembles that of the *tang kui* [see 46].

Cust. Med., No. 247 :—川芎 *ch'uan kung, Pleurospermum*, sp., or *Conioselinum univittatum*, Turcz. (Umbellif.)

Cust. Med., p. 66 (13):—*Ch'uan kung* exported 1885 from Han kow 7,180 piculs,—p. 58 (6), from I chang 1,368 piculs.—*Ibid.*, p. 66 (21), *si kung* exported from Han kow 77 piculs,—p. 342 (62), from Canton 50 piculs.

So moku, V, 2 :—芎藭 or 川芎 umbelliferous plant not identified by FRANCHET. The *Phon zo* [IX, 4] represents the large nodose roots of the same plant. See also *Kwa wi*, 34, 川芎, *sen kiu*.

So moku, V, 2, 3 :—大葉川芎 (large-leaved), *Angelica refracta*, Fr. Schm.

SIEB., *Pl. œcon.*, 254 :—*Selinea?* Sinice 芎藭, jap. *sen kjo*. Colitur in usum officinarum. SIEBOLD'S *senkjo* is, according FRANCHET, *Angelica decursiva*, Miq.

48.—蘪蕪 *mi wu*. *P.*, XIVa, 9. *T.*, CXLIII.

Pen king :—*Mi wu*. Leaves officinal. Taste pungent. Nature warm. Non-poisonous.

For other ancient names see *Bot. sin.*, II, 89, 412.

Pie lu:—Other names: 薇蕪 *wei wu*, 江蘺 *kiang li*. The leaves of the *kung k'iung* plant are the *mi wu* [v. 47]. The *mi wu* is brought from the marshes of Yung chou [Mid Shen si, App. 424] and Yüan kü [in Shan tung, App. 415]. The leaves are gathered in the 4th and 5th months and dried in the sun.

T'ao Hung-king:—The plant is produced in Li yang [in An hui, App. 186] where it is much cultivated by the people. The leaves are fragrant, resemble those of the *she ch'uang* (*Cnidium*). The plant is frequently mentioned by poets but seldom used in medicine.

Su Kung [7th cent.]:—There are two sorts, both fragrant. One has the leaves of the *k'in* (Celery), the other resembles the *she ch'uang* (*Cnidium*).

Li Shi-chen quotes an ancient author who states that the name *kiang li* is derived from the name of the provinces situated on the (lower) Kiang (Yang tsz' kiang), where the plant grows. The *Pie lu* gives *kiang li* as a synonym for *mi wu*, but Sz' ma Siang ju (the celebrated poet, who lived in the 2nd cent. B.C.) in one of his poems keeps them apart. The tender young leaves of the plant are called *mi wu*. The same plant, after the roots have developed, is called *kung k'iung*. The *mi wu* has small leaves resembling the *she ch'uang*. The *kiang li* is a variety of it. It has large leaves resembling the *k'in* (Celery).

49.—蛇牀 *she ch'uang.* P., XIVa, 10. T., CLXVII.

Pen king:—She ch'uang (serpent's bed), 蛇粟 *she su* (serpent's millet), 蛇米 *she mi* (serpent's rice). The seeds are officinal. Taste bitter. Nature uniform. Non-poisonous.

For other ancient names see *Bot. sin.*, II, 157.

Pie lu:—Other names: 牆蘼 *ts'iang mi,* 思益 *sz' i,* 繩毒 *sheng tu,* 棗棘 *tsao ki.* The *she ch'uang* grows in Lin tsz' [in Shan tung, App. 194] in river valleys and fields. The fruit (seeds) is gathered in the 5th month and dried in the shade.

T'AO HUNG-KING :—It is common in fields. Flowers and leaves resemble those of the *mi wu* [see 48].

HAN PAO-SHENG [10th cent.] :—The leaves resemble those of the small-leaved *kung k'iung* [see 47]. White flowers. Seed like millet, yellowish white. The plant grows in low, moist places. The best kinds are produced in Yang chou [Kiang su and Che kiang, App. 400] and Siang chou [in Hu pei, App. 305].

SU SUNG [11th cent.]:—The plant grows two or three feet high. Fine leaves, like those of the *hao* (*Artemisia*). Flowers white, arranged at the end of the stalks like an umbrella, more than a hundred together, forming a nest [the author means to describe an umbelliferous inflorescence] like the *ma k'in* [see *Bot. sin.,* II, 38]. The seeds are light, of a grayish yellow colour, like millet.

The *she ch'uang* is *Cnidium Monnieri.* For further particulars see *Bot. sin.,* II, 157.

Cust. Med., p. 372 (419):—Seeds of *she ch'uang* exported from Canton 7 piculs,—p. 296 (334), from Amoy 0.3 picul.

50.—藁本 *kao pen. P.,* XIVa, 12. *T.,* CXLIX.

Pen king:—*Kao pen,* 鬼卿 *kui k'ing,* 鬼新 *kui sin.* The root is officinal. Taste pungent. Nature warm. Non-poisonous.

Pie lu:—It is also called 微莖 *wei heng*. The *kao pen* grows in the mountain valleys of Sung shan [in Ho nan, App. 317]. The root is dug up in the 1st and 2nd months and dried in the sun. The fruit (seeds) are likewise used in medicine.

Su Kung [7th cent.]:—The *kao pen* in its stem, leaves and root, and also in taste, is very much like the *kung k'iung.* The best is produced in Tang chou [in Kan su, App. 331].

Su Sung [11th cent.]:—The plant grows in Si ch'uan [in Kan su, App. 296], Ho tung [Shan si, App. 80], Yen chou [in Shan tung, App. 404], Hang chou [in Che kiang, App. 58]. The leaves resemble those of the *pai chi hiang* [see 51] and the *kung k'iung,* but are smaller. In the 5th month it bears white flowers, in the 7th or 8th month it produces seed. The root is of a purple colour.

Li Shi-chen:—The *kao pen* grows in the mountain recesses of Kiang nan [Kiang su, An hui, App. 124]. The root resembles that of the *kung k'iung,* but is lighter and less juicy. It is unfit for making a beverage (decoction, for which purpose the *kung k'iung* is used). In ancient times it was used as a perfume and called *kao pen hiang.*

Tatar., *Cat.,* 26 :—*Kao pen.* Rad. *Conii* seu *Cicutæ?* Erroneous identification.—P. Smith, 62.

In Japan the above Chinese name is applied to *Nothosmyrnium japonicum,* Miq. (Umbellifera). For further particulars see *Bot. sin.,* II, 413.

Cust. Med., p. 342 (62):—*Kao pen* exported 1885 from Canton 50 piculs,—p. 168 (417), from Shang hai 0.65,—p. 186 (46), from Ning po 0.53 picul.

In Hu peh the drug *kao pen* is derived from *Ligusticum sinense,* Oliv. See Dr. Henry's note in Hooker's *Icones. Plant.,* tab. 1958.

14

51.—白芷 *pai chi*.   *P.*, XIVa, 14.   *T.*, CLVII.

Pen king:—*Pai chi*, | | 香 *pai chi hiang*, 白茝 *pai ch'i*, 芳香 *fang hiang* (fragrance).   The root is officinal. Taste pungent.   Nature warm.   Non-poisonous.

Pie lu:—Other names: 澤芬 *tse fen* (marshes' fragrance), 苻蘺 *fu li*.   The *pai chi* grows in Ho tung [S.W. Shan si, App. 80], in river valleys and marshes.   The root is dug up in the 2nd and 8th months.   The leaves are likewise used in medicine.

T'AO HUNG-KING:—It is a common plant in Eastern China.   The leaves are used as a perfume.

SU SUNG [11th cent.]:—It is common in the country of Wu [Kiang su, App. 389].   Root more than one foot long, coarse or slender, of a white colour.   Leaves opposite as broad as three fingers.   Yellowish white flowers.   The best is produced in Huang tse [unknown.   App. 96].

TATAR., *Cat.*, 4:—*Pai chi*.   Radix *Umbelliferæ* (*Angelicæ*). —GAUGER, 4:—Same drug figured and described.   Umbellifera.—P. SMITH [120] identifies it erroneously with *Iris florentina*.

LOUREIRO, *Fl. cochin.*, 114:—*Dorstenia chinensis* (a dubious plant unknown to botanists), sinice *pe chi*.   Habitat in provinciis borealibus imperii Sinensis.   Radix aromatica in usum medicum.   It seems that LOUREIRO described the plant partly from a Chinese drawing.

Cust. Med., No. 940:—*Pai chi*.   Root of *Angelica anomala*, Pall.

Cust. Med., p. 68 (45):—*Pai chi* exported 1885 from Han kow 1,825 piculs,—p. 142 (78), from Shang hai 550 piculs,—p. 58 (16), from I chang 337 piculs.

For further particulars regarding the *pai chi*, an umbelliferous plant, see *Bot. sin.*, II, 410.

52.—芍藥 *shao yo.* *P.,* XIV*a*, 18. *T.,* CXV.

Pen king:—*Shao yo.* The root is official. Taste bitter. Nature uniform. Non-poisonous.

Pie lu:—Other names: 犁食 *li shi,* 餘容 *yü yung,* 莛 *yen.* The *shao yo* grows on the Chung yo mountain [in Ho nan, App. 33] in river valleys. The root is dug up in the 2nd and 8th months and dried in the sun.

Other ancient names: 將離 *tsiang li,* 離草 *li ts'ao.*

T'AO HUNG-KING:—The best sorts are found on the Pai shan and Tsiang shan mountains [both unknown. App. 238, 351] and on the Mao shan [in Kiang su, App. 218]. The root is white and about a foot long. The plant is also found in other localities, but that is for the greater part the red sort, which is of an inferior quality.

MA CHI [10th cent.]:—There is a white and a red sort (according to the colour of the root). The flowers are also white or red.

SU SUNG [11th cent.]:—It is a common plant. The best drug comes from Huai nan [An hui, Kiang su, App. 90]. The young sprouts are of a red colour. Leaves on the top of the stem, three branches, five leaves (the author means biternate leaves) resembling the leaves of the *mou tan* [*Pæonia mou tan.* See 53], but they are longer and narrower. The plant is from one to two feet high. Its flowers are white, red or purple. The fruit resembles that of the *mou tan,* but is smaller. According to the *Ku kin chu* [4th cent.] there are two kinds of *shao yo,* the 草 | | *ts'ao* (herbaceous) *shao yo* and the 木 | | *mu* (tree) *shao yo.* The latter has large flowers of a deep (red) colour and is commonly called 牡丹 *mou tan* [see 53]. According to other authors the *mu shao yo* is a name for the purple

shao yo, (the root of which) is poor and fibrous, whilst the
shao yo with a white, fat root is called 金 | | *kin* (gold)
shao yo.

CH'EN CH'ENG [11th cent.] says, that the *shao yo*,
mentioned as a wild plant by the ancient authors, is now
much cultivated by the people.

LI SHI-CHEN :—In ancient times Lo yang [the ancient
capital of China, in Ho nan, App. 201] for the cultivation
of its *mou tan* flowers and Yang chou [in Kiang su, App.
400] for its *shao yo*. For medical use now the drug
obtained from the *shao yo* cultivated in Yang chou is
generally employed. There are more than thirty varieties
of the cultivated *shao yo*, single and double flowered. The
root of the single flowered is used in medicine. It is white
or red according to the colour of the flowers.

The *shao yo* is *Pæonia albiflora*, Pall. For further
particulars see *Bot. sin.*, II, 403.

LOUR., *Fl. cochin.*, 419 :—*Pæonia officinalis* [LOURRIRO
describes under this name *P. albiflora*], sinice *ro yo* (*sho yo*).
Varietates flore albo et rubro (radice rubescente). Habitat
culta spontaneaque per totum imperium Sinense, maxime
in provinciis borealibus. Virtus radicis, imprimis rubræ,
nervina, cephalica, emmenagoga.

TATAR., *Cat.*, 15 :—赤 | | *ch'i* (red) *shao yo*, Rad.
Pæoniæ rubræ.—P. SMITH, 169.

Pæonia albiflora, Pall., is common in the mountains of
North China and also much cultivated in gardens under the
name of *shao yo*. It has the same Chinese name in Hu pei
[see HENRY, *Chin. pl.*, 393] and in Japan.

Cust. Med., p. 122 (44) :—*Pai* (white) *shao yo* exported
1885 from Chin kiang 7,388 piculs,—p. 68 (46), from Han
kow 2,134 piculs,—p. 58 (17), from I chang 327 piculs,—
p. 24 (43), from Tien tsin 5 piculs,

Ibid., p. 22 (10):—*Ch'i* (red) *shao yo* exported from Tien tsin 2,075 piculs,—p. 2 (6), from New chwang 211 piculs,—p. 44 (7), from Che foo 2 piculs.

53.—牡丹 *mou tan*. *P.*, XIVa, 22. *T.*, CCLXXXVII to CCXCII.

Pen king:—*Mou tan* (the male red) 鼠姑 *shu ku*, 鹿韭 *lu kiu*. The bark of the root is officinal. Taste pungent. Nature cold. Non-poisonous.

Pie lu:—The *mou tan* grows in the mountain valleys of Pa [E. Sz ch'uan, App. 235] and in Han chung [S. Shen si, App. 54]. Root dug up in the 2nd and 8th months, and dried in the shade.

T'AO HUNG-KING:—Now this plant is also found in the eastern provinces of China. The red sort is good.

SU KUNG [7th cent.]:—It grows in Han chung and Kien nan [W. Sz ch'uan, App. 136]. The plant has the appearance of the *yang t'ao* [see *Bot. sin.*, II, 493]. In summer it puts forth white flowers, followed in autumn by roundish green fruit which becomes red in winter and does not fall off. The root resembles that of the *shao yo* [see 52]. It has white flesh and a red rind. The local name is 百兩金 *po liang kin* (hundred taels gold). In Ch'ang an [in Shen si, App. 6] it is known under the name 吳 | | *wu mou tan* (*mou tan* from Wu), which is the true *mou tan*.

SU SUNG [11th cent.]:—Now the drug from Ho chou [in Sz ch'uan, App. 69, b.] is considered the best. Those from Ho chou and Süan chou [both in An hui, App. 71, 315] are also of a good quality. The *mou tan* grows in a wild state in the mountains of Tan chou and Yen chou [both in Shen si, App. 327, 403], in Ts'ing chou [in Shan tung, App. 363], Yüe chou [in Che kiang, App. 418], in

Ch'u chou and Ho chou [both in An hui, App. 25, 71].
Its flowers are of different colours—yellow, purple, red,
white—and appear in the 3rd month. The flowers and leaves
of the wild-growing plant are the same as in the cultivated
sorts, but the wild *mou tan* produces only single flowers.
In the 5th month it produces fruit of a black colour,
resembling a cock's head, with large seeds. The root is of
a yellowish white colour, from five to seven inches long, of
the thickness of a pencil-holder.

K'OU TSUNG-SHI [Sung dynasty]:—The rind of the root
of the mountain *mou tan* is that which is used in medicine.
The cultivated plant produces also dark red and pale blue
flowers.

LI SHI-CHEN :—From ancient times the *mou tan* flower
has been called 花王 *hua wang* (king of flowers). OU YANG-
SIU [Sung dynasty] enumerates more than thirty cultivated
varieties of it. The *Hua pu* (a treatise on flowers, Sung
dynasty) records that to the west of Tan chou and Yen chou
[in Shen si, *v. supra*] the *mou tan* is so common that the
country people use its wood for fuel like the *king* (*Vitex*)
and *ki* (*Zizyphus*).

The *mou tan* is the China Tree Pæony, *Pæonia Moutan*,
Sims., a favorite garden-flower of the Chinese, which they
have cultivated from a remote period. In ancient times Lo
yang, the old capital of China, in Ho nan, was famed for its
mou tan flowers [see sub 52].

A good drawing of the plant is found in the *Ch.* [XXV,
18].

TATAR., *Cat.*, 39 :—*Mou tan p'i* (rind). Radix *Pæoniæ
moutan.*—GAUGER [28] figures and describes the drug. In
the drug-shops it is simply called 丹皮 *tan p'i.*—P. SMITH,
169.

Cust. Med., p. 104 (87):—*Tan p'i* exported 1885 from
Wu hu 1,606 piculs.

Amœn. exot., 862 :—牡丹 *bo tan.* Pæonia major stirpe ligneo surrecto, folio ramoso, laciniis inæqualiter divisis.

Phon zo, IX, 13, 14 :—Same Chinese name, *Pæonia Moutan.*

54.—木香 *mu hiang.* P., XIVa, 24. T., CXVII.

Pen king :—*Mu hiang* (wood perfume). The root is officinal. Taste pungent. Nature warm. Non-poisonous.

Pie lu :—Other name 蜜香 *mi hiang* (honey perfume). The *mu hiang* grows in the mountain valleys of Yung ch'ang [in W. Yün nan, App. 426].

T'AO HUNG-KING :—This drug (spoken of in the *Pie lu*) is the 青木香 *ts'ing* (green) *mu hiang*, which now however is not brought from Yung ch'ang. The *mu hiang* now employed in China is all brought by foreign ships. People say that it is produced in Ta Ts'in [the Roman Empire in Asia and Europe]. It is not used as a medicine, but only as a perfume.

Su KUNG [7th cent.]:—There are two kinds of *mu hiang*. The best comes from K'un lun [see App. 171]. That from Si hu [see App. 299] is of an inferior quality. The leaves of the *mu hiang* resemble those of the *yang t'i* [*Rumex.* See 193] but are longer and larger. The flowers resemble those of the *kü hua* (*Chrysanthemum*). The fruit is yellowish black and likewise officinal. The *mu hiang* is much used in medicine. T'AO HUNG-KING is wrong in stating that it is only employed as a perfume.

CHEN KUAN [7th cent.]:—According to the *Nan chou i wu chi* [3rd cent.] the *ts'ing mu hiang* comes from T'ien chu (India). It is the root of an herbaceous plant and has the appearance of the *kan ts'ao* (*Liquorice*).

SU SUNG [11th cent.]:—The drug *mu hiang* is brought
in ships from Kuang chou (Canton), but is not produced
there.   Large wrinkled root like that of the *k'ie tsz'*
(*Solanum melongena*).   The leaves resemble those of the
yang t'i [*v. supra*] but are longer and larger.   They are also
like those of the *shan yao* (*Dioscorea*).   Large root.   Purple
flowers.   The buds of the root used in medicine.   The
mu hiang root looks like a rotten bone.   That which is of a
bitter taste and sticks to the teeth is of a good quality.
There is a sort of *mu hiang* which grows in Kiang and Huai
[Kiang su and An hui, App. 124, 89], and is called 土青
木香 *t'u* (native) *ts'ing mu hiang*, which is not much
used in medicine.   The *Shu pen ts'ao* [10th cent.] states
that in the garden of the prince MENG CH'ANG the *mu hiang*
was cultivated.   It was a plant from three to five feet high,
leaves eight or nine inches long, wrinkled, soft, and covered
with hair.   Yellow flowers.   This was probably the *t'u mu
hiang*.   In Buddhist books the *mu hiang* is called 矩瑟佗
kü-se-t'o (probably *kush tam* is intended, *Costus*).

K'OU TSUNG-SHI [12th cent.]:—The *ts'ing mu hiang* is
found beyond the frontier [west of] Min chou [in Kan su,
App. 223].   The plant has leaves like the *niu p'ang* [*Arctium
Lappa*. See 91] but they are narrower and longer.   The
stem is from two to three feet high and bears one yellow flower
resembling the *kin ts'ien* (*Inula*).   The fresh root is fragrant
and has a pungent taste.

CH'EN CH'ENG [11th cent.]:—The *mu hiang* is brought
to China from foreign countries, as has already been stated
by T'AO HUNG-KING and SU SUNG.   But the *mu hiang*
which is produced in Ch'u chou [in An hui, App. 25] and
Hai chou [in Kiang su, App. 48] is the root of a plant
called 馬兜鈴 *ma tou ling* (horse's bell), which is also used
in medicine.

Lɪ Sʜɪ-ᴄʜᴇɴ notices that the above name *mi hiang* is also applied to the *ch'en hiang* [*Aloewood.* See 307] and 木 香 *mu hiang* to a kind of Rose [the fragrant *Rosa Banksia.* Comp. 171].

Ch., XX, 21 :—*Ma tou ling* or *t'u ts'ing mu hiang*, and [XXI, 2] *t'u ts'ing mu hiang*, good drawings representing an *Aristolochia.* See also *Kiu huang*, XLVl, 15 :—*Ma tou ling.* The latter name (horse's bell) refers to the shape of the fruit.—*Ch.*, XXV, 11 :—Three miserable drawings of the *ts'ing mu hiang*, produced in Ch'u chou and Hai chou, and of the *mu hiang* from Canton.

Tᴀᴛᴀʀ., *Cat.*, 40 :—*Mu hiang*, and [27] *kuang* (Canton) *mu hiang*, *Costus amarus.*—Gᴀᴜɢᴇʀ, 23 :—*Kuang mu hiang* described and figured. The root has a violet-like smell. It seems to belong to a plant of the *Composite* order and resembles the root of *Inula Helenium.*

Hᴀɴʙ., *Sc. pap.*, 257 :—Root *mu hiang* received from Shang hai. It was the root of *Aucklandia Costus*, Falc. (*Aplotaxis Lappa*, Dcne. *Composita*). *Costus* root or *Putchuk.*—Wɪʟʟɪᴀᴍs, *Chin. Commerc. Guide*, p. 100.— P. Sᴍɪᴛʜ, 29.

In Dʏᴍᴏᴄᴋ's *Vegetable Mat. Med. of W. India* [p. 372] this plant is called *Aplotaxis auriculata*, DC., in Sanscrit *kushta*, in Arabic and Persian *kust*, in Bengal *patchak.* The root is collected in large quantities in the highlands of Kashmeer and exported to Punjab. It is much shipped to China. P. Sᴍɪᴛʜ is wrong in stating that *putchuk* is a Canton name for the drug.—Gᴀʀᴄɪᴀs ᴀʙ Oʀᴛᴏ [middle of the 16th cent.] in his Indian Pharmacopœia [Clusius, Exot. 204] says :—Costus in Malacca, ubi ejus plurimus est usus *pu cho* dictus et inde vehitur in Sinarum regionem.

As to the *mu hiang* or *ts'ing mu hiang* produced in China and called there also *t'u ts'ing mu hiang* or *ma tou ling*, this is, according to TATAR. [*Cat.*, 12] Rad. *Aristolochiæ.* See also HANB., *Sc. pap.*, 259.—TATAR., *Cat.*, 38 :—*Ma tou ling.* Fructus *Aristolochia contortæ* (a Peking species).— HANB., *l.c.*, 239 :—*Ma tou ling.* Fruits of *Aristolochia Kæmpferi.*—P. SMITH, 22.

According to the Customs' Report on Trade for 1867 [p. 42] and 1868 [p. 51], the native *puchuk* grown in the neighbourhood of Ning po is a common garden creeper, an *Aristolochia.* Some years later Dr. HANCE examined this plant. It proved to be a new species—*A. recurvilabra.* See *Journ. Bot.*, 1873, p. 72.

HENRY, *Chin. pl.*, 294 :—*Ts'ing mu hiang* in Hu pei, *Aristolochia*, sp.

It is not quite clear whether the *ts'ing mu hiang* in the *Cust. Med.* is the foreign or the native drug. It is stated to have been exported 1885 from Han kow [p. 66 (9)] to the extent of 34 piculs,—and [p. 338 (20)] imported into Canton 24 piculs. *Ibid.*, p. 28 (88):—*Ma tou ling* exported from Tien tsin 27 piculs.

Dr. HENRY states [in HOOKER's *Icones. Plant.*, tab. 1975] that *Inula racemosa*, Hook. fil., is cultivated in the mountains of Hu peh as a substitute for *putchuk.*

FRANCHET refers the drawings in the *Phon zo* [IX, 14, 15] sub 木香, and in the *So moku* [XVII, 3, 4] sub 土木香, to *Inula Helenium*, L.—*Phon zo*, XXVI, 4-6 :—馬兜鈴 *Aristolochia Kæmpferi*, Willd.

55.—杜若 *tu jo.* P., XIVa, 30. T., CXLVIII.

Pen king :—*T'u jo*, 杜衡 *tu heng.*[19] The root is officinal. Taste pungent. Nature slightly warm. Non-poisonous.

---

[19] Regarding this synonym *see* 41.

Pie lu:—Other names : 杜連 *tu lien,* 若芝 *jo chi,* 白芩 *pai k'in,* 白連 *pai lien.* The *tu jo* grows in the marshes of Wu ling [in Hu nan, App. 394] and in Yüan kü [in Shan tung, App. 415]. The root is dug up in the 2nd and 8th months and dried in the sun. In the *Kuang ya* it is called 楚衡 *ch'u heng.*

T'AO HUNG-KING :—It is a common plant. Its leaves resemble *kiang* (ginger) leaves and are veined. The root resembles the *kao liang kiang* (*Galanga*), but is smaller, of a pungent taste and fragrant. It is also very much like the root of the *süan fu* [*Calystegia.* See 169] and is confounded with it, but the leaves are different. The *tu jo* is mentioned as a fragrant plant in the Elegies of Ts'u [4th cent. B.C.].

SU KUNG [7th cent.]:—The plant is common in Kiang and Hu [Mid China, App. 124, 83]. It grows in shady places. The plant resembles the *lien kiang* [a *Zingiberacea.* P., XlVa, 29], the root the *kao liang kiang.*

HAN PAO-SHENG [10th cent.]:—The plant resembles the *shan kiang* [*Alpinia.* See 56]. Yellow flowers, red fruit, as large as small jujubes. Inside the fruit resembles the *tou k'ou* [*Cardamom.* See 58]. That produced in Ling nan [S. China, App. 197] and Hia chou [in Hu pei, App. 64] is the best. The *Fan tsz' ki jan* says that the *tu heng* and the *tu jo* are produced in the southern prefectures and in Han chung [S. Shen si, App. 54].

LI SHI-CHEN :—There is in the mountains of Ch'u [Hu kuang, App. 24] a plant which the people call 茛薑根 *liang kiang ken* (root). It resembles ginger and is of a pungent taste. This is the plant which CHEN KUAN [7th cent.] notices under the name 獠子薑 *chao tsz' kiang.*[20] SU SUNG [11th cent.] calls it 山薑 *shan kiang* (mountain ginger) and states that it is produced in Wei chou [in Ho nan,

---

[20] *Chao tsz'* a barbarian tribe in the S.W. of China.

App. 391], that it has purple flowers and no seeds. The root used in medicine. All these names according to LI SHI-CHEN refer to the *tu jo*. The larger sort is called *kao liang kiang*, the smaller *tu jo*. In the T'ang period the *tu jo* was brought as tribute from Hia chou [*v. supra*].

Ch., XXV, 9 :—*Tu jo.* Representation of a *Zingiberacea*, probably an *Alpinia*, 䓞 薑 *liang kiang* is given as a synonym.

So moku, VII, 13 :—杜 若 *Pollia japonica*, Hornst. (*Commelinaceæ*).

56.—山 薑 *shan kiang.* P., XIVa, 31. T., XLIII.

As we have seen, LI SHI-CHEN takes the *shan kiang* or mountain ginger to be the same as the *tu jo*, but in the next article he describes it as a distinct plant, of which the root, the flowers and the seed are officinal.

T'AO HUNG-KING :—The eastern people (East China) call it *shan kiang*. In the south it is called 美 草 *mei ts'ao* (beautiful plant).

CHEN KUAN [7th cent.]:—The root and the whole plant of the *shan kiang* much resemble ginger, but it (the root) is larger, has the smell of camphor - wood. The southern people eat it. There is one sort which is called *chao tsz' kiang*. It is of a yellow colour, very pungent, acrid and strong [compare above, 55].

SU SUNG [11th cent.]:—The *shan kiang* is produced in Kiu chen [in Cochinchina, App. 154] and Kiao chi [Cochinchina, App. 133], but it is also found in Min [Fu kien, App. 222] and Kuang [Kuang tung and Kuang si, App. 160]. The *Ling piao lu i* [T'ang dynasty] says, regarding this plant :—The stem and the leaves all resemble the ginger-plant, but the root is not much eaten. The flowers resemble those of the *tou k'ou* [*Cardamom*. See 58], but are smaller. The flowers are arranged in spikes and appear between the leaves. The flower-buds (?) are like wheat-grains, small and of a red colour. In the south the unopened flowers are called 含胎花 *han t'ai hua*. They are prepared with salt water and mixed with sweet dregs. In winter they then become like amber in colour and are of a pleasant, fragrant and pungent taste.

Li Shi-chen :—The *shan kiang* grows in the south (of China). Its leaves resemble those of ginger. The flowers are red, very pungent. The fruit (or seeds) is like *Cardamom* [*ts'ao tou k'ou*, see 58]. The root resembles the *tu jo* [see 55] and the *kao liang kiang* [see 57]. The seeds are substituted for the *ts'ao tou k'ou*, but are very hot and strong.

Ch., XXV, 53 :—*Shan kiang*. Rude drawing, perhaps *Alpinia*.

Loui., *Fl. cochin.*, 13 :—*Canna indica*, L., sinice *san kiam* (*shan kiang*).[21]

So moku, I, 11 :—山薑 *Alpinia japonica*, Miq.

Cust. Med., p. 372 (416):—*Shan kiang* seeds exported 1885 from Canton 116 piculs.—The same exported also from Han kow. See *Hank. Med.*, p. 35.

57.—高良薑 *kao liang kiang*. P., XIVa, 32. *T.*, CLXXVIII.

Pie lu :—*Kao liang kiang*. The root and the fruit officinal. Taste pungent. Nature very hot. Non-poisonous. It is produced in the district of Kao liang [in Kuang tung, App. 117]. The root is dug up in the 2nd and 3rd months.

---

[21] *Canna indica* is cultivated at Peking under the name of 美人蕉 *mei jen tsiao*. It does not seem that the *shan kiang* in the *P.* refers to this plant.

It resembles in shape and in odour the *tu jo* [see 55], but the leaves are like those of the *shan kiang* [see 56].

T'AO HUNG-KING :—This is a kind of ginger produced in the district of Kao liang, whence the name.

LI SÜN [8th cent.]:—紅豆蔲 *hung tou k'ou* (red nutmeg) is the name for the fruit of the *kao liang kiang*. It is a common plant in Nan hai [Kuang chou fu, App. 228]. It looks like a reed. The leaves resemble ginger-leaves. The flowers are veined with red and arranged in a spike which is at first enclosed in a spathe. The young flowers are prepared with salt.

SU SUNG [11th cent.]:—The *kao liang kiang* is a common plant in Ling nan (South China), also in K'ien and Shu [N. Kui chou and Sz ch'uan, App. 141, 292]. It is also found in Central China, but this sort is not much used in medicine. The plant grows from one to two feet high. The leaves resemble ginger-leaves. Purplish red flowers like those of the *shan kiang*.

FAN CH'ENG-TA [12th cent.] in his description of the southern provinces of China, says that the *hung tou k'ou* is a plant with leaves like a reed. It shoots forth one stem bearing a large spathe which bursts and then a drooping spike of beautiful pale red flowers appears. The flowers resemble peach or apricot flowers.

LI SHI-CHEN states that the plant is also known under the name of 蠻薑 *man kiang* (ginger of the Southern Barbarians).

Ch., XXV, 39 :—*Kao liang kiang.* The drawing seems to represent a *Zingiberacea.* The plant is said to grow in Yün nan. Yellow flowers.

LOUR. [*Fl. cochin.*, 7] gives *cao leam kiam* as the Chinese name for *Amomum Galanga,* L., the *Galanga major* of Rumphius, Galanga root, = *Alpinia Galanga*, Sw. Pale yellow flowers. Root and seeds used in medicine by the Chinese.

TATAR., *Cat.*, 26, 34 :—*Kao liang kiang* or *liang kiang, Galanga.* WILLIAMS [*Chin. Commerc. Guide* 120] has 良薑 *liang kiang*, Galangal, the root of *Alpinia Galanga.* The seeds of the same plant are used as aromatic medicine under the name of *hung tou k'ou.* *Ibid.* [p. 84] the same seeds are called *kao liang kiang tsz'.*—HANB. [*Sc. pap.*, 107, 252] describes and figures the fruit capsules received from Shanghai under the name of *kao liang kiang tsz'* or *hung tou k'ou.* They proved to belong to *Alpinia Galanga.*— P. SMITH, 9, 10.

Another kind of Galanga, the lesser or Chinese Galanga of commerce, the *Galanga minor* of Rumphius, is referred in the *Flora Hongk.* [349] to *Alpinia chinensis*, Rosc., a plant of smaller stature than the *A. Galanga*, known from Canton more than one hundred years ago. But in 1873 Dr. HANCE described [in the *Journ. Linn. Soc.*, XIII] a plant which had been presented to him by TAINTOR as growing wild and cultivated in the island of Hai nan and called *liang kiang* by the Chinese. HANCE named it *Alpinia officinarum*, and believes that this yields the true Chinese Galanga. It has white flowers, veined with dull red.

It would seem from the ancient Chinese accounts above translated regarding the *tu jo* and the *kao liang kiang*, that the first is the *Galanga minor*, the second the *Galanga major.* But probably the above names were applied to different species of *Alpinia* in various parts of China. MARCO POLO [YULE's 2nd edition, II, 207, 208] mentions the *galingale* produced in immense quantities in the kingdom of Fu ju (province of Fu kien), and also in Java [II, 254]. Dr. FR. HIRTH thinks [*China Review*, II, 97] that the name *Galanga* has been derived from the Chinese *kao liang kiang*. It seems however more naturally to trace it in *kulanjana*, the Sanscrit name for Galangal.

So moku, I, 10 :—高艮薑 *Alpinia* allied to *A. chinensis.*—*Phon zo*, IX, 20, 21 :—Same Chinese name, same identification.

58.—豆蔻 *tou k'ou.　P.*, XlVa, 35.　*T.*, CXLVII.

Pie lu :—The *tou k'ou* grows in Nan hai [Southern Sea, App. 228]. Seeds and flowers used in medicine.

In the *Nan fang i wu chi* it is called 漏 | *lou k'ou.*

Su Kung [7th cent.] :—The plant resembles the *shan kiang* [see 56]. The flowers are yellowish white. The root and the seeds resemble the *tu jo* [see 55].

Su Sung [11th cent.] :—The 草豆蔻 *ts'ao* (herbaceous) *tou k'ou* is a common plant in Ling nan [S. China, App. 197]. It grows like a reed. The leaves resemble those of the *shan kiang*. The root is like the root of the *kao liang kiang* [see 57]. The flowers[22] open in the 2nd month, they are in spikes at the bottom of the stem. The young leaves are rolled up. The flowers are of a reddish colour, darker at the end of the spike. Gradually the leaves become larger and the flowers paler. The flowers are sometimes of a yellowish white colour. The southern people collect the flowers and salt them. The fruit resembles the *lung yen* (*Nephelium longan*) but is pointed, and the rind (capsule) is not squamous. The seeds within the capsule resemble those of the pomegranate. They ripen in summer and are then gathered and dried in the sun. The root and all parts of the plant exhale an odour which recalls camphor-wood and are of a pungent taste.

[22] This account is taken from an earlier work, the *Nan fang ts'ao mu chuang* [3rd cent.], *tou k'ou hua.*

LI SHI-CHEN :—The *ts'ao tou k'ou* and the 草果 *ts'ao kuo* are not the same, as some believe. There are differences. Now the *tou k'ou* produced in Kien ning [in Fu kien, App. 138] has a fruit as large as the *lung yen*, but a little longer. It (the capsule) has a yellowish white thin rind with prominent ridges. The seeds are as large as the *su sha* (*Amomum villosum*, Lour.), pungent and fragrant. But the *ts'ao kuo* which grows in Tien [Yün nan, App. 338] and in Kuang [Kuang tung and Kuang si, App. 160] has a large oblong fruit resembling the *ho tsz'* (*Terminalia chebula*). The rind (of the capsule) is black and thick, the ridges are close together. The seed is coarse, pungent and of an unpleasant odour recalling that of Cantharides. The people use it as tea or in various other ways as a spice. The people of Kuang take the fresh *ts'ao kuo* and steep it in the juice of the *mei* fruit (*Prunus mume*) mixed with salt. After it has become red it is dried in the sun and offered with wine. This is called *hung yen* (red salt) *ts'ao kuo*. The small unripe fruit is called 鸚哥舌 *ying k'o she* (parrot's tongue). In the time of the Mongol dynasty the *ts'ao kuo* was much valued as a spice.

LI SHI-CHEN quotes from Buddhist books the Sanscrit name of the *tou k'ou*, being 蘇乞迷羅細 *su-ki-mi-lo-si.*—*Sukmil* is the Tibetan name for *Cardamom* [see further on].

Ch., XXV, 30 :—*Tou k'ou.* Representation of an *Amomum* with large leaves and small, wrinkled capsules.

LOUR., *Fl. cochin.*, 6 :—*Amomum globosum.* Sinice *tsao keu* (*ts'ao tou k'ou*). Corolla supera, albo-rubra . . . . Pericarpium globosum . . . . cortice tenui fragili.

TATAR., *Cat.*, 5 :—*Ts'ao tou k'ou, Cardamomum.*

This is the *Large Round Chinese Cardamom* figured and described sub *ts'ao* (*tou*) *k'ou* in HANB., *Sc. pap.*, 95, 96, 248.—P. SMITH, 14.

16

Lour., *Fl. cochin.*, 5 :—*Amomum medium*, sinice *tsao quo* (草果). Pericarpium oblongum, striatum, crassum, coriaceum.—Tatar., *Cat.*, 5 :—*Ts'ao kuo.* Fructus *Amomi medii.*—This is the *Ovoid China Cardamom* figured and described in Hanb., *Sc. pap.*, 105, 106, 250 :—P. Smith, 14.

Cust. Med., p. 372 (433):—*Ts'ao (tou) k'ou* exported 1885 from Canton 0.2 picul.

Ibid., p. 372 (434):—*Ts'ao kuo* exported from Canton 653 piculs,—p. 406 (164), from Kiung chow 428 piculs,—p. 424 (132), from Pak hoi 402 piculs.

The drawing in the *Phon zo* [IX, 21, 22] sub 草豆蔻 represents, it seems, Loureiro's *Amomum globosum*, of which only the fruits are known to European botanists.

白豆蔻 *pai tou k'ou* (white *Cardamom*) is the Chinese name for the *Cardamom* imported from foreign countries. *P.*, XIVa, 37. *T.*, CXLVII. The seeds are used in medicine. It does not seem to be mentioned in Chinese works before the 8th cent.

Ch'en Ts'ang-k'i [8th cent.]:—The *pai tou k'ou* is produced in the country of Ka-ku-lo, and is called there 多骨 *to ku*. The plant resembles the *pa tsiao (Musa, Banana)*. The leaves resemble those of the *tu jo* [*Alpinia.* See 55]. They are from eight to nine feet long, shining, evergreen. Flowers of a pale yellow colour. The fruit is produced in clusters, hanging down like grapes. They are at first green but become white when ripe. They are gathered in the 7th month.

Su Sung [11th cent.]:—This plant is now grown in Kuang chou (Kuang chou fu) and in I chou [in Kuang si, App. 103], but the drug is inferior in value to that brought by foreign ships.

Li Shi-chen:—The fruit (capsule) of the *pai tou k'ou* is globular, as large as that of the *k'ien niu* [*Pharbitis.* See 168]. Its outer skin is thick and of a white colour. The seeds are like the *su sha* [*v. supra*]. To prepare it for medical use the skin is taken off and the seeds are roasted.

Ch., XXV, 64 :—*Pai tou k'ou.* Rude drawing. The *Cardamom* plant seems to be intended.

Lour., *Fl. cochin.,* 4 : — *Amomum Cardamomum,* L. Sinice *pe teu keu.* Flores albo-lutei. Capsula 3 gona rotunda. Semina cortice laevi, albicante.

The *Amomum Cardamomum* of Linnæus is the *Round* or *Cluster Cardamom,* a native of Cambodja, Siam, Java. The *pai tou k'ou* is still much imported into China from Cochinchina, Siam and Malabar. It seems that the *Malabar Cardamom, Elettaria Cardamomum,* the seeds of which are very similar in odour and taste to those of the *Cluster Cardamom,* go also under the name of *pai tou k'ou.* The *pai tou k'ou* which I obtained from a Tibetan apothecary's shop at Peking was *Malabar Cardamom.* The Tibetans call it *sukmil* [comp. above the Sanscrit name *su-ki-mi-lo-si*].

Rheede [*Malab.,* XI, p. 10], in describing the *Elettaria,* says :—In aprico fructus exsiccatur solo, ubi cortex, qui primo crassus, viridisque, extenuatur et ex ruffo albescit.

The country Kakulo, mentioned in the above Chinese account as producing the *pai tou k'ou,* is unknown to me. I may however observe, that *kakula* is the Arabic name for *Cardamom* [Roxbg., *Fl. ind.,* 1874, p. 24].

The *Round* or *Cluster Cardamom* is also known under the name of 東坡 | | *Tung p'o tou k'ou,* probably after the celebrated poet Su Tung-p'o, who, towards the end of the 11th century, lived for some years in the island of Hai nan

and wrote notices of useful plants. Comp. HANB., *Sc. pap.*,
253.—See also WILLIAMS, *Commerc. Guide*, p. 84.—
P. SMITH, 14.

肉豆蔻 *jou* (fleshy) *tou k'ou* is the Chinese name for
Nutmegs, the nuts of *Myristica moschata*. Mace, the arillus
of the nutmeg, is called 肉豆花 *jou tou hua* (flower). It
seems improbable that nutmegs were known to the Chinese
before the 8th century. *P.*, XIV*b*, 45. *T.*, CXLVII.

CH'EN TS'ANG-K'I, the first Chinese author who mentions
the *jou tou k'ou*, states that it is brought by ships from
foreign countries, where it is called *ka-kù-le* (probably
intended for *kakula*, which, however, as we have seen, is
Cardamom).

SU SUNG [11th cent.] reports that the *jou tou k'ou* is
also cultivated in South China.

LI SHI-CHEN :—The *jou tou k'ou* in its flowers and fruit
resembles the *ts'ao tou k'ou*. The difference is that the
latter (is a capsule) in which the seeds are contained, whilst
the *jou tou k'ou* is solid (a solid nut), the outer skin of
which is covered with wrinkled lines, and the inner substance
is reticulated and mottled like the betelnut.

Ch., XXV, 63 :—*Jou tou k'ou.* Rude, incorrect drawing.
But the *Phon zo* [IX, 27, 28] sub 肉豆蔻 gives a good
figure of *Myristica moschata*.

WILLIAMS [in his *Commercial Guide*, 98, 95] gives
[erroneously, it seems] 豆蔻 *tou k'ou* as the Chinese name
for nutmegs. As we have seen above, the original meaning
of *tou k'ou* is *Cardamom*.

TATAR., *Cat.*, 64 :—P. SMITH, 156, 141.

59.—莎草 *so ts'ao*, 香附子 *hiang fu tsï'*. *P.*, XIV*b*,
58. *T.*, CVII.

For other ancient names compare *Rh ya*, 97.

Pie lu:—The *so ts'ao* grows wild. It (evidently the root) is gathered in the 2nd and 8th months. Taste sweet. Nature slightly cold. Non-poisonous.

The leaves and the flowers are likewise officinal.

T'AO HUNG-KING :—This plant is mentioned in the *Shi king*. It is no longer used as a medicine. There is a medicinal plant 鼠 | *shu so*, but that is different.

SU KUNG [7th cent.]:—The root of the *so* is called *hiang fu tsz'* [*hiang* = fragrant, *fu tsz'* properly the small tubers of *Aconite*. See 143] also 雀頭香 *tsio t'ou hiang* (sparrow-head fragrance). The stem and the leaves of the plant resemble the *san leng* (triangular grass, *Scirpus, Cyperus*). It (the root, tubers) is used as a perfume.

SU SUNG [11th cent.]:—It is a common plant, which in its leaves resembles the *hiai* [*Allium*. See 242] but is weaker. The root resembles the head of a chopstick. In a topographical work of the T'ang period the 水香稜 *shui hiang leng* (cornered fragrant water-plant) is spoken of as growing in the ponds and marshes of Po p'ing [in Shan tung, App. 260]. Its root is called 莎結 *so kie* (*so* knot) also 草附子 *ts'ao fu tsz'*. In Ho nan and Huai nan [An hui, Kiang su, App. 90] it is known by the name of 水 | *shui so*, in Lung si [in Kan su, App. 216] they call it 地頼根 *ti lai ken*, in Shu (Sz ch'uan) it is 頼根 *su ken* (attached root tubers) also 水巴戟 *shui pa ki*. The plant now grows abundantly in Fou tu [in Sz ch'uan, App, 42] and is called there 三稜草 *san leng ts'ao* (triangular grass). It is used for making shoes. The whole plant, and especially the root (tubers), is used in medicine.

K'OU TSUNG-SHI [12th cent.]:—*Hiang fu tsz'* is the name for the tubers which are frequently found attached to the root of the *so* plant. These tubers have a thin, chapped skin of a purplish black colour and are very hairy. After the skin has been removed the white flesh becomes apparent.

LI SHI-CHEN :—The leaves of the *so* plant are like *Allium* leaves—hard, shining, sharp on the margins. The stem is hollow, triquetrous. Green flowers in spikes. The roots are fibrous. Beneath the radical fibres small tubers are produced. These are of the size of a small jujube, pointed at both ends, and covered with fine black hair. They are much used in medicine. In Sanscrit books the plant is called 目苯哆 *mu ts'ui ch'e* (or *ta*).

Ch., XXV, 35 :—*So ts'ao.* Representation of a *Cyperus* with oblong tubers.

As has been stated in *Bot. sin.*, II, 97, the names *so ts'ao* and *hiang fu tsz'*, given in the *P.* as synonyms, were applied in ancient times to two distinct cyperaceous plants,— *so* to a *Scirpus*, the culms of which were used for making shoes, umbrellas, rain-cloaks, hats, and *hiang fu tsz'* to the fragrant tubers of a *Cyperus*.

LOUR., *Fl. cochin.*, 53 :—*Cyperus rotundus*, L. Ubique in Cochinchina et China. Radix tuberibus ovatis, parvis, odoratis, pilosis. Sinice *hiam phu cu.*

TATAR., *Cat.*, 45 :—*Hiang fu tsz'.* Radix *Cyperi.*— GAUGER [39] describes and figures these tubers, which he refers to *C. rotundus.*

P. SMITH, 81 :—*So ts'ao, hiang fu tsz', Cyperus esculentus,* and [51] *so ts'ao, Carex hirta* [arbitrary identification].

Comp. also HENRY, *Chin. pl.*, 144.

Cust. Med.; p. 210 (15) :—*Hiang fu tsz'* exported 1885 from Wen chow 76 piculs,—p. 340 (41), from Canton 75 piculs,—p. 186 (31), from Ning po 58 piculs.—Smaller quantities exported also from Shang hai, Amoy, Swatow.

I may observe, regarding the name *mu ts'ui ta*, as given in Chinese Buddhist books, that *musta* or *mustuka* is a Sanscrit name for *Cyperus rotundus*, L.

Regarding the Japanese cyperaceous plants, to which the above Chinese names are applied, see *Bot. sin.*, II, 97.

There is another cyperaceous plant with officinal tubers which is described in the *P.* [XIV*b*, 55] under the name 荆三稜 *san leng*, from the country of King (Hu pei). *San leng* (triquetrous) is a general name for several cyperaceous plants. See the drawing *Ch.*, XXV, 55.— *T.*, CLXXVII.

TATAR., *Cat.*, 44 :—*San leng ts'ao.* Rad. *Cyperi* seu *Scirpi.*—GAUGER [37] describes and figures the *san leng.* Tuber about one inch in diameter.—P. SMITH, 82 :—*King san leng, Cyperus rotundus.*

Cust. Med., p. 70 (53) :—*San leng* exported 1885 from Han kow 109 piculs,—p. 130 (147), from Chin kiang 60 piculs,—p. 188 (76), from Ning po 29 piculs,—p. 92 (70), from Kiu kiang 5 piculs.

Comp. *Phon zo*, IX, 33 :—荆三稜 *Cyperacea.*

60.—薰草 *hün ts'ao.* P., XIV*b*, 72. T., CVIII.

Pie lu :—The *hün ts'ao*, which is also called 蕙草 *hui ts'ao*, grows in low, marshy places. It is gathered in the 3rd month and dried in the shade. That with the joints taken off is good.—The same *Pie lu* says also :—The | 實 *hün shi* (fruit) grows in marshes in Lu shan [in Ho nan, App. 203]. The whole plant is officinal. Taste sweet. Nature uniform. Non-poisonous.

T'AO HUNG-KING :—According to the *Yao lu* [attributed to T'UNG KÜN, a minister of Emperor HUANG TI], the *hün ts'ao* has leaves resembling those of the *ma* (hemp) and each two

standing opposite. The *Shan hai king* states :—On the Fou
shan (mountain) there grows a plant with leaves like the *ma*.
It has a square stem, red flowers, black fruit. It smells like
the *mi wu* [see *Rh ya*, 89] and is called *hūn ts'ao*. It is
good for curing ulcers. Now it is commonly called 燕草
yen ts'ao. Some say it has the appearance of the *mao*
(*Imperata*) and is fragrant, but that is not the *hūn ts'ao*
which the people cultivate. The plant *hui* [*Bot. sin.*, II,
406], frequently mentioned by poets in ancient times, is
unknown to T'AO HUNG-KING, as he says.

CH'EN TS'ANG-K'I [8th cent.]:—The *hūn ts'ao* is the
same as the 零陵香 *ling ling hiang* (fragrance). *Hūn* is
the name for the root of the *hui* plant.

MA CHI [10th cent.]:—The *ling ling hiang* grows in the
mountain valleys of Ling ling [in Hu nan, App. 196]. Its
leaves resemble those of the *lo le* (*Ocimum basilicum*). The
Nan yüe chi [5th cent.] says that the local name of the
plant is *yen ts'ao* [*v. supra*]. It is also called *hūn ts'ao* or
hiang ts'ao (fragrant herb). This is the *hūn ts'ao* of the
Shan hai king.

SU SUNG [11th cent.]:—The *ling ling hiang* is now a
common plant in Hu kuang [App. 83] where it grows in
marshy places. Its leaves resemble those of hemp, each two
standing opposite. Square stem. In the 7th month it produces
very fragrant flowers. In ancient times it was called *hūn ts'ao*.
The people of Southern China dry it by artificial heat till
it assumes a yellow colour. It is also found in Kiang and
Huai [An hui and Che kiang, App. 124, 89] and used
as a perfume. But this is less valued than that from Hu
kuang and Ling nan. The fragrance increases when the
plant decays. In ancient times the *hūn ts'ao* was used in
medicine, and the name *ling ling hiang* was unknown. But
now the people use it only as a perfume added to cosmetics.

FAN CH'ENG-TA, in his account of the southern provinces [end of the 12th century], states that the *ling ling hiang* is a common plant in I chou and in Yung chou [both in Kuang si, App. 103, 430] and other places. The plant is used for making mats, pillows and matresses. The locality Ling ling [spoken of by earlier authors as producing this plant] is now called Yung chou [in Hu nan, App. 425], but this plant does not grow there.

LI SHI-CHEN :—In ancient times this plant was burned to make the spirits descend. Now the people of Wu [Kiang su, App. 389] cultivate it for sale. It is also termed 廣 l l l *kuang ling ling hiang* and 黃陵草 *huang ling ts'ao.*

Regarding the identification of the *hün ts'ao* or *ling ling ts'ao,* which seems to be *Ocimum basilicum,* see *Bot. sin.,* II, 406, 407.

61.—蘭草 *lan ts'ao.* *P.,* XIV*b*, 75. *T.,* LXXXI and LXXXII.

Pen king :—*Lan ts'ao,* 水香 *shui hiang* (water perfume). The leaves are officinal. Taste pungent. Nature uniform. Non-poisonous.

Pie lu :—The *lan ts'ao* grows in the ponds and marshes of T'ai Wu. It (the leaves) is gathered in the 4th and 5th months.

T'AO HUNG-KING :—It is not used now in prescriptions. T'ai Wu is the kingdom of Wu [Kiang su, App. 389] where T'ai Po[23] lived. There is now in Tung men [Eastern Gate. Unknown to me] a plant used for making fragrant oil[24] and which is called 蘭香 *lan hiang.* This is the *lan ts'ao.*

---

[23] The founder of the state of Wu. See MAYERS' *Chin. Read. Man.,* 243.

[24] 煎澤草.

Li Tang-chi [3rd cent.]:—The *lan ts'ao* is the same plant as that which the people now cultivate under the name of 都 梁 香 *tu liang hiang* [Fragrance from Tu liang, in Hu nan, App. 370]. The *tse lan* [see the next] is also called *tu liang hiang.*

Su Kung [7th cent.]:—The *lan*, a fragrant plant, is the same as the *tse lan*. It has a round stem, a purple receptacle of flowers. In the 8th month the flowers are white. It is commonly called *lan hiang* (fragrance) and grows by the sides of rivulets. It is also much cultivated as an ornamental plant.

Han Pao-sheng :—The *lan ts'ao* grows in low, damp places. Its leaves resemble those of the *tse lan*, but are longer, pointed and coarsely toothed. Flowers red and white, fragrant.

Ch'en Ts'ang-k'i :—The *lan ts'ao* and the *tse lan* are two distinct plants. The *lan ts'ao* grows by the sides of marshes, its leaves are glabrous, succulent. The root is small and of a purple colour. It is gathered in the 5th and 6th months and dried in the shade. This is the *tu liang hiang*. Women mix it (it seems the leaves) with oil to dress their hair.—The *tse lan* has pointed, slightly hairy leaves, not glabrous, and is succulent. Square stem, purple joints. This is the plant regarding which Su Kung states that it bears white flowers in the 8th month.

Li Shi-chen :—The *lan ts'ao* and the *tse lan* are two species of the same genus. Both grow on the borders of water-courses or in swamps. They have perennial roots, purple, branched stems with red joints, opposite leaves issuing from the joints, slightly serrated. But the *lan ts'ao* has a round stem, long joints (internodes), glabrous leaves, whilst the *tse lan* has a nearly square stem, short joints and leaves covered with hair. The young leaves of both are gathered and worn (in satchels) on girdles. In the 8th or 9th month the plants are from three to four feet high. The flowers are in spikes like those of the *ki su* (a *Labiata*). The flowers are red and white (or perhaps reddish white). Small seeds. The plant which in the *P'ao chi lun* is called *ta* (great) *tse lan* is the same as the *lan ts'ao*, and the *s'ao* (small) *tse lan*, there is what we call *tse lan*.

For the identification of the *lan ts'ao* see the next.

62.—澤蘭草 *tse lan ts'ao.* P., XIV*b*, 78. *T.*, LXXXII.

Pen king:—*Tse lan ts'ao* (marsh *lan*), 虎 | *hu* (tiger) *lan*, 龍棗 *lung tsao* (dragon jujube). The leaves are officinal. Taste bitter. Nature slightly warm. Non-poisonous.

Pie lu:—The *tse lan* grows on the margins of all the great lakes or swamps in Ju nan [in Ho nan, App. 110]. It (the leaves) is gathered on the 3rd day of the 3rd month, and dried in the shade.

The descriptions of the *tse lan* as given by the authors quoted in the *P.* are not characteristic and much confused. Some compare it to plants of the Labiate order, from other descriptions it would seem that it is a *Composita*. According to LI SHI-CHEN the roots are eaten and called 地筍 *ti sun.* The seeds are also used in medicine.

I have already pointed out [*Bot. sin.*, II, 405] that the fragrant plant 蘭 *lan* mentioned in the Classics, and by early Chinese poets, was most probably a fragrant orchid. The figure in the *T.* [*l.c.*] under the name of *lan* is without doubt intended for a plant of this order. LI SHI-CHEN observes that this *lan* of the Classics and poets is probably called *lan hua* (*lan* flower). It has leaves like the *mai men tung* (*Ophiopogon*) and is not to be confounded with the *lan ts'ao*, which is quite different,

China is very rich in orchids. In our days one of the most favorite of them among the Chinese in the south is the 吊蘭花 *tiao* (suspended) *lan hua,* called also 風 | *feng* (air) *lan,* the *Aërides odorata* of LOUREIRO, *Fl. cochin.,* 642.—BRIDGM., *Chin. Chrest.,* p. 452 (5).—*Amœn. exot.,* 864 :—*Fu ran,* cum icone. [*V. infra,* sub 202].

The *Cust. Med.* [p. 160 (319)] notices 30 piculs of 蘭花米 *lan hua mi,* classed among seeds, as imported to Shang hai.

Ibid., p. 152 (201):—佩蘭葉 *pei lan ye* (*lan* leaves worn on the girdle) exported from Shang hai 1.15 picul. Said to come from Sz ch'uan. *Ibid.,* p. 194 (163):—The same imported to Ning po 1.10 picul.

Ibid., 360 (283):—*Tse lan* exported 1885 from Canton 22 piculs,—p. 288 (222), from Amoy 5 piculs.

The figures of the *lan ts'ao* and the *tse lan ts'ao,* in the *Ch.* [XV, i and 13], seem both to be intended for species of *Eupatorium,* order of *Compositæ.* Dr. HANCE states that in S. China *Eupatorium stœchadosmum* is cultivated on account of the fragrance of its flowers. See *Ind. Fl. sin.,* I, 405.

In Japan the Chinese names 蘭草 and 澤蘭草 are both applied to *Eupatorium.* For particulars see *Bot. sin.,* II, 405.

63.—香薷 *hiang ju.* P., XIV*b*, 81. T., CL.

Pie lu :—Only the name *hiang* (fragrant) *ju* and medical properties noticed. It seems the leaves are officinal.

T'AO HUNG-KING :—It is commonly eaten raw as a vegetable. It is also gathered in the 10th month and dried.

The *Shi liao Pen ts'ao* [7th cent.] calls it 香菜 *hiang jou* and 香茸 *hiang jung,*

Su Sung [11th cent.]:—The plant is cultivated but rarely in the north. It resembles the *pai su* [*Perilla*. See 67] but the leaves are smaller. It is produced in Shou ch'un and Sin an [both in Che kiang, App. 291, 310]. There is one kind which is called 石 | | *shi hiang ju*. It grows on rocks, is slender, of a yellow colour, pungent and fragrant and much valued.

K'ou Tsung-shi [12th cent.]:—The *hiang ju* grows wild in the mountains of North and South King Hu [Hu nan, App. 147]. In Pien and Lo [both in Ho nan, App. 248, 201] the people cultivate it in gardens and eat it as a vegetable during the hot season.

Li Shi-chen:—There are the wild-growing *hiang ju* and the cultivated one. The latter is called 香菜 *hiang ts'ai* (fragrant vegetable). There is a large-leaved and a small-leaved sort, the first is the best. The plant has a square stem, incised leaves like the *huang king* (*Vitex*) but smaller. In the 9th month purple flowers in spikes, followed by small seeds. There is one sort with more slender leaves like those of the *lo chou* (*Kochia*) and which grows only a few inches high. This is the *shi hiang ju*. Another name for the *hiang ju* is 蜜蜂草 *mi feng ts'ao* (bee plant).

Ch., XXV, 32:—*Hiang ju*. Representation of a *Labiata*, probably *Elsholtzia*.

Tatar., *Cat.*, 46:—*Hiang ju*, *Elsholtzia cristata*. This is a common plant in the Peking mountains. Debeaux [*Flor.* Shang hai, 48, Tien tsin, 36] saw it cultivated in Chinese gardens.—P. Smith, 94.

Cust. Med., p. 80 (202):—*Hiang ju* exported 1885 from Han kow 173 piculs,—p. 374 (464), from Canton 31 piculs,—p. 92 (84), from Kiu kiang 31 piculs,—p. 300 (390), from Amoy 8 piculs.

So moku, XI, 16:—香薷 *Elsholtzia cristata*, Willd.

64.—爵牀 *tsio chuang*.   P., XIV*b*, 84.   *T.*, CLXVIII.

Pen king:—*Tsio chuang*.   Leaves and stem officinal.
Taste saltish.   Nature cold.   Non-poisonous.

Pie lu:—Other name: 香蘇 *hiang su* (fragrant *su*).
The *tsio chuang* grows in Han chung [S. Shen si, App. 54],
in river-valleys and fields.

Wu P'u [3rd cent.] calls it 爵麻 *tsio ma*.

Su Kung [7th cent.]:—This plant grows in marshes
and corn-fields and by way sides.   It resembles the *hiang ju*
[*Elsholtzia*.   See 63], but the leaves are longer and larger.
They resemble the *jen* [*Perilla*.   See 67] but are smaller.
Its vulgar name is 赤眼老母草 *ch'i yen lao mu ts'ao*
(red-eyed old mother's herb).

Li Shi-chen:—It is a common plant in the plain and
in waste places.   Square stem with joints.   It resembles the
large-leaved *hiang ju* [see 63].   But when rubbed [the
leaves] between the fingers the latter is fragrant, whilst the
tsio chuang exhales a somewhat unpleasant odour.

Ch., XXV, 23:—*Tsio chuang*.   The drawing represents
a labiate plant.

So moku, XI, 17:—爵狀 *Mosla punctata*, Maxim.
Same as *Ocymum punctatum*, Thbg., *Fl. japon.*, 249.—Order
Labiatæ.

65.—假蘇 *kia su*.   P., XIV*b*, 85.   *T.*, LVIII.

Pen king:—*Kia su* (*Pseudo-Perilla*), 鼠蓂 *shu ming*.
The whole plant, especially the flower-spikes, used in
medicine.   Taste pungent.   Nature warm.   Non-poisonous.

Pie lu:—Other name: 薑芥 *kiang kie* (*kiai*) or ginger-
mustard.   The *kia su* grows in the marshes of Han chung
[S. Shen si, App. 54].

Wu P‘u [3rd cent.] calls the plant also 荆芥 *king kie* (*kiai*) and states that it has·leaves like those of the *lo li* (*Chenopodium*) but smaller. The plant is eaten in Shu [Sz ch‘uan, App. 292].

T‘AO HUNG-KING :—The *kia su* is not used now in medical prescriptions.

Su KUNG [7th cent.] classes it among the vegetables.

LI SHI-CHEN :—The *king kie* grows wild and is also much cultivated. The young plants are fried and eaten. The taste is pungent and fragrant. The plant has a square stem, small leaves resembling those of the *tu chou* [*Kochia*. See 111] but narrower and smaller, of a pale yellowish green colour. In the 8th month it opens its small flowers, arranged in spikes like those of the *su* [*Perilla*. See 67]. The seeds are small like those of the *t‘ing li* [see 114].

Ch., XXV, 22 :—*Kia su* or *king kie*. It seems *Salvia plebeja*, R. Br., is intended by the drawing. This plant at Peking is called *king kie*, but the same name is also applied to *Nepeta tenuifolia*, Bth.

TATAR., *Cat.*, 58 :—*King kie, Salvia plebeja.*—P. SMITH, 192.

PARKER, *Canton pl.*, *king kie* = *Salvia plebeja*, also *Moslea lanceolata*, Maxim. But LOUREIRO [*Fl. cochin.*, 453] gives *quam tum kim kiai* (*king kie* of Canton) as the Chinese name for *Origanum creticum*, L. [*O. vulgare*. See *Ind. Fl. sin.*, II, 282].

HENRY, *Chin. pl.*, 70 :—The name *king kie* applied in Hu pei to various plants : *Phtheirospermum, Mosla, Elsholtzia, Melampyrum.*

Cust. Med., p. 132 (172) :—*King kie* exported 1885 from Chin kiang 463 piculs,—p. 202 (264), from Ning po 200 piculs,—p. 92 (80), from Kiu kiang 123 piculs.—The *Hank. Med.* [p. 6] mentions it also as exported from Han kow.

So moku, **XI**, 31 :—假蘇 or 荆芥 *Nepeta japonica*, Maxim., and [32] same Chinese names, *N. tenuifolia*, Benth.

66.—積雪草 *tsi süe ts'ao.　P.*, **XIV***b*, 92.　*T.*, **CXXXVI.**

Pen king :—Tsi süe ts'ao (snow plant).　Stem and leaves officinal.　Taste bitter.　Nature cold.　Non-poisonous.

Pie lu:—The *tsi süe ts'ao* grows in the river-valleys of King chou [Hu kuang, App. 146].

T'AO HUNG-KING explains the name (snow plant) by the cooling properties of the plant.　But it was then not used in medical prescriptions.

SU KUNG [7th cent.]:—This plant has leaves resembling the round Chinese copper coins, and therefore the people of King Ch'u [Hu kuang, App. 145] call it 地錢草 *ti ts'ien ts'ao* (ground coin herb).　It has a slender but strong stem, creeps on the ground.　It grows near rivulets.

SU SUNG [11th cent.]:—It is a common plant, which is also called 連錢草 *lien ts'ien* (connected coins) *ts'ao*. According to the collection of prescriptions of the T'ien pao period [9th cent.] it grows in Hien yang [Shen si, App. 65] in low, marshy places, also in Lin tsz' [in Shan tung, App. 194] and Tsi yang [in Shan tung, App. 346] in ponds and marshes.　It is very fragrant, has round leaves, resembles the *po ho* (*Mentha*) and is also called 胡薄荷 *hu po ho*. It is very common in Kiang tung, Wu Yüe and Tan yang [all in An hui and Kiang su, App. 124, 389, 328], where the people eat it.　In Liu ch'eng situated in the province of Ho pei [Chi li, App. 198] it is called 海蘇 *hai su*.

LI SHI-CHEN :—In An hui and Che kiang, where the people used to drink an infusion of the leaves, the plant is called 新羅 丨 丨 *sin lo po ho* [Sin lo = S. Corea, App. 311].　It is also found in Hu kuang and in Min (Fu kien).

Ch., XXV, 24 :—*Tsi süe tsao.* The figure shows only
leaves. Probably *Nepeta Glechoma*, Benth. (*Glechoma hederacea*, L.), our Ground Ivy, which is a common plant in
China, is intended. The above descriptions in the *P.* agree
in a general way.

Amœn. exot., 887 :—積雪 *sakusetz, kakidoro, tsubogusa.*
Herba repens Hederæ terrestris facie ac folio, flosculis
hexapetalis, purpureis, etc. According to THBG. [*Fl. japon.*,
116] this is *Hydrocotyle asiatica*, L. But in the *So moku*
[XI, 2] 積雪 is *Nepeta Glechoma.*

67.—蘇 *su. P.*, XIV*b*, 94. *T.*, LVIII.

Pie lu :—*Su.* Stem, leaves and seeds used in medicine.
Taste pungent. Nature warm. Non-poisonous.

The *su* is mentioned in the *Rh ya* [64].

T'AO HUNG-KING :—The *su* has its leaves purple underneath. They are very fragrant. Another sort, the leaves
of which are not purple coloured, and which resembles the
jen, is called 野蘇 *ye* (wild) *su.* It is not much used. The
same author says, in another work quoted in the *Ry ya i* :—
The 荏 *jen* resembles the *su*, but it grows higher, is white
(downy, not purple-coloured leaves), and not very fragrant.
The seeds are oily, and by pressure oil is obtained from
them. In Kiang tung [An hui, App. 124] the people call it
蔫 *yü.*

SU SUNG [11th cent.] :—The *su* is the 紫蘇 *tsz'* (purple)
su. The best sort has the leaves purple coloured on both
sides. The stem and the leaves are gathered in summer,
the seeds in autumn. There are several kinds of *su*, the
shui (water) *su* [see 68], the *yü* (fish) *su*, etc. All these are
kinds of *jen* [*v. supra*].

18

LI SHI-CHEN :—There is the 紫 | *tsz'* (purple) *su* and
the 白 | *pai* (white) *su*. The seeds of both are sown in
the 2nd and 3rd months. They grow also spontaneously
from seeds left on the ground. Square stem, roundish,
pointed, toothed and serrated leaves. In a rich soil it (the
tsz' su) has leaves of a purple colour on both sides, but
in a poor soil they are green on the upper side, purple only
underneath. The *pai* or white *su* has its leaves white
(downy) on both sides. This is the 荏 *jen*. The young
leaves of the purple *su* are eaten as vegetable food salted
or pickled together with the *mei* fruit [*Prunus Mume.*
See 272]. In summer they make a beverage of the leaves,
which are very fragrant. The root is also used. In the
8th month it opens its small purple flowers, arranged in
spikes, and afterwards capsules are formed like those of the
king kie [*Salvia.* See 65]. In the 9th month, when the
plant is half withered, the seeds are gathered. These are
small, like mustard seeds, of a yellowish red colour. They
yield an oil like the oil obtained from the *jen*.

The *su* and the *jen* are species of *Perilla*. For further
particulars, see *Bot. sin.*, II, 64.

Cust. Med., p. 373 (422):—*Su tsz'* (seeds) exported
1885 from Canton 24 piculs,—p. 164 (360), from Shang hai
3 piculs,—p. 200 (241), from Ning po 2 piculs.

Ibid., p. 298 (360):—*Tsz' su* from Amoy 2.5 piculs.

Ibid., p. 130 (142):—*Pai su tsz'* from Chin kiang
9.7 piculs,—p. 110 (164), from Wu hu 6 piculs.

Ibid., p. 360 (275, 276):—*Su* stalks and leaves, from
Canton 24 piculs.

68.—冰蘇 *shui su.   P.*, XIV*b*, 97.   *T.*, LVIII.

Pen king:—*Shui* (water) *su*. The stem and the leaves
are officinal. Taste pungent. Nature slightly warm. Non-
poisonous.

Pie lu:—Other names: 芥蒩 *kie tsu,* 芥苴 *kie tsü.* The *shui su* grows in Kiu chen [App. 154], in ponds and marshes. It is gathered in the 7th month.

WU P'U [3rd cent.] calls it 雞蘇 *ki su* (chicken *su*).

T'AO HUNG-KING:—It is not used in medical prescriptions. *Kiu chen* is a distant place which has not been identified.

SU KUNG [7th cent.]:—This kind of *su* grows in marshes and by the sides of water-courses. It resembles the *süan fu* [*Inula.* See 81]. The leaves stand in twos opposite and are very fragrant. In Ts'ing, Ts'i [both in Shan tung, App. 363, 348], and in Ho kien [in Chi li, App. 75] the people call this plant *shui su* [as above], in Kiang tso [S. An hui, App. 124] it is known by the name 荠苧 *tsi ning,* in Wu Hui [in Kiang su, Che kiang, App. 391] it is called *ki su* [as above].

HAN PAO SHENG [10th cent.]:—Leaves like those of the *pai wei* [*Vincetoxicum.* See 44], in twos opposite. Violet flowers coming out between the joints. Taste pungent, aromatic.

SU SUNG [11th cent.]:—The *shui su* is a common plant by the sides of water-courses. It is much eaten in the south as a vegetable. It is also frequent north of the Kiang, but the people there do not eat it. In Kiang tso [*v. supra*] the *ki su* is not the same as the *shui su.* The *tsi ning* [*v. supra*] is also a different plant. The leaves of the *shui su* are toothed, fragrant, of a pungent taste, those of the *tsi ning* are narrow and longer, covered with hair, and exhale an unpleasant odour.

WU SHUI [Mongol period] says that the *shui su* is also called 龍腦薄荷 *lung nao po ho* (Camphor mint), but from an earlier account, quoted by SU SUNG, it would seem that this plant resembles an *Artemisia.*

Ch., XXV, 20 :—*Shui su.* Figure of a *Labiata*, as also under the same Chinese name in the *Phon zo*, XII, 14.

The 薺薴 *tsi ning* is described in the *P.* [XIV*b*, 99] as a distinct plant, called also 臭蘇 *ch'ou* (stinking) *su.* Judging from the drawings in the *Ch.* [XXV, 51] and in the *Phon zo* [XII, 15] it is a *Labiata*. The drawing under the above Chinese name in SIEB., *Icon. ined.*, VI, is *Calamintha?*

69.—菊 *kü.* *P.*, XV, i. *T.*, LXXXVII-IX.

Pen king :—*Kü*, in the Index 菊華 *kü hua*, also 節華 *tsie hua*. Flowers and other parts of the plant officinal. Taste bitter. Nature uniform. Non-poisonous.

For other ancient names see *Bot. sin.*, II, 130, 404.

Pie lu :—Other names : 女節 *nü tsie*, 女華 *nü hua*, 女莖 *nü heng*, 日精 *ji tsing*, 更生 *keng sheng*, 傳延年 *fu yen nien*, 陰成 *yin ch'eng*, 周盈 *chou ying*. The *kü hua* grows in river-valleys and fields in Yung chou [in Shen si, App. 424]. The root is dug up in the 1st month, the leaves are gathered in the 3rd, the stem in the 5th, the flowers in the 9th, the seeds in the 11th month, and dried in the shade.

According to TS'UI SHI [Han period] the names *nü tsie, nü hua* refer to the flowers. In the *Pao p'o tsz'* [3rd and 4th cent.] it is stated that the above names *ji tsing, keng sheng,* and *chou ying* in Taoist prescriptions to promote longevity, are applied to the root, stem, flower and seeds of the *kü* plant.

T'AO HUNG-KING :—There are two kinds of *kü.* One has a purple stem, is fragrant and of a sweet taste. The leaves are used in soups. This is the genuine *kü.* The other, with a green stem, is larger and has the smell of the *hao* and *ai* (both *Artemisia*). It is of a bitter taste and not much eaten. This is the *pseudo kü* called also 苦薏 *k'u* (bitter) *i*. The leaves in both are about the same. The *kü* grows plentifully in Li hien, in the prefecture of Nan yang [in Ho nan, App. 183, 231] and is also common in other places. It is much cultivated. There is also a variety called *pai kü* with white flowers.

Su Sung [11th cent.]:—The *kü* is a common plant. That produced in Nan yang [*v. supra*] is the best. It flowers in autumn and bears seed in winter. There are many varieties, with large and small flowers. Some have flowers with a yellow disk and white ray flowers, others are entirely yellow.

Wu Shui [Mongol period]:—That with large, fragrant flowers is the 甘菊 *kan* (sweet) *kü*, that with small yellow flowers is the 黃 | *huang kü*, that with small flowers of an unpleasant odour is the 野 | *ye* (wild) *kü*.

Li Shi-chen :—There are a hundred varieties of the *kü*. The flowers are of various colours, single or double. The *kan* (sweet) *kü* is used in medicine. It is much cultivated, and grows also wild in the mountains. Its leaves are eaten.

Kü is a general name for many plants of the order *Compositæ*: *Chrysanthemum*, *Aster*, etc.—but the *kü* par excellence, and which the ancient Chinese authors above quoted call the true or sweet *kü*, is the *Chrysanthemum sinense*, Sab., the favourite winter-flower of the Chinese, who have cultivated it from time immemorial, it seems, in numerous varieties. It is also common in a wild state in the mountains of North China and also in other parts of the empire. The wild plant is about one foot high, and blossoms late in autumn. Small flower-heads. Florets of the disk yellow, those of the circumference rose coloured. I suspect that the *huang* (yellow) *kü* of the ancient authors is the *Chrysanthemum indicum*, L., likewise a common wild plant all over China.

It has small flower-heads, yellow florets in the disk as well
as in the circumference. At Peking it is called *siao ye kü
hua* (small, wild *Chrysanthemum*).

Kiu huang, LIII, 20 :—*Kü.* Rude drawing. Small
flower-heads.

Ch., XI, i :—*Kü hua.* Two figures. One represents a
Chrysanthemum with large double flowers, the other a plant
with small flower-heads ; probably the wild form is intended.

LOUR., *Fl. cochin.*, 610 :—*Chrysanthemum indicum*, L.
Late cultum ob pulchritudinem floris in Cochinchina et
China. Sinice *ta kio hua* (large *kü hua*). LOUREIRO'S *Chr.
indicum* is the *Chr. sinense.*

Ibid.:—*Chr. procumbens* [LOUR. describes under this
name *Chr. indicum*]. Spontaneum, cultumque in Cochinchina et China. Inveniuntur multæ varietates :—(1) flore
pleno, integre ligulato, flavo,—(2) flore radiato, disco et
radio flavis,—(3) disco flavo, radio albo,—(4) flosculis
omnibus albis. [Comp. above the varieties according to the
Chinese authors.]

P. SMITH [62] erroneously identifies the *pai kü hua*
with *Chrysanthemum album*, and [145] *ye kü hua* with
Matricaria Chamomilla. Both these plants have not been
recorded from China. *Ibid.*, 19 :—*Huang kü, Anthemis.*

Cust. Med., p. 74 (113) :—*Kü hua* exported 1885 from
Han kow 315 piculs,—p. 30 (108), from Tien tsin 270
piculs,—p. 196 (173), from Ning po 210 piculs.

Ibid., p. 362 (293):—*Kan kü* exported from Canton
45 piculs.

Ibid., p. 128 (102):—*Huang kü hua* exported from
Chin kiang 20 piculs,—p. 324 (150), from Swatow 2.3 piculs.

Ibid., p. 324 (155):—*Pai kü hua* exported from Swatow
147 piculs,—p. 128 (107), from Chin kiang 42 piculs.

Amœn. exot., 875 :—菊 *kik, kikf* vel *kikku,* i.e. *Matricaria.* Cujus cum sylvestris tum præcipue hortensis, plurimæ sunt varietates.

So moku, XVII, 18:—菊花 *Pyrethrum (Chrysanthemum) sinense.*

Ibid., 21:—冬菊 (Winter *Chrysanthemum*) *Pyr. (Chrys.) indicum.*

Ibid., 22 :—野菊花 *Pyretrum seticuspe,* Maxim. Small, yellow flower-heads.

70.—卷薗 *an lü.   P., XV, 5.   T.,* CXLVI.

Pen king:—An lü. Seeds used in medicine. Taste bitter. Nature slightly cold. Non-poisonous.

*Pie lü:—*The *an lü* seed is produced in Yung chou [in Shen si, App. 424] in river-valleys, also in Shang tang [S.E. Shan si, App. 275] by waysides. The seeds are gathered in the 10th month and dried in the shade.

T‘AO HUNG-KING :—It has the appearance of the *hao* and the *ai* (both *Artemisia*), and is a common plant in Middle China. The Taoists use it. People cultivate it. Snakes dislike it.

SU SUNG :—It grows in Kiang and Huai [Kiang su, An hui, App. 124, 89], from two to three feet high, resembles *Artemisia* in its leaves, flowers in the 7th and bears seed in the 8th month.

LI SHI-CHEN :—The leaves of the *an lü* do not resemble the *ai* (*Artemisia vulgaris*) but rather the *kü* (*Chrysanthemum*), and are thinner and much divided into narrow segments. Leaves green on both sides. The stem grows from four to five feet high, is white, like that of the *ai,* ·and rough. In the 8th or 9th month it opens its small, pale

yellow flowers. Fruit (seed) like that of the *ai*, small.
The old plant is used for thatching roofs, whence it is also
called 罷 菌 *fou lü*.

Ch., XI, 3 :—*An lü*. Representation of an *Artemisia*.

So moku, XVI, 21 :—菴 閭 *Artemisia Keiskiana*, Miq.
(known only from Japan, E. Manchuria and Corea).—
HOFFM. & SCHLT. [548] identify the same Chinese name
with *Siphonostegia chinensis*, Benth., but this seems to be a
mistake. Comp. *infra*, 86.

71.—蓍|*shi*. *P.*, XV, 5. *T.*, C.

Pen king:—Shi. Fruit receptacles with the achenes
used in medicine. Taste bitter and acid. Nature uniform.
Non-poisonous.

This is the Chinese divining plant, about which see
Bot. sin., II, 428, *Achillea sibirica*. Comp. LEGGE's *Yi king*,
Appendix, V, p. 422 :—" Anciently when the sages made the
" Yi in order to give mysterious assistance to the spiritual
" intelligences, they produced the rules for the use of the
" divining plant *shi*."

Pie lu:—The *shi* fruit is produced in the mountain
valleys of Shao shi [in Ho nan, App. 281]. It is gathered
in the 8th and 9th months, and is dried in the sun.

SU KUNG [7th cent.]:—The stem of this plant is used
in divination (筮).

SU SUNG [11th cent.]:—The *shi* is found growing near
the sacrificial hall of the white tortoise at Shang ts'ai hien
in the prefecture of Ts'ai chou [in Ho nan, App. 276, 342].
It has the appearance of the *hao* (*Artemisia*), grows from
five to six feet high. From thirty to fifty stems spring up
from one root. Late in autumn purple flowers appear at
the end of the branches. They resemble the *kü hua* (*Chrysanthemum, Aster*). Fruit like those of the *ai* (*Artemisia*).

LI SHI-CHEN says that the *shi* is a kind of *hao* (*Artemisia*), a divine plant.

For further particulars see *Bot. sin.*, II, 428.

Cust. Med., p. 8 (54, 55):—*Shi* exported 1885 from New chwang 760 piculs.[25]

72.—艾 *ai*. *P.*, XV, 8. *T.*, CI.

Comp. also *Bot. sin.*, II, 77, 429.

Pie lu:—The *ai* is also called 醫草 *i ts'ao* (vulnerary herb). The plant which yields the *ai* leaves [used for cauterizing] grows in the fields. The leaves are gathered on the 3rd day of the 3rd month, and dried in the sun. Taste bitter. Nature slightly warm. Non-poisonous.

SU SUNG [11th cent.]:—The *ai* is a common plant. The best sorts are produced in 複道 Fu tao (elevated road) and 四明 Sz' ming (name of a monastery in Che kiang). The *ai* leaves are used for cauterizing, and therefore the plant is also called 灸草 *kiu ts'ao* (moxa). It is a kind of *hao* (*Artemisia*). The leaves are white (downy) underneath. The leaves must be gathered on the 3rd day of the 3rd month or on the 5th of the 5th month.

LI SHI-CHEN:—In the Sung period that from Fu tao in T'ang yin [in Ho nan, App. 335] and Sz' ming was considered the best. The first was called 北艾 *pei* (northern) *ai*, the other 海丨 *hai* (sea) *ai*. Since the Ch'eng hua period (1465-1488) the drug from K'i chou [in Hu pei, App. 121] is much valued and known under the name of 蕲艾 *k'i ai*. This plant is common on mountain plateaux. Perennial root, straight, white stem, four or five feet high. The leaves resemble those of the *hao*, are five-lobed with

---

[25] A mistake: the drug exported from New chwang is a kind of *huang-ch'i-shih;* in the Customs List 蓍 is a misprint for 耆 *ch'i* (*i.e.* without the 140th radical).—A. HENRY.

small points, green on the upper side, white and downy underneath, soft and thick. In the 7th or 8th month flower-spikes like those of the *ch'e ts'ien* [*Plantago*. See 115] with small flowers come out between the leaves. Small seeds.

The *ai* is the *Artemisia vulgaris*, L., very common in N. China, both wild and cultivated. Good figure sub *ai* in the *Ch.* [X, 81] also in the *Kiu huang* [XLVIII, 25] sub 野艾蒿 *ye* (wild) *ai hao*.

Ch., XIV, 65 :—千年艾 *ts'ien nien ai* (a thousand years' *ai*) or 蘄艾 *k'i ai*. It is said there that this plant grows wild on the To ho shan mountains in Hu pei. This is *Tanacetum chinense*, A. Gray. It is cultivated under the name of *k'i ai* at Peking and its downy leaves are used for moxa.

LOUR., *Fl. cochin.*, 600 :—*Artemisia vulgaris.* Sinice *ngai ye* (*ai* leaves). Ibidem *Artemisia chinensis.* Sinice *khi ngai*. Ex plantæ hujus foliis exsiccatis et contusis fit *moxa* seu cauterium actuale. It seems that the plant LOUR. describes as *A. chinensis* is *Tanacetum chinense*.

TATAR., *Cat.*, 1 :—*Ai tse'*, *Artemisia indica.*—P. SMITH, 25.

HENRY, *Chin. pl.*, 7 :—*Ai hao* in Hu pei is *A. indica* (a variety of *A. vulgaris*).

Cust. Med., p. 360 (278):—*Ai* (large-leaved) exported 1885 from Canton 3 piculs,—p. 378 (511), *ai jung* (moxa punk) from Canton 8 piculs.

Ibid., p. 356 (224):—*K'i ai* exported from Canton 22 piculs,—p. 286 (181) from Amoy 3 piculs.

The Chinese mode of cauterizing by burning the down of *Artemisia vulgaris* or *Tanacetum chinense* upon the skin, seems to be of very ancient date. Its invention is ascribed to the Emperor HUANG TI. In the History of the Sui dynasty, in the section on Literature, there is the title of a work 黃帝鍼灸經 *Huang ti chen kiu king* or Emperor HUANG TI's work on Acupuncture and Cauterizing. The *Rh ya* [77] gives 冰臺 *ping t'ai* (ice turret) as a synonym for *ai*. The *Po wu chi* [3rd cent.] explains the character *ping* (ice) in the name by the fact that the *ai* leaves (or moxa) were ignited by means of a piece of ice cut into a roundish form which collected the sun-beams.

Amœn. exot., 897 :—艾 *gai*, vulgo *jamogi*. *Artemisia vulgaris major*; quæ junior vocatur *futz*, ex qua fit *Moxa*, celebris stupa pro cauterio actuali.

SIEB., *Œcon.*, 213 :—*Artemisia chinensis*, Moksa japon. 艾. Ex herba præparantur moksa celebrata.—*Ibid.*, 376 :— *Artemisia ibuki jomogi* (*A. vulgaris*). Ad præparandam moksam.

So moku, XVI, 16 :—艾 *Artemisia vulgaris*. Japonice *yomogi*, *ibuki yamogi*.—GUIBOURT, in his *Hist. naturelle des drogues* [III, 52], says, as many other authors did before him, that *moxa* is a Chinese and Japanese word. But this is an error which has already been refuted, 200 years ago, by RUMPHIUS, who, in his *Herbarium Amboinense* [V, 261, 262,¹ sub *Artemisia latifolia, baru tschina*], writes :—H.e sinensis fomes igniarius. Hic fomes vulgo *moxa* vocatur, per longum autem tempus detegere non potui quænam vox moxa esset, quumque animadvertebam a Sinensibus hic habitantibus non intelligi, hinc concludi, Portugallicam esse corruptam vocem, derivatam a *mocho* h.e. muco, cum quo forte Portugalli siccatam hujus herbæ fomitem comparant.

BAILLIÈRE [*Dict. de Médecine*, etc., 1873] suggests that moxa may be derived from the Greek Μύκες, a fungus which in a dried state was used by the ancient Greeks to burn the skin, and which is mentioned by Hippocrates.

73.—茵蔯蒿 *yin ch'en hao.*   *P.,* XV, 14.   *T.,* CXLIV.

Pen king:—Yin ch'en hao. Leaves and stem used in medicine. Taste bitter. Nature uniform, slightly cold. Non-poisonous.

*Pie lu:—*The *yin ch'en* grows in T'ai shan [in Shan tung, App. 322] on mountain slopes. It is gathered from the 5th month to the beginning of autumn, and dried in the shade.

In the *Kuang ya* [3rd cent.] it is called 因塵 *yin ch'en.*

T'AO HUNG-KING :—It is a common plant and resembles the *p'eng hao* [see *Bot. sin.,* II, 436] but the leaves are smaller and denser. The stem does not die in winter and in spring thrives again.

HAN PAO-SHENG [10th cent.]:—Its leaves resemble those of the *ts'ing hao* [see 74] but are white (downy) underneath.

SU SUNG [11th cent.]:—It is a common plant in Mid China. The best sort comes from T'ai shan [in Shan tung, App. 322]. It resembles the *p'eng hao,* but the leaves are smaller (finer). It has neither flowers nor seed. One kind, the *shan* (mountain) *yin ch'en* is used like the *ai hao.* It is different from the cultivated *yin ch'en.*

LI SHI-CHEN :—The *yin ch'en hao* was much cultivated in ancient times as a vegetable. The *shan* (mountain) *yin ch'en,* which is used in medicine, is different. Now the people of Huai and Yang [An hui and Kiang su, App. 89, 400] on the 2nd day of the 2nd month gather the leaves of the wild *yin ch'en,* mix them with flour, and prepare cakes which they call | | 餅 *yin ch'en ping.*

Ch., XI, 22 :— *Yin ch'en hao.* Representation of an *Artemisia.*

LOUR., *Fl. cochin.,* 598 :— *Artemisia abrotanum* [the plant LOUREIRO describes is not this species. See *DC, Prodr.,*

VI, 108 : certe diversa].   Sinice *yin chin hao*.   Habitat incultum, cultumque in Cochinchina, China.

TATAR., *Cat.*, 30 :—*Yin ch'en hao*.   Flores *Artemisiæ*.— P. SMITH, 25 :—*Art. abrotanum*.

Cust. Med., p. 74 (112):—*Yin ch'en* exported 1885 from Han kow 135 piculs,—p. 90 (52), from Kiu kiang 103 piculs,—p. 126 (99), from Chin kiang 1.7 picul,—p. 324 (145), from Swatow 3 piculs,—p. 30 (107), from Tien tsin a small quantity.

Amœn. exot., 897 :—茵 蔯 *intsjin*, vulgo *fki jamogi & kawara jamogi*.   Abrotanum campestre.   THBG. [*Fl. jap.*, 309] refers this to his *Artemisia capillaris* (with capillary leaves).— *So moku*, XVI, 28:—Same Chinese name *A. capillaris*.

74.—青 蒿 *ts'ing hao*.   P., XV, 16.   T., LXII.

Pen king :—*Ts'ing hao* (green *Artemisia*), 草 | *ts'ao* (herbaceous) *hao*, 方潰 *fang hui*.   Leaves, stem, root and seed used in medicine.   Taste bitter.   Nature cold.   Non-poisonous.

For other ancient names see *Bot. sin.*, II, 13.

Pie lu :—The *ts'ao hao* grows in the marshes of Hua yin [in Shen si, App. 87].

T'AO HUNG-KING :—This is a common plant, now generally called *ts'ing hao*.   The people eat it mixed with fragrant vegetables.

HAN PAO-SHENG [10th cent.]:—The *ts'ao hao* in Kiang tung [App. 124] is called 犰 | *sin hao*, for its smell resembles that of the beast called *sin*.   In the north it is called *ts'ing hao*.   The young leaves are pickled in vinegar. The leaves resemble those of the *yin ch'en* [see 73] but are not white underneath [whence the name *ts'ing* or green *hao*]. The plant grows more than four feet high.   The leaves are used in medicine. This plant is mentioned in the *Shi king* [the author refers to *Shi king*, 246, hao. See *Bot. sin.*, II, 431].

Su Sung [11th cent.]:—The *ts'ing hao* has very fine (small) leaves, which are eaten. Late in autumn it bears pale, yellow flowers. The seeds are like millet and are gathered in the 9th month. Root, stem, leaves and seeds are all used in medicine.

K'ou Tsung-shi [12th cent.]:—The *ts'ing hao* is also called 香 | *hiang hao* (fragrant *Artemisia*). It has a red root and fragrant leaves.

Li Shi-chen:—The *ts'ing hao* has a coarse, succulent stem of the thickness of a finger. The stem and the leaves are of a dark green colour. The leaves resemble those of the *yin ch'en* but are dark green on both sides. The root is white and firm. The flowers appear in the 7th or 8th month. They are small, yellow and slightly fragrant. The fruit resembles that of hemp and contains small seeds.

Ch., XI, 93, sub *ts'ing hao* or *ts'ao hao*, representation of an *Artemisia* with capillary leaves.

Tatar., *Cat.*, 12 :—*Ts'ing hao*, *Artemisia*.—P. Smith, 25 :—*Ts'ing hao*, *Artemisia Dracunculus*. He says it is abundant in Hu peh, and sometimes eaten as a vegetable.

Debeaux, *Fl. de Shang hai*, 38, *Fl. de Tien tsin*, 28 :— *Artemisia Dracunculus*, sinice *tsin kao*, cultivated as a vegetable.

Cust. Med., p. 356 (231) :—*Ts'ing hao* exported 1885 from Canton 18 piculs.

Hank. Med., 6, 7 :—The same mentioned as exported from Han kow.

So moku, XVI, 25 :—青蒿 *Artemisia apiacea*, Hance (*A. abrotanum*, Thbg.),

75.—白蒿 *pai hao*. *P.*, XV, 20. *T.*, LXII.

Pen king:—*Pai hao* (white *Artemisia*). Leaves, root and seeds used in medicine. Taste sweet. Nature uniform. Non-poisonous.

Pie lu:—The *pai hao* grows in the marshes of Chung shan [in Chi li, App. 31]. It is gathered in the 2nd month.

T'AO HUNG-KING states that the *pai hao* is not used in medicine.

SU KUNG [7th cent.]:—The 蘩 *fan* or 皤蒿 *pai hao* of the *Rh ya* [12] is the 白 | *pai hao*. The leaves are covered with a white down, whence the name.

SU SUNG [11th cent.]:—In ancient times the people used the leaves of the *pai hao* for food. Now they employ for this purpose the 蔞蒿 *lou hao*, which some authors have erroneously identified with the *pai hao*.

LI SHI-CHEN:—The *pai hao* is a common plant. There are two kinds of it, one growing in water, the other in dry land. The first is the *pai hao* of the *Pie lu*. It is fragrant and pleasant, whilst the dry land plant is pungent and unpleasant. The *pai hao* of the *Pie lu* is without doubt the same as the *lou hao*.

The *pai hao* as well as the *lou hao* are species of *Artemisia*. For further particulars see *Bot. sin.*, II, 433, 430.

76.—馬先蒿 *ma sien hao*. *P.*, XV, 22. *T.*, LXII.

Pen king:—*Ma sien hao*, 馬矢蒿 *ma shi hao*. Taste bitter. Nature uniform. Non-poisonous.

Pie lu:—Other names: 練石草 *lien shi ts'ao*, 爛石草 *lan shi ts'ao*. The *ma sien hao* grows in Nan yang [W. Ho nan, App. 231], in marshes.

Su Kung [7th cent.]:—Leaves as large as those of the *ch'ung wei* [*Leonurus*. See 78]. The flowers are red and white. Stem and leaves gathered in the 2nd and 8th months, the seed ripens in the 8th and 9th months. This plant is also called 虎麻 *hu ma* or 馬新蒿 *ma sin hao*.

This is the *Incarvillea sinensis*, Juss. For further particulars see *Bot. sin.*, II, 432.

77.—牡蒿 *mou hao*.　P., XV, 23.　T., LXII.

Comp. *Bot. sin.*, II, 14, 432. Lu ki confounds it with the *ma sien hao* [76].

Pie lu:—*Mou hao* (male *Artemisia*). Leaves used in medicine. Taste bitter, slightly sweet. Nature warm. Non-poisonous. This plant grows in the fields. It is gathered in the 5th and 8th months.

T'ao Hung-king :—It is not used in medicine.

Su Kung [7th cent.]:—It is also called 齊頭蒿 *ts'i t'ou hao*. It has leaves resembling those of the *fang feng* [*Stenocœlium*. See 31] but finer and thinner, not glaucous.

Li Shi-chen :—Its leaves are flat, narrow at the base, broad and lobed at the end. The young leaves can be eaten. Deer are fond of the plant. In autumn it bears small, yellow flowers. The fruit is as large as that of the *ch'e ts'ien* [*Plantago*. See 115] and contains minute seeds, hardly distinguishable, wherefore the ancients asserted that the plant has no seeds, and called it the male southernwood.

The *mou hao* is an *Artemisia*. In Japan this Chinese name is applied to *Art. japonica*. See *Bot. sin.*, II, 432.

78.—茺蔚 *ch'ung wei*.　P., XV, 24.　T., CXXIX.

Pen king:—*Ch'ung wei*, 益母 *i mu*, 益明 *i ming*, 火杴 *huo hien*. The seeds are officinal. Taste sweet. Nature slightly warm. Non-poisonous.

For other ancient names see *Bot. sin.*, II, 25, 444.

Pie lu:—Other name : 貞蔚 *chen wei.* The *ch'ung wei* grows near the seashore and on the margins of pools and marshes. It (the seeds) is gathered in the 5th month.

T'AO HUNG-KING :—This plant is found everywhere. Its leaves resemble those of the *jen* [*Perilla.* See 67]. The stem is square. Small, oblong triangular seeds. Seldom used in medicine.

CH'EN TS'ANG-K'I [8th cent.]:—The popular name of the *ch'ung wei* is 臭草 *ch'ou ts'ao* (stinking plant).

TA MING [10th cent.]:—Stem, leaves and root likewise used in medicine.

SU SUNG [11th cent.]:—The plant is common in gardens and in waste places. It is mentioned in the *Rh ya* [25]. The seeds are black and resemble those of the *ki kuan* (*Celosia cristata*). Square stem.

K'OU TSUNG-SHI [12th cent.]:—The young plant can be used for food.

LI SHI-CHEN :—The *ch'ung wei* grows abundantly near water-courses, in damp places. The young plant in spring resembles a young *hao* (*Artemisia*). In summer it is from three to four feet high. It has a square stem. The leaves resemble *ai* leaves (*Artemisia vulgaris*) but are green (not downy) underneath. They are trilobed with long points. The small flowers are arranged (in a whorl) around the joints, and are of a red colour tinged with white. Each calyx contains four seeds as large as those of the *t'ung hao* (*Chrysanthemum Roxburghii*), triangular and of a gray colour. The living plant has an unpleasant odour. The root is white. The descriptions given by ancient authors—as, for instance, that the leaves resemble *Perilla* leaves, that the seeds are black, etc.—are incorrect. There are two kinds of *ch'ung*

20

wei, one with purple and the other with white flowers. The last is the *i mu*, the purple-flowered is called 野天麻 *ye t'ien ma*. Other authors say that the purple-flowered is the true *i mu*. The name *i mu* (mother's help) is explained by its seeds being useful in women's diseases.

Ch., XI, 25 :—*Ch'ung wei* or *i mu ts'ao*. Good drawing representing *Leonurus sibiricus*, L. This plant has red flowers and a disagreeable odour.

TATAR., *Cat.*, 29 :—*I mu ts'ao, Leonurus sibiricus.*— According to PARKER, the same Chinese name is applied to the same plant in Sz ch'uan [*China Review*, X, 169] and at Canton *Leonurus sibiricus* is *i mu ai.*—P. SMITH, 132 :— *Ch'ung wei* or *i mu, Leonurus sinensis* [the species name is purely imaginary on his part].

Cust. Med., p. 376 (470) :—*I mu ts'ao* exported 1885 from Canton 18 piculs,—p. 300 (398), from Amoy 2 piculs,— p. 40 (231), from Tien tsin *i mu* plaster 0.4 picul.

So moku, XI, 41 :—益母 and 芜蔚 *Leonurus sibiricus.*

79.—薇銜 *wei hien*.   *P.*, XV, 30.   *T.*, CLI.

Pen king :—*Wei hien,* 糜銜 *mi hien*. Stem and leaves officinal. Taste bitter. Nature uniform. Non-poisonous.

Pie lu :—Other names 承膏 *ch'eng kao*. The *wei hien* grows in Han chung [S. Shen si, App. 54], in marshes, also in Yüan kü [in Shan tung, App. 415] and Han tan [in Chi li, App. 56]. In the 7th month the stem and the leaves are gathered and dried in the shade.

Other names noticed by WU P'U [3rd cent.]:—無心草 *wu sin ts'ao* (plant without a heart), 無顛 *wu tien*, 承膿 *ch'eng ki*.

SU KUNG [7th cent.]:—The southern people call the plant 吳鳳草 *wu feng ts'ao*. In Ch'u [Hu kuang, App. 24] they distinguish two sorts, the greater and the smaller *wu feng*. Another name is 鹿 銜 草 *lu* (deer) *hien ts'ao*. People say that deer, when sick, have recourse to this plant and then recover. This plant grows in a bushy manner, resembles the *ch'ung wei* [*Leonurus*. See 78] and also the *pai t'ou weng* [see 24]. Its leaves are covered with hair. The stem is red.

HAN PAO-SHENG [10th cent.] adds that it has yellow flowers and a reddish black root.

LI SHI-CHEN :—The *Shui king chu* [5th cent.] states that the *wei hien* plant grows plentifully in the Si shan mountains in Wei hing [in Shen si, App. 384]. The plant is said not to move by wind. It moves only when the air is still.

In the *Phon zo* [XIV, 12, 13] 薇 銜 is given as the Chinese name for various species of *Senecio*, viz.: *S. nikoensis*, Miq., *S. palmatus*, Pall., *S. nemorensis*, L.

80.—夏枯草 *hia ku ts'ao*. *P.*, XV, 31. *T.*, CXXXVI.

Pen king :—*Hia ku ts'ao* (plant withering in summer), 夕 句 *si kü*, 乃 東 *nai tung*. Stem and leaves officinal. Taste bitter and pungent. Nature cold. Non-poisonous.

Pie lu :—Other names : 燕 面 *yen mien*. The *hia ku ts'ao* grows in the river-valleys of Shu [Sz ch'uan, App. 292]. It is gathered in the 4th month.

SU KUNG [7th cent.] :—It is found everywhere in marshes, and grows till late in winter. Its leaves resemble those of the *süan fu* [*Inula*. See 81]. In the 3rd or 4th month it bears purplish white flowers, in spikes, resembling those of the *tan shen* [*Salvia*. See 20], then produces seed, and withers in the 5th month.

LI SHI-CHEN :—It is a common wild plant, grows from one to two feet and higher. Its stem is nearly square,

Leaves opposite on the joints, resembling those of the *süan fu*, but longer, serrated on the margin, downy underneath, and finely veined. Small pale purple flowers in spikes which issue from the top of the stem. Small seeds, four together. It is also called 鐵色草 *t'ie se ts'ao* (plant of the colour of iron).

Ch., XI, 66:—*Hia ku ts'ao.* The figure seems to represent *Prunella vulgaris,* L., our common self-heal. The drawing under the same name in the *Kiu huang* [XLVII, 23] is indistinct. It is said there that the *hia ku ts'ao* grows in Ho nan, Che kiang, An hui, especially in Ch'u chou [App. 25].

D'INCARVILLE, *Peking plants* (MS.):—*Hia kou tsao,* Brunelle. On en voit chez les droguistes à Pékin. Elle vient des provinces méridionales.

LOUR., *Fl. cochin.,* 203:—*Celosia margaritacea.* Sinice: *hia khu ts'ao.*

TATAR., *Cat.,* 45:—*Hia ku ts'ao, Lophanthus.*—P. SMITH [138] *Lophanthus.*

PARKER [*China Rev.,* X, 162]:—*Prunella vulgaris, hia ku ts'ao,* a common roadside plant in Sz ch'uan.

HENRY, *Chin. pl.,* 161:—The name *hia ku ts'ao* in Hu pei is applied to *Prunella vulgaris,* also to *Ajuga decumbens,* Thbg.

The drug (dried leaves and flowers) *hia ku ts'ao* which I received from Wen chou seemed to belong to *Prunella.* The Chinese name " iron coloured plant " refers probably to the brown coloured bracts and calyces of *Prunella.*

Cust. Med., p. 202 (266):—*Hia ku ts'ao* exported 1885 from Ning po 157 piculs,—p. 374 (463), from Canton 85 piculs.

Amœn. exot., 897:—夏枯草 vulgo *utsu bogusa* et *urukki.* Brunella major, folio non dissecto,

So moku, XI, 4:—Same Chinese name, *Ajuga genevensis,* L.—*Ibid.,* 8:—滁州夏枯草 [i.e. *hia ku ts'ao* from Ch'u chou, *v. supra*], *Prunella vulgaris.*

81.—旋覆花 *süan fu hua.* P., XV, 35. T., CXX.

Pen king:—*Suan fu hua,* 金沸草 *kin fu ts'ao.* Flowers officinal. Taste saltish. Nature warm. Slightly poisonous.

For other ancient names see *Bot. sin.,* II, 139.

Pie lu:—Other name: 戴椹 *tai shen.* The *süan fu* grows in low marshes and valleys. The flowers are gathered in the 5th month and dried 20 days. The root is also used in medicine.

T'AO HUNG-KING:—It grows in Mid China in low, damp places, and resembles the *kü hua* (*Chrysanthemum*) but is larger. It must not be confounded with the 旋葍 *süan fu,* the root of which plant is used in medicine. [*Ealystegia.* See 169.]

SU SUNG [11th cent.]:—This plant grows along the edge of the water. It resembles the *hung lan* (*Carthamus tinctorius*), but is not prickly. It grows from one to two feet high. Leaves like willow-leaves. Stem slender. In the 6th month it opens its flowers, which resemble the *Chrysanthemum.* They are as large as a small copper coin, and of a deep yellow colour. The people of Shang tang [S.E. Shan si, App. 275], where it grows in fields, call it 金錢花 *kin ts'ien hua* (gold coin flower). In Mid China it is much cultivated in gardens.

LI SHI-CHEN:—The wild plant, which grows on the margins of water-courses, has small single flowers resembling the *kü.* But when cultivated the flowers become large. The root is small and white. It is also called 夏菊 *hia kü* (summer *Chrysanthemum*) and 滴滴金 *ti ti kin* (dripping gold).

Ch., XI, 68 :—*Süan fu hua.* Good drawing of an *Inula.*
See also *Kiu huang,* XLVI, 16.

Tatar., *Cat.,* 48 :—*Süan fu hua.* Flores *Inula chinensis.*

P. Smith [119] states that *Inula chinensis* was introduced
into China in the 6th century. His assertion is evidently
based upon a statement in the *Yu yang tsa tsu,* reproduced
in the *P.,* that a plant *kin ts'ien hua* (gold coin flower)
was introduced into China during the Liang dynasty, from
a foreign country where its original name was *p'i-shi-sha.*
But this plant was certainly not *Inula chinensis,* which is a
common indigenous plant in N. China. It has beautiful
yellow flowers.

Cust. Med., p. 362 (301) :—*Süan fu hua* exported 1885
from Canton 11 piculs,—p. 154 (218), from Shang hai
1.3 picul,—p. 290 (234), from Amoy 1.3 picul,—p. 214
(64), from Wen chou 1 picul.

Amœn. exot., 877 :—旋覆 *sen fuki,* vulgo *oguruma,*
Aster luteus. Figured in Kæmpf., *Icon. sel.,* 30. This is
Inula japonica, Thbg.

So moku, XVII, 5 :—Same Chinese name, *Inula japonica.*

82.—青葙 *ts'ing siang.* P., XV, 37. T., CLVI.

Pen king :—*Ts'ing siang,* 草蒿 *ts'ao hao,*[26] 萋蒿 *ts'i
hao.* Name of the seeds : 草決明 *ts'ao küe ming.*[27] The
leaves, the stem and the seeds officinal. Taste bitter. Nature
slightly cold. Non-poisonous.

Pie lu :—The *ts'ing siang* grows in the plains, by roadsides. Stem and leaves gathered in the 3rd month, the
seeds in the 5th and 6th months.

---

[26] Comp. also 74 *Artemisia,*   [27] Comp. 110 *Cassia,*

T'AO HUNG-KING :—It is a common plant. It resembles a flowering wheat-ear. The seeds are very small.

SU KUNG [7th cent.]:—This plant is about one foot high. It has small, soft leaves. Flowers whitish purple. The fruit forms a horn. The seeds are black, flat, shining, and look like the seeds of the *hien* [*Amarantus Blitum*. See 256] but are larger. The plant grows in low, damp places. The people of King and Siang [both in Hu pei, App. 146, 305] call it 崑崙草 *k'un lun ts'ao.*

SU SUNG [11th cent.]:—The plant is common in Kiang and Huai [Kiang su and An hui, App. 124, 89] and in Mid China. It grows from three to four feet high. The leaves are broad, resemble willow-leaves, but are softer. The stem is like that of the *hao* (*Artemisia*), greenish red. Its flowers appear in the 6th or 7th month, they are red in the upper part (of the ear), white in the lower part. Seeds black, shining and flat, resembling those of the *lang tang* [*Scopolia*. See 139]. The root is like *Artemisia* root.

LI SHI-CHEN :—The *ts'ing siang* grows wild. The young plant resembles the *hien* [*v. supra*], and can be eaten. In its leaves, flowers and seeds the *ts'ing siang* resembles the *ki kuan* (*Celosia cristata*, cockscomb), only the flower-spikes of the latter are larger and flat. The *ts'ing siang* is therefore also called 雞冠莧 *ki kuan hien* or *ye* (wild) *ki kuan*. SU KUNG is wrong in saying that the fruit has the shape of a horn.

Kiu huang [XLVII, 18] and *Ch.* [XI, 46] sub *ts'ing siang*, representations of *Celosia argentea*, L. The description in the *P.* agrees.

LOUR., *Fl. cochin.*, 203 :—*Celosia argentea.* Sinice *tsim siam tsu.*

TATAR., *Cat.*, 13 :—*Ts'ing siang tsz'.* Semina *Celosiæ argenteæ.*—P. SMITH, 57.—I have also seen seeds received under the above Chinese name from Wen chou. They were small, black, shining and undoubtedly belonged to the above-mentioned plant.

Cust. Med., p, 366 (343) :—*Ts'ing siang* exported 1885 from Canton 5.68 piculs,—p. 292 (272), from Amoy 0.85 picul.

So moku, III, 61 :—菁 菥, japonice *nokeito*, *Celosia argentea.*

SIEB., *Œcon.*, 127 :—Same Chinese and Japanese names, *Celosia argentea.* Herba tenera edulis.

83.—大 薊 *ta ki* and 小 薊 *siao ki.* P., XV, 43. T., CVIII.

Pie lu :—The *ta ki* and the *siao ki* (the great and little *ki* or thistle) are gathered in the 5th month. Of both the leaves and the root are officinal. Taste sweet. Nature warm. Non-poisonous.

FAN WANG [4th cent.] uses the name 馬 | *ma ki* (horse thistle).

T'AO HUNG-KING :—The great *ki* is also called 虎 | *hu* (tiger) *ki*, the little one 貓 | *mao* (cat) *ki*. These plants resemble each other in their leaves, which are very spiny. They grow abundantly in a wild state. Not much used in medicine.

SU KUNG [7th cent.]:—The great and the little *ki* resemble each other, but the medical virtues of the drugs are different. The great *ki* grows in the mountains, the root is beneficial in ulcers and abscesses,—the little *ki* grows in low marshes, it does not accelerate the bursting of abscesses, but it breaks the blood.

In the *Ji hua Pen ts'ao* [10th cent.] we met with the names 山 牛 蒡 *shan niu p'ang* (mountain burdock) and 刺 薊 *ts'z' ki* (spiny thistle) applied to these plants.

Su Sung [11th cent.]:—The *siao ki* is a common plant. Its vulgar name is 青剌薊 *ts'ing ts'z' ki* (green spiny thistle). The young leaves together with the roots, in spring, when the plant is from two to three inches high, are used for food and considered very palatable. In the 4th month the plant is about one foot high and very spiny. The flower-heads issue from the heart of the plant and resemble those of the *hung lan* (*Carthamus tinct.*). They are of a greenish purple colour. The people in the north call it 千針草 *ts'ien chen ts'ao* (thousand needles plant). For medical use the stem and the leaves are gathered in the 4th month, the root in the 9th, and dried in the shade.—The greater *ki* in its leaves and root resembles the lesser, but the plant is fatter and larger.

K'ou Tsung-shi [12th cent.]:—The greater and the lesser *ki* resemble each other. Flowers like tufts of hair. The greater grows from four to five feet high, has wrinkled leaves, the lesser is only one foot high. Leaves not wrinkled. It is used as a vegetable.

Li Shi-chen gives as synonyms the names 雞頂草 *ki ting ts'ao* (cockscrown), 野紅花 *ye hung hua* (wild *Carthamus*). Some authors refer the *ta ki* to *Rh ya*, 137, *ki* or *kou tu*.

Ch., XI, 86 :— *Ta ki.* Two drawings, apparently *Cnicus* is intended. One of the plants figured shows a tuberous root. It is said there that the root of the *ta ki* in Kiang si is known to the native physicians under the name of *t'u jen shen* (native ginseng).—Comp. also *Kiu huang*, XLVI, 3, *ta ki.* Leaves eaten. Root said to be poisonous.

Lour., 589 :—*Carduus tuberosus*, Canton. Sinice *thu gin sen.* The tuberous root is officinal.

Tatar., *Cat.*, 19 :— *Ta ki.* Radix *Cardui* seu *Dipsaci.*— In the Peking mountains the people apply the name *ta ki* to

21

Cnicus pendulus, a thistle-like *Composita*, from five to six feet high, very spiny, with enormous purple flower-heads.

HENRY, *Chin. plants*, 515 :—*Ta ki* in Hu pei, *Cnicus japonicus*, DC. var.

So moku, XV, 37 :—大薊 *Cnicus spicatus*, Maxim.

Ch., XI, 85 :—*Siao ki.* The drawing represents a *Cnicus* with small spiny leaves, small flower-heads. See also *Kiu huang*, XLVI, 2.

LOUR., *Fl. cochin.*, 588 :—*Carduus lanceolatus*, L. [= *C. chinensis*, *DC. Prodr.*, VI, 629], sinice *siao ky* or *la di ts'ao*.

TATAR., *Cat.*, 47 :—*Siao ki*, *Carduus* seu *Dipsacus.*— Comp. P. SMITH, 51, 64, sub *Carduus* and *Cirsium*.

HENRY, *l.c.*, *siao ki* = *Cnicus segetum*, Bge.

Amœn. exot., 897 :—薊 *kei*, vulgo *asami*. *Carduus pratensis latifolius*.

So moku, XV, 36 :—小薊 *Cnicus japonicus*, DC. (*vide* Maxim.).

Phon zo, XV, 13 :—Same Chinese name, *Cnicus purpuratus*, Maxim., and *Cn. Hilgendorfii*, Franch. But in the *So moku* [XV, 42] the latter appears with the Chinese name 鷄頂草 [*v. supra*, LI SHI-CHEN].

SIEB., *Œcon.*, 210 :—*Carduus acaulis*, sinice 小薊, japonice *noa sami*. Radix atque herba edules. According to MAXIMOWICZ this is *Cnicus japonicus*.

Cust. Med., p. 350 (139) :—*Ta ki* exported 1885 from Canton 0.2 picul.—*Ibid.*, p. 344 (44), *siao ki* exported from Canton 1.68 picul,—p. 276 (36), from Amoy 0.07 picul.

84.—續斷 *su tuan*. P., XV, 45. T., CXXXIII.

Pen king :—*Su tuan*, 屬折 *su che*. (Both these names mean : to join together what is broken). The root is officinal. Taste bitter. Nature slightly warm. Non-poisonous.

Pie lu:—Other names : 接骨 *tsie ku* (join together broken bones), 龍豆 *lung tou*. The *su tuan* grows in the mountain-valleys of Ch'ang shan [in Chi li, App. 8]. It is gathered in the 7th and 8th months and dried in the shade.

Wu P'u [3rd cent.]:—It is produced in Liang chou [in Ho nan, App. 187].

The descriptions of the *su tuan* given by authors of various times are confused and contradictory, and no conclusion can be drawn from them ; some compare it to the *ch'u ma* (*Boehmeria*), others to a thistle.

The drawing in the *Ch.* [XI, 33] sub *su tuan* may be intended for *Dipsacus*.

TATAR., *Cat.*, 49 :—*Su tuan*. Rad. *Cardui* seu *Dipsaci*. In the Peking mountains *su tuan* is *Dipsacus japonicus*, Miq.

P. SMITH, 64 :—*Su tuan* or 川獨 *Ch'uan tuan, Cirsium lanceolatum*, an imaginary identification.

HENRY, *Chin. pl.*, 164 :—In Hu pei *su tuan* = *Dipsacus asper*, Wall.

Cust. Med., p. 340 (46):—*Su tuan* exported 1885 from Canton 240 piculs.—The *Cust. Med.* mentions the *Ch'uan* [Sz ch'uan] *su tuan* as imported into several Chinese ports.[28]

In Japan 續斷 is *Lamium album*, L. *So moku*, XI, 11.

Comp. also *Bot. sin.*, II, 118.

---

[28] The Hankow list of medicines.—*Customs Med.*, p. 66 (14) mentions it (1,610 piculs) as *Ch'uan tan* 川胆 . See *Alphabetical Index of Customs Med.*, No. 474, for the various popular names given to *Hsü-tuan.*— A. HENRY.

It is stated there that it comes from Han kow and Shang hai, but neither the *Ch'uan su tuan* or the *su tuan* are mentioned as articles of export in the Han kow list, and in the Shang hai list we find, p. 188 (21), that 1,810 piculs of *Ch'uan su tuan* were *imported* to Shang hai from Han kow and other ports.

85.—苦芙 *k'u yao.* *P.,* XV, 47. *T.,* CLXI.

Pie lu:—*K'u* (bitter) *yao.* Leaves used in medicine. Taste bitter. Nature slightly cold. Non-poisonous. It can also be eaten.

T'AO HUNG-KING:—The *k'u yao* is a common plant. The people of Ch'eng [Ho nan, App. 17] eat the stem raw.

HAN PAO-SHENG:—It grows in low, damp places. The stem is round, not spiny, and can be eaten raw. Seeds like those of the *mao ki* [cat's thistle. See sub 83].

LI SHI-CHEN:—This is the plant *kou* or *yao* of the *Rh ya* [62]. It (the stem) is of the thickness of a finger, and hollow inside. At the top of the stem is a flower-head like that of the *ki* (thistle). The young plant can be eaten. The people of Che tung [Che kiang, App. 10] at the *ts'ing ming* feast (in spring) gather the young plants and eat them. It is believed that then they will not be afflicted with sores for a year. The juice of the plant is mixed with rice for food. In the *Tsao hua chi nan* (an alchemistic work) the plant is called 苦板 *k'u pan,* the larger kind 苦藉 *k'u tsie.* Its leaves resemble those of the *ti huang* [Rehmannia. See 100], and are of a bitter taste. When young they are downy. In summer the stem developes. It is covered with hair. The flowers are white and numerous, the fruit is small. That kind which bears neither flowers nor fruit is called 地膽草 *ti tan ts'ao* (ground gall plant). Its juice is bitter, like gall. This plant is common in damp places.

Ch., XIV, 6:—*K'u yao.* The drawing represents leaves of a spiny plant.

So moku, XV, 41:—苦芙 *Cnicus nipponicus,* Maxim.

86.—漏盧 *lou lu.* *P.,* XV, 47. *T.,* CXXXIII.

Pen king:—*Lou lu,* 野蘭 *ye lan.* Root and leaves officinal. Taste saltish. Nature cold. Non-poisonous.

Pie lu:—The *lou lu* grows in the valleys of the K'iao shan mountain [in Shen si, App. 134]. The root is dug up in the 8th month and dried in the shade.

T'AO HUNG-KING:—The K'iao shan is Emperor HUANG TI's burial-place. It lies in Shang kün [in Shen si, App. 273]. The leaves and the root are used in medicine. The latter is commonly called 鹿驪根 *lu li ken.* Ground with bitter wine it is useful in curing itch-sores.

SU KUNG [7th cent.]:—The popular name of this drug is 茭蒿 *kia hao* (*Artemisia* with pods). The stem and the leaves resemble those of the *pai hao* [*Artemisia.* See 75]. Yellow flowers. The pod is as long as that of the *si ma* (small hemp, unknown to me, perhaps *Sesam*) and as thick as a quill. It has four or five divisions (cells) and becomes black in the 7th or 8th month. By this (*i.e.* by producing a pod) it is distinguished from all the *hao* (*Artemisia*) plants. The stem and the leaves, also the seeds, are used in medicine, but not the root. The name *lou lu* is also applied to several other plants.

MA CHI [10th cent.]:—The *lou lu* has a stem like a quill, from four to five feet high. The fruit is a capsule, like that of the *yu ma* (oil-hemp or *Sesamum*) but smaller. The people of Kiang tung [Kiang su, etc., App. 124] use the leaves in preference to the root. The drug from Kiang ning [Nan king, App. 129] and Shang tang [S.E. Shan si, App. 275] is considered the best.

CH'EN TS'ANG-K'I [8th cent.]:—The people in the south use the leaves, the people in the north the root.

HAN PAO-SHENG [10th cent.]:—Its leaves resemble those of the *kie hao* (*Incarvillea*). It is a common plant in the marshes of Ts'ao chou and Yen chou [both in Shan tung, App. 344, 404]. In the 6th and 7th months the stem is gathered and dried in the sun. It then becomes blacker than any other dried plant.

The *Ji hua Pen ts'ao* [10th cent.] calls this plant 鬼油麻 *kui yu ma* (devil's *Sesam*).

There are descriptions of the plant by several other Chinese authors, but they are confused and contradictory. Evidently they confound several plants.

LI SHI-CHEN observes, that there is in Min [Fu kien, App. 222] a plant called *lou lu*. It has a stem six or seven feet high resembling that of the *Sesam* plant. In autumn, when it withers, it becomes black, like varnish. This is the true *lou lu*. This latter name means "black."

Ch., XI, 36:—*Lou lu*. Rude drawing. Comp. also *Kiu huang*, XLVI, 26.

Cust. Med., No. 756:—*Lou lu*, root of an herb (*Serratula* sp. ?)

Cust. Med., p. 122 (34):—*Lou lu* exported 1885 from Chin kiang 1.38 picul,—p. 344 (82), from Canton 0.1 picul.

Phon zo, XV, 16:—洞盧 *Siphonostegia chinensis*, Benth. (*Scrophularineæ*).—The *So moku* [XI, 62] figures the same plant under the Chinese name 鬼油麻. This plant, which is common in China, becomes indeed black in autumn or when dried, but it is certainly not the plant figured in Chinese works sub *lou lu*. *Siphonostegia chinensis* seems to be intended by the 陰行草 *yin hing ts'ao* figured in *Ch.*, X, 21.

TATAR. [*Cat.*, 35] gives 劉寄奴 *liu ki nu* as the Chinese name for *Siphonostegia chinensis*. See also P. SMITH, 198. But judging by the description of the *liu ki nu* in the *P.* [XV, 32] and *T.* [CLXXI] this is a plant of the order *Compositæ*. It is first spoken of by SU KUNG [8th cent.], and vaguely described. LI SHI-CHEN says that it has yellow flowers resembling those of a small *kü* (*Chrys-*

anthemum). The *Ch.* [XIV, 20, 21] has sub *liu ki nu* two rude figures, one of them seems to refer to a *Composita.* According to the *Cust. Med.*, p. 374 (452), 202 (275), 286 (183), this drug is exported from Canton, Ning po, Amoy, resp. 7, 2, 1 piculs.

So moku, XVII, 40 :—劉寄奴 *Solidago virgo aurea,* L. (our common Golden Rod) and *ibid.*, 43, same Chinese name, *Senecio palmatus,* Pall.

87.—飛亷 *fei lien.* *P.,* XV, 50. *T.,* CXXXIV.

Pen king :—Fei lien. Root and flowers officinal. Taste bitter. Nature uniform. Non-poisonous.

*Pie lu :—*Other names : *lou lu* [see 86], 木禾 *mu ho,* 飛雉 *fei ki* (flying cock), | 輕 *fei k'ing* (to fly and not heavy), 伏兎 *fu t'u,* 伏豬 *fu chu,* 天薺 *t'ien tsi.* The *fei lien* grows in Ho nei [in Ho nan, App. 77]. In the 1st month the root is dug up, and in the 7th and 8th months the flowers are gathered and dried in the shade.

T'AO HUNG-KING :—It is a common plant and much resembles the *k'u yao* [*Cnicus.* See 85], but its leaves are much incised (or sinuated or laciniated). The stem has skins (raised lines) which issue from the base of the leaves and continue down the stem, resembling the wings of an arrow. Purple flowers. This drug is not used now in medicine, only Taoists consider the stem and the branches as securing longevity.

SU KUNG [7th cent.] :—There are two kinds of this plant. One grows in low marshes and is the plant described by T'AO HUNG-KING. The other grows on the summits of mountains. Its leaves resemble those of the first but are not sinuated, and very downy. Its stem is red, not winged. The root goes straight down, has no lateral branches, the rind (of the root) is black, the flesh white with black veins. When dried in the sun it becomes black, like the root of the *yüan shen* [see 18].

Ch., **XI**, 37 :—*Fei lien.* The drawing is probably intended for *Carduus crispus*, L., with which the above Chinese description of the *fei lien* agrees. *Carduus crispus* is a common plant in the Peking mountains.

So moku, **XV**, 40 :—飛廉 *Carduus crispus.* See also *Kwa wi*, p. 9.

88.—苧麻 *ch'u ma.* *P.*, **XV**, 51. *T.*, **XXXIX**.

This is the grass-cloth plant, *Boehmeria nivea.* See *Bot. sin.*, II, 391.

The *Pie lu* gives only the name *ch'u ma*, and specifies its medical virtues. Root and leaves used in medicine. Taste sweet. Nature cold. Non-poisonous.

T'AO HUNG-KING says, the *ch'u* is a kind of hemp, used for spinning thread.

SU SUNG [11th cent.]:—The ancient authors do not record where the *ch'u ma* is produced. Now it grows plentifully in Min [Fu kien, App. 222], Shu [Sz ch'uan, App. 292], Kiang [Kiang su, etc., App. 124] and Che [Che kiang, App. 10]. From its fibres cloth is woven. The plant grows from seven to eight feet high. Its leaves resemble the leaves of the *ch'u* [*Broussonetia papyrifera.* See 333], but are not lobed. They are green on the upper side, white underneath, covered with short hairs. In summer and autumn the plant bears spikes of small green flowers. Root yellowish white, not heavy.

LI SHI-CHEN :—The *ch'u* occurs wild and is also much cultivated. The people distinguish the 紫 | *tsz'* (purple) *ch'u*, with purple leaves, and the 白 | *pai* (white) *ch'u* with its leaves green on the upper side and white underneath. The leaves can be prepared for food. The seeds are of a brownish gray colour. Perennial root.

The plant has the same Chinese name in Japan. For further particulars see *Bot. sin.*, II, 391.

89.—大青 *ta ts'ing.   P., XV, 54.   T., CV.*

Pie lu:—*Ta ts'ing* (great green). Its stem is gathered in the 3rd or 4th month and dried in the shade.

T'AO HUNG-KING :—This plant is found in East China and in the border provinces. It has a purple stem about one foot high. Stem and leaves official.

SU SUNG [11th cent.]:—It is produced in all the prefectures of Kiang tung [E. of the Kiang, App 124], in King nan [Hu nan, App. 148], in Mei, Shu [Sz ch'uan, App. 219, 292] and in Hao [in An hui, App. 59]. Stem purplish green, resembling the *shi chu* (*Dianthus*). Flowers purplish red, resemble those of the *ma liao* (*Polygonum*) and the *yüan hua* [*Daphne*. See 156]. The root is yellow.

LI SHI-CHEN :—It is a common plant, and grows from two to three feet high. Stem round. Leaves from three to four inches long, green on the upper side, paler underneath, placed in pairs at the joints. In the 8th month small red flowers in corymbs. Fruit at first green, of the size of the *tsiao* (*Zanthoxylon*) fruit. In the 9th month it becomes red. LI SHI-CHEN explains the name *ta ts'ing* (great green) by the dark green colour of the stem.

Ch., XI, 88 :—*Ta ts'ing.* A plant figured which is unknown to me.

TATAR., *Cat.*, 18 :—*Ta ts'ing.* Caules et folia *Polygoni tinctorii.*—P. SMITH, 175.

22

Cust. Med., 360 (277) :—*Ta ts'ing* leaves exported 1885 from Canton 1.15 picul,—p. 350 (140), *ta ts'ing* root 0.1 picul. But *ta ts'ing* is identified there with *Indigofera tinctoria*, whether correctly, I am not prepared to say.

HOFFM. & SCHLT., 312 :—大青 *Justicia crinata*, Thbg. (*Acanthaceæ*). This is, according to *DC. Prodr.*, XI, 493 = *Peristrophe tinctoria*, Nees. = *Justicia purpurea*, Lour., *Fl. cochin.*, 31.—But, strange to say, [in *DC. Prodr.*, XI, 485] THUNBERG's *Just. crinata* is identified with another acanthaceous plant, the *Dicliptera crinita*. This latter is the same as *Dianthera japonica*, Thbg. [*Fl. jap.*, 21, tab. 4]. I may observe that LOUREIRO describes also a *Justicia tinctoria* : Folia viridi colore saturata, eodem telas pulchre imbuunt. The *ta ts'ing* in Chinese works does not seem to refer to one of these acanthaceous plants.

The *P.* [XV, 55] notices also a plant called 小青 *siao ts'ing* (little green). The *T'u king Pen ts'ao* [11th cent.] says only a few words regarding this plant. It is said to grow in Fu chou [in Fu kien, App. 46]. It flowers in the 3rd month. The people there employ the leaves.

Ch., IX, 16 :—*Siao ts'ing*. Figure of a plant with pinnate leaves and a large root. Unknown to me. *Ibid.*, XIV, 48 :—*Siao ts'ing*. This figure seems to represent *Ardisia japonica*, Bl. Comp. HENRY, *Chin. pl.*, 13.

90.—蠡實 *li shi*. *P.*, XV, 57. *T.*, CLXV.

Pen king :—*Li shi*, 劇草 *k'i ts'ao*, 豕首 *shi shou* (pig's head). The fruit is official. Taste sweet. Nature uniform. Non-poisonous.

For other ancient names see *Bot. sin.*, II, 36, 467.

Pie lu :—Other name : 荔實 *li shi.* The *li shi* grows in Ho tung [in Shan si, App. 80] in river-valleys. The fruit is gathered in the 5th month and dried in the shade.

Su Kung [7th cent.]:—The *li shi* is also called 馬藺子 *ma lin tsz',* and mentioned in the *Yüe ling* (*Li ki*) under the name of 荔 *li.*

Su Sung [11th cent.]:—The plant is common in Shen si [App. 284]. It is found also in Ting chou and Li chou [both in Hu nan, App. 341, 185] and frequent near Pien [in Ho nan, App. 248]. Its leaves resemble those of the *hiai* [*Allium.* See 242] but are longer and thicker. In the 3rd month it opens its blue flowers. The fruit is formed in the 5th month. It is a horn (capsule). The seeds are as large as hemp seed, red and angulous. The root is fibrous and long, of a yellow colour. The people use it for brushes. Kao Yu, of the Han dynasty, says that it grows in the marshes of Ho pei [in Shan si, App. 78]. The people of Kiang tung used to cultivate it in front of their houses and called it 旱蒲 *han p'u.*

Li Shi-chen refers to the *Rh ya* [36] and states that the people south and north of the Yellow River call the plant 鐵掃帚 *t'ie sao chou* (iron besom). Another name for it is 三堅 *san kien.*

Ch., XI, 42 :—*Li shi.* The drawing represents an *Iris.*

At Peking *ma lin* is a common name for *Iris oxypetala,* Bge. The name is sometimes erroneously written 馬蘭 *ma lan.*[29] Tatar., *Cat.,* 38, has *ma lan hua.* Flores *Iridis oxypetalæ.*

Amœn. exot., 872 :—馬藺 *farin,* vulgo *buran.* Iris hortensis, alba, germanica.—According to Thbg. [*Fl. jap.,* 33] this is *Iris sibirica.*

---

[29] *Ma lan* is an *Aster.* See *P.,* XIV*b,* 80, and *Ch.,* XXV, 50.— *So moku,* XVI, 44 :—*Aster trinervius,* Roxb.

So moku, II, 8 :—Same Chinese name, *Iris ensata*, Thbg.
—*Phon zo*, XVI, 2 :—盍寶, japonice *haran*, *Iris ensata*.
The same Chinese name is applied there to *I. sibirica*.

鐵掃帚 *t'ie sao chou*, which in the *P*. is given as a
synonym for *li shi* or *Iris*, appears in the *So moku* [XIV, 22]
as the Chinese name for *Lespedeza juncea*, a leguminous
plant, and the drawings under the same name, in the *Kiu
huang* [L, 20] and *Ch.* [XII, 45], although rude figures,
seem to refer to the same plant.

91.—惡寶 *wu shi*.   *P*., XV, 60.   *T*., CLXVI.

Pie lu:—*Wu shi* (evil fruit), 鼠粘 *shu nien*, 牛蒡
niu p'ang. The *wu shi* grows in Lu shan [in Ho nan, App.
203], in marshes. The seeds, the stem and the root are
officinal. Taste of the seeds pungent. Nature uniform.
Non-poisonous.—Taste of the root and the stem bitter.
Nature cold. Non-poisonous.

Su Kung [7th cent.]:—Lu shan [also the name of a
mountain] lies to the north-east of Teng chou [in Ho nan,
App. 337]. The leaves of the *wu shi* plant are as large as
those of the *yü* [*Colocasia*. See 261]. The involucrum of
the fruit has the appearance of the [spiny husk] of the
chestnut. The seeds are small and long like those of the
ch'ung wei [*Leonurus*. See 78].

Su Sung [11th cent.]:—The *wu shi*, also called *niu
p'ang*, is a common plant. It has large leaves like the *yü*
[*v. supra*], but longer. The seeds resemble the seeds of the
grape, and are of a gray colour. The involucrum of the fruit
is like the husk of the chestnut, but smaller, of the size of
the end of a finger. It is covered with spines. The root,
which is sometimes very large, is used for food. The seeds

are officinal. The spines of the involucrum of the fruit lay hold of the rats who pass near it and stick to them, whence the name *shu nien* (rat and to stick).

Li Shi-chen:—In ancient times the people cultivated the *niu p'ang* plant in a rich soil, and the leaves were eaten as a vegetable. The root was likewise prepared for food. It is very nourishing, but is now seldom eaten. The plant grows from three to four feet high. In the 4th month it opens its pale purple flowers, which are crowded together. The fruit resembles that of the *feng* tree (*Liquidambar formosana*) but is smaller. The calyces of the flowers have small spines. More than a hundred flowers form a globular head. The root is of the thickness of an arm, nearly a foot long and of a gray colour. The seeds are gathered in the 7th month, the root is taken up in the 10th. The plant (flower-head) is provided·with hooked spines, whence the name *wu shi* (evil fruit). Other vulgar names are 牛菜 *niu ts'ai* (ox-vegetable), 大力子 *ta li tsz'*, 便牽牛 *pien k'ien niu*, 夜叉頭 *ye ch'a t'ou*, 蝙蝠剌 *pien fu ts'z'* (bat spine), 勞翁菜 *p'ang weng ts'ai*.

Kiu huang [LIII, 4] and *Ch.* [XI, 84], sub *niu p'ang tsz'*, good drawings of *Arctium Lappa*, L., which is a common plant in North and Central China. In Peking it is called *niu p'ang tsz.* It is known under the same name or as *ta li tsz'*, in Sz ch'uan and Hu pei [PARKER in *China Rev.*, XI, "Names of Sz ch'uan Plants," and HENRY, *l.c.*, 312].

TATAR., *Cat.*, 41:—*Niu p'ang tsz'*. Semina Lappæ.

Cust. Med., p. 78 (182):—*Ta li tsz'* exported 1885 from Han kow 636 piculs,—p. 16 (129), from New chwang 520 piculs,—p. 34 (174), from Tien tsin 5.50 piculs. *Ibid.*, p. 34 (160), *niu p'ang tsz'* from Tien tsin 1.89 picul.

Amœn. exot., 828:—牛蒡 *yobo, umma bufuki.* Bardana major: quæ hic in terra pulla colitur, ob radicem, ante caulium eruptionem, culinis destinandam.

So moku, XV, 34 :—Same Chinese name, *Lappa major* Gært.   Japonice *gobo*.

SIEB., *Œcon.*, 211 :—*Arctium Lappa*, sin. 牛蒡, japon. *koboo*.   Radices obsonium vulgatum.

92.—枲耳 *si rh*.   *P.*, XV, 64.   *T.*, CLVI.

Pen king :—*Si rh*, 胡枲 *hu si*, 地葵 *ti k'ui* (ground mallow).   The fruit, the stem and the leaves are officinal. Taste of the fruit bitter.   Nature warm.   Slightly poisonous. The stem and the leaves are bitter and pungent.   Nature slightly cold.   Slightly poisonous.

For other ancient names of the plant in the Classics see *Bot. sin.*, II, 184, 438.

Pie lu :—The *si rh* grows in the river-valleys of An lu [in Hu pei, App. 1] and in Ta an [in Shan si, App. 320] in neglected places.   The ripe fruit is gathered.

T'AO HUNG-KING :—The people of Ch'eng [Ho nan, App. 17] eat it and call it 常思菜 *ch'ang sz' ts'ai*.   The leaves are used for dyeing clothes a yellow colour.   The plant is seldom employed in medicine.

SU SUNG [11th cent.]:—In the *Shi king* it is called 卷耳 *küan rh*, in the *Rh ya* the name reads 蒼耳 *ts'ang rh*. [This is an error.   The character *ts'ang* was not used in the classical period to designate this plant.]   The *Po wu chi* [3rd cent.] reports that in Lo [in Ho nan, App. 201] there was a man who drove sheep to Shu (Sz ch'uan).   The fruit of the 胡枲 *hu si* plant, being covered with prickles, adhered to the fleeces of the sheep, and when they returned they brought the seeds of the plant to China, whence it is also called 羊負來 *yang fu lai* (it came carried on the back of sheep).   Another popular name is 道人頭 *Tao jen t'ou* (Taoist head).

Kiu huang, LII, 22:—蒼耳 *ts'ang rh* (green ear). Rude drawing representing *Xanthium strumarium,* L. It is said there that the fruit resembles a mulberry and is covered with prickles. The leaves are eaten.

Ch., XI, 50:—*Si rh.* Good drawing of *Xanthium strumarium.* This is a very common weed all over China, called *ts'ang rh* at Peking.

TATAR., *Cat.,* 5:—*Ts'ang rh.* Fructus *Xanthii strumarii.*—HANB., *Sc. pap.,* 233.—P. SMITH, 233.

According to PARKER [*Canton Plants*] *ts'ang rh* is also the Chinese name for this plant at Canton.

Cust. Med., p. 372 (427):—*Ts'ang rh* exported 1885 from Canton 9.7 piculs,—p. 298 (353), from Amoy 1.23 picul,—p. 350 (156), from Canton 0.25 picul of the root.

Amœn. exot., 892:—蒼耳 *sooni, namome.* Xanthium. Lappa minor.

So moku, XX, 25:—Same Chinese name, *Xanthium strumarium.*

But 卷耳 [*So moku,* VIII, 71] is *Cerastium glutinosum,* Fries.

93.—天名精 *t'ien ming tsing. P.,* XV, 68. *T.,* CLXXIV.

Pen king:—*T'ien ming tsing,* 麥句薑 *mai kü kiang,* 蝦蟆藍 *hia ma lan* (frog blue), 豕首 *shi shou* [pig's head. Comp. 90]. Leaves and root officinal. Taste sweet. Nature cold. Non-poisonous.

Comp. also *Bot. sin.,* II, 35.

Pie lu:—Other names: 天蔓菁 *t'ien man ts'ing,* 天門精 *t'ien men tsing,* 壼松 *ti sung,* 玉門精 *yü men tsing,* 蟾蜍蘭 *shan chu lan* (toad *lan*), 覭顛 *chi lu* (pig's head). The *t'ien ming tsing* grows in marshes in the plain. It is gathered in the 5th month.

The accounts given by the ancient authors of this plant,
which has so many names, are confused and not characteristic. Probably several plants are confounded. Su Kung
[7th cent.] says, that the *t'ien ming tsing* is also called
活鹿草 *huo lu ts'ao*. The people in the south call it
地菘 *ti sung* (ground cabbage), also *t'ien man tsing* (heavenly
rape), for the leaves resemble cabbage or rape-leaves, and
are of a sweet, pungent taste.

Li Shi-chen:—The young leaves of the *t'ien men tsing*
are wrinkled like cabbage-leaves or leaves of the mustard
plant. They smell somewhat of foxes, but can be eaten when
cooked. The plant has small yellow flowers like small
Asters. The fruit resembles that of the *t'ung hao* (*Chrysanthemum Roxburghii*). The seeds stick to people's clothes.
They have at first an unpleasant, fox-like smell, but after
heating become fragrant. The root is white. Other names
of the plant: 皺面草 *tsou mien ts'ao* (wrinkled leaf),
毋豬芥 *mu chu kie* (sow mustard). The fruit (seed) is
called 鶴蝨 *ho shi* (crane's louse), the root 土牛膝 *t'u
niu si.*

The *ho shi* (crane's louse) or seed of the *t'ien ming tsing*,
according to the authors of the T'ang and Sung periods, is
an important medicine. It is of a bitter and pungent taste,
slightly poisonous, and reputed to destroy insects.

Cust. Med., p. 366 (358):—*Ho shi* exported 1885
from Canton 12 piculs,—p. 158 (296), from Shang hai
5.49 piculs,—p. 198 (204), from Ning po 1.15 picul.

Ch., XI, 16:—*T'ien ming tsing.* The drawing may
perhaps be intended for *Carpesium* [*v. infra*].

Tatar., *Cat.*, 56:—*T'ien ming tsing. Semen Amaran-
thaceæ.*—P. Smith, 12.

In Japan the name 天名精 is applied to *Carpesium
abrotanoides*, L. See *So moku*, XV, 62. This plant, of the

order *Compositæ*, yellow flowers, is common in South and Mid China. Three other species of *Carpesium* are known from China.

94.—蘆 *lu.* *P.*, XV, 76. *T.*, CIX.

This is the common reed, *Arundo phragmites*, L. (*Phragmites communis*), also called 葦 *wei* and 葭 *kia.* Comp. *Bot. sin.*, II, 210, 211, 213, 455.

Pie lu:—Lu. The root, the young sprouts, the stem and the leaves are officinal. Taste of the root, the stem and the leaves, sweet. Nature cold. Non-poisonous. The sprouts are slightly bitter.

Su Kung [7th cent.]:—The *lu* root is produced in low marshes. The stem and the leaves of the *lu* resemble the bamboo, the flowers are like those of the *ti* [a smaller reed. See *Bot. sin.*, II, 455] and are called 蓬 蕽 *p'êng nung.* The root is dug up in the 2nd and 8th months and dried in the sun.

Su Sung [11th cent.]:—The *lu* resembles the bamboo, its leaves clasp the stem, which is not branched. White flowers in panicles resembling those of the *mao* [*Imperata.* See 37]. The root is also like that of the bamboo, but the joints are at a greater distance apart. That part of the root which is below the water (in the mud) is sweet and pungent, that which is in the water is not good for use. The tender sprouts of the *lu* are much used for food, like bamboo-sprouts.

Kiu huang, LIII, 13:—*Lu sun* (*lu* sprouts), and *Ch.*, XIV, 10:—*Lu* or *wei*, figures of *Arundo phragmites.*— P. Smith, 171.

Henry, *Chin. pl.*, 258:—蘆 柴 *lu ch'ai* (fuel) in Hu pei — *Phragmites Roxburghii*, Kth.

23

Cust. Med., p. 344 (83):—*Lu ken* (root) exported 1885 from Canton 3.43 piculs.

Amœn. exot. :—葦 *i*, vulgo *assi, jussi.* Arundo palustris vulgaris, foliis latioribus, calamis firmis, ex quibus puto penicilla scriptoria confici.

According to MIQUEL [*Prol. Fl. jap.*, 166] *josi* is the Japanese name for *Phragmites Roxburghii*, Nees. The common reed *Phr. communis* is also found in Japan. MIQUEL does not mention its Japanese name. One of these two species is figured in the *Phon zo*, XVI, 9, 10, sub 蘆.

95.—甘蕉 *kan tsiao.* P., XV, 59. T., CLXXXV.

Pie lu:—Only the name is given and the medical properties of the root are noticed.

Kan tsiao is the plantain or Banana (*Musa*), of which detailed and good descriptions by ancient authors are given in the *P.* I defer the translation of these accounts to another part of the *Botanicon sinicum*.

96.—蘘荷 *jang ho.* P., XV, 81. T., CXLIII.

Pie lu :—*Jang ho,* 覆葅 *fu tsü,* | 草 *jang ts'ao.* It grows in the mountain valleys of Huai nan [in An hui, Kiang su, App. 90]. The root is officinal. Taste pungent. Nature warm, slightly poisonous. It is used as a vermifuge. The leaves also are employed.

The *Shuo wen* [beginning of 2nd cent.] says that the *jang ho* is also called 蒩葅 *fu tsu.* According to the *Sou shen ki* [4th cent.], quoted in the *Ts'i min yao shu*, the *jang ho* is the plant mentioned under the name of 嘉草 *kia ts'ao* (excellent herb) in the *Chou li.* Comp. BIOT's translation, II, 386 :—" Le cuiseur d'herbes chargé d'expulser les animaux vénimeux .... il les attaque par des *plantes excellentes*, qu'il fait cuire."

In the poems of Sz' Ma Siang ju [† B.C. 120] this plant is called 猲直 *po tsü.*

The *Ku kin chu* [4th cent.] says, the *jang ho* resembles the *fu tsu* [*v. supra.* Other reading *pa tsiao,* Banana] but is white. The flowers issue from the root. Its unopened flowers are eaten. The leaves and the root are like ginger.

T'ao Hung-king :—There are two sorts. The red sort (I suppose red root) is now called *jang ho,* whilst the white is *fu tsü.* The first is good to eat, the white is used in medicine. The leaves in both sorts are the same.

Su Sung [11th cent.]:—The *jang ho* is much cultivated in King and Siang [both in Hu pei, App. 145, 305] and in Kiang and Hu [Mid China, App. 124, 83]. It is also found in the northern provinces. The leaves resemble those of the banana. The root is like ginger-shoots but more succulent. The leaves wither in winter. The root is much pickled. The best is that grown in the shade of trees.

Ch., III, 68 :—*Jang ho.* Henry [*Chin. pl.,* 359] refers this drawing to *Lilium giganteum,* Wall. But the descriptions of the *jang ho* in the *P.* seem rather to point to *Zingiber.*

Amœn. exot., 826 :—蘘荷 *Dsjooka* vulgo *Mjoga.* Zingiber edule, sapore molli . . . . . bulbo florigero ex radice circa caulem in terræ superficie nascente . . . . .

This is the *Amomum mioga,* Thbg. [*Fl. jap.,* 14] figured in Banks' *icon. sel.* Kæmpf., tab. 1.—Same Chinese name [*So moku,* I, 9] *Zingiber Mioga,* Rosc.

Sieb., *Œcon.,* 101 :—*Zingiber Mioga,* same Chinese name. Soboles juniores quæsitum obsonium.

The above Chinese descriptions of the *jang ho* seem to agree with *Zingiber Mioga.* This plant however has not been recorded from China. Hance [in *Journ. Bot.,* 1882, p. 80] described *Z. integriloba,* a new species from Hong kong, which is allied to *Z. Mioga.*

97.—麻黃 *ma huang*.   *P.*, XV, 83.   *T.*, CLIV.

Pen king:—*Ma huang* (hemp yellow), 龍沙 *lung sha*. The stem is officinal. Taste bitter. Nature warm. Non-poisonous.

Pie lu:—Other names: 卑相 *pei siang*, 卑鹽 *pei yen*. The *ma huang* grows in Tsin [Shan si, App. 353] and Ho tung [S.W. Shan si, App. 80]. The stem is gathered in the beginning of spring and dried in the shade until it assumes a green colour.

T'AO HUNG-KING :—This plant now grows in Ts'ing chou [E. Shan tung, App. 363], in P'eng ch'eng [in Kiang su, App. 247]. In Yung yang and Chung mou [both in Ho nan, App. 427, 30] the best sorts are produced. It is of a dark green colour and has much foam (?).[30] The drug from Shu (Sz ch'uan) is of an inferior quality.

SU KUNG [7th cent.]:—It grows plentifully in Cheng chou [in Ho nan, App. 15], Lu t'ai [unknown, App. 205], in Kuan chung [Shen si, App. 158], in Sha yüan [in Shen si, App. 267], on the banks of the Yellow River and on sandy islets.

The *Yu yang tsa tsu* [8th cent.] states that the small yellow flowers of the *ma huang* appear in cymes at the top of the stem. Its fruit resembles the *fu p'en tsz* [raspberry. See 166] and is edible.

SU SUNG [11th cent.]:—The *ma huang* is common near Pien king [K'ai feng fu, App. 248]. The best drug comes from Yung yang and Chung mou [*v. supra*]. The plant grows about one foot high. Yellow flowers at the top. The fruit is small, resembles the scaly bulb of a lily, and is of a sweet taste. Its smell recalls the *ma huang* (hemp yellow, pollen

[30] 色青而多沫.

of hemp?). The outer skin of the fruit is red. It contains black seeds. The root is purplish red. The people distinguish the female and the male *ma huang* plant. The male produces neither flowers nor fruit.

Li Shi-chen :—Its root, which is about one foot long, has a yellowish red skin.

Ch., XI, 51 :—*Ma huang.* The drawing seems to be intended for *Equisetum.*

Lour., *Fl. cochin.*, 823 :—*Equisetum arvense.* Sin. *ma huam.*

Tatar., *Cat.*, 37 :—*Ma huang, Ephedra.*—P. Smith, 93 :—*Ephedra flava.*—The drug which I received under the name of *ma huang* from an apothecary's shop in Peking— slender yellow or green stems or branches of a plant, cut into small pieces—proved to belong, on microscopical examination, to *Ephedra. E. vulgaris* is a common plant in N. China and Mongolia. The description of the *ma huang* in the *P.* agrees with *Ephedra.*

Phon zo, XVI, 18 :—麻黃 *Equisetum arvense.*

Ephedra (order *Gnetaceæ*), with its leafless branches, has a slight resemblance to *Equisetum*, which is a cryptogamous plant. *Ephedra* has yellow flowers, male and female flowers on different plants, and produces red, edible berries.

Cust. Med., p. 12 (91) :—*Ma huang* exported 1885 from New chwang 966 piculs,—p. 36 (199), from Tien tsin 213 piculs,—p. 74 (107), from Han kow 66 piculs,—p. 194 (161), from Ning po 31 piculs.

Cust. Med., No. 801 :—*Ma huang, Ephedra vulgaris*, Rich. var. *helvetica*, Hook. & Thom.

98.—石龍芻 *shi lung ch'u. P.*, XV, 90. *T.*, CLXXVI.

Pen king :—*Shi lung ch'u* (stone dragon grass), 龍鬚 *lung sû* (dragon's moustaches), 龍珠 *lung chu*, 草續斷 *ts'ao su tuan.* Stem and root officinal. Taste sweet. Nature uniform. Non-poisonous.

The *Shan hai king* mentions a plant 龍修 *lung siu.* Kuo P'o comments that this is the plant 龍須 *lung sü* which grows in rock-holes, and the stem of which is used for making mats.

The *Ku kin chu* [4th cent.] relates that when Emperor Huang ti soared up to Heaven on a dragon, his ministers seized the dragon's moustaches, which dropped to the earth and produced the plant so called.

Other legends are found in the *Shu i ki* [6th cent.] in which this plant is mentioned in connection with Mu wang's eight famous horses and the fabulous Si wang mu or western royal mother. The plant is also called 西王母簪 Si wang mu's hair-pin. In Kiang tung the mats woven of this plant were known under the name of mats of Si wang mu.

Some ancient authors refer the *lung sü* grass to *Rh ya,* 16.

Pie lu:—Other names: 龍華 *lung hua,* 懸莞 *hüan huan,* 方賓 *fang pin.* The *shi lung ch'u* grows in Liang chou [in Ho nan, App. 187] in mountain valleys and marshes. The stem is gathered in the 5th and 7th months and dried in the sun. That with nine joints, and which has a strong taste, is the best.

T'ao Hung-king:—It has a slender green stem with joints, red fruits. It is found in Mid China, growing in water between stones, and is employed for making mats.

Ch'en Ts'ang-k'i:—It grows in Fen chou, Ts'in chou, Shi chou [all in Shan si, App. 38, 362, 286].

Li Shi-chen:—The *lung sü* grass grows in tufts. It resembles the *tsung sin ts'ao* [see the next] and the *fu ts'z* [*Eleocharis.* See *Rh ya,* 59]. The plant rises straight, has neither branches nor leaves. In summer there appear on the top of the stem small flowers in spikes, followed by small fruits (seeds). The people of Wu [Kiang su, App. 389] cultivate it much and use it for matting. It is rarely met with there in a wild state. It is also known under the name of 箐雲草 Tsin yün ts'ao. LI SHI-CHEN observes that Tsin yün is the name of a district in the prefecture of Ch'u chou in Che kiang, where this grass is produced on the Sien tu shan mountain. [See Bot. sin., I, p. 226 (42).]

I have already noticed [Bot. sin., II, 455] that the fine mats made of the lung sü grass are still known in China and much prized. But the grass which furnishes the material is unknown to botanists.

Ch., XI, 39:—Shi lung ch'u or lung sü ts'ao, and Phon zo, XVI, 22:—石 龍 芻 representations of fine-leaved grasses.

99.—龍常草 lung chang ts'ao. P., XV, 91. T., CLXIV.

Pie lu:—It grows by the sides of rivers in summer as well as in winter, and resembles the lung ch'u [see 98]. The stem is used in medicine. Taste saltish. Nature warm. Non-poisonous.

LI SHI-CHEN thinks that this is the plant mentioned in the Rh ya [16]. It is also called 棕心草 tsung sin ts'ao.

Ch., XIV, 13:—Lung shang ts'ao.—Phon zo, XVI, 21:—龍常草 figure of a Graminea.

100.—地黃 ti huang. P., XVI, i. T., CXXXVII.

Pen king:—Ti huang (earth yellow), 地髓 ti sui (earth marrow). The root is official. Taste sweet. Nature cold. Non-poisonous.

For other ancient names of the plant see Bot. sin., II, 180.

Pie lu:—The *ti huang* grows in the marshes of Hien
yang [in Shen si, App. 65]. That which is produced in a
loamy soil is the best. The root is dug up in the 2nd and
9th months and dried in the shade. The drug is used in
this dried state, 乾地黃 *kan* (dry) *ti huan*, or the fresh
root is used, 生地黃 *sheng* (fresh) *ti huang*.

T‘AO HUNG-KING:—Hien yang is the same as Ch‘ang
an [the ancient capital of China, in Shen si, App. 6]. The
ti huang grows in Wei ch‘eng [same as Hien yang, App.
382]. The fruit looks like a wheat-grain. Now the dried
ti huang from P‘eng ch‘eng [in Kiang su, App. 247] is the
best. That from Li yang [in An hui, App. 186] is of
second quality. The drug from Pan k‘iao in Kiang ning
[Nan king, App. 240] is also prized. The juice, pressed
from the root, is likewise used in medicine.

SU SUNG [11th cent.]:—It is a common plant. The
best drug comes from T‘ung chou [in Shen si, App. 378].
The leaves of the *ti huang* appear in the 2nd month. They
resemble the leaves of the *ch‘e ts‘ien* [*Plantago.* See 115]
and are all on the ground (radical leaves). They are
wrinkled, rough, and veined. The scapes are from three or four
inches to one foot high. They bear flowers resembling those
of the *yu ma* (*Sesamum*), of a purple colour. Sometimes
the flowers are yellow. The fruit is a capsule, like that of
the *lien k‘iao* [*Hypericum.* See 120]. Small grayish brown
seeds. The root is like a man's hand with fingers, varies
as to size, and is yellow throughout. The plant is not always
raised from seed, frequently it is propagated from the root.
In a fat soil the root attains considerable dimensions and
becomes very juicy. The extract obtained by pressing and
steaming the root is called 熟地黃 *shu ti huang*.

LI SHI-CHEN gives a similar description of the plant.
Leaves and scapes covered with hair. Flowers red and

yellow. The young leaves can be eaten. The drug from Huai k'ing fu (in Ho nan) is considered the best.

The *P.* [XVI, 10] notices after the *ti huang* a plant called 胡面莽 *hu mien mang*, regarding which the *Pen ts'ao shi i* [8th cent.] says, that it grows in Ling nan [S. China, App. 197]. Leaves like those of the *ti huang*. It (the root) is of a sweet taste and used in medicine.

The above descriptions of the *ti huang* agree well with *Rehmannia glutinosa*, Libosch., a common plant in N. China. At Peking, where it is not cultivated, it is called *ti huang*. Four more species of *Rehmannia* are known from other parts of China.

Ch., X, 8:—*Ti huang*, two good drawings of *Rehmannia*, one of them represents a plant with a very large fleshy root. See also *Kiu huang*, LIII, 3.

According to Father Cibot [*Mém. conc. Chin.*, V, 498] the root of the *ti huang* furnishes a yellow dye. The *P.* says nothing about its being employed for tinctorial purposes.

Tatar., *Cat.*, 20 :— *Ti huang*. Radix *Rehmanniæ chinensis.*—Gauger [15, 16] figures and describes the root of the *ti huang.*—P. Smith, 184, 69, 99.

Cust. Med., p. 26 (52):—*Sheng* (fresh) *ti huang* exported 1885 from Tien tsin 22,549 piculs,—p. 70 (58), from Han kow 8,365 piculs,—p. 46 (28), from Chefoo 638 piculs,— p. 124 (52), from Chin kiang 96 piculs.

Ibid., p. 280, 318, 348:—*Shu ti huang* (extract) exported in small quantities from Amoy, Swatow and Canton.

So moku, XI, 64, and *Phon zo*, XVII, 2:—地黃 *Rehmannia lutea*, Maxim., a Japanese species.—*Rehmannia glutinosa* is depicted in the *So moku* [XI, 63] and *Phon zo* [XVII, 3] under the Chinese names 地筋 and 胡面莽 [v. supra, *hu mien mang*]. See also *Kwa wi*, 50.

Both species are depicted in Sieb., *Icon. ined.* [VI], *R. glutinosa* with purple flowers and brownish root,—*R. lutea* with yellow flowers and yellow root.

101.—牛膝 *niu si*.   *P.,* XVI, 10.   *T.,* CLXVI.

Pen king :—*Niu si* (ox knee), 百倍 *po pei*. The root
is officinal. Taste bitter and acid. Nature uniform. . Non-
poisonous.

Pie lu :—The *niu si* grows in Ho nei [S.E. Shan si,
App. 77] in river-valleys, also in Lin k'ü [in Shan tung,
App. 193]. The root is dug up in the 2nd and 8th months
and dried in the shade.

In the *Kuang ya* [3rd cent.] it is called 牛莖 *niu heng*.

WU P'U [3rd cent.]:—Its leaves resemble those of the
hia lan (summer blue. Unknown to me). The stem at its
beginning is red.

T'AO HUNG-KING :—It is produced in Mid China. The
drug from Ts'ai chou [in Ho nan, App. 342] is the best,
it is large, long and succulent. The stem has joints like the
knee of an ox, whence the name. That with large purple
joints is the male, that with small green joints is the female
plant. The male is the best.

SU SUNG [11th cent.]:—It is found in Kiang and in
Huai (Kiang su and An hui), in Min [Fu kien, App. 222]
and Yüe [S. China, App. 419], also in Kuan chung [in
Shen si, App. 158]. The genuine drug comes from Huai
k'ing [in Ho nan, App. 94] and is considered the best.
The stem grows from two to three feet high, is of a greenish
purple colour, and has joints like the knee of a crane or an ox.
The leaves are pointed, round, resemble a spoon in shape,
and come out from the joints in pairs opposite. Flowers in
spikes. In autumn it produces very small fruits (or seeds).
The root attains a length of three feet and is succulent.
Stem and leaves also used in medicine.

LI SHI-CHEN :—The *niu si* is a common plant. The
wild-growing is not so much used for food as that cultivated

in N. China and in Sz ch'uan, which is of a superior quality. It has a square stem with large joints. Leaves opposite, resembling those of the *hien* [*Amarantus blitum.* See 256], whence the plant is also called 山莧菜 *shan hien ts'ai.* The seeds resemble the small vermin found on rats, and are hirsute. They are attached along the stem and hang down. Root, leaves and stem used in medicine, the leaves are eaten.

Ch., XI, 20:—*Niu si.* Two drawings. *Achyranthes* probably intended.—*Kiu huang*, XLVI, 4:—*Shan hien ts'ai* or *niu si.*

LOUR., *Fl. cochin.*, 124:—*Cyathula geniculata.* Caulis herbaceus . . . geniculatus. Cochinchina. Sinice: *nieu si.* In Sinis inveniuntur duæ plantæ, quæ eodem nomine dignoscuntur. Has ego non vidi, nisi pictas in herbario sinensi. Una illarum similem habitum portat, videturque tam genere, quam specie cum cochinchinensi convenire.

GAUGER [30] describes and figures the root of the *niu si.* He means that it belongs to *Achyranthes aspera*, L. (allied to *Cyathula.* Order *Amarantaceæ*).

TATAR., *Cat.*, 41:—*Niu si.* Radix *Pupaliæ* (*Cyathulæ*) *geniculatæ ?*—P. SMITH, 180.

HENRY, *Chin. pl.*, 314, 315:—*Niu si* in Hu pei, *Achyranthes bidentata*, Bl. The same name also applied to *Polygonum filiforme*, Thbg.

The description of the *niu si* in the *P.* agrees in a general way with *Achyranthes*.

Cust. Med., p. 24 (39):—*Niu si* exported 1885 from Tien tsin 1,954 piculs,—p. 68 (41), 66 (12), from Han kow 612 piculs, besides this 1,030 piculs of *Ch'uan* (Sz ch'uan) *niu si*,—p. 46 (23), from Che foo 42 piculs,—p. 58 (14), from I chang 12 piculs.

Amœn. exot., 863 :—牛 膝 *goositz*, vulgo *ino kadsitz*.
Amarantus siculus spicatus Bocconi, flosculis pentapetaloidibus
albis ; semine fusco.—The Amar. sicul. spic. Bocconi is
Achyranthes argentea, Lam., not *Celosia argentea*, to which
THUNBERG [*Fl. jap.*, 106] refers the plant noticed by
KÆMPFER. But [*Amœn. exot.*, 911] the same Chinese name
is [probably erroneously] referred to a *Thlaspi*.

So moku, IV, 2 :—牛 膝 *Achyranthes bidentata*, Bl.
(= *A. aspera*, Thbg., non L.).

Phon zo, XVII, 4 :—牛 膝 *Achyranthes bidentata* and
[3] 川 牛 膝 [from Sz ch'uan] *A. lactea.*—See also *Kwa
wi*, 43.

102.—紫 菀 *tsz' yüan*.   P., XVI, 13.   T., CLIX.

Pen king :—*Tsz'* (purple) *yüan*. The root is officinal.
Taste bitter. Nature warm. Non-poisonous.

Pie lu :—Other names : 青 | *ts'ing yüan*, 紫 蒨 *tsz'
ts'ien*. The *tsz' yüan* grows in Han chung (S. Shen si) and
in Fang ling [in Hu pei, App. 36], in mountain-valleys,
also in Chen ting [in Chi li, App. 11] and Han tan
[S. Chi li, App. 56]. The root is dug up on the 3rd day of
the 2nd month and dried in the shade.

T'AO HUNG-KING :—It is a common plant in Mid China.
It covers the ground. Purple flowers. Lower part covered
with white hair. Fibrous, pliable root. The white sort is
called 白 | *pai yüan* [see the next].

SU SUNG [11th cent.] :—This plant is found in Yao
chou [in Shen si, App. 401], in Ch'eng chou [in Kan su,
App. 18], in Sz' chou and Shou chou [both in An hui,
App. 319, 290], in T'ai chou [in Che kiang, App. 326],
in Meng chou [in Ho nan, App. 220], in Hing kuo [in
Hu pei, App. 67]. Leaves two or four together. Flowers

yellow, white and purple [probably the author has in view an *Aster* with a yellow disk and purple or white radiate florets]. The seed is black.

CH'EN TSZ'-MING [13th cent.]:—The best *tsz' yüan* comes from Lao shan [App. 179]. The root resembles that of the northern *si sin* [*Asarum*. See 40]. It is also met with in I chou and Yen chou [both in Shan tung, App. 106, 404] and eastward.

LI SHI-CHEN:—The *Shuo wen* writes the name of the plant 茈菀 *ts'z' yüan*. Another ancient name [Taoist?] is 返魂草 *fan hun ts'ao*. It is also called 夜牽牛 *ye k'ien niu* [comp. 168].

Cust. Med., p. 124 (66):—*Tsz' yüan* exported 1885 from Chin kiang 347 piculs,—p. 282 (140), from Amoy 0.24 picul.

In the *So moku* [XVI, 59] 紫菀 is *Aster trinervius*, Roxbg.—*Ch.*, XI, 53:—*Tsz yüan*. Only leaves represented. It does not seem that an *Aster* is intended.

P. SMITH [71] identifies the *tsz' yüan* arbitrarily with *Convolvulus*.

103.—女菀 *nü yüan*. P., XVI, 14. T., CXXX.

Pen king:—*Nü yüan*. Root officinal. Taste pungent. Nature warm. Non-poisonous.

Pie lu:—Other names: 白 | *pai* (white) *yüan*, 織女 | *chi nü yüan*. The *nü yüan* grows in Han chung [S. Shen si, App. 54], in mountain-valleys, also in Shan yang [in Shan tung, App. 270]. It (the root) is gathered in the 1st and 2nd months and dried in the shade.

In the *Kuang ya* [3rd cent.] it is called 女復 *nü fu*.

The Chinese authors do not describe the *nü yüan*, but LI SHI-CHEN says that it is allied to the *tsz' yüan* [102].

The *Ch.* [XI, 54] figures sub *nü wan* a *Labiata.* In the Peking mountains this name is applied to *Plectranthus glaucocaly.r*, Maxim. (*Labiatæ*).

So moku, XVI, 61:—女菀 *Aster fastigiatus,* Fish & Mey.

104.—麥門冬 *mai men tung.* P., XVI, 16. T., CLXXIV.

Pen king:—Mai men tung. The root (tubers) is officinal. Taste sweet. Nature uniform. Non-poisonous.

For other ancient names see *Bot. sin.,* II, 108.

*Pie lu:—*Other names : 羊韭 *yang kiu* (sheep *Allium*), 禹餘糧 *Yü yü liang* [Emperor Yü's extra ration. See also 179]. The leaves of the *mai men tung* resemble those of the *kiu* [*Allium odorum.* See 240] and are green the whole year round. The plant grows in Han ku [*v. infra*] in river-valleys, on declivities, in a fat soil, between stones. The root is dug up in the 2nd, 8th and 10th months and dried in the shade.

Wu P'u [3rd cent.]:—It is known also under the names: 禹韭 *Yü kiu* (Emperor Yü's *Allium*), 忍凌 *jen ling,* 不死草 *pu sz' ts'ao* (undying plant).

T'ao Hung-king :—Han ku [mentioned in the *Pie lu*] is the same as Ts'in kuan [the barrier of Ts'in. In N.W. Ho nan, App. 55, 359]. The *mai men tung* is a common plant. It bears green (or blue) globular fruits in winter. The root (tubers) resembles the *kung mai* (barley), whence the name *mai men tung* (*mai* = wheat).

Ch'en Ts'ang-k'i [8th cent.]:—The drug produced in Kiang ning [Nan king, App. 129] is small but succulent, that from Sin an [in Che kiang, App. 310] is large and white. The larger sort has leaves like the *lu ts'ung* (stag onion), the smaller like the *kiu* [*v. supra, Allium*]. There are three or four species. All these have nearly the same medical properties. The fruit is globular and blue.

Su Sung [11th cent.]:—The *mai men tung* has green, persistent leaves resembling those of the *so ts'ao* [*Cyperus.* See 59] and about a foot long. The root is yellowish white, fibrous, with roundish tubers. In the 4th month the plant opens pale red flowers, resembling those of the *hung liao* (*Polygonum*). The fruits are globular, blue. That sort which grows in Kiang nan [South of the Yang tsz, App. 124] has large leaves. The best drug is produced in Wu [Kiang su, App. 389].

Li Shi-chen :—In ancient times the people used [the tubers of] the wild-growing plant. Subsequently they began to cultivate it for medical use. The tubers can also be eaten, whence the name *Yü yü liang.* It is popularly called 門冬 *men tung.* Other names 僕壘 *pu lei,* 隨脂 *su chi.*

At Peking *Ophiopogon spicatus,* Ker., is cultivated under the name of *mai men tung.* Under the same Chinese name this plant is represented in the *Ch.* [XI, 10]. The description in the *P.* agrees. See also *Kiu huang,* LI, 6.

Lour., *Fl. cochin.,* 50 :—*Commelina medica* (*Aneilema medicum,* R. Br., an obscure plant). Sinice : *me muen tum.* Tubera in frequenti usu medico. Sapor subdulcis, odor gratus.

Tatar., *Cat.,* 38 :—*Mai men tung.* Radix *Aneilematis medici.*

Gauger [27] describes and figures the oblong tubers of the *mai men tung.*—Hanb. [*Sc. pap.,* 257] figures the same tubers, and identifies the *mai men tung* with *Ophiopogon japonicus,* Ker.—P. Smith, 162, 17, 194.

In the *Rep. on Trade, Chin. M. Cust.,* for 1880 [p. 141], there are interesting details regarding the cultivation of the *mai men tung* near Ning po and other places.

Cust. Med., p. 186, 188 (49-53):—*Mai men tung* exported 1885 from Ning po 2,431 piculs,—p. 58 (12), from I chang 482 piculs,—p. 78 (161), from Han kow 454 piculs.

Amœn. exot., 823 :—門冬 *mondo*, *biakf mondo*, vulgo *rjuno fige*. Gramen flore hexapetaloide spicato, radice fibrosa et tuberosa. Cum icone, p. 824. Detailed description of the plant, which is *Ophiopogon japonicus*, Ker. KÆMPFER adds : Usus radicum in medicina familiaris est ; tubera saccharo condita Sinensis ægrotis commendant.—Altera datur species, literatis *temondo* dicta, in prov. Satzuma frequens, quæ quod radicem et tubera habeat pinguiora, priori antiferri solet.

So moku, VI, 46 :—小葉麥門冬 (small-leaved), *Ophiopogon japonicus*. *Ibid.*, 44, 45 :—大葉麥門冬 (large-leaved), *Oph. spicatus*.

Both these species—the larger, *O. spicatus*, and the smaller, *O. japonicus*—are recorded from China by our botanists, and, as we have seen, they are correctly distinguished by the ancient Chinese authors.

105.—葵 *k'ui*. *P.*, XVI, 22. *T.*, LXXXV.

Pen king :—*K'ui*, 冬葵子 *tung k'ui tsz'* [this name is in the Index of the *Pen king*]. Leaves, root and seeds officinal. Taste of the leaves sweet. Nature cold and mucilaginous. Non-poisonous. The root is sweet and cold. The seeds are sweet, cold and mucilaginous.

Regarding the *k'ui*, or Mallow of the Classics, see *Bot. sin.*, II, 368.

Pie lu :—The *tung k'ui tsz'* grows on the Shao shi shan mountain [in Ho nan, App. 281].

T'AO HUNG-KING:—The *kui* which is sown in autumn grows during the winter and bears seed in spring. It is therefore called *tung k'ui* (winter *k'ui*). It is used in medicine and is very mucilaginous. The 春 | *ch'un* (spring) *k'ui* is likewise mucilaginous, but not much used.

SU KUNG [7th cent.]:—This is the *k'ui* which is commonly used for food. There are many sorts of it. They are not employed as medicines.

SU SUNG [11th cent.]:—The *k'ui* is a common plant. The young plants and the leaves are used as a vegetable, which is sweet and palatable. In ancient times the seeds of the *tung k'ui* were commonly used in medicine. There are many sorts of *k'ui*: the 蜀 | *Shu k'ui* (*Althœa rosea*), the 錦 | *kin k'ui* (*Malva sylvestris*), the 黃 | *huang* (yellow) *k'ui* (*Hibiscus Manihot*, also *II. Abelmoschus*), the 終 | *chung k'ui* [*Basella*. See *Rh ya*, 148], the 莵 | *t'u k'ui* [*Anemone*. See *Rh ya*, 115]. All these plants are useful.

LI SHI-CHEN:—The *k'ui* was a common food in ancient times and was considered the first of the five vegetables, but nowadays it is not much eaten. It was also called 露葵 *lu* (dew) *k'ui*. Now the people call it also 滑菜 *hua ts'ai* (mucilaginous vegetable), but it is rarely cultivated. Two sorts are distinguished, according to the colour of the stem, which is purple or white. The latter is preferred. It has large leaves, small purple and yellow flowers. That sort with very small flowers is called 鴨脚 | *ya kio k'ui* (duck's-foot mallow). The fruit (of the *k'ui*) is of the size of the end of a finger and flattened. Its skin is thin. The seeds within are light and resemble the seed-vessels of the elm. From that sown in the 4th or 5th month seed may be left. That sown in the 6th or 7th month is called 秋葵 *ts'iu* (autumn) *k'ui*, that sown in the 8th and 9th months is 冬 | *tung* (winter) *k'ui*, that sown in the 1st

25

month is 春 | *ch'un* (spring) *k'ui*. Thus the plant can be utilized all the year.

Ch., III, 1:—*Tung k'ui*. The figure represents *Malva verticillata*, L. (*M. pulchella*, Bernh.). See also *Kiu huang* LVIII, 32.

Lour., *Fl. cochin.*, 514 :—*Malva verticillata*. Sinice : *tung quei tsu*. Habitat culta Cantone Sinarum.

Tatar., *Cat.*, 22 :—*Tung k'ui tsz'*. Semina *Hibisci.*— P. *Smith*, 112 :—*Tung k'ui tsz'*, *Hibiscus Abelmoschus.*— According to Ford & Crow ["Notes on Chin. Mat. Med.," in *China Rev.*, XVI, p. 8] *tung k'ui tsz'* at Hong kong is *Abutilon indicum*, G. Don. But what I obtained under this Chinese name in an apothecary's shop in Peking were the seeds of *Malva*. Comp. also *Henry*, *Chin. pl.*, 156.

Cust. Med., p. 80 (192):—*Tung k'ui tsz'* exported 1885 from Han kow 6.40 piculs,—p. 298 (359), from Amoy 0.1 picul.

Amœn. exot., 858:—葵 *ki* vulgo *awoi*, in specie *Kara awoi* (Kara = China) Malva hortensis sive rosea, Malva arborescens. *Thbg.* [*Fl. jap.*, 271] identifies this with *Malva mauritiana*.

So moku, XII, 56 :—冬葵 *Malva pulchella*.

106.—酸漿 *suan tsiang*. P., XVI, 33. T., CLX.

Pen king:—*Suan tsiang* (sweet juice), 醋漿 *ts'u* (vinegar) *tsiang*. The leaves, the stem and the root are officinal. Taste bitter. Nature uniform. Non-poisonous. The fruit is also used in medicine. Taste acid. Non-poisonous.

Comp. also *Rh ya*, 55.

Pie lu:—The *suan tsiang* grows in King Ch'u [Hu kuang, App. 145] in marshes, also in fields and gardens. Gathered in the 5th month and dried in the shade.

T'ao Hung-king :—The *suan tsiang* is a common plant.
It resembles the *shui* (water) *k'ie* (*Solanum nigrum*) but is
smaller. The leaves are edible. The fruit is of the size
of a plum, of a yellowish red colour, and is enclosed within
a capsule (bladder, inflated calyx). Children eat it.

Su kung [7th cent.]:—This plant is also called 燈籠草
teng lung ts'ao (lantern plant). The stem is branchy and from
three to four feet high. The red fruit is contained within a
red flower (bag) which has the appearance of a lantern.
It is very handsome. All parts of the plant are used in
medicine.

Han Pao-sheng [10th cent.]:—The *suan tsiang* is the
same as the 苦蔵 *k'u chen* [see *Rh ya*, 55]. The root
resembles the *tsü k'in* (pickled celery), is of a white colour
and very bitter.

Chang Yü-hi [11th cent.]:—The 苦䏌 *k'u tan* grows
in neglected places from two to three feet high. The fruit
is globular, red when ripe, and enclosed within a kind of
bag. The people of Kuan chung [in Shen si, App. 158]
call it 洛神珠 *lo shen chu*. Other names are 王母珠
wang mu chu, 皮弁草 *p'i pien ts'ao* (skin bonnet plant).
There is a small variety which is called 苦蘵 *k'u chi*. The
author refers to *Rh ya*, 144. Comp. also *supra*, 34.

K'o Tsung-shi [12th cent.]:—The *suan tsiang* or *k'u tan*
resembles the *t'ien k'ie tsz'* (*Solanum nigrum*). Small white
flowers. Fruit like a cherry, red, and enclosed in a bladder.
It contains small seeds like those of the *lo su* (*Solanum
melongena*). Its taste is like green grass.

Li Shi-chen gives a similar description of the plant:
White flowers, five cleft corolla, fruit a pentagone pendent
bladder like a lantern. Other name 天泡草 *t'ien p'ao ts'ao*
(Heaven bladder).

The plant thus described by the Chinese authors is the *Physalis Alkekengi*, L., our common winter cherry, which is also very common in China. The Chinese descriptions are quite correct.

Ch., XI, 48 :—*Suan tsiang*. A rude drawing, but recognizable ; *Physalis*. The same plant is figured in the *Kiu huang* [LII, 23] s.n. 紅姑娘 *hung ku niang* (red girl), which is also the vulgar name for *Ph. Alkekengi* at Peking. Also 豆 | | *tou ku niang*.

LOUR., *Fl. cochin.*, 164 :—*Physalis Alkekengi*. Sin. : *soan tsiam*.

HENRY, *Chin. pl.*, 465, 466 :—The name *t'ien p'ao ts'ao* is applied in Hu pei to two of species of *Physalis*—*Ph. minima*, L. ? and *Ph.* aff. *Ph. angulatæ*—but also to several species of *Solanum*.

Amœn. exot., 785 :—酸漿 *san sjo* vulgo *foo dsukki*. Solanum vesicarium, vulgo Alkekengi.

So moku, III, 43 :—酸漿 *Physalis Alkekengi*. The 苦蘵 in the *So moku* [III, 45] and *I'hon zo* [XVII, 25] is *Physalis angulata*, L.

SIEB., *Œcon.*, 196 :—*Physalis Alkekengi*. Japonice *hoosuki*, sinice 酸漿. Fructus edulis ac pro nugis habetur venalis.

The drawing in the *Ch.* [XI, 80] sub 王不留行 *wang pu liu hing* agrees exactly with that of *Physalis angulata* in the Japanese works. The Chinese name *wang pu liu hing* here is, I suspect, a mistake for *k'u chi*, for *wang pu liu hing* is a *Silene*, and such a plant is figured in the *Ch.* [XI, 76] sub *wang pu liu hing*. The confusion arose probably from *Silene* having likewise a persistent, inflated calyx which encloses the fruit.

107.—蜀羊泉 *Shu yang ts'üan.* *P.,* XVI, 36.
T., CLXXVI.

Pen king:—*Shu* (Sz ch'uan) *yang ts'üan.* The leaves
are officinal. Taste bitter. Nature slightly cold. Non-
poisonous.

Pie lu:—Other names: 羊泉 *yang ts'üan,* 羊飴 *yang i.*
This plant grows in Shu (Sz ch'uan) in mountain-valleys.

T'AO HUNG-KING:—It is not employed in medicine now.

SU KUNG [7th cent.]:—The vulgar name of this plant
is 漆姑 *ts'i ku.* Its leaves resemble those of the *kü* (*Chrysanthemum, Aster*). Purple flowers. The fruit resembles
that of the *kou k'i* [*Lycium.* See 345]. The root is like
that of the *yüan chi* [*Polygala.* See 16]. It grows in shady
moist places. The leaves are gathered in the 3rd and 4th
months and dried in the shade.

LI SHI-CHEN states that the name *ts'i ku* is also applied
to another plant.

Ch., XI, 26:—*Shu yang ts'üan.* Rude drawing,
Perhaps a *Solanum* is intended. The drawing is taken from
the *Kiu huang* [XLVII, 2], where this plant is figured
under the names of 青杞 *ts'ing k'i* or *Shu yang ts'üan.*

P. SMITH, 200:—*Shu yang ts'üan, Solanum dulcamara*
[arbitrary identification].

So moku, III, 51:—蜀羊泉 *Solanum lyratum.* Thbg.—
Ibid., VIII, 32:—漆姑草 *Sagina maxima,* A. Gray
(*Caryophyllaceæ*).

108.—敗醬 *pai tsiang.* *P.,* XVI, 37. *T.,* CLXI.

Pen king:—*Pai tsiang.* The root is officinal. Taste
bitter. Nature uniform. Non-poisonous. It is also called
鹿腸 *lu ch'ang.*

of Kiang hia [in Hu pei, App. 126]. The root is gathered in the 8th month and dried in the shade.

T‘AO HUNG-KING : — The root has the smell of old, spoiled *tou tsiang* (soy), whence the name *pai tsiang* (spoiled soy). It is a common plant in Mid China. Its leaves resemble those of the *hi lien* (*Siegesbeckia orientalis*). The root resembles the *ch‘ai hu* [*Bupleurum*. See 29].

SU KUNG [7th cent.]:—It is a mountain plant. Yellow flowers, purplish (brown) root, having the colour of old soy. The leaves do not resemble those of the *hi lien*.

LI SHI-CHEN :—It is a common wild plant. Its vulgar name is 苦菜 *k‘u ts‘ai* (bitter vegetable). The savages eat it. The people of Kiang tung [Kiang su, Che kiang, App. 124] gather it. In spring, when the plant begins to grow, the leaves cover the ground. They resemble cabbage-leaves but are narrower and longer, serrated and toothed. The leaves are dark green on the upper side, paler underneath. The stem attains a height of from two to three feet, and has joints. The leaves are four in a whorl like an umbrella. The flowers are white, and appear on the top in umbels like those of the *k‘in* (Celery) or the *she ch‘uang* [*Cnidium*. See 49]. Small fruits. The root is purplish white, resembling that of the *ch‘ai hu* [*v. supra*]. The plant is sometimes also called *k‘u chi*, which is properly a name for the *suan tsiang* [see 106]. The people in the south eat the young plants steamed, as a vegetable. It has a slightly bitter taste and the smell of spoiled soy.

From the above description it would seem that the *pai tsiang* is an umbelliferous plant. The drawing of it in the *Ch.* [XI, 47] shows only leaves.

So moku, II, 21, 22 :—敗 醤 *Patrinia villosa*, Juss., and *P. scabiosœfolia*, Link. Order *Valerianeœ.*—See also *Kwa wi*, 56.

109.—欸冬花 *k'uan tung hua.* P., XVI, 38. T., CXIX.

Pen king:—*K'uan tung hua* (flowers which like the winter), 橐吾 *t'o wu*, 虎鬚 *hu sü* (tiger's moustaches). The flowers are officinal. Taste pungent. Nature warm. Non-poisonous.

Comp. *Rh ya*, 160.

Pie lu:—Other name : 氏冬 *shi tung.* The *k'uan tung* grows in Ch'ang shan [in Chi li, App. 8], in mountain-valleys, also in Shang tan [in S.E. Shan si, App. 275] by river-sides. The flowers are gathered in the 11th month and dried in the shade.

T'AO HUNG-KING :—The best is produced in Ho pei [in Shan si, App. 78]. It resembles the *shun* [*Brasenia.* See 199]. The best is that with its flowers not yet opened. The next in quality comes from Kao li [Corea, App. 116] and Po tsi [S. Corea, App. 261]. Its flowers resemble the great *kü* (*Chrysanthemum*). A third sort is produced in Pei pu, in Shu [Sz ch'uan, App. 244], and in Tang ch'ang [in Kan su, App. 330]. The author says it is unknown to him that this plant grows in winter under the ice. [He seems to refer to a statement found in the *Shu cheng ki* [5th cent.] that the *k'uan tung* blossoms in the depth of winter in the ice of the Lo river [in Ho nan] whence the name *k'uan tung*]. It is gathered in the 12th and 1st months.

SU KUNG [7th cent.]:—It is found in Yung chou [in Shen si, App. 424], in the Nan shan mountains [S. Shen si, App. 230], also in the mountain-valleys of Hua chou

S.E. Shen si, App. 85]. Leaves like those of the *k'ui*
(*Mallow*) but larger. The flowers issue directly from the
root.

Su Sung [11th cent.]:—Now it is also found in Kuan
chung [in Shen si, App. 158]. The root is of a purple
colour. The leaves resemble those of the *pei hiai* [*Dioscorea.*
See 178]. It blossoms in the 12th month. Yellow flowers.
Greenish purple receptacle (*involucrum ?*). These flowers are
from one to two inches above the ground. There is one kind
with red flowers and large leaves like those of *Nelumbium
speciosum.*

K'ou Tsung-shi [12th cent.]:—The *k'uan tung* is the
only plant that is independent of frost and snow, for it
flowers long before spring, wherefore it is also called 鑽 凍
tsuan tung (piercing the cold).

Ch., XI, 44 :—*K'uan tung hua.* Flowers and leaves
figured. Perhaps *Tussilago.* See also *Kiu huang* [XLVI, 5].
Only leaves figured.

The above Chinese descriptions of the *k'uan tung hua*
agree in a general way with *Tussilago Farfara,* L.,—Colt's
foot, which sets forth, early in spring, its short flower-stalks ;
each bearing at its summit a single large yellow radiated
flower-head.

Lour., *Fl. cochin.,* 614 :—*Tussilago Farfara,* L. Habitat
inculta in China boreali. Sinice *koan tum hoa.* Ibidem
Tuss. anandria, sin. : *Lu chau koan tum hoa.* China borealis
[perhaps ancient Lu chou in Shan si. See App. 204].

Tatar., *Cat.,* 27 :—*K'uan tung hua.* Flores *Farfaræ.*—
P. Smith, 68, sub Colt's foot.

In Williams' *Chin. Dict.* [477] we read that *k'uan
tung hua* are the flowers of *Eriobotrya japonica.* Henry
[*Chin. pl.,* 124] states that at I chang the dried flowers of
Eriobotrya japonica are called *k'uan tung hua.*

About ten years ago I sent to my late friend MAXIMOWICZ the drug *k'uan tung hua* obtained from a Peking drug-shop. He found that it was the flower-buds of a *Tussilago*, and observed that no species of this genus has hitherto been gathered in China, but an allied genus—*Petasites*—is represented in Japan and China by *P. japonicus*, Miq. (*Tussilago petasites*, Thbg.).[31]

Cust. Med., 74 (122):—*K'uan tung hua* exported 1885 from Han kow 540 piculs,—p. 30 (115), from Tien tsin 99 piculs,—p. 62 (48), from I chang 37 piculs.

Amœn. exot., 831 :—款冬 *ro* vulgo *fuki sabuki*. Petasites vulgaris. Caules hic inter olera recipiuntur.

So moku, XVII, 25, 26 :—Same Chinese name, *Petasites japonicus*, Miq.

款吾 *T'o wu*, given in the *P.* as a synonym for *k'uan tung hua*, is in Japan applied to another *Composita*.

Amœn. exot., 827 :— *Tswa* [no Chinese characters]. Doronicum radice tuberoso, folia Petasitidis, floribus luteis Chrysanthemi. Caules et pediculi inter olera recipiuntur.— This plant, figured in the *Icones Kœmpf.* [sel. 27, 28] is the *Tussilago japonica*, L., and THBG., *Fl. jap.*, 313, = *Senecio Kœmpferi*, DC. Prod., VI, 363, = *Ligularia Kœmpferi*, S. & Z. *Fl. jap.*, I, p. 77, tab. 35. Nomen japonicum *tswa*, nomen sinicum *takgo*. In SIEB., *Icon. ined.*, the Chinese characters 款吾 are given. It flowers late in autumn and in winter. It is the *Farfugium Kœmpferi*, Benth., *Fl. hongk.*, 191.

---

[31] I obtained specimens of *tung-hua* growing wild in T'ang district, Hupeh, whence the drug is sent to Hankow for sale, and these were identified at Kew as *Tussilago Farfara*, L. This is an excellent example of the accuracy and extent of knowledge of LOUREIRO, who found out the correct facts 100 years ago.—A. HENRY.

26

The *Ligularia gigantea*, S. & Z. [*Fl. jap.*, I, 79, tab. 36] sinice 大葉橐吾 (the large-leaved) is the *Farfugium grande*, Lindl., introduced from China into Europe by R. FORTUNE. This is probably only a variety of *F. Kœmpferi*.

See also *So moku*, XVII, 27.

110.—決明 *küe ming*.   *P.*, XVI, 41.   *T.*, CXXXI.

Pen king:—*Küe ming.* The seeds are officinal and are employed in diseases of the eye. Taste saltish. Nature uniform. Non-poisonous.

Comp. *Rh ya*, 56.

Pie lu:—The *küe ming* grows in Lung men [*v. infra*], in marshes. The seeds are gathered on the 10th day of the 10th month and dried in the shade for a hundred days.

T'AO HUNG-KING:—Lung men lies north of Ch'ang an [the ancient capital of China, in Shen si, App. 211]. The *küe ming* is a common plant. Its leaves resemble those of the *kiang mang* [*Cassia?* See further on], the seeds resemble a horse's hoof, hence the name 馬蹄決明 *ma ti* (horse's hoof) *küe ming*. A different plant is the *ts'ao küe ming* which is the *ts'i hao* [*Celosia*. See 82].

SU SUNG [11th cent.]:—This plant is now much cultivated in gardens. It grows from three to four feet high. The root is tinged with a purple colour. The leaves resemble those of the *mu su* [*Medicago*. See 255] but are larger. It blossoms in the 7th month. Yellow flowers. The fruit is a pod resembling that of the *lü tou* (green bean, *Phaseolus Mungo*) but pointed. The seeds are gathered in the 10th month. This plant is mentioned in the *Rh ya* [56]. There is another kind which is called *ma t'i küe ming*. Its leaves resemble those of the *kiang tou* [*v. infra*], the seeds resemble a horse's hoof.

LI SHI-CHEN:—There are two sorts of *küe ming*. One is the *ma t'i küe ming*. It grows from three to four feet high. The leaves (leaflets) are larger than those of the *mu su* [*v. supra*], narrow at the base, broader at the top. They stand in pairs opposite (pinnate leaves). In daytime they are open, at night (the leaflets) all fold together. It blossoms in autumn. Flowers of a pale yellow colour with five petals. The fruit looks like a young pod of the *kiang tou* (*Dolichos sinensis*), is from five to six inches long, and contains a number of irregularly shaped, dark green seeds resembling a horse's hoof. They are very useful in diseases of the eye.—The other sort is called 茳芒決明 *kiang mang küe ming*. In the *Kiu huang Pen ts'ao* [LVII, 3] it is called 山扁豆 *shan pien tou*. It resembles the *ma t'i küe ming*, only the leaves (leaflets) are pointed at the top. These leaves resemble very much the leaves of the *huai* [*Sophora japonica*. See 322]. They do not fold together at night. It blossoms in autumn. Dark yellow flowers with five petals. The pod is of the size of a little finger, two inches or more in length. The seeds within are in rows, and resemble the seeds of the *huang k'ui* [*Hibiscus*. See sub 105], but flattened, of a gray colour, sweet taste and mucilaginous. From the leaves of both kinds yeast can be prepared. The leaves of the *kiang mang* as well as its flowers and seeds can be eaten boiled, but the *ma t'i küe ming* is bitter and not edible. There is also one sort which is called 石 | | *shi* (stone) *küe ming*.

P., XVI, 43 :—The *kiang mang* is again spoken of in a special article, where a short account of the plant by CH'EN TS'ANG-K'I [8th cent.] is given. A fragrant beverage is prepared from the leaves.

Ch., XI, 30 :—*Küe ming*, also 望江南 *wang kiang nan*. The drawing represents a *Cassia*, as also the figure sub *wang kiang nan* in the *Kiu huang* [LIII, 22]. The above

descriptions in the *P.* likewise refer doubtlessly to *Cassia.*
At Peking *wang kiang nan* [I suspect the same as *kiang man*
in ancient works] is the common name for *Cassia Sophera,* L.

LOUR., *Fl. cochin.,* 324 :—*Cassia sophera.* Sinice *xy tsi
tau, kiue mim tsu.*

TATAR., *Cat.,* 5 :—*Ts'ao kŭe ming.* Semina *Cassiæ
Toræ.* [As we have seen, the *P.* admits *ts'ao kŭe ming* only
as a name for *Celosia*].—HANB., *Sc. pap.,* 231 :—*Kŭe ming
tsz'.* Received from Shang hai. Seeds of *Cassia Tora.*
They are of a cylindrical form, from two to three lines long,
pointed at one extremity, rounded at the other, of a dark
brown colour, etc.—P. SMITH, 54.

PARKER, *Chinese Names of Canton Plants,* 54 :—*Cassia
occidentalis,* 石 | | *shi kŭe ming,* also *ye* (wild) *pien tou.*

The *shan pien tou* given by LI SHI-CHEN as a synonym
for the *kiang mang kŭe ming* is figured in the *Kiu huang*
[LVII, 3]. A rude drawing. It seems *Cassia mimosoides,* L.,
is intended. Comp. also *Ch.,* II, 10.

Cust. Med., p. 372 (429) :—*Ts'ao kŭe ming* exported
1885 from Canton 17 piculs,—p. 164 (370), from Shang hai
7.72 piculs,—p. 302 (433), from Amoy 0.35 picul.

So moku, VIII, 2 :—决明 *Cassia Tora,* L. Leaflets
broadest at the top. According to MIQUEL [resp. SIEBOLD]
introduced into Japan from China.

Ibid., 3 :—望江南 *Cassia occidentalis,* L. Leaflets
pointed at the top.

Ibid., II, 39 :—山扁豆 *Cassia mimosoides,* L. Japonice
kawara kets mei.—SIEB., *Œcon.,* 297 :—*Sooja nomame,* or
jawara kets mai. Sinice 山扁豆 Plantæ sponte crescentis
folia adhuc tenera pro potu Theæ colliguntur. The first
Japanese name seems to be a mistake, for it means the
Soy-bean, but *jawara kets mai* is the Japanese name for *Cassia
mimosoides.*

111.—地膚 *ti fu.*　*P.,* XVI, 44.　*T.,* CXXXVIII.

Pen king:—*Ti fu,* 地葵 *ti k'ui.* The seeds are officinal. Taste bitter. Nature cold. Non-poisonous.

Comp. *Rh ya,* 9.

Pie lu:—Other name: 地麥 *ti mai.* The *ti fu* grows in King chou [Hu pei, App. 146] in marshes and fields. The seeds are gathered in the 8th and 10th months and dried in the shade. The leaves are also used in medicine.

T'AO HUNG-KING:—It is a common plant in fields, and is employed for making besoms. Seeds very small and used in medicine.

SU KUNG [7th cent.]:—The peasants call it *ti mai ts'ao* [*v. supra*]. In the north it is known under the name of 涎衣草 *ts'ien i ts'ao* (plant which spits on clothes). Small leaves, red stem, very pliable and prostrate. The plant is used for besoms.

The *Yao sing Pen ts'ao* [7th cent.] calls it 益明 *i ming.* Seeds useful in the treatment of diseases of the eye.

TA MING [10th cent.]:—The *ti fu* is also called 落帚子 *lo chou tsz'.* The seeds are green and very small.

SU SUNG [11th cent.]:—It is common in Shu ch'uan [Sz ch'uan, App. 292] and Kuan chung [in Shen si, App. 158] and the adjoining provinces. It is also called 獨帚 *tu chou* and 鴨舌草 *ya she ts'ao* (duck's-tongue weed, from the form of the leaves).

LI SHI-CHEN:—The young tender leaves of the *ti fu* are eaten. It grows in a bushy manner, and is much cultivated in gardens. The old plant is good for besoms. Other names: 白地草 *pai ti ts'ao,* 千心妓女 *ts'ien sin ki nü* (thousand hearts' courtesan).

Ch., XI, 31:—*Ti fu.* The figure represents *Kochia scoparia,* Schr. (*Chenopodium scoparia,* L.). See also *Kiu huang* [XLVII, 7] sub 獨帚 *tu chou.* Rude drawing.

The common Chinese name at Peking for *Kochia scoparia* is 掃帚菜 *sao chou ts'ai* (besom vegetable).

TATAR., *Cat.*, 20 :— *Ti fu tsz'*, *Semina Kochiæ?*— P. SMITH, 128.

Cust. Med., p. 298 (348) :— *Ti fu* seeds exported 1885 from Amoy 3.84 piculs,—p. 372 (425), from Canton 3 piculs.

Amœn. exot., 885 :—地膚 *tsisu*, *fawa kingi*, *fookigusa*, etc. Scoparia sive Belvedere Italorum. Insigne Japonibus medicamentum præstat.

So moku, IV, 41 :—*Kochia scoparia*, same Chinese name.

SIEB., *Œcon.*, 117 :— *Kochia scoparia*. Hookigusa, 地膚 Pro scopis caules adhibentur. Herba tenera edulis ac adulta medico agricolis usui.

112.—瞿麥 *k'ü mai*. P., XVI, 46. T., CXX.

Pen king :—K'ü mai, 巨句麥 *kü kü mai*. It seems the whole plant is officinal. Taste bitter. Nature cold. Non-poisonous.

Comp. *Rh ya*, 125.

*Pie lu :—*Other name : 大蘭 *ta lan*. The *k'ü mai* grows in the mountain-valleys of T'ai shan [in Shan tung, App. 322]. It is gathered in the beginning of autumn.

T'AO HUNG-KING :—It is a common plant in Mid China. Small leaves, handsome purplish red flowers. The plant is cut [for medical use] together with leaves and fruits (capsules). The seed resembles wheat (*mai*), whence the name. There are two sorts. One has rather large flowers and the margin of the petals incised. The other, with smaller flowers, is more generally used. The leaves are covered with hair, the flowers are late and of a reddish carnation colour.

The *Ji hua Pen ts'ao* [10th cent.] gives 石竹 *shi chu* (stone bamboo) as a synonym for *k'ü mai*.

Su Sung [11th cent.]:—It is a common plant, and grows about one foot high. Leaves small, narrow and pointed. Root of a purplish black colour, and resembles a small rape. The flower is purplish red and resembles the *ying shan hung* [*Rhododendron*. See 155]. It blossoms from the 2nd to the 5th month, and in the 7th it produces fruit in racemes (or fascicles). The seed has a slight resemblance to wheat. The plant produced in Ho yang [in Ho nan, App. 81] and Ho chung fu [S.W. Shan si, App. 74] is good for [medical] use. There is one kind grown in Huai [An hui, App. 89] with a slender (fibrous) root. The country people use it (the root) for making brushes. The *k‘u mai* is mentioned in the *Rh ya* [125].

Li Shi-chen:—The 石竹 *shi chu* (stone bamboo) has leaves resembling those of the *ti fu* [*Kochia*. See 111]. They are small and narrow like young bamboo-leaves, whence the name. The stem is one foot and more in height and has joints. That which grows wild in the fields has purplish red flowers as large as a copper coin. That sort which the people cultivate in gardens has smaller flowers of a red or pink colour, sometimes striped, and very handsome. It is commonly called 洛陽花 *Lo yang hua* [Lo yang, the ancient capital of China, in Ho nan]. The fruit (capsule) resembles that of the *yen mai* (swallow wheat. *Avena?*). It contains small black seeds. The young plant is eaten cooked.

Kiu huang, XLVI, 8 :—*Shi chu* and *Ch.* XI, 55 :—*K‘u mai*, representations of *Dianthus*.

Shi chu in N. China is a vulgar name for *Dianthus chinensis*, L., and *D. superbus*, L., both common plants.

Tatar., *Cat.*, 13 :—*K‘u mai*, *Dianthus Fischeri* (same as *D. chinensis*). *Ibid.*, 54 :—*Shi chu*, *Dianthus et Commelyna* [*Commelyna* is *chu ye ts‘ai*, or vegetable with bamboo-leaves].—P. Smith, 86.

The *Cust. Med.* mentions the *k'ü mai* as imported into several ports, but it is not stated there from what place it is exported. In the *Hank. Med.* [12] it is noticed as a drug exported from Han kow.

Amœn. exot., 910 :—石竹 *seki tsiku*, vulgo *nadesko* et *tokunatz.* Caryophyllus hortensis simplex, flore majore.

So moku, VIII, 20 :—瞿麥 *Dianthus superbus.*—*Ibid.*, 21 :—洛陽花 [comp. same name in *P.*], a large double *Dianthus.*—*Ibid.*, 22 :—石竹 *Dianthus Seguieri* (same as *D. chinensis*).

113.—王不留行 *wang pu liu hing.* P., XVI, 48. T., CLXXX.

Pen king, Index :—*Wang pu liu hing.*

Pie lu:—The *wang pu liu hing* grows in the mountain-valleys of T'ai shan [in Shan tung App. 322]. It is gathered in the 2nd and 8th months. The leaves and the seed officinal. Taste bitter. Nature uniform. Non-poisonous.

Wu P'u [3rd cent.] writes the name 不留行 *pu liu hing.*

In the *Ji hua Pen ts'ao* [10th cent.] it is called 剪金花 *tsien kin hua* and 禁宮花 *kin kung hua.* The meaning of the latter name is "forbidden in the palace," and the original name *wang pu liu hing* means, as Li Shi-chen explains, about the same. The descriptions of the plant given by the ancient authors are confused and incorrect, for they confound it with *Physalis* [see 106].

Li Shi-chen says :—The *wang pu liu hing* is frequently met with in corn-fields, it grows from one to two feet high. In the 3rd or 4th month it opens its small flowers resembling little bells, of a reddish white colour. The fruit (capsule), of the size of a bean, and is within a bladder like a lantern with five angles. Small globular seeds like cabbage-seeds, at first white, when ripe black.

Kiu huang [LII, 25] and *Ch.* [XI, 76], sub *wang pu liu hing*, rude drawings, but it seems *Silene* is intended.

The description given by LI SHI-CHEN agrees with *Silene*, which genus is characterized by a persistent inflated calyx which encloses the capsule. But *Saponaria*, another genus of *Caryophyllaceæ*, has also an inflated calyx.

TATAR., *Cat.*, 62 :—*Wang pu liu hing*, *Silene.*— P. SMITH, 197.

At Peking *Silene aprica*, Turcz., is called *wang pu liu hing*. Four species of *Silene* are known from China. See *Ind. Fl. sin.*, I, 64.

Cust. Med., p. 282 (141):—*Wang pu liu hing* exported 1885 from Amoy 2.33 piculs,—p. 368 (387), from Canton 4 piculs.

So moku, VIII, 27 :—王不留行 *Saponaria vaccaria*, L. —*Phon zo*, XVIII, 19 :—The same Chinese name applied to *Saponaria vaccaria*, [fol. 20, 21] to *Melandrium firmum* and *Polemonium cœruleum*, L.

114.—葶藶 *t'ing li*.   P., XVI, 51.   T., CXLI.

Pen king :—*T'ing li*, 大室 *ta shi*, 大適 *ta shi*. Seeds used in medicine. Taste pungent. Nature cold. Non-poisonous.

Comp. *Rh ya*, 78.

Pie lu :—Other names : 丁歷 *ting li*, 狗薺 *kou tsi* (dog shepherd's purse), grows in Kao ch'eng [in Chi li, App. 115], in marshes and fields. The seeds are gathered in summer.

T'AO HUNG-KING:—The best is produced in P'eng ch'eng [in Kiang su, App. 247]. This is the 公荠 *kung* (male) *tsi*. The *mu* (female) *tsi* grows in Mid China. Small yellow seeds and very bitter. They are boiled for use.

Su SUNG [11th cent.]:—It is common in all the prefectures of Pien tung [Ho nan, App. 250], Shen si [App. 284] and Ho pei [S. Chi li and W. Shan tung, App. 78]. The drug from Ts'ao chou [in Shan tung, App. 344] is the best. The plant grows from six to seven inches high and resembles the *tsi* [*Capsella*. See 251]. The root is white. It flowers in the 3rd month. Yellowish flowers. The fruit is a horn (capsule, silique). The seeds are small, slightly oblong, flattened, yellow, and resemble millet. The plant is mentioned in the *Yüe ling* of the *Li ki* under the name of 蘼草 *mei ts'ao*. [LEGGE, *Li ki*, I, 271, translates " delicate herbs." The Chinese commentator CHENG HÜAN says that it is a kind of *tsi* (*Capsella*) or *t'ing li*.]

LI SHI-CHEN :—There are two sorts of *t'ing li*—the sweet and the bitter. The first is also called 狗芥 *kou kiai* (dog mustard).

Cust. Med., p. 164 (367):—*T'ing li tsz'* exported 1885 from Shang hai 0.78 picul,—p. 130 (156), from Chin kiang 0.69 picul.—*Hank. Med.*, 45 :—Exported also from Han kow.

Ch., XI, 63 :—*T'ing li*. The figure represents a cruciferous plant.

For the identification of the *t'ing li*, see *Bot. sin.*, 78, *Sisymbrium*, *Draba*, etc.

115.—車前 *ch'e ts'ien*. P., XVI, 54. T., CLXII.

Pen king:—*Ch'e ts'ien* (cart-track plant), 當道 *tang tao*. The seeds are officinal. Taste sweet. Nature cold. Non-poisonous.

For other ancient names comp. *Rh ya*, 200, *Classics*, 439.

Pie lu :—Other names : 牛遺 *niu i* (ox track), 蝦蟆衣 *hia ma i* (frog's cloak). The *ch'e ts'ien* grows in Chen ting [in Chi li, App. 11] in marshes, also in the mountains and on roads. It is gathered on the 5th day of the 5th month and dried in the shade.

T'ao Hung-king :—It is a common plant about houses and by waysides.

Su Kung [7th cent.]:—Now the best comes from K'ai chou [in Sz ch'uan, App. 113].

Su Sung [11th cent.]—It is a common plant in Kiang, Hu and Huai [Mid China, App. 124, 83, 89] near cultivated land, also in Pien [in Ho nan, App. 248] and in North China. The leaves are all on the surface of the ground (radical leaves), grow the whole year, attain the length of one foot and more and are ladle-shaped. Several stems rise from the centre of the leaves, each bearing at the top a spike of small green flowers tinged with red, like a rat's tail. Brown seeds resembling those of the *t'ing li* [see 114]. It is also cultivated in gardens for its seeds. That from Shu (Sz ch'uan) is much valued. The leaves are gathered in the 5th month, the seeds in the 7th and 8th. In North China the people sell the root as a substitute for the *tsz' yüan* [*Aster.* See 102].

The *Kiu huang* [XLVI, 11] figures the plant under the name of 車輪菜 *ch'e lun ts'ai* (cart-wheel vegetable). This figure and that in the *Ch.* [XI, 28] sub *ch'e ts'ien* represent *Plantago*.

Ch'e tsien is the common name of *Plantago major*, L., at Peking.

Lour., *Fl. cochin.*, 90 :—*Plantago major*. Sinice : *che tsien tcao*. Decocto seminum maxime atuntur indigenæ at diuresim promovendam. Succo foliorum expresso, melle apum condito medentur tenesmo et fluxui sanguineo.

TATAR., *Cat.*, 14 :—*Ch'e ts'ien tsz'*. Semina *Plantaginis*.
—P. SMITH, 14.

HENRY, *Chin. pl.*, 20 :—*Ch'e ts'ien ts'ao*, in Hu pei
Plantago major. · The seeds enter into the composition of
liang fen, a jelly used in summer.

Cust. Med., p. 12 (98) :—*Ch'e ts'ien* seeds exported 1885
from New chwang 366 piculs,—p. 74 (125), from Han kow
254 piculs,—p. 90 (58), from Kiu kiang 98 piculs,—p. 30
(118), from Tien tsin 57 piculs,—p. 198 (193), from Ning po
3.34 piculs,—p. 374 (450), from Canton 1.7 picul.

Amœn. exot., 912 :—車前 *sjaden*, vulgo *obacko*. Plantago major vulgaris, latifolia.

So moku, II, 27 :—Same Chinese name, *Plantago
asiatica*, L.

116.—馬鞭草 *ma pien ts'ao*. P., XVI, 57. T., CLXV.

Pie lu :—*Ma pien ts'ao* (whip herb). Only the name
given and the medical virtues explained. The leaves are
officinal. Taste bitter. Nature cold. Non-poisonous. Also
the root, which is said to be pungent and harsh.

T'AO HUNG-KING :—It is a common plant near villages,
on burial wastes and on pathways. The stem is like the
si sin [*Asarum*. See 40]; purple flowers somewhat resembling those of the *p'eng hao* [given in the *P.*, XXVI, 54, as
a synonym for *t'ung hao* which is *Chrysanthemum Roxburghii*.
Comp. also *Bot. sin.*, II, 436].

SU KUNG [7th cent.]:—The leaves of this plant resemble
those of the *lang ya* [*Potentilla?* See 134] and also those
of the *ch'ung wei* [*Leonurus*. See 78]. It sends up three
or four spikes of small flowers like those of the *ch'e ts'ien*
[*Plantago*. See 115]. This spike resembles a whip, whence
the name. It does not resemble the *p'eng hao*. Another
name is 鳳頸草 *feng king ts'ao*.

HAN PAO-SHENG [10th cent.] says that the flowers are white.

SU SUNG [11th cent.]:—It is common in Heng shan [in Hu nan, App. 61], in Lŭ shan [in Kiang si, App. 209] and in Kiang and Huai [Kiang su and An hui, App. 124, 89]. It resembles the *i mu* [*Leonurus.* See 78], but the stem is round, from two to three feet high.—Regarding the 龍牙草 *lung ya ts'ao* (dragon's tooth, which some believe to be the same as the *ma pien ts'ao*), SU SUNG says that it grows in Shi chou [in Hu pei, App. 288] and that the root is used in medicine.

LI SHI-CHEN:—The *ma pien* is a common plant in low places. It has a square stem, leaves resembling those of the *i mu* (*Leonurus*) and standing opposite. In autumn small purple flowers in spikes like those of *Plantago.* The seeds resemble those of the *p'eng hao* [*v. supra*] and are small. The root is small and white. T'AO HUNG-KING and SU SUNG describe the plant incorrectly.

Ch., XIV, 8:—*Ma pien ts'ao.* By the plant figured *Verbena* may be intended. The description in the *P.* agrees in a general way.

LOUR., *Fl. cochin.,* 33 :—*Verbena officinalis,* L. Sinice: *ma pien tsao.*

PARKER, *Canton pl.,* 266 :—*Verbena officinalis, ma pien ts'ao.* In Sz ch'uan the same plant is called *t'ie ma pien* [PARKER in litt.]. See also HENRY, *Chin. pl.,* 457.— P. SMITH, 226.

Cust. Med., p. 376 (481):—*Ma pien ts'ao* exported 1885, from Canton 481 piculs,—p. 424 (141), from Pakhoi 0.56 picul,—p. 288 (218), *t'ie ma pien* from Amoy 1.3 picul.— In the *Hank. Med.* the *ma pien ts'ao* is mentioned as exported from Hankow.

So moku, XI, 42:—馬 鞭 草 *Verbena officinalis.*—
Ibid., IX, 9:—龍 牙 草 *Agrimonia viscidula,* Bge. This
drawing agrees with that sub *lung ya ts'ao* in *Ch.,* XII, 29.
See also *Kiu huang,* LII, 17.

117.—蛇含 *She han.*   P., XVI, 59.   T., CLXVII.

Pen king:—*She han,* 蛇 衔 *she hien* (snake's bridle). The
leaves are officinal. Taste bitter. Nature slightly cold. Non-
poisonous.

Pie lu:—The *she han* grows in I chou [Yün nan, App.
102] in mountain-valleys. It is gathered in the 8th month
and dried in the shade.

T'AO HUNG-KING:—The *she hien* is a common plant.
There are two kinds, both grow on stony ground also in
loamy soil. That generally used has small leaves and yellow
flowers.

SU SUNG [11th cent.]:—It grows in Hing chou [in
Shen si, App. 66] and in other places. It grows on stony
ground, also in damp places. In Shu (Sz ch'uan) the people
cultivate this plant. Snakes dislike it. There are two kinds
of this plant, one with 5 leaves, the other with 7, on the same
footstalk. The root is gathered in the 8th month and dried
in the shade.

CH'EN JI-HUA [author of the Sung dynasty] and SU SUNG
[11th cent.]:—The stem and the leaves used in medicine.
There is one kind in Shu (Sz ch'uan) which is called 紫背
龍牙 *tsz' pei lung ya* [leaves purple underneath].

LI SHI-CHEN:—There are two kinds of this plant. That
with small leaves is the *she hien,* that with large leaves is
龍衔 *lung hien* (dragon's bridle). It is used for plasters. The
purple *she hien,* which is smaller and has the back (of the leaf)
of a purple colour is called the 小龍牙 *siao* (small) *lung ya.*

This is the *tsz' pei lung ya* of SU SUNG. [Regarding *lung ya* see 116.]

Ch., XI, 65 :—*She han.* Rude drawing, only leaves. Perhaps *Potentilla* is intended.

So moku, IX, 35 :—蛇含 *Potentilla inclinata*, Vill.

Phon zo, XIX, 7 :—蛇含 *Geum dryadoides*, S. & Z. *Ibid.*, 7, 8 :—紫背龍牙草 *Geum strictum*, Ait.

118.—女青 *nü ts'ing*. *P.*, XVI, 60. *T.*, CXXX.

Pen king :—*Nü ts'ing*, 雀瓢 *tsio p'iao* (sparrow's calabash). The root is officinal. Taste pungent. Poisonous.

Pie lu :—The *nü ts'ing* is the root of the *she hien* [see 117]. It grows in Chu yai [App. 22], is gathered in the 8th month and dried in the shade.

T'AO HUNG-KING objects to the statement of the *Pei lu* that the *nü ts'ing* is the root of the *she hien*. He considers it to be a distinct plant which has a root like the *su tuan* [see 84] and very bitter leaves. The *nü ts'ing* root is produced in King chou [Hu pei, App. 146].

SU KUNG [7th cent.] :—The *nü ts'ing* or *tsio p'iao* grows in marshes. Its leaves resemble those of the *lo mo* (*Metaplexis Stauntoni*) and are opposite. The fruit has the appearance of a calabash, whence the name. It is about the size of a jujube. The root resembles the *pai wei* [*Vincetoxicum.* See 44]. The stem and the leaves have an offensive odour. The *she hien* is a different plant.

The name *nü ts'ing* is not found in the *Ch.*, but [*Ch.*, V, 7] *tsio p'iao* is given as another name for *ti shao kua*, which is *Vincetoxicum sibiricum*, a small plant with linear lanceolate leaves and a large edible fruit.

So moku, IV, 39 :—女青 *Pæderia fœtida*, L. This may be the *nü ts'ing* described by SU KUNG.

In South China, however, *Pœderia fœtida* is known by
the name 鷄屎藤 *ki shi t'eng* (chicken's excrement creeper).
See LOUR., *Fl. cochin.*, 213, sub *Gentiana scandens*, PARKER,
Canton plants, 189. It has, according to PARKER, the same
name in Sz ch'uan. The drawing of the *ki shi t'eng* [*Ch.*,
XIX, 55] is very rude.

119.—鼠尾草 *shu wei ts'ao.* P., XVI, 62. T., CLXVIII.

Comp. *Rh ya*, 17.

Pie lu :—The *shu wei ts'ao* (rat's-tail herb) grows in
marshes. In the 4th month the leaves are gathered, and in
the 7th the flowers. Taste bitter. Nature slightly cold. Non-
poisonous.

WU P'U [3rd cent.] calls this plant 山 陵 翹 *shan ling
k'iao.*

T'AO HUNG-KING :—It is a common wild-growing plant.
The people use its juice for dyeing a black colour.

In the *Pen ts'ao shi i* [8th cent.] the synonyms 烏草
wu ts'ao (black herb) and 水菁 *shui ts'ing*. The plant has
purple flowers. The stem and the leaves are used for dyeing
black.

HAN PAO-SHENG [10th cent.]:—It grows in damp
places. In K'ien chung [Kui chow, App. 142] the people
collect the plant for medical use. The leaves resemble those
of the *hao* (*Artemisia*). In the summer four or five flower-
spikes, like those of the *ch'e ts'ien* [*Plantago*. See 115], spring
from the top of the stem. The flowers are red or white.

This is probably a *Salvia.* For further particulars see
Bot. sin., II, 17.

Cust. Med., p. 288 (215):—*Shu wei* exported 1885 from
Amoy 0.1 picul.

120.—連翹 *lien k'iao.*　*P.,* XVI, 65.　*T.,* CXXXI.

Comp. *Rh ya,* 120.

Pen king :—Lien k'iao. The Index of the *Pen king* has 翹根 *k'iao ken* (*k'iao* root). Taste of the root sweet. Nature cold and uniform. Slightly poisonous.

CHANG CHUNG-KING [2nd cent.] calls the root of this plant 連軺 *lien yao.*

Pie lu :—Other names : 三廉 *san lien,* 竹根 *chu ken* (bamboo-root). The *lien k'iao* grows in the mountain-valleys of T'ai shan [in Shan tung, App. 322]. It is gathered in the 8th month and dried in the shade.

WU P'U [3rd cent.] calls it 蘭華 *lan hua* [which is properly a name applied to several orchideous plants, see 62].

T'AO HUNG-KING :—It is a common plant. Now the stem, together with the flowers and fruit, is used in medicine.

The *Yao sing Pen ts'ao* [7th cent.] calls it 旱蓮子 *han lien tsz'.*[33]

SU KUNG [7th cent.]:—There are two kinds of this plant, the large *k'iao* and the small *k'iao.* The larger grows in low, damp places. Leaves narrow and long. Handsome yellow flowers. The fruit is a peculiarly opening capsule resembling that of the *ch'un* tree (*Cedrela*).

The small *k'iao* grows on elevated plains. It resembles the first in its leaves, flowers and capsules, but is smaller. In Shan nan [S. Shen si, App. 268] both drugs are used, but in Ch'ang an [ancient capital of China, App. 6] they use only the fruit of the larger sort and do not employ the stem and the flowers.

---

[33] In the *P.* [XVI, 63] 旱蓮草 *han lien ts'ao* is given as a synonym for 鱧腸 *li ch'ang,* which in Japan is *Eclipta. Cust. Med.,* 374 (461), *han lien ts'ao* exported 1885 from Canton 3,89 piculs. BRAUN [*Hank. Med.,* 14] identifies *han lien ts'ao* with dried lilies.

Sư Sung [11th cent.]:—It is now common in Pien king [in Ho nan, App. 248], in Ho chung [S.W. Shan si, App. 74], in Kiang ning [Nan king, App. 129] and many other places in North and Mid China. After this follows a description of the plant similar to that given by Su Kung.

The name *lien k'iao*, in China as well as in Japan, is applied to *Forsythia* and *Hypericum*. For further particulars see *Bot. sin.*, II, 120.

TATAR., *Cat.*, 34 :—*Lien k'iao*, Fructus *Forsythiæ*. P. SMITH, 98.

Cust. Med., p. 72 (95):—*Lien k'iao* exported from Hankow 2,864 piculs,—p. 32 (150), from Tien tsin 789 piculs,—p. 48 (44), from Chefoo 35 piculs.

121.—陸英 *lu ying*. P., XVI, 67. T., CXLVII.

Pie lu :—The *lu ying* grows in Hiung rh [in N.W. Ho nan, App. 69], in river-valleys, also in Yüan kü [in Shan tung, App. 415]. It is gathered in the beginning of autumn. Taste bitter. Nature cold. Non-poisonous.

Sư Kung [7th cent.] states that the *lu ying* is the same as the *so t'iao* [see the next]. But later authors [MA CHI, SU SUNG, LI SHI-CHEN] keep them apart, proving that the *so t'iao* is a different plant, and is poisonous. The description of the *lu ying* is vague. SU SUNG refers to *Rh ya*, 222. The flowers are officinal.

For identification see the next.

122.—蒴藋 *so t'iao*. P., XVI, 68. T., CXLVII.

Pie lu :—*So t'iao*, 堇草 *kin ts'ao*, 芨 *ki*. The *so t'iao* grows in fields. The leaves are gathered in spring and summer, the stem and the root in autumn and winter. Taste sour. Nature warm. Poisonous.

T'AO HUNG-KING :—It is a common plant near fields, in burial wastes and near villages.

Su KUNG [7th cent.] says, the *so t'iao* is the same as the *lu ying*, and refers to *Rh ya*, 134.

K'OU TSUNG-SHI [12th cent.]:—The *so t'iao* has white flowers. The fruit is at first green, like the *lü tou* (green bean, *Phaseolus Mungo*), arranged in clusters, several hundreds together, like a shallow cup. In the 10th month these berries become red.

LI SHI-CHEN adds :—Every leaf consists of 5 leaflets. It is also known by the name of 接骨草 *tsie ku ts'ao* (plant which joins together [broken] bones).

Ch., XI, 75 :—*Lu ying* or *so t'iao*. Representation of a plant with pinnate leaves, berry-like fruits. HENRY [*Chin. pl.*, 80] means that it is *Sambucus Chinensis*, Lindl.

There is in the *P.* [XXXVI, 74] a short account, from the *T'ang Pen ts'ao* [7th cent.], of the 接骨木 *tsie ku mu* or 續骨木 *su ku mu* [both these names mean wood (tree) which joins broken bones] also called 木蒴藋 *mu* (tree) *so t'iao*. SU KUNG says, the leaves and the flowers of the *tsie ku mu* resemble those of the *lu ying*, but the *tsie ku mu* is a tree from 10 to 20 feet high [the *lu ying* is herbaceous]. Its wood is very light and empty, it has no heart. It is much cultivated. The *tsie ku mu* has the property of healing broken bones, whence the name. [It seems a decoction of the leaves is administered internally to that effect.] The skin of the root is also employed in medicine.—CH'EN TS'ANG-K'I [8th cent.] says that the *tsie ku mu* is slightly poisonous.

The *tsie ku mu* is figured in the *Ch.* [XXXVIII, 35], only leaves, rude drawing, and again XXXV, 15. HENRY [*Chin. pl.*, 81] means that it may be *Sambucus Sieboldiana*, Bl. (*S. racemosa*, L.).

Cust. Med., p. 298 (374):—*Tsie ku ts'ao* exported 1885 from Amoy 1.45 picul.

So moku, V, 45 :—蒴 藋 *Sambucus Thunbergiana,* Bl. (shrub).

Phon zo, XCII, 23, 24 :—接 骨 木 *Sambucus racemosa,* L. (tree). *Kwa wi,* 93, *tatzu noki.*

SIEB., *Œcon.,* 239 :—*Sambucus pubescens* (a variety of *S. racemosa*), japonice : *tadsu* ; sinice : 接 骨 木 Folia adhibentur in vulneribus.

123.—藍 *lan.* *P.,* XVI, 70. *T.,* CV.

Comp. *Rh ya,* 178, *Classics,* 392.

Pen king :—*Lan.* The Index of the *Pen king* has 藍 實 *lan shi* (fruit). The fruit (seed) used in medicine. Taste bitter. Nature cold. Non-poisonous.

Pie lu :—The *lan shi* grows in Ho nei (S.E. Shan si, App. 77] in marshes. Stem and leaves used for dyeing blue.

T'AO HUNG-KING :—This is the plant now employed for dyeing a dark blue (紺 碧) colour. That with pointed leaves is the best.

SU KUNG [7th cent.] :—There are two kinds of *lan.* One has round leaves, 2 inches in diameter and 2 to 4 *fen* thick. It yields a dark blue colour. It grows in Ling nan [S. China, App. 197] and in the Court of Sacrifices[33] it is called 木 藍 子 *mu* (tree) *lan tsz'.* The plant of which T'AO HUNG-KING speaks is the 菘 | *sung* (Cabbage) *lan.* From the juice [of its leaves] the 澱 *tien* (Indigo) is prepared. The *lan* referred to in the *Pen king* are the seeds of the 蓼 | *liao* (*Polygonum*) *lan.* From this plant also Indigo is prepared.

SU SUNG [11th cent.] :—The *lan* is a common plant which is much cultivated in gardens. It grows from 2 to 3 feet high, and has leaves like the *shui liao* [Water *Polygonum.* See *Rh* ya, 65], rose-coloured flowers. Seeds also like those of the liao (*Polygonum*) but larger and black. It is used for dyeing a blue colour, but it is not fit preparing *tien* (Indigo). It is called *liao lan* [*v. supra*]. This is the plant used in medicine. Another kind is termed *mu lan* (*v.s*). It grows in Ling nan [S. China], is not officinal. The *tien* or Indigo is also prepared from the *sung lan* (*v. s*). The 馬 | *ma lan* is mentioned in the *Rh ya* [178]. In Yang chou [in Kiang su, App. 400] there is one kind of *ma lan* the leaves of which resemble those of the *k'u* (bitter) *mai* [*Lactuca* or *Sonchus*. See 257]. The people there use the root in medicine. The 吳 | *Wu lan* which is produced in Kiang ning [Nan king, App. 129] resembles the *hao* (Artemisia) and has white flowers.

For further particulars and the identification of the Chinese Indigo plants see *Bot. sin.*, II, 392. The *mu lan* is *Indigofera tinctoria*, L., and other species yielding the true Indigo.— The *sung lan* is *Isatis tinctoria*, L.—The *liao lan* is *Polygonum tinctorium.*

124.—蓼 *liao.*   *P.*, XVI, 76.   *T.*, LVIII.

Pen king :—*Liao.* The Index of the *Pen king* has 蓼實 *liao shi* (fruit, seeds). The seeds are officinal. Taste pungent. Nature warm. Non-poisonous.

Pie lu :—The *liao shi* is produced in Lei tse [App. 182], in marshes. The leaves also used in medicine.

T'AO HUNG-KING :—There are many kinds of *liao*. Three of them are used for food. The first is the 青 | *ts'ing* (green) *liao*, which is much employed by the people. Its leaves are round or pointed. The round-leaved is the best. The second is the 紫 | *tsz'* (purple) *liao*. It resembles the first but is of a purple colour. The third is the 香 | *hiang* (fragrant) *liao*. All these sorts are not very pungent in taste and are good to eat.

Han Pao-sheng [10th cent.] enumerates seven kinds of *liao*, viz. :—

1. *ts'ing liao*
2. *hiang liao* } these two have thin leaves.

3. 水 | *shui* (water) *liao*.
4. 馬 | *ma* (horse) *liao*, also called 大 | *ta* (large) *liao*.

} Both have large, broad leaves covered with black spots. The flowers of the *ta liao* are red and white. The seeds resemble Sesam-seeds, are of a brown colour, flattened and pointed. [Comp. 125.]

5. *tsz'* (purple) *liao*
6. 赤 | *ch'i* (red) *liao* } both have small, narrow leaves.

7. 木 | *mu* (tree) *liao*, also called 天 | *t'ien liao*. A creeper or twiner. Its leaves resemble the leaves of the *che* (*Cudrania triloba*). Flowers yellowish white.

The above names refer to various species of *Polygonum*. For further particulars see *Bot. sin.*, II, 366.

The 蓼 *liao* seems to be *Polygonum hydropiper*. *Ch.*, XI, 57.—*Amœn. exot.*, 891.—*So moku*, VII, 57. But according to Lour. [*Fl. cochin.*, 296], *leao xi* (*shi*) at Canton is *Pol. barbatum*, L.—Sieb., *Œcon.*, 104:—*Polygonum barbatum*, sinice 蓼 *Herba edulis*.

The 水 | *shui liao* is represented *Ch.*, XIV, 19. *So moku*, VII, 62 :—*Polygonum*, not determined. Hoffm. & Schlt., 451 (resp. Siebold), refer the above Chinese name to *Polygonum perfoliatum*, L.—Lour., *Fl. cochin.*, 295 :—*Polygonum hydropiper*. Sinice: *xuei leao* (*shui liao*). Virtus acris, stimulans, urens. Habitat in paludibus et infra ripas fluminum in Cochinchina et China, ubi ex illo formantur sepes ad olera aquatica continenda.

馬 | *ma liao*, *Ch.*, XI, 59. *So moku*, VII, 63, under this Chinese name, *Polygonum persicaria*, L.

The 香 | *hiang liao* (fragrant *Polygonum*) is perhaps the *Polygonum odoratum.* LOUR., *Fl. cochin.*, 299. The *Polygonum* figured under the above Chinese name in the *Phon zo* [XX, 3] has not been determined.

The 毛 | *mao* (hairy) *liao* is noticed by CH'EN TS'ANG-KI [8th cent.], *P.*, XVI, 80, and figured *Ch.*, XIV, 30, *Phon zo*, XX, 6 [not identified].—*Phon zo*, XX, 2 :—赤 | and 紫 | *Polygonum* [not identified].

Comp. also TATAR., *Cat.*, 34, *liao*,—P. SMITH, 175,—HENRY, *Chin.*, *pl.*, 239.

125.—葒草 *hung ts'ao.*   *P.*, XVI, 79.   *T.*, LVIII.

Comp. *Rh ya*, 102, *Classics*, 426.

Pie lu :—Other names : 石龍 *shi lung*, 天蓼 *t'ien* (heavenly) *liao*, 火 | *ta* (great) *liao* [for the last two names comp. 124]. The *hung* grows on the banks of water-courses, and resembles the *ma liao* [see 124] but is larger. In the 5th month the seeds are gathered. Taste salty. Nature slightly cold. Non-poisonous.

T'AO HUNG-KING :—It grows abundantly in damp places, and much resembles the *ma liao*, but grows very high. It is mentioned in the *Shi king* and in the *Rh ya*.

SU SUNG [11th cent.]:—The *hung* is also called 水 | *shui* (water) *hung*. It resembles the *liao* [*Polygonum*. See 124]. Large leaves, red and white [flowers]. It attains a height of 10 feet and more.

LI SHI-CHEN :—It has a coarse stem as thick as a finger, covered with hair. Its leaves are as large as those of the *shang lu* (*Phytolacca.* See 131). Pale red flowers in spikes. Seeds flat, brown, and resemble the kernels of the *suan tsao* (small *Jujube.* See 336] but are smaller. The flesh within is white, not very pungent in taste. The seeds are cooked for food.

The seeds and the flowers used in medicine. The plant is also known by the name 鴻䔮 *hung kie*.

The *hung* is the *Polygonum orientale*, L. See *Bot. sin.*, II, 426. *Ch.*, XI, 90.—HENRY, *Chin.*, *pl.*, 239.—*So moku*, VII, 76.

SIEB., *Œcon.*, 105 :—*Polygonum orientale.* Japon.: *oketade*; sinice : 莊, *Herba edulis.*

126.—虎杖 *hu chang.* *P.*, XVI, 83. *T.*, CLXIV.

Comp. *Rh ya*, 19.

The *Pie lu* gives only the name, *hu chang* (tiger's staff), and notices the medical virtues of the drug. The root is used.

T'AO HUNG-KING :—It grows abundantly in fields. It resembles the *ta ma liao* [*Polygonum.* See 124]. The stem is variegated, the leaves are round.

In the *Yao sing Pen ts'ao* [6th and 7th cent.] it is called 大蟲杖 *ta ch'ung chang*,—in the *Pen ts'ao shi i* [8th cent.] it is 苦杖 *k'u chang.*

HAN PAO-SHENG [10th cent.]:—It grows in low, damp places, like a tree, ten feet and more high. Red stem. Yellow root. The root is gathered in the 2nd and 3rd months and dried in the sun.

In the *Ji hua Pen ts'ao* [10th cent.] it is called 斑杖 *pan chang* (variegated staff).

SU SUNG [11th cent.]:—It now grows in Fen chou [in Shan si, App. 38], in Yüe chou [in Che kiang, App. 418], in Ch'u chou [in An hui, App. 25], where it is a common plant. The young plants resemble bamboo-sprouts. It is covered with red spots. Leaves like small apricot-leaves. It flowers in the 7th month, and bears seed in the 9th.

The skin of the root is black, but the root is yellow within, like the root of the willow. The plant is more than ten feet high.

According to LI SHI-CHEN the plant is also called 酸杖 *suan chang*.

Ch., XI, 91:—*Hu chang.* Rude drawing. Perhaps a *Polygonum* is intended.

So moku, VII, 78:—虎杖 *Polygonum cuspidatum,* S. & Z.

127.—萹蓄 *pien ch'u.* P̣., XVI, 85. T., CXLIV.

Comp. *Rh ya,* 54.

Pen king:—*Pien ch'u.* It seems the whole plant is used in medicine. Taste bitter. Nature uniform. Non-poisonous.

Pie lu:—The *pien ch'u* grows in Tung lai [in Shan tung, App. 373] in mountain-valleys. It is gathered in the 5th month and dried in the shade.

WU P'U [3rd cent.] calls it 扁辨 *pien pien* and 扁蔓 *pien man.*

T'AO HUNG-KING:—It is a common plant which covers the ground. White flowers between (around) the joints. Small green leaves. Its vulgar name is 扁竹 *pien chu.*

SU SUNG [11th cent.]:—It grows by roadsides, covering the ground. The young plant resembles the *k'ü mai* [*Dianthus.* See 112]. Leaves small, green, like bamboo-leaves. Red stem. Very small greenish yellow flowers. The root looks like *Artemisia* root. The leaves and the stems are gathered in the 4th or 5th month, according to others in the 2nd month. Mentioned in the *Rh ya.*

LI SHI-CHEN:—The plant has small leaves like those of the *lo chou* [*Kochia.* See 111], but they are not pointed. Slender stem which lies on the ground with the joints near

29

each other. It flowers in the 3rd month. Small red flowers
like those of the *liao lan* [*Polygonum tinctorium*. See 123].
The plant is also called 粉節草 *fen tsie ts'ao* (flour joint
plant), for the stem is covered with a [white] powder. As
the plant grows by waysides it is also called 道生草 *tao
sheng ts'ao* (way plant).

This is probably a small *Polygonum*. According to
TATAR. [*Cat.*, 3], *pien ch'u* is *Polygonum hydropiper*, L.
P. SMITH [175] means that it is *P. aviculare*, L. For
Chinese and Japanese drawings see *Bot. sin.*, II, 54.

Cust. Med., p. 332 (253):—*Pien ch'u* exported 1885
from Swatow 1.72 picul.

128.—蓋草 *tsin ts'ao*. *P.*, XVI, 86. *T.*, CVII.

Comp. *Rh ya*, 10, *Classics*, 461.

Pen king:—*Tsin ts'ao*. It seems the whole plant is
officinal. Taste bitter. Nature uniform. Non-poisonous.

Pie lu:—The *tsin ts'ao* grows in Ts'ing i [in Sz ch'uan,
App. 364] in river-valleys. It is gathered in the 9th and
10th months. The plant is fit for dyeing a gold-yellow
colour.

WU P'U [3rd cent.] calls it 黃草 *huang ts'ao* (yellow
herb) and states that it grows in the mountain-valleys of
T'ai shan [Shan tung, App. 322].

SU KUNG [7th cent.]:—The locality Ts'ing i [mentioned
in the *Pie lu*] lies west of I chou [See App. 102]. The
tsin ts'ao is a common plant in marshes and on the banks
of rivulets. The leaves resemble bamboo-leaves but are
small and tender. Stem slender and round. The people of
King and Siang [both in Hu pei, App. 146, 305], by boiling
the plant, prepare an excellent yellow dye. The vulgar name
of the plant is 菉薽 *lü ju* or | 竹 *lü chu*.

LI SHI-CHEN :—This plant is green, but it can be used for dyeing a yellow colour. The ancient dictionary *Shuo wen* notices a plant 蒚 *li* (*lei*) which dyes a sort of yellow. This is the plant under review. In the History of the Earlier Han it is stated that the feudal princes had a golden seal with a green ribbon (鑾綬). A commentator of the 4th century says that the plant which yielded this colour is called 鑾草 *li ts'ao* and grows in Lang ye and P'ing ch'ang [both in Shan tung, App. 178]. It resembles the *ai* [*Artemisia*. See 72] and is used for dyeing. This is also the *tsin ts'ao.*

For the identification of this plant see *Bot. sin.*, II, 461 (*Phalaris*). The *Phon zo* [XX, 15] figures sub 蕛草 a *Graminea.* But two other figures with the same Chinese name, on the same plate, seem to be intended for *Polygonum.*

129.—蒺藜 *tsi li.* *P.*, XVI, 86. *T.*, CXLI.

Comp. *Rh ya*, 90, *Classics*, 427.

Pen king:—Tsi li, 旁通 p'ang t'ung, 屈人 k'ü jen, 止行 chi hing, 休羽 hiu yü. The fruit (seed) is official. Taste bitter. Nature warm. Non-poisonous.

Pie lu:—The *tsi li* grows in Feng i [in Shen si, App. 40] in marshes and by roadsides. The fruit is gathered in the 7th and 8th months and dried in the sun.

T'AO HUNG-KING :—The plant grows abundantly on roads and walls. The leaves cover the ground. The fruit resembles the *ling* [*Trapa*. See 296], but is smaller. It is provided with spines. The plant is very common in Ch'ang an [the ancient capital of China in Shen si]. When the people walk it sticks to their wooden shoes. In war *tsi li* made of iron are used to defend a passage. [It seems a kind of *chevaux de frise* is meant]. The *tsi li* is mentioned in the *I king* [LEGGE's *Yi king*, p. 162 (3). He translates *tsi li* by thorns] and in the *Shi king*.

SU SUNG [11th cent.]:—It (the fruit) is gathered in winter. It is of a yellowish white colour. The plant is mentioned in the *Rh ya*. There is a peculiar kind which is called 白 | | *pai* (white) *tsi li* and which is produced in the district of 沙苑 *Sha yüan* in the prefecture of T'ung chou [in Shen si, App. 267, 378] on pasture grounds. It is also found in Mid China. It creeps over the sand, and flowers in the 7th month. The flowers are yellow and purple, and resemble pea-flowers but are smaller. In the 9th month it is in fruit. The fruit is a pod. The seeds are grayish green, of a sweet taste and a somewhat strong smell.

K'OU TSUNG-SHI [12th cent.]:—There are two kinds of *tsi li*. One is called 杜 | | *tu tsi li*. This is the common plant which covers the ground by waysides. It has small yellow flowers and a spiny fruit. The other is the *pai tsi li*, which grows in Sha yüan [*v. supra*]. It has reniform seeds of the size of millet, and is used in complaints of the kidneys.

The common *tsi li* is the *Tribulus terrestris*, L. For ancient Chinese names and further particulars see *Bot. sin.*, II, 427.

LOUR., *Fl. cochin.*, 331:—*Tribulus terrestris*. Sinice *Cie li tsu*.

TATAR., *Cat.*, 57:—*Tsi li*, *Tribulus terrestris*.—P. SMITH [221] is wrong in identifying the *pai tsi li* with *Tribulus*, for this Chinese name seems to refer to a leguminous plant.

Cust. Med., p. 80 (190):—*Tsi li* exported 1885 from Han kow 20 piculs,—p. 34 (45), *sha yüan tsi li*, from Tien tsin 45 piculs.

The *Phon zo* [XX, 16] represents sub 沙苑蒺藜 a leguminous plant. *Vicia?*

130.—大黃 *ta huang.* *P.*, XVIIIa, 1.   *T.*, CLV.

Pen king:—*Ta huang* (great yellow), 黃良 *huang liang* (yellow excellent), 將軍 *tsiang kün* (captain general). The root is officinal. Taste bitter. Nature cold. Non-poisonous.

Pie lu:—The *ta huang* grows in Ho si [west of the Yellow River. See App. 79] in mountain-valleys, also in Lung si [in Kan su, App. 216]. The root is taken up in the 2nd and 8th months and dried by fire.

Wu P'u [3rd cent,]:—Other names: 火參 *huo shen* (fire Ginseng) and 膚如 *fu ju.* The *ta huang* grows in Shu [Western Sz ch'uan, App. 292], in Pei pu [in Sz ch'uan, App. 244], also in Lung si. In the 2nd month the young leaves are rolled up and are of a yellowish red colour. The plant grows three feet and more high. The leaves are arranged four together opposite around the stem. In the 3rd month it opens its yellow flowers, in the 5th it bears black fruits (seeds). In the 8th month the root, which contains a yellow juice, is dug up, cut up in slices and dried in the shade. Emperor SHEN NUNG and LEI KUNG [the latter is said to have lived in the days of Emperor HUANG TI] considered the *ta huang* to have poisonous properties. Other ancient authors say it is not poisonous. [The *P.* classes it among the poisonous drugs.]

T'AO HUNG-KING:—The name *ta huang* refers to the yellow colour of the root, the name *tsiang kün* to the excellent and prompt effect of the drug. The drug which is now brought from I chou [Sz ch'uan, App. 102], namely from Pei pu [*v. supra*] and Wen shan [in Sz ch'uan, App. 388] and the western mountains, is not equal in quality to that from Ho si and Lung si [*v. supra*], it is darker in colour and of a very bitter and harsh taste. That from Si ch'uan [in Kan su, App. 296], which is dried in the shade, is of a

good quality. In Pei pu they dry it in the sun. That
dried by fire appears slightly charred and is not eaten by
worms.

Su Kung [7th cent.]:—This plant in its stem, leaves
and seeds resembles the *yang ti* [*Rumex*. See 193], but
its stem grows from six to seven feet high. It is easily
broken and of a sour taste. The stalks are much eaten in
a raw state. The leaves are coarse, long and thick. The
root is red and resembles that of an old *yang ti* (*Rumex*)
root. In shape it is like a bowl, two feet long. Its nature
is soft and moist, and it is easily destroyed by worms. That
dried by artificial heat is durable. It is dried by means
of heated stones on which are placed the roots cut in slices.
Being thus heated for a day, a hole is made in each piece,
through which they are strung together. Now the best
drug comes from Tang chou [in Sz ch'uan or Kan su,
App. 331], Liang chou [in Kan su, App. 189], Si Kiang
[Kukonor, App. 300], Shu [Sz ch'uan, 292]. It is also
found north of Yu [present Peking, App. 411], but this is
smaller in size, and in point of strength inferior to that
from Shu. What T'AO HUNG-KING says about the *ta huang*
from Shu being inferior to that of Lung si is incorrect.

Su Sung [11th cent.]:—The *ta huang* is now produced
in all the prefectures of Shu ch'uan [Sz ch'uan, App. 292],
Ho tung [Shan si, App. 80] and Shen si [present Shen si
and Kan su, App. 284]. But the drug from Shu, with
fine veins, is the best. Next comes that from Ts'in [in
Kan su, App. 358] and Lung [in Shen si, App. 215], but
perhaps Lung si [in Kan su, App. 216] is meant. This
is called 土番大黃 *T'u fan* (Tibetan) *ta huang*. The *ta
huang* plant begins to grow in the 1st month. Its leaves
resemble those of the *pi ma* (*Ricinus*) and are as large as a
fan. The root resembles that of the *yü* [*Colocasia*. See
261], the largest is of the size of a bowl and one or two feet

long. It flowers in the 4th month. The flowers are yellow, green and red, resembling those of buck-wheat. The stem is like a bamboo, of a green and purple colour. The root is taken up in the 2nd and 8th months and, the black skin which covers it being taken off, it is cut in slices and dried by fire. The *ta huang* from Shu is flattened like the tongue of an ox and is therefore called 牛舌大黄 *niu she* (ox tongue) *ta huang*. There is a sort of *ta huang* produced in Kiang and Huai [Kiang su and An hui, App. 124, 89] which is called 土大黄 *t'u* (native) *ta huang*. It flowers in the 2nd month and produces small fruits.

SUNG K'I [11th cent.], in his description of I chou [Sz ch'uan, App. 102], states that the *ta huang* grows abundantly in the high mountains of Shu. It has a red stem and large leaves. The root is so large that in the drug markets they use it as a pillow. The interior is beautifully veined with purple or brown.

In the days of LI SHI-CHEN the best *ta huang* was brought from Chuang lang [in Kan su, App. 27], and the author observes that this agrees with the localities noticed in the *Pie lu* as producing this drug.

Ta huang is still the common name in China for Rhubarb. The drawing of the plant in the *Ch.* [XXIV, i] is rude and incorrect. Of the species of *Rheum* which furnish this Chinese drug two are known to our botanists. Seeds of the true Rhubarb plant, procured from China by way of Kiakhta, were first received in St. Petersburg, in 1750, and distributed by the Russian government to the Horticular Societies of England, Scotland and Germany, and from that time the plant was much cultivated in Europe. LINNÆUS named it *Rheum palmatum*. It was for a long time doubted whether this was really the genuine Rhubarb, till the late General [then Captain] PRZEWALSKY, in 1872, visited the province of Kan su, where he observed the plant which yields the much-valued Kiakhta Rhubarb, so called because it was imported through Kiakhta. The plants raised in the Botanic Gardens, St. Petersburg, from the seeds he had brought home, proved to be the well-known *Rheum palmatum.* Another species from Sz ch'uan and S.E. Tibet, from which a great part of the best Chinese Rhubarb is derived, was obtained in 1867 by the French missionaries, and sent to Paris, where it was cultivated and described by BAILLON as *Rheum officinale.*

That which the natives in North China call *t'u ta huang* (native Rhubarb) is the root of *Rheum rhaponticum,* L., and the variety *compactum,* frequently seen in the Peking mountains.

P. SMITH, 185.—HENRY, *Chin. pl.,* 438.

Cust. Med., p. 70 (61) :—*Ta huang* exported 1885 from Han kow [probably Sz ch'uan Rhubarb] 5,650 piculs,— p. 58 (22), from I chang 2,123 piculs,—p. 26 (53), from Tien tsin 1,093 piculs [probably Kan su Rhubarb].

So moku, VII, 91, 92 :—漢種大黃 (Chinese Rhubarb, cultivated), *Rheum undulatum,* L.—*Ibid.,* 28 :—土大黃 *Rumex aquaticus.*

SIEB., *Œcon.,* 111 :—*Rheum tataricum.* Japonice : *Too dai woo.* Colitur in usum medicum, Chinensi vero longe inferior radice.

Ibid., 112 :—*Rheum palmatum.* Rarius cultum.

131.—商陸 *shang lu.* P., XVIIa, 8. T., CXXXI.

Comp. *Rh ya,* 112.

Pen king :—Shang lu, 夜呼 *ye hu.* The root is officinal. Taste pungent. Nature uniform. Poisonous.

In the *Kuang ya* [3rd cent.] it is called 馬尾 *ma wei* (horse's tail).

Pie lu :—The *shang lu* grows in Hien yang [in Shen si, App. 65] in mountain-valleys. It (the root) has the shape of a man and has spiritual (divine) power (神).

LEI HIAO [5th cent.]:—There are two kinds of this plant which resemble each other in their leaves. One is called 赤昌 *ch'i* (red) *ch'ang*. It is not edible, but is injurious to man. The other has white flowers [and is called the white *ch'ang*. See further on]. It is cultivated, and the *sien jen* (immortals, Taoists) use it for food.

SU KUNG [7th cent.]:—There are two sorts—the red and the white. The white is used in medicine, the red is possessed of evil power and is very poisonous.

HAN PAO-SHENG [10th cent.]:—The plant has large, thick, succulent leaves resembling in shape the tongue of an ox. That with red flowers has also a red root. The root of the white flowered is white. The root is dug up in the 2nd and 8th months and dried in the sun.

SU SUNG [11th cent.]:—The popular name of the plant is 章柳根 *chang liu ken*. It is much cultivated in gardens. It grows from three to four feet high. The leaves resemble the tongue of an ox, but are longer. Stem green and red and soft. It flowers in summer and in autumn. Purplish red flowers in clusters. The root resembles a radish, but is longer. It is mentioned in the *Rh ya*. The flowers are also used in medicine.

In the *K'ai Pao Pen ts'ao* [10th cent.] the plant is also called 當陸 *tang lu* and 白昌 *pai* (white) *ch'ang* [*v. supra*].

LI SHI-CHEN :—In ancient times the *shang lu* was cultivated as a vegetable. The sort with a white root as well as the purple sort are propagated by planting the root cut in pieces. It can also be raised from seeds. The root, as

20

well as the leaves and the stem [of the white sort] can be eaten when cooked. But that of a red and yellow colour is not edible, for it is poisonous.

Ch., XXIV, 3:—*Shang lu.* *Phytolacca.* Good drawing. See also *Kiu huang*, LI, 5, sub *chang liu ken.*

TATAR., *Cat.*, 51:—*Shang lu.* Rad. *Phytolaccæ octandræ.* —GAUGER [33] describes and figures the root of the *shang lu.*—P. SMITH, 171.

Cust. Med., p. 348 (130):—*Shang lu* exported 1885 from Canton 1 picul,—p. 280 (103), from Amoy 1.25 picul.— According to *Hank. Med.*, p. 36, exported also from Hankow.

The plant cultivated in Peking under the name of *shang lu*, and which is found also wild in the mountains, is *Phytolacca acinosa*, Roxbg.

It has the same Chinese name in Japan. See *Bot. sin.*, 112.

SIEB., *Œcon.*, 128 :—*Phytolacca octandra* (*Ph. acinosa*). Japonice: *jama goboo*; sinice : 商陸, Radix habetur venenata. Herba agricolis remedium diureticum, ac adhuc tenera inter obsonia posita.

132.—狼毒 *lang tu.* P., XVIIa, 11. T., CLXV.

Pen king:—*Lang tu* (wolf poison). The root is officinal. Taste pungent. Nature uniform. It is very poisonous. [34]

[34] The *Po wu chi* [3rd cent.] quotes a passage from the *Shen nung Pen ts'ao* in which it is stated that among medicines there are five poisonous drugs, *viz.* :—

1. The 狼毒 *lang tu.* Counter poison the *chan sz'* [see 356].
2. The 巴豆 *pa tou* [*Croton Tiglium.* See 331]. Counter poison the 藿汁 *huo chi* [the juice expressed from the leaves of the soybean].
3. 黎蘆 *li lu* [*Veratrum.* See 142]. Counter poison 湯 *t'ang* (broth).
4. The 天雄 *t'ien hiung* and the 烏頭 *wu t'ou* [*Aconite.* See 144, 146]. Counter poison the soy-bean.
5. 斑茅 *pan mao* [*Cantharides*]. Counter poison stone salt.

Pie lu:—The *lang tu* grows in Ts'in t'ing [in Kan su, App. 361] in mountain-valleys, also in Feng kao [in Shan tung, App. 41]. The root is gathered in the 2nd and 8th months and dried in the shade. That which is old and heavy and sinks in water is good.

T'AO HUNG-KING :—This drug is also produced in Tang ch'ang [in Kan su, App. 330], but it is rare there, for certain vipers are said to eat the root. That from T'ai shan [in Shan tung, App. 322] is more generally used. The drug from Han chung [S. Shen si, App. 54] and Kien p'ing [in W. Hu pei, App. 139] resembles very much the root of the *fang k'ui* [an umbelliferous plant. See 133], but the latter does not sink in water.

Su KUNG [7th cent.]:—This drug is now produced in Ts'in chou and Ch'eng chou [both in Kan su, App. 358, 18]. The plateau of Ts'in t'ing [mentioned in the *Pie lu*] lies on the border of these two prefectures. Su KUNG refutes T'AO HUNG-KING's statements regarding the *lang tu.*

MA CHI [10th cent.]:—The leaves of the *lang tu* resemble those of the *shang lu* and the *ta huang* [*Phytolacca* and *Rhubarb.* See 131, 130]. Leaves and stem covered with hair. The skin of the root is yellow, the flesh is white. The drug of a good quality must be succulent and heavy. Ts'in t'ing [*v. supra*] lies in Lung si [in Kan su, App. 216]; Feng kao [likewise mentioned in the *Pie lu*] is a district at the foot of the T'ai shan mountain [in Shan tung]. There are six drugs which are called the 六陳 *liu ch'en* or six old drugs, *viz.* the *lang tu,* the *ma huang* [*Ephedra.* See 97], the *kü pi* [orange-peel. See 281], the *pan hia* [*Pinellia tuberifera.* See 150], the *chi shi* [fruit of *Citrus trifoliata.* See 334] and the *Wu chu yü* [*Boymia.* See 291].

Su SUNG [11th cent.]:—The *lang tu* is now found in all prefectures of Shen si [modern Shen si and Eastern Kan su], also in Liao chou and Shi chou [both in Shan si, App. 190, 286]. MA CHI's description is correct.

LI SHI-CHEN:—The *lang tu* is produced in [ancient] Ts'in and Tsin [Kan su, Shen si and Shan si, App. 358, 353]. The people frequently confound this drug with the *lü ju* [*Euphorbia*. See 135].

Ch., XXIV, 6:—*Lang tu*. Figure of a plant with a large root. *Solanea?* P. SMITH, 232:—*Lang tu*, Wolf's bane.

Cust. Med., p. 344 (74):—*Lang tu t'ou* exported 1885 from Canton 2.29 piculs.

Phon zo, XXI, 7, 8:—狼毒. The drawing is perhaps intended for *Mandragora*.

133.—防葵 *fang k'ui*. P., XVIIa, 13. T., LXXXV.

Pen king:—*Fang k'ui*, 梨蓋 *li kai*. The root is officinal. Taste pungent. Nature cold. Non-poisonous.

Pie lu:—Other name 房苑 *fang yuan*. The *fang kui* grows in Lin tsz' [in Shan tung, App. 194] in river-valleys. It is likewise found in Sung kao [in Ho nan, App. 317], in T'ai shan [in Shan tung, App. 322] and in Shao shi [see App. 281]. The root is dug up on the 3rd day [probably of the 3rd month] and dried in the sun.

WU P'U [3rd cent.]:—Other names 利茹 *li ju*, 爵離 *tsio li*, 方蓋 *fang kai*, 農果 *nung kuo*. The stem and the leaves resemble the *k'ui* [*Malva*. See 105]. It is dark yellow in its upper part. The root is of the size of the *kie keng* root [*Platycodon*. See 6]. The flesh is of a reddish white colour. It flowers in the 6th month, white flowers, and bears a white fruit in the 8th month. The root is dug up in the 3rd month.

CHEN KUAN [6th cent.] says that the root is slightly poisonous. On account of this statement, probably, LI SHI-

CHEN classes the *fang k'ui* with the poisonous drugs. The other ancient authors consider it non-poisonous.

SU KUNG [7th cent.]:—The root and the leaves resemble those of the *k'ui* (*Malva*). Flowers, seeds and root are fragrant. The odour [or taste] resembles that of the *fang feng* [*Siler* or *Peucedanum*. See 31], hence the name *fang k'ui*. The plant is found east of the Wang ch'u shan mountain, which lies in the prefecture of Siang yang [in Hu pei. App. 380, 306], but sparsely. It grows also in Hing chou [in Shen si, App. 66] and westward and southward in the country of Shu [Sz ch'uan]. The drug from Hing chou is the best.

SU SUNG [11th cent.]:—Now this plant is found in Siang yang. The author knows nothing about its growing in the other localities [mentioned by SU KUNG]. The leaves of the *fang k'ui* resemble mallow-leaves. Three leaves are on the same stalk (petiole). A number of branches spring from the same point. Between them a stalk pushes upward which at the top bears flowers in the mode of the onion and the *king t'ien* [*Sedum*. See 205]. These flowers are white and open in the 6th month. Afterwards the fruit is produced. The root resembles that of the *fang feng* [*v. supra*] in its fragrance and taste. When it is dug up in the proper season it sinks in water. Only the rotten root floats on the surface. T'AO HUNG-KING's statement regarding the resemblance of the *fang k'ui* root to the *lang tu* [see 132] is wrong.

LI SHI-CHEN agrees with SU SUNG. In the time of the T'ang dynasty the *fang k'ui* was an article of tribute brought from Lung si and Ch'eng chou [both in Kan su, App. 216, 18].

Ch., VII, 34 :—*Fang k'ui*. Rude drawing. Perhaps an umbelliferous plant is intended. The description of the plant there seems to refer to an umbelliferous plant.

So moku, V, 13 :—防葵, *Peucedanum japonicum*, Thbg.
Comp. also *supra*, 31, sub *fang feng*.

SIEB., *Œcon.*, 250 :—*Peucedanum japonicum*. Thbg.
Japonice : *booki ;* sinice : 防葵. Herba tenera edulis.

134.—狼牙 *lang ya*. *P.*, XVIIa, 15. *T.*, CLXV.

Pen king :—*Lang ya* (wolf's tooth). 牙子 *ya tsz'*. The
root is officinal. Taste bitter. Nature cold. Poisonous.
[Other ancient authors say non-poisonous.]

Pie lu :—Other names : 狼齒 *lang ch'i* (wolf's tooth),
狼子 *lang tsz'*. The *lang ya* grows in Huai nan [An hui,
Kiang su, App. 90] in river-valleys, also in Yüan kü [in
Shan tung, App. 415]. The root is dug up in the 8th month
and dried in the sun. When moist and rotten internally and
mouldy it is a deadly poison.

Wu P'u [3rd cent.]:—Other names : 火牙 *ta ya* (great
tooth), 抱牙 *pao ya*. The root is yellowish red. The plant
flowers in the 6th or 7th month, in the 8th it produces black
fruit [or seed]. Root dug up in the 1st and 8th months.

LI TANG-CHI [3rd cent.] calls it 支蘭 *chi lan*.

T'AO HUNG-KING :—It (the root) resembles the tooth of
an animal, hence the above names.

HAN PAO-SHENG [10th cent.]:—The leaves of this plant
resemble those of the *she mei* [*Fragaria*. See 167] but are
thicker, larger, and dark green. The root is black and has
the shape of an animal's tooth.

SU SUNG [11th cent.]:—This plant is common in all the
prefectures of Kiang tung [Kiang su, etc. App. 124] and
Pien tung [in Ho nan, App. 250].

Ch., XXIV, 7 :—*Lang ya*. Rude drawing representing
a plant with a large root.

So moku, IX, 36:—狼 牙, *Potentilla cryptotænia*, Maxim.

135.—藺 茹 *lü ju.*　P., XVIIa, 16.　T., CXLV.

Pen king:—Lü ju. The root is officinal. Taste pungent. Nature cold. Slightly poisonous.

Pie lu:—Other name 離 蔞 *li lou.* The *lü ju* grows in Tai [in Shan si, App. 321] in river-valleys. The root is dug up in the 5th month and dried in the shade. That with a black head is the best.

Wu P‘u [3rd cent.]:—It is an herbaceous plant from 4 to 5 feet high. Round, yellow leaves standing four together and opposite. It flowers in the 4th month and bears black fruit in the 5th month. Root yellow and contains a yellow sap. The leaves and the stem are gathered in the 3rd month, the root is dug up in the 5th.

T‘ao Hung-king:—Now the best sort comes from Kao li [Corea, App. 116]. It is of a yellow colour. When broken it discharges a yellow sap which after hardening becomes black like varnish, whence it is called 漆 頭 *ts‘i t‘ou* (varnish head). An inferior sort is produced in Mid China. This is called 草 ｜ ｜ *ts‘ao* (herbaceous) *lü ju.* It is white, but by heating it on iron the head becomes black.

Su Sung [11th cent.]:—Now this plant grows also in Ho yang [in Ho nan, App. 81], in Tsz‘ chou and Ts‘i chou [both in Shan tung, App. 367, 348]. Leaves like those of the *ta ki* [*Euphorbia*. See 136]. Yellow flowers. The root resembles a radish, its skin is reddish yellow, the flesh white. When broken it discharges a sap which in hardening becomes black like varnish. Flowers, pale red or yellow, appear in the 3rd month. It does not bear fruit.

LI SHI-CHEN :—The name was originally written 藘茹 *lü ju.* Another name is 掘据 *küe kü,* also pronounced *kie kü.* An ancient work says that the *lü ju* is produced in Wu tu [in Kan su, App. 395]. The yellow is the best. The *ts'ao lü ju* grows in Kien k'ang [Nan king, App. 137]. It is white. LI SHI-CHEN says it is a common mountain plant, it grows from 2 to 3 feet high and has a large long root like a radish or a rape, sometimes forked, with a yellowish red skin and white flesh, containing a yellow sap. Stem and leaves resemble those of the *ta ki* [*v. supra*], but the leaves are longer and broader and not very pointed. When broken they discharge a white sap. There are shorter stems clasping leaves (floral leaves) standing opposite. From the midst of the leaves spring dichotomous or trichotomous small branches (umbels) which in the 3rd month bear small purple flowers. The fruit is of the size of a pea. It consists of three parts united into one body, is at first green and becomes black when ripe. The kernels within are white like the seeds of the *su sui tsz'* [*Euphorbia lathyris. So moku,* IX, 23]. The root of the *lü ju* is frequently confounded with that of the *lang tu* [see 132]. But the latter has leaves resembling those of the *shang lu* (*Phytolacca*) and Rhubarb, and the root is not replete with sap.

Ch., XXIV, 12 :—*Lü ju.* The drawing undoubtedly represents an *Euphorbia* with a large root. The description in the *P.* agrees.

So moku, IX, 11 :—草藘茹, *Euphorbia palustris,* L.— *Ibid.,* 12 :—漆頭藘茹, *Euphorbia adenochlora,* Morr. & Den.

In the *Phon zo* [XXI, 11, 12] we have 藘茹, *Euphorbia Sieboldiana,* Morr. & Den.,—12, 13 :—草藘茹, the same species [this is also depicted in SIEB., *Icon. ined.,* VII, with a peculiar root],—13 :—白藘茹. Not identified by FRANCHET.

136.—大戟 *ta ki.*   *P.,* XVIIa, 17.   *T.,* CLXII.

Pen king:—*Ta ki* (great lance). The root is officinal. Taste bitter. Nature cold. Slightly poisonous. [Other ancient authors say "very poisonous."]

Comp. *Rh ya,* 186.

Pie lu:—The *ta ki* grows in Ch'ang shan [in Chi li, App. 8]. The root is dug up in the 12th month and dried in the shade.

HAN PAO-SHENG [10th cent.]:—The plant in its leaves resembles the *kan sui* [*Euphorbia* or *Passerina.* See 138], but it grows higher. The leaves contain a white juice. Yellow flowers. The root resembles a small *k'u shen* root [*Sophora angustifolia.* See 34]. Its skin is yellow, the flesh yellowish white. The leaves and the stem are gathered in the 2nd month, the root is dug up in the 8th month.

SU SUNG [11th cent.]:—It is a common plant in Mid China. The sprouts which appear in spring are of a red colour. The plant grows to the height of one foot. Leaves like young willow-leaves but smaller and crowded. Yellowish purple flowers in the 3rd or 4th month, agglomerated, resembling apricot flowers or the *wu i* [see *Rh ya,* 57]. There are several sorts.

LI SHI-CHEN:—The *ta ki* is a common plant in marshes. It has an upright stem, from two to three feet high, hollow inside. When broken it discharges a white juice. Leaves long and narrow like willow-leaves, the upper leaves in a whorl. The purple *ta ki* of Hang chou [in Che kiang, App. 58] is the best. The 土 | | *t'u* (native) *ta ki* of Kiang nan is inferior in quality. There is in the northern regions a sort called 綿大戟 *mien* (floss-like) *ta ki,* which is of a white colour. The skin of the root is soft like floss.

31

It is very dangerous, and injurious to life. The root of the
ta ki is of a bitter, acrid taste, causes scratching in the
throat.   A popular name for it is 下 馬 仙 *hia ma sien*.

Ch., XXIV, 13 :—*Ta ki.*   An *Euphorbia* figured.

In the *Cust. Med.* the drug *ta ki* is noticed as imported
into several ports,—p. 342 (55), *hung ya ta ki* (*ta ki* with
red sprouts) exported 1885 from Canton 6 piculs.

So moku, IX, 17-20 :— 大 戟 *Euphorbia lasiocaula*,
Boiss.—*Kwa wi*, 40.

137.—澤漆 *tse ts'i*.   P., XVIIa, 20.   T., CXL.

Pen king :—*Tse ts'i* (marsh varnish), 漆 莖 *ts'i heng*
(varnish stalk).   The stem and the leaves are officinal.
Taste bitter.   Nature cold.   Non-poisonous.   [Later authors
say "slightly poisonous."]

Pie lu :—*Tse ts'i* is a name applied to the stem and the
leaves of the *ta ki* [see 136].   The plant grows in T'ai shan
[in Shan tung, App. 322]. in marshes.   Its stem and leaves
are gathered on the 3rd day of the 3rd month or on the
7th day of the 7th month and dried in the shade.

T'AO HUNG-KING :—*Tse ts'i* is a name for the leaves and
the stem of the plant *ta ki*.   The fresh plant is of a purple
colour and replete with a white, acrid juice.

TA MING [10th cent.]:—*Tse ts'i* consists of the flowers of
the *ta ki*.   It grows in marshes.   Small yellow flowers.   The
young leaves eaten as a vegetable.

SU SUNG [11th cent.]:—This plant grows in Ki chou
[in Chi li, App. 119], in Ting chou [in Hu nan, App. 341],
in Ming chou [in Che kiang, App. 224] and other localities
in Mid China.

LI SHI-CHEN :—All the above-quoted authors are wrong
in stating that the names *tse ts'i* and *ta ki* refer to the
same plant.   The leaves of the *ta ki* are not edible.   In the

T'u su Pen ts'ao [a Taoist Materia Medica] and in other works the *tse ts'i* is called 猫兒眼睛草 *mao rh yen tsing ts'ao* (cat's pupil [iris] herb), also 綠葉綠花草 *lü ye lü hua ts'ao* (herb with green leaves and green flowers) and 五鳳草 *wu feng ts'ao*. It is a common plant in the plains and marshes of Kiang and Hu [Mid China, App. 124, 83]. The leaves (floral leaves) are round and yellow, resembling the pupil of a cat's eye. Flower-stalks five-branched. Small green flowers. The whole plant contains a white viscid juice. The root is of a white colour and hard like a bone. The *ta ki* root is not the same as some have asserted. In the 5th month the juice is collected and prepared for medical use. This preparation is called *tse ts'i* (marsh varnish).

Ch., XXIV, 15 :—*Tse ts'i*. The plant figured seems to be *Euphorbia helioscopia*, L. But the *tse ts'i* figured in the *Kiu huang* [XLVI, 19] is another plant, which is stated there to be used as a vegetable.

TATAR., *Cat.*, 38 :—猫眼草 *mao yen ts'ao*. Caules et folia *Euphorbiæ lunulatæ*, Bge., and [57] *tse ts'i*, *Leguminosa?* —P. SMITH, 95.

Amœn. exot., 896:—澤漆 *takusitzu*, vulgo *totaigusa*. Esula vulgaris minor. Tithymalus arvensis latifolius Germanicus C. Bauh. P.—KÆMPFER means *Euphorbia platyphyllos*, L.

So moku, IX, 16 :—澤漆 *Euphorbia helioscopia*, L.

138.—甘遂 *kan sui*. P., XVIIa, 22. T., CLX.

Pen king:—*Kan sui*. The root is officinal. Taste bitter. Nature cold. Poisonous.

Pie lu:—Other names : 甘藁 *kan kao*, 陵澤 *ling tse*, 重澤 *chung tse*, 主田 *chu t'ien*. The *kan sui* grows in the river-valleys of Chung shan [in Chi li, App. 31]. The root is gathered in the 2nd month and dried in the shade.

Wu P'u [3rd cent.] says it is gathered in the 8th month, and gives the following synonyms: 陵藁 *ling kao*, 甘澤 *kan tse*, 白澤 *pai tse*, 鬼醜 *kui ch'ou*, 苦澤 *k'u tse*.

T'AO HUNG-KING explains that Chung shan lies in Tai kün [in Shan si. Comp. App. 321]. The best drug is brought from T'ai shan [in Shan tung, App. 322] and Kiang tung [Kiang su, etc., App. 124]. That with a red skin is of a good quality. The white-skinned is inferior in quality. It is also called 草 | | *ts'ao* (herbaceous) *kan sui*.

Su KUNG [7th cent.]:—The *kan sui* in its stem and leaves resembles the *tse ts'i* [137]. The root has a red skin and white flesh. It forms tubers. The *ts'ao kan sui* [*v. supra*] is a quite different plant and the same as the *tsao hiu* [*Paris.* See 151], which is commonly called *ch'ung t'ai* and which has leaves resembling those of the *kui kiu* [see 152] and the *p'i ma* (*Ricinus*) and its root has a white skin.

Su SUNG [11th cent.]:—This plant (the *kan sui*) now grows in Shen si [App. 284] and Kiang tung [Kiang su, etc., App. 124] and resembles the *tse ts'i* [137], but the stem is shorter, more slender, the leaves contain a juice, the root has a red skin and a white flesh forms tubers of the size of the end of a finger.

Ch., XXIV, 31:—*Kan sui.* HENRY [*Chin. pl.*, 282] means that the figure is intended for a *Wickstrœmia* (order *Thymelaceœ*).

TATAR., *Cat.*, 25:—*Kan sui.* Radix *Passerinœ?* (*Thymelaceœ*).—GAUGER [22] describes and figures the *kan sui* root. Cylindrical or elliptical tubers which smell like ginger. —P. SMITH, 168.

Cust. Med., p. 68 (33):—*Kan sui* exported 1885 from Han kow 3.30 piculs,—p. 278 (54), from Amoy 0.07 picul.

So moku, IX, 13:—甘遂 *Euphorbia Sieboldii*, Morr. & Dcn. [*E. corraloides*, Thbg., *Fl. jap.*, 197].

139.—莨菪 *lang tang.*   *P.*, XVIIa, 26.   *T.*, CXLVI.

Pen king:—*Lang tang,* 横唐 *heng t'ang.* The seeds and the root are officinal. Taste of the seeds bitter. Nature cold. Non-poisonous. Taste of the root bitter and acrid. Poisonous.

Pie lu:—The *lang tang* grows in the river-valleys of Hai pin [in Chi li, App. 49], also in Yung chou [in Shen si, App. 424]. The seeds are gathered in the 5th month.

T'AO HUNG-KING:—It is a common plant. The seeds somewhat resemble the kernels of the *wu wei* [*Schizandra.* See 164], but are very small.

HAN PAO-SHENG [10th cent.]:—Its leaves resemble those of the *sung lan* [*Isatis.* See 123]. The whole plant is covered with fine hairs. White flowers. The covering of the seed (fruit) looks like a jar [perhaps a persistent calyx]. The seeds are small, flattened, as large as a millet-grain, of a greenish yellow colour. The seeds are gathered in the 6th and 7th months and dried in the sun.

SU SUNG [11th cent.]:—This plant is also called 天仙子 *t'ien sien tsz'.* It is common and grows from two to three feet high. Leaves like those of the *ti huang* [*Rehmannia.* See 100], the *wang pu liu hing* [*Silene,* also *Physalis.* See 113] etc., as broad as three fingers. Its purple flowers open in the 4th month, in the 5th the fruit is formed. The covering of the fruit is like a jar, the fruit is like a small pomegranate, it is a capsule and contains very small greenish white seeds, like millet. The plant is covered with white hairs.

LI SHI-CHEN:—The name of the plant is also written 蔏蔏 *lang tang.* It is also called 行唐 *hing t'ang.* The seeds when eaten cause one to become mad.

Ch., XXIV, 49:—*Lang tang.* Rude drawing which does not permit of identification.

So moku, III, 18 :—莨菪 *Scopolia japonica,* Maxim.
Order *Solanaceæ.* Fruit unknown. This plant hitherto not
observed in China. The *lang tang* of the Chinese authors
may perhaps be *Hyoscyamus niger*[35] or *H. physaloides,* both
common plants in North China. The calyx enlarges as the
fruit ripens.

140.—雲實 *yün shi.* P., XVIIa, 30. T., CXXXVI.

Pen king :—*Yün shi* (cloud fruit). The fruit (seeds)
and the flowers are officinal. Taste of the fruit pungent.
Nature warm. Non-poisonous. [In the *P.,* however, this
plant is classed with the poisonous plants.]

Pie lu :—Other names : 員實 *yüan shi,* 雲英 *yün ying.*
The *yün shi* grows in Ho kien [in Chi li, App. 75] in river-
valleys. It is gathered in the 10th month and dried in the
sun.

Wu P'u [3rd cent.]:—Other name : 天豆 *t'ien tou*
(heavenly bean). The plant grows from four to five feet
high, stem hollow inside, large leaves like hemp-leaves
standing in pairs opposite. It flowers in the 6th month, and
bears fruit in the 9th. Fruit gathered in the 10th month.

T'ao Hung-king :—It is a common plant. The seeds are
small and black, like those of the *t'ing li* [*Sisymbrium.* See
114]. The fruit resembles that of the *lang tang* [see 139].

Su Kung [7th cent.]:—The *yün shi* is of the size of the
shu (*Panicum*) and hemp-seed, of a yellowish black colour.
It resembles also a bean, whence the name *t'ien tou* (heavenly
bean). It grows on the borders of marshes, from five to six
feet high. The leaves are like small *huai* (*Sophora*) leaves
or like those of the *mu su* [*Medicago.* See 255]. Spines
in the axils of the twigs. Popular name 草雲毋 *ts'ao yün*

---

[35] *Lang-tang* cultivated in a mountain garden in Hupeh proved to be
Hyoscyamus niger, L.—A. Henry.

mu. T‘AO HUNG-KING is wrong in saying that it resembles the *t‘ing li* [SU KUNG is himself wrong in likening the seeds to millet or hemp seed].

HAN PAO-SHENG [10th cent.]:—Similar description of the plant as above given. He adds: Yellowish white flowers. The fruit is a pod. Seed greenish yellow, and resembles hemp-seed.

SU SUNG [11th cent.]:—Its leaves are like those of the *huai* (*Sophora*) but narrower and longer. The branches are spiny. The stem with the leaves are also called 臭草 *ch‘ou ts‘ao* (stinking plant), also 羊石子草 *yang shi tsz‘ ts‘ao* [probably meaning sheep's-dung plant]. The fruit [or seed] is 馬豆 *ma tou* (horse bean).

LI SHI-CHEN :—This plant is common in the mountains. Popular name 粘剌 *nien ts‘z‘* (viscid spines). The stem is red, hollow inside, scandent and prickly. Leaves like *Sophora* leaves. Its flowers are yellow, in racemes, and open in the 3rd month. The pod is three inches long, and resembles that of the *fei tsao* [*Cæsalpinia*. See 325]. It contains five or six seeds of the size of the *ts‘io tou* (magpie bean), slightly pointed at both ends, very hard, with a thick, dark coloured, variegated skin, white flesh and of an unpleasant odour. The root is also used in medicine.

Ch., XXIV, 17 :—*Yün shi*. The figure seems to represent a *Cæsalpinia*. According to HENRY [*Chin. pl.*, 501], *Cæsalpinia sepiaria*, Roxb.

Phon zo, XXI, 22, 23 :—雲實 *Cæsalpinia sepiaria*, Roxb. (*C. japonica*, S. & Z.). This climbing shrub is found in Japan as well as in Central China. LI SHI-CHEN'S description of the *yün shi* agrees well.

141.—常山 *ch‘ang shan*. *P.*, XVIIa, 36. *T.*, CXXXVIII.

Pen king:—*Ch‘ang shan*, 蜀漆 *Shu ts‘i* (Sz ch‘uan varnish), 互草 *hu ts‘ao*. The root is official. Taste bitter. **Nature cold. Poisonous.**

Pie lu:—The *ch'ang shan* grows in the river-valleys of I chou [Yün nan or Sz ch'uan, App. 102], also in Han chung [S. Shen si, App. 54]. The root is gathered in the 2nd and 8th months and dried in the shade. The *Shu ts'i* grows in the valleys of the Kiang lin mountains [App. 127]. The above name in Shu Han [Sz ch'uan, App. 293] is applied to the stem and the leaves of the *ch'ang shan.*

T'AO HUNG-KING:—The *ch'ang shan* is produced in I tu and in Kien p'ing [both in Hu pei, App. 104, 139]. That with a small fruit and yellow is the best. It is called 雞骨常山 *ki ku* (chicken's bones) *ch'ang shan.* *Shu ts'i* is the name for the stem and the leaves of the *ch'ang shan* [according to the *Pie lu*]. The Kiang lin mountains [*v. supra*] are the same as the Kiang yang mountains in I chou [Sz ch'uan, App. 130].

SU KUNG [7th cent.]:—The *ch'ang shan* grows in mountain-valleys. It has a round stem with joints, not higher than three or four feet. Leaves like *ming* (tea) leaves, but longer, narrower, standing opposite, and in pairs. It flowers in the 2nd month. White flowers, green in the centre. In the 5th month it bears fruit, green, round capsules with three seeds. This plant when dried in the sun keeps a pale green colour. It is much used. When dried in the shade it becomes black and is easily spoiled.

SU SUNG [11th cent.]:—It is a common plant in Mid China and has been correctly described by [the above mentioned] previous authors. There is one sort produced in Hai chou [in Kiang su, App. 48] which has leaves like those of the *tsiao* [*Zanthoxylon.* See 280], reddish white flowers in the 8th month, green [or blue] fruits resembling the *shan lien tsz'* [*Melia.* See 321], but smaller. Another kind, which grows on the T'ien t'ai shan mountain [in Che kiang, App. 340], is called 土常山 *t'u* (native) *ch'ang shan.* The leaves, which are very sweet, are used for preparing a sweet beverage.

LI SHI-CHEN observes that *ch'ang shan* is properly the name of a mountain which is also called Heng shan, *ch'ang* and *heng* having the same meaning (perpetual). It was also the name of a prefecture in Chi li [see App. 8] where this drug is produced.

Ch., XXIV, 10 :—*Ch'ang shan.* Rude drawing.— *Ch.*, X, 7, 8, 9, sub *t'u ch'ang shan*, three drawings, one of them [8] seems to represent a *Hydrangea.*

LOUR., *Fl. cochin.*, 369 :—*Dichroa febrifuga* (order *Hydrangeæ*). Sinice : *cham chan* (*ch'ang shan*). Frutex arboreus. Corolla exterius alba, intus cœrulea, sicut etiam stamina. Virtus foliorum et radicis febrifuga.

TATAR., *Cat.*, 14 :—*Ch'ang shan.* Radix *Lysimachiæ.*— P. SMITH [141] says that TATARINOV's identification is doubtful. The drug, shoots and coarse roots are used in the treatment of ague.

Cust. Med., p. 66 (2) :—*Ch'ang shan* [root] exported 1885 from Han kow 450 piculs,—p. 184 (3), from Ning po 17 piculs,—p. 210 (1), from Wen chow 2 piculs,—p. 356 (223), from Canton *ch'ang shan* leaves 8.67 piculs.

HOFFM. & SCHLT., 126 :—*Celastrus orixa*, S. & Z. (*Orixa japonica*, Thbg.). Nom chinois de la racine 常山, nom des feuilles 蜀漆.

Comp. the drawing under the same Chinese names, *Phon zo*, XXII, 2. Not identified by FRANCHET. *Ibid.*, 3 :—土常山, a *Hydrangea.*

142 :—藜蘆 *li lu.* *P.*, XVIIa, 41. *T.*, CXLVI.

Pen king:—*Li lu.* The root is officinal. Taste acrid. Nature cold. Poisonous.

Pie lu :—Other names : 山 葱 *shan ts'ung* (mountain onion), 葱 苒 *ts'ung jan,* 葱 葵 *ts'ung t'an.* The *li lu* grows in T'ai shan [in Shan tung, App. 322] in mountain-valleys. The root is dug up in the 3rd month.

Wu P'u [3rd cent.]:—The plant has large leaves and a small root. Li Tang-chi [3rd cent.] says it is very poisonous. It is also called 葱 葵 *ts'ung k'ui,* 豐 盧 *li lu.*

T'ao Hung-king :—It is a common plant in Mid China. The root in its lower part resembles very much the root of the *ts'ung (Allium fistulosum).* It is covered with hairs (radical fibres) which are scraped off before use, and then the root is slightly roasted.

Su Sung [11th cent.] :—It is common in Shen si [App. 284] and in North and Mid China. The best sort is produced in Liao chou, Kie chou [both in Shan si, App. 190, 135] and in Kün chou [in Hu pei, App. 172]. The plant first begins to grow in the 3rd month and the leaves then resemble the opening heart of the *tsung* [*Chamærops.* The author seems to have in view the spathe which incloses the flowers as a sheath]. The [developed] leaf resembles that of the *ch'e ts'ien* [*Plantago.* See 115], the stem that of the onion. The stem is of a pale green colour tinged with purple, from five to six inches high. There is a black skin (sheath) like a palm spathe which envelops the stem. Flesh-coloured flowers. The root resembles the *ma ch'ang ken* [horse's bowels root. *P.,* XVIIa, 43. Unknown to me]. It is from four to five inches and more long, of a yellowish white colour. It is dug up in the 2nd and 3rd months and dried in the shade. There are two kinds of *li lu.* One is called 水 ‖ *shui* (water) *li lu.* It grows on stones near water-courses. The root has many rootlets. More than a hundred stems (?) It is not used in medicine. The other sort, which is officinal, is called 葱 白 ‖ *ts'ung pai li lu* [Onion *li lu.* Comp. regarding *ts'ung pai,* 241].

The root has but few rootlets. From twenty to thirty stems (?) That growing on elevated mountains is the best. In Kün chou [v. supra] it is called 鹿葱 lu ts'ung (deer onion).

Li Shi-chen says that the name li lu means "black stem" and refers to the black sheath which envelops the stem. In North China it is also called 憨葱 han ts'ung (silly onion, i.e. which causes insanity), in the south they call it lu ts'ung [v. supra].

In the Cust. Med. the li lu is mentioned as a drug imported to Shanghai and Canton [p. 142, 344], and exported [p. 294 (313)] only from Amoy in a small quantity. According to the Hank. Med. [24] the li lu is an article of export in Hankow.

Ch., XXIV, 8 :— Li lu. The drawing seems to be intended for Veratrum (order Liliaceæ).

Tatar., Cat., 35 :— Li lu. Folia et Radix Veratri nigri.—P. Smith 226.

Veratrum nigrum, L., is common in the Peking mountains and known there under the name of li lu. The descriptions in the P. agree in a general way.

In Loudon's Encycl. of plants it is stated :— Veratrum is said by Lemery to be so called because its root is vere-atrum (truly black).

That which Kaempfer [Amœn. exot., 785] describes under the Chinese name 藜蘆, japonice : kiro, rirjo, vulgo omotto, comp. also Kaempf., Icon. sel., 12,—is Rhodea japonica, Roth., (Liliaceæ). In the So moku [VII, 17], however, this plant is figured under the Chinese name 萬年青, and the figure agrees with that under the same Chinese name in the Ch. [XV, 24].

The drawings sub 藜蘆 in the So moku [XX, 64] and Phon zo [XXII, 6-8] and Kwa wi [24] represent Veratrum nigrum and album.

143.—附子 *fu tsz'*.   *P.*, XVIIa, 44.   *T.*, CXXVII.

Pen king:—*Fu tsz'*. The root is officinal. Taste acrid.
Nature warm. Very poisonous.

Pie lu:—The *fu tsz'* is produced in Kien wei [in Sz
ch'uan, App. 140] in mountain-valleys, also in Kuang Han
[in Sz ch'uan, App. 161]. That root which is dug up in
the winter months is called *fu tsz'*, that taken up in the
spring is 烏頭 *wu t'ou* (crow's head). It is of a sweetish
taste.

T'AO HUNG-KING explains that *fu tsz'* and *wu t'ou* are
names applied to the root of the same plant. That taken up
in the 8th month is called *fu tsz'*, and that with eight horns is
the best. The root dug up in spring, when the stem begins to
rise up, is called *wu t'ou*, from its resembling a crow's head in
shape. It shows two protuberances (or branches). That with
a pedicle like an ox-horn [perhaps he means the tail into
which the root tapers] is called 烏喙 *wu hui* (crow's beak).
The inspissated juice is called 射罔 *she wang*. The 天雄 *t'ien
hiung* [comp. 144] resembles the *fu tsz'* but is more slender,
from three to four inches long. The 側子 *tse tsz'* [comp.
145] is a large lateral horn of the *fu tsz'*. All these names
refer to the root of the same plant. The *Pen king* [he means
the *Pie lu*], however, considers them to be applied to different
plants, each of them growing in a different locality [see
further on, 144-146].

SU KUNG [7th cent.]:—The drugs *t'ien hiung, fu tsz'* and
wu t'ou all come from the province of Shu [Sz ch'uan], the
best sorts from Mien chou and Lung chou [both in Sz ch'uan,
App. 221, 210]. The drugs produced in Kiang nan [Fu
kien, Kiang su, Che kiang, App. 124] are not much used.

TA MING [10th cent.]:—The *t'ien hiung* is of a large size
and long. It has but few pointed horns, and is of a solid
structure,—the *fu tsz'* is large and short, solid, its horns are

rounded.—The *wu hui* resembles the *t'ien hiung*.—The *wu t'ou* stands near to the *fu tsz'*.—The *tse tsz'* is smaller than the *wu t'ou*. That drug which consists of agglomerated masses is called 虎掌 *hu chang* [tiger's paw. This name is properly applied to an *Arisœma*. See 148]. All these drugs are various forms of the root produced by the same plant.

LI SHI-CHEN explains that *wu t'ou* is the mother of the *fu tsz'* [*fu* properly means an appendix. Here we have to understand "younger tubers appended to the old root"]. There are two kinds of *wu t'ou*. That which grows in Chang ming [in Sz ch'uan, Lung an fu, Chang ming hien] is commonly called 川烏頭 *Ch'uan* (Sz ch'uan) *wu t'ou*. The root which is dug up in the spring, and which then has not yet produced the small lateral tuber (子), is called *wu t'ou*: that taken up in the winter, with a small lateral tuber, is *fu tsz'*. The names *t'ien hiung*, *wu hui*, *tse tsz'*, all refer to the variously shaped root with small tubers. The drug produced in Kiang tso [An hui and Kiang su], Shan nan [in Shen si and Ho nan, App. 268] is the *wu t'ou* of the *Pen king* [see 146]. It is now commonly called 草烏頭 *ts'ao* (herbaceous) *wu t'ou*.

All the above Chinese names refer to *Aconite*. For further particulars, see 146.

144.—天雄 *t'ien hiung*. *P.*, XVIIb, 1. *T.*, CXXVIII. Comp. 143.

Pen king :—*T'ien hiung*, 白幕 *pai mo* (*mu*). Root. Taste acrid. Nature warm. Very poisonous.

Pie lu :—The *t'ien hiung* grows in Shao shi [App. 281] in mountain-valleys. The root is dug up in the 2nd month and dried in the shade.

T'AO HUNG-KING :—Now the drug for medical use is dug up in the 8th month. The *t'ien hiung* resembles the *fu tsz'* but is more slender, from three to four inches and more long. As the *t'ien hiung* with the *fu tsz'* and *wu t'ou* form three sorts of the same drug, which is produced in Kien p'ing [in Sz ch'uan and Hu pei, App. 139], they are also known under the name 三 建 *san kien* (the three *kien*). Now the drug from Lang shan in I tu [in Hu pei, App. 177, 104] is much valued and called 西 建 *si* (western) *kien*. That from Ts'ien t'ang [in Che kiang, App. 352] is called 東 建 *tung* (eastern) *kien*. This is less potent.

LI SHI-CHEN :—There are two kinds of *t'ien hiung*. One is produced by the *fu tsz'* tubers planted by the people of Shu (Sz ch'uan). It (the root) grows very long and sometimes assumes the shape of the cultivated *yü* [*Colocasia*. See 261]. The other kind grows wild in the same country and is a kind of *ts'ao wu t'ou* [see 146].

The *t'ien hiung* is likewise Aconite. See 146.

145.—側 子 *tse tsz'*.   *P.*, XVII*b*, 3.   *T.*, CXXVII. Comp. above 143.

Pie lu:—Tse tsz' [this name means "lateral tuber"]. Taste acrid. Nature very hot. Very poisonous.

The *Shuo wen* [1st cent.] writes 萴 子 *tse tsz'*, and gives as synonym 烏喙 *wu hui* [*v. supra*].

T'AO HUNG-KING :—*Tse tsz'* are large lateral horns coming out from the *fu tsz'*. They are cut off and used in the treatment of rheumatism of the legs.

SU KUNG [7th cent.]:—From the principal root of the *wu t'ou* spring lateral tubers. The smaller ones are called *tse tsz'*, the larger are *fu tsz'*.

LI SHI-CHEN adds that the smallest lateral tubers are called 漏藍子 *lou lan tsz'*.

146.—烏 頭 *wu t'ou*.   *P.*, XVII*b*, 4.   *T.*, CXXVIII. Comp. above 143,

Pen king:—*Wu t'ou* (crow's head), 烏喙 *wu hui* (crow's beak), 兩頭尖 *liang t'ou tsien* [means "pointed at both ends"], 奚毒 *hi tu*. Taste acrid. Nature warm. Very poisonous.

In the *Kuang ya* [3rd cent.] we find *hi tu* [*v. supra*] and 蔠 *cho* given as synonyms for *fu tsz'*.

Pie lu:—The *wu t'ou* or *wu hui* grows in Lang ling [in Ho nan, App. 176], in mountain-valleys. The root is dug up in the first and second months and dried in the shade. It is three inches long. The best is the *t'ien hiung* [see 144].

Wu P'u [3rd cent.] gives the following synonyms: 耿子 *keng tsz'*, 毒公 *tu kung* (respectable poison), 帝秋 *ti ts'iu* (Emperor's autumn).

Ta Ming [10th cent.] notices the 土附子 *t'u* [native] *fu tsz'*, the inspissated juice of which, called 射罔 *she wang*, is used by archers to poison their arrows.

Li Shi-chen:—This *wu t'ou* [mentioned in the *Pen king*. Comp. also above 143] is a wild-growing species and is commonly called 草烏頭 *ts'ao* (herbaceous) *wu t'ou*, also 竹節烏頭 *chu tsie* (bamboo-joint) *wu t'ou*. That which grows in Kiang pei (north of the Yang tsz') is called 淮｜｜ *Huai wu t'ou*. This is the *t'u fu tsz'* mentioned by Ji Hua (Ta Ming). The 烏喙 *wu hui* is that with two protuberances. It is now commonly called 兩頭尖 *liang t'ou tsien* (pointed at both ends). The *ts'ao wu t'ou* is a common plant. Its root, leaves, fruit, all resemble those of the *Ch'uan wu t'ou* [see 143]. It grows wild [the other is cultivated]. The root has a black skin, is white within, shrivelled. It is very poisonous.

The Chinese drugs noticed from 143 to 146 are the roots of several species of *Aconitum*, and the above descriptions by the ancient authors are quite correct. The root of the European *Aconitum napellus*, as described in Flückiger and

HANBURY's *Pharmacographia*, is more or less conical or
tapering, enlarged and knotting at the summit, which is
crowned with the base of the stem. Numerous branched rootlets spring from its sides. If dug up in the summer it will be
found that a second or younger root [occasionally a third] is
attached to it near its summit by a very short branch, and
is growing out of it on one side. This second root (*fu tsz'* of
the Chinese) has a bud at the top which is destined to
produce the stem of the next season.

Ch., XXIV, 21:—*Fu tsz'*. Good drawing of an *Aconite*.

TATAR., *Cat.*, 24:—*Fu tsz'*, Radix *Aconiti chinensis*.—
Ibid., 52:—*Sheng fu tsz'*, Radix *Aconiti chinensis* cruda.—
Ibid., 63:—*Wu t'ou*, Radix *Aconiti*.—*Ibid.*, 5:—*Ts'ao wu
t'ou*, Radix *Aconiti*.

HANBURY, *Sc. pap.*, 258:—川 烏 *Ch'uan wu* (*t'ou*),
described and figured. Root of *Aconitum*. This figure may
serve to explain the Chinese name "crow's head" for the
root of *Aconite* and the "horns" in the ancient Chinese
descriptions of the drug.

Ibid.:—草 烏 *ts'ao wu* (*t'ou*), figured and described.
Tubers of *Aconitum japonicum*, Thbg. (=*A. Lycoctonum*, L.,
floribus ochroleucis).

P. SMITH, 2, 3.—

The *Index Fl. sin.* [I, 20] enumerates eight species of
Aconitum recorded from China.

Father DAVID [*Journ. Trois. voy.*, I, 367] mentions an
Aconitum (*Napellus?*) cultivated for medical use in Southern
Shen si and Sz ch'uan.

HENRY, *Chin. pl.*, 534:—烏獨 *wu tu*, *Aconitum Fischeri*.
Rich, ? This species occurs wild in the mountains (of Hu
pei) and is used as a drug. HENRY's native collector explained that the tuberous root of the first year's growth is
known as *wu tu*; a secondary tuber, which comes in the
second year, is called 附 子 *fu tsz'*; and a smaller tuber,
which it is rare to find, appearing in the third year, is 天雄

t'ien hiung.—Large quantities of *Aconite* are exported from Sz ch'uan under the names 川附 *Ch'uan fu* and 附片 *fu p'ien* (slices).

Cust. Med., p. 58 (5, 7):—*Ch'uan fu* and *fu p'ien* exported 1885 from Ichang 6,341 piculs.—According to the *Hank. Med.* [9] the same drug also exported from Hankow. Comp. also *Rep. on Trade,* Ch. Mar. Customs, for 1879, p. 3, Hankow.

The *t'ien hiung,* according to the *Cust. Med.,* imported to many ports, but it is not clear from which port the drug is brought.

Cust. Med., p. 8 (60):—*Ts'ao wu* exported 1885 from New chwang 76 piculs,—p. 70 (70), from Hankow 62 piculs.

So moku, X, 24:—草烏頭, *Aconitum uncinatum,* L. Blue flowers. Known also from the Peking mountains.

Phon zo, XXII, 10:—附子, *Aconitum Fischeri.* Reich. Blue flowers. [Japan, North and Mid China].—*Ibid.,* 11, 12:—川烏頭, *Aconitum* [not identified by FRANCHET]. Blue flowers. Root with lateral tubers as described by the Chinese authors.—*Ibid.,* 12:—烏頭, *Aconitum,* violet flowers and [13], same Chinese name, *A.* with rose-coloured flowers and [13], *A.* with green flowers, *A. Lycoctonum* (Franchet). *Ibid.,* 14, 15:—Same Chinese name, various species of *Aconitum* with blue or yellow flowers.

SIEB., *Icon. ined.,* I:—草烏頭, *Aconitum chinense.* SIEB. (=*Aconitum Fischeri,* Reich.).

147.—白附子 *pai fu tsz'. P.,* XVII*b,* 11. *T.,* CXXVII.

Pie lu:—*Pai* (white) *fu tsz'.* It grows in Shu (Sz ch'uan). Root officinal, dug up in the third month. Taste acrid and sweet. Nature very warm. Slightly poisonous.

33

Su Kung [7th cent.]:—This drug came originally from Kao li [Corea, App. 116]. Now it is produced west of Liang chou [in Kan su, App. 189]. It is not met with in Shu. It grows on sand-hills and in low damp places. A solitary stem resembling that of the *shu wei ts'ao* [*Salvia*. See 119]. Small leaves in a whorl between the flower spikes. The root resembles the *t'ien hiung* [*Aconite*. See 144].

Li Sün [8th cent.]:—The *Nan chou i wu chi* [earlier than the 6th cent.] says that the *pai fu tsz'* grows in the Eastern Sea, in the kingdom of Sin lo [S. Corea, App. 311] and Liao tung [S. Manchuria, App. 191].

Li Shi-chen:—Its root looks exactly like that of the *ts'ao wu t'ou* [*Aconite*. See 146] but is smaller, about one inch long. The dried drug is shrivelled and knobby.

Lour., *Fl., cochin*, 718 :—*Jatropha janipha*, L. [*Manihot Loureiri*, DC. *Prodr*., XV, 2, 1073]. Sinice : *pe fu tsù*. Planta fruticosa, caule recto simplicissimo. Folia palmata. Radix tuberosa, fasciculata, tuberibus ovato-oblongis, carnosis, intus et extra albis, sapore subdulci, subardente.

Tatar., *Cat*., 2 :—*Pai fu tsz'*, Radix *Aroideœ*.—Gauger [1] describes and figures these tubers.—P. Smith, 23.—

Cust. Med., No. 944 :—*Pai fu*, *Arisœma*, sp.

Cust. Med., p. 6 (36):—*Pai fu tsz'* exported 1885 from Newchwang 130 piculs,—p. 280 (87), from Amoy 0.03 picul.—According to *Hank. Med.* [31] the *pai fu tsz'* is also exported from Hankow.—Loureiro's *pai fu tsz'* is hardly the same drug as that exported under this name from Manchuria.

148.—虎掌 *hu chang*. *P*., XVII*b*, 13.   *T*., CLXXIII.

Pen king:—*Hu chang* (tiger's paw). The root is officinal. Taste bitter. Nature warm. Very poisonous.

Pie lu:—The *hu chang* grows in Han chung [S. Shen si, App. 54] in mountain-valleys, and in Yüan kü [in Shan tung, App. 415]. It is dug up in the 2nd and 8th months and dried in the shade.

T'AO HUNG-KING:—It is also found in Mid China. It (the root) resembles the *pan hia* [*Pinellia*. See 150] but is larger and has four lateral tubers which make it resemble a tiger's paw.

SU KUNG [7th cent.]:—This is the old root of the 由跋 *yu po* [see 149]. The plant has one stem (stalk) with a forked leaf at the end. The root varies in size from that of a fist to that of a hen's egg. In shape it resembles a flattened persimmon. On the four sides are round protuberances which give the root the appearance of a tiger's paw. The young root is called *yu po*. It is twice or thrice as large as the *pan hia* and has no lateral protuberances.

HAN PAO-SHENG [10th cent.]:—At the top of the (common) stalk are from 8 to 9 leaves (pedate leaves). The flowers come out between the stalks.

In the *Ji hua Pen ts'ao* [10th cent.] this plant is called 鬼蒟蒻 *kui kü jo.*

CH'EN TS'ANG-K'I [8th cent.] notices a plant 天南星 *t'ien nan sing* (southern cross of heaven) which grows in the mountain-valleys of An tung [in Kiang su, App. 2]. Its leaves resemble those of the *ho* [*Nelumbium speciosum*. See 295]. Solitary stem. Root used [in medicine].

SU SUNG [11th cent.]:—The *t'ien nan sing* is the same as the *hu chang* of the ancient authors. The smaller kind is called *yu po* [*v. supra*]. The *hu chang* is now found in Ho pei [S. Chi li and W. Shan tung, App. 78]. The root when it first begins to grow is not larger than a bean ; afterwards, when developing itself, it resembles the *pan hia* [see 150], but

is flattened [the tubers of the *pan hai* are globular].   After a
year the root becomes spherical and is then as large as a hen's
egg and shows from 3 to 6 protuberances.   The leaves shoot
forth in the 3rd or 4th month.   The plant grows about one
foot high.   The leaf is at the top of the stalk from 5 to 6
cleft.   The stalk which bears the flower-spike is like a rat's
tail and is enclosed in a spathe which resembles a ladle.   The
flowers are of a greenish gray colour, the seeds as large as
hemp-seed.   The people in Ki chou [in Chi li, App. 119]
cultivate it in gardens under the name of *t'ien nan sing*.
Another account says :—The *t'ien nan sing* is also a common
wild plant in marshes, it grows about one foot high.   The
leaves resemble those of the *k'u jo*.[36]   They stand opposite and
clasp the stem.   It flowers in the 5th month.   The flowers
are yellow and resemble [the spadix] the head of a snake.
The seeds are produced in the 7th month, they are of a red
colour, resemble pomegranate seeds, and are arranged in a
spike (spadix).   The root which is gathered in the 2nd and
8th months resembles the *yü* [*Colocasia.* See 261], is
spherical, flattened, and resembles the root of the *kü jo*
[*v. supra*] with which it is frequently confounded.   But the
kü jo plant is distinguished by having a spotted stem and
purple flowers.   The root of the *t'ien nan sing* is small, soft and
succulent.   When roasted it bursts and splits.   The *nan sing*
(or southern cross) is the *hu ch'ang* of the *Pen king*.   The
larger roots have on the margin protuberances (secondary
tubers) which are generally cut off from the fresh root.

LI SHI-CHEN:—The larger root is called *hu ch'ang* or
nan sing, the smaller is *yu po*.   Both belong to the same
species.   The larger kind is sometimes erroneously called
鬼臼 *kui kiu*.   But this is a different plant [see 152].
Another name for the *hu ch'ang* is 虎膏 *hu kao*.

蒟蒻 In Japan this Chinese name is applied to *Conophallus konjak*,
Schott, Order *Aroideæ.*   Comp. *P.*, XVII*b*, 17.

The *Ch.* [XXIV, 23-26] figures sub *t'ien nan sing* four aroidaceous plants with variously shaped leaves : palmate, pedatisect, peltatosect.

LOUR., *Fl. cochin.*, 652 :—*Arum pentaphyllum*, L. Sinice : *tien nan sin*.

TATAR., *Cat.*, 40, 56 :—*Nan sing* or *t'ien nan sing*, Radix *Ari pentaphylli.*—GAUGER [29] describes and figures this root. It has indeed a resemblance to a star or a tiger's paw.—HANB., *Sc. pap.*, 263.—P. SMITH, 26.

In the Peking mountains the name *t'ien nan sing* is applied to *Arisæma Tatarinowii*, Schott. Peltatosect leaves.

Amœn. exot., 786 :—南星 *nan soo*, vulgo *jamma konjakf*, item *osomi*, Medicis *ten nan sio* dictus, Dracunculus minor trifolius, etc.—This is *Arum triphyllum*, Thbg. [*Fl. jap.*, 233] and *Arisæma ringens*, Schott.

So moku, XIX, 16, and *I'hon zo*, XXII, 18, 19 :—天南星, *Arisæma japonicum*, Bl. Comp. also *Kwa wi*, 58.

Cust. Med., p. 78 (165):—*T'ien nan sing* exported 1885 from Han kow 220 piculs,—p. 34 (159), from Tien tsin 10 piculs,—p. 130 (154), from Chin kiang 6.8 piculs.

149.—由跋 *yu po. P.*, XVII*b*, 14. *T.*, CXXXII.

Pen king:—*Yu po.* The root is officinal. Taste acrid and bitter. Nature warm. Poisonous.

SU KUNG [7th cent.]:—The *yu po* is the young root of the *hu chang* [see 148]. It is double the size of the *pan hia* tuber [see 150]. There are no secondary lateral tubers. The old root is the *hu chang.* The plant mentioned by T'AO HUNG-KING under the name of *yu po* as cultivated in Shi hing [App. 289] is not this plant but the *yuan wei* [an *Iris.* See 154].

CH'ENG TS'ANG-K'I [8th cent.]:—The *yu po* grows in forests. It is from one to two feet high and resembles the *kü jo* [see sub 148]. Root of the size of a hen's egg.

HAN PAO-SHENG [10th cent.]:—It sends up in spring one stem (petiole) at the top of which are eight or nine leaves (pedatisect leaf). The root is round and flattened, and its flesh is white.

LI SHI-CHEN :—The *yu po* is a small *t'ien nan sing*. It is seldom used in medicine.

Ch., XXIV, 27 :—*Yu po*. The drawing represents a plant with peltatosect leaves, *Arisæma*.

150.—半夏 *pan hia*. P., XVII*b*, 20. T., CXXXVI.
Comp. *Classics*, 422.

Pen king:—*Pan hia* (midsummer plant), 守田 *shou t'ien* (guardian of the field), 水玉 *shui yü*, 和姑 *ho ku*. The root (tubers) is officinal. Taste acrid. Nature uniform. Poisonous.

Pie lu:—Other name: 地文 *ti wen*. The *pan hia* grows in Huai li [in Shen si, App. 95] in river-valleys. The root is dug up in the 5th and 8th months and dried in the sun.

T'AO HUNG-KING :—The place Huai li lies in Fu feng [in Shen si, App. 44]. Now the best sort is brought from Ts'ing chou [in Shan tung, App. 363]. It is also found in Wu [Kiang su, etc., App. 389]. That of a good quality has a white flesh. The old drug is rejected.

SU KUNG [7th cent.]:—It grows still in the above-mentioned localities. That sort which grows in marshes is called 羊眼半夏 *yang yen* (sheep's eye) *pan hia*. The best sort is globular and white. That from Kiang nan [Kiang si, etc., App. 124] is large [the tuber], one inch in diameter,

Su Sung [11th cent.]:—The best sort comes from Ts'i chou [in Shan tung, App. 348]. The leaves are trifoliate, of a light green colour, somewhat resembling bamboo-leaves. The sort produced in Kiang nan has leaves resembling those of the *shao yo* [*Pæonia albiflora*. See 52]. The tuber has a yellow skin and white flesh. The small tubers are called *yang yen pan hia* [*v. supra*].

Ch., XXIV, 28, 30 :—*Pan hia*. The drawings seem to represent *Pinellia tuberifera*, Ten. Ternate leaves.

Lour., *Fl. cochin.*, 652 :—*Arum triphyllum*, L. [this is *Arisæma Loureiri*, according to Blume. Species dubia]. Sinice : *puon hia*. Bulbus subrotundus albus.—*Ibid.*, 651 :—*Arum dracontium*, L. [Blume calls it *Arisæma cochinchinense*]. Sinice : *puon hia*. Folia pedata. Bulbus subrotundus albus.

Tatar., *Cat.*, i :—*Pan hia*. Radix *Ari macrouri*, Bge. [same as *Pinellia tuberifera*].—Gauger, 2 :— *Pan hia* described and figured. Small spherical, white tubers.

Hanbury, *Sc. pap.*, 262 :—生 半 夏 *sheng* (fresh) *pan hia*. Tubers of *Pinellia tuberifera* described and figured.— P. Smith, 22, 26, 149.

At Peking the name *pan hia* is applied to *Pinellia tubifera* as well as to *P. pedatisecta*, Schott.

China Review, X, 380 [Parker's "Travels in Sz ch'uan"]:—A root drug called *pan hia*, looking like round pellets of bone, was drying in the sun.

Cust. Med., p. 78 (171):—*Pan hia* exported 1885 from Han kow 1,200 piculs,—p. 130 (144), from Chin kiang 58 piculs,—p. 188 (72), from Ning po 55 piculs,—p. 58 (19), from I chang 39 piculs.

So moku, XIX, 2, and *Phon zo*, XXII, 23, 24 :—半 夏, *Pinellia tuberifera*.—*So moku*, XIX, 4 :—大 半 夏 (large *pan hia*), *P. tripartita*, Schott.

151.—蛋休 *tsao hiu.* *P.*, XVIIb, 28. *T.*, CXXX.

Pen king:—*Tsao hiu.* The root is official. Taste bitter. Nature slightly cold. Poisonous.

Pie lu:—Other name: 蚩 休 *ch'i hiu.* The *tsao hiu* grows in Shan yang [in Shan tung, App. 270] in river-valleys, and in Yüan kü [in Shan tung, App. 415].

SU KUNG [7th cent.]:—It is now called 重樓金線 *chung lou kin sien,* also 重臺 *chung t'ai.* In the south it is known by the name 甘遂 *kan sui* [comp. 138]. The plant has a solitary stem bearing at the top from six to seven leaves [arranged in a whorl] in two or three rows, like those of the *wang sun* [*Paris.* See 22], the *kui kiu* [see 152] or the *pi ma* (*Ricinus*). The root is white and delicate with fine fibres. It resembles a large, succulent *ch'ang p'u* root [*Acorus.* See 194].

HAN PAO-SHENG [10th cent.]:—Its leaves resemble those of the *kui kiu* [see 152] and *mou meng* [*Paris.* See 22]. The root resembles that of the *tsz' shen* [*Polygonum bistorta.* See 21]. It has a yellow skin and white flesh. It is dug up in the 5th month and dried in the sun.

SU SUNG [11th cent.]:—The *tsao hiu* is also called 紫河車 *tsz' ho ch'e.* This plant is now found in Ho chung [in Shan si, App. 74], in Ho yang [in Ho nan, App. 81] in Hua chou and Feng chou [both in Shen si, App. 85, 39], Wen chou [in Kan su, App. 386] and in Kiang and Huai [Kiang su and An hui, App. 124, 89]. The leaves resemble those of the *wang sun, kui kiu,* etc. They form two or three rows. It flowers in the 6th month. The flowers are yellow and purple, the stamens [or anthers] are of a reddish yellow colour and run out into gold-coloured drooping filaments. The fruit, which appears in autumn, is red. The root is like a succulent ginger-root and has a red skin and white flesh.

LI SHI-CHEN:—The *chung lou kin sien* is also called 三眉草 *san ts'ang ts'ao* and 白甘遂 *pai kan sui* [in these names *chung lou* means "many storied," *san ts'ang* = three-storied, both terms referring to the rows of the leaves. *Kin sien* = gold thread]. It is a common plant which grows in mountain recesses in moist, shady places. It has a solitary stem on which the leaves, which resemble peony leaves, are arranged in two or three rows (whorls), each row consisting of seven leaves. The flowers appear in the summer at the top of the stem, each flower has seven petals. The flowers are provided with gold-coloured filaments, from three to four inches long. On the Wang wu shan mountain [in Ho nan. *Bot. sin.*, I, p. 228 (65)] there grows one species which has its leaves in from five to six rows. The root resembles the root of the *kui kiu* or that of the *ts'ang shu* [*Atractylis.* See 12]. It has a purple skin and white flesh.

Ch., XXIV, 34:—*Tsao hiu.* Good drawing representing a *Paris.* From seven to eight leaves in a whorl. The descriptions in the *P.* seem also to refer to one or several species of *Paris* or the allied genera *Trillium* and *Trillidium.* Order *Liliaceæ.*

Phon zo, XXIII, 1, 2:—蚕休 *Trillidium japonicum,* Franchet. Represented with a large root.—*Ibid.*, 3:—Same Chinese name, *Trillium erectum*, L.

Kwa wi, 26:—Same Chinese name, *Paris hexaphylla,* Cham.

152.—鬼臼 *kui kiu. P.*, XVII*b*, 30.　*T.*, CXXVI.

Pen king:—*Kui kiu* (devil's mortar), 九臼 *kiu kiu,* 㿜犀 *tsio si,* 馬目毒公 *ma mu tu kung.* The root is officinal. Taste acrid. Nature warm. Poisonous.

Pie lu:—Other names : 天臼 *t'ien kiu* (heaven mortar),
解毒 *kie tu* (counter poison). The *kui kiu* grows in Kiu
kü [App. 156], in mountain-valleys, also in Yüan kü [in
Shan tung, App. 415]. The root is dug up in the 2nd and
8th months.

T'AO HUNG-KING :—The root resembles the roots of the
she kan [*Pardanthus.* See 153], the *shu* [*Atractylis.* See
12] and the *kou wen* [see 162]. There are two sorts. One is
produced in Ts'ien t'ang [in Che kiang, App. 352] and in
Mid China. It is of a sweet taste and covered with dense
hair (radical fibres). This is the best. The other, which
comes from Hui ki and Wu hing [both in Che kiang, App.
98, 390], is larger, of a bitter taste, not covered with hair, and
less potent. The *ma mu tu kung* has a root which resembles
the *huang tsing* [*Polygonatum.* See 7]. It has excavations
(臼 *kiu*, properly mortar) resembling horse's eyes, is tender
and succulent. It is less frequently used in medicine than the
genuine *kui kiu.*

SU KUNG [7th cent.]:—The *kui kiu* grows in the depths
of the mountains, in shady places. The leaves resemble
those of the *pi ma* (*Ricinus*), are bi-lobed, arranged in
rows at the top of a solitary stem. The root sends up one
stem every year, and when the stem decays it leaves an
excavation on the root. Thus after 20 years the root shows
20 excavations. The name *kiu kiu* (nine excavations) is
also derived from this peculiarity of the root. In its skin,
flesh and hairs (radical fibres) the root resembles the *she kan*
[*v. supra*], which is frequently substituted for it. The
kui kiu is now an article presented as tribute in the district
Tang yang in the prefecture of King chou, in the district
of Yüan an in the prefecture of Hia chou, and in the district of King shan in the prefecture of Siang chou [all
these localities are in Hu pei, App. 333, 146, 417, 64, 149,
305]. It grows there in the mountains, but is very rare.

Su Sung [11th cent.]:—The drug is now produced in Kiang ning fu, Ch'u chou, Shu chou [all in An hui, App. 129, 25, 294], Shang chou [in Shen si, App. 278], Ts'i chou [in Shan tung, App. 348], Hang chou [in Che kiang, App. 58] and in the localities mentioned by Su Kung. Red flowers which appear in the 3rd month in the stem (spathe?), afterwards fruits. The leaves are lobed and arranged like an umbrella at the top of the stem.

Li Shi-chen enumerates the following synonyms applied to this plant, viz.: 唐婆鏡 t'ang p'o king, 羞天花 siu t'ien hua, 八角盤 pa küe p'an (eight-horned dish. This name refers to the large, lobed leaves), 尤律草 shu lü ts'ao, 害母草 hai mu ts'ao, 鬼藥 kui yao, 瓊田草 k'iung t'ien ts'ao, 山荷葉 shan ho ye, 旱荷 han ho. It is also called 獨荷草 tu ho ts'ao and 獨脚蓮 tu kio lien. Some authors say that its leaf resembles the leaf of *Nelumbium speciosum* and is palmately lobed, to which some of the above names allude. Li Shi-chen states that the plant is common in South China. In North China it is found only on the Lung men shan and Wang wu shan mountains [App. 212, and *supra* sub 151]. It has only one stem, which is hollow and produces at the top seven round leaves resembling those of *Nelumbium*, but smaller and palmately divided, purple underneath. The flowers appear beneath the leaves.

The ancient Chinese authors probably confound under the above enumerated names several different plants.

Henry [*Chin. pl.*, 323] states:—八角蓮 pa küe lien, *Diphylleia?* sp.[37] *nova*. This curious plant appears in shaded places in the mountains. It is the 鬼臼 kui kiu of books, and is figured in the *Ch.* [XXIV, 35] where the name used at I chang is given as a synonym. P. Smith [46] wrongly

---

[37] This plant, so first identified from imperfect specimens, is now found from better specimens to be *Podophyllum versipelle*, Hance.—A. Henry.

says it is *Caladium*. This error probably arose from the fact that 獨 脚 蓮 *tu kio lien*, which is given also as a synonym of *kui kiu*, in Hu peh signifies *Arisœma*.

Diphylleia and *Podophyllum* are both genera of the order *Berberideœ*. Some of the above descriptions of the *kui kiu* by the Chinese authors agree. In Northern America *Diphylleia cymosa* is called "the umbrella plant" [comp. *supra* Su Sung's description].

Ch., VIII, 61:—獨 脚 蓮 *tu küe lien*. Rude drawing. Perhaps an *Aroidea* is intended. The plant is said to grow in Fu chou.

Tatar., *Cat.*, 22:—*Tu küe lien*. Radix *Caladii xanthorizi?* —P. Smith, 46.—The plant *tu küe lien*, wild and cultivated at Peking, and which was raised in the Botanic Gardens, St. Petersburg, from seeds I had procured in Peking, proved to be *Typhonium giganteum*, Engl. (order *Araceœ*).

Henry, *l.c.*, 476:—*Tu küe lien* in Hu pei = *Arisœma heterophylla*, Bl.—But according to Ford and Crow ["Notes on Chin. Mat. Med.," in *China Review*, XVI, 7], *tu küe lien* at Canton is *Podophyllum versipelle*, Hance.

Comp. Hooker's *Icon. Plant.*, tab. 1996, *Podophyllum versipelle*, Dr. Henry's note.

Phon zo, XXIII, 5, 6:—鬼 臼, *Diphylleia.—So moku*, VII, 25:—山 荷 葉, *Diphylleia Grayi*, Fr. Schm.

153.—射 干 *she kan*. P., XVII*b*, 32. T., CLXII.

This plant is mentioned in Sz' Ma siang ju's poems [† B.C. 120].

Pen king:—*She kan*, 烏 扇 *wu shan*, 烏 蒲 *wu p'u*. The root is officinal. Taste bitter. Nature uniform. Poisonous.

Pie lu:—Other names: 烏 翣 *wu sha* (black feathers), 烏 吹 *wu ch'ui*, 草 薑 *ts'ao kiang*. The *she kan* grows in

Nan yang [in Ho nan, App. 231] in mountain-valleys and in fields. The root is dug up on the 3rd day of the 3rd month and dried in the shade.

Wu P'u [3rd cent.] calls it 黃遠 *huang yüan.*

T'ao Hung-king:—It has a yellow root, and is frequently cultivated. There is one sort with white flowers. It has a long stem. This accounts for the name *she kan* (lance).

Su Kung [7th cent.]:—The *she kan* resembles in its leaves the *yüan wei* [*Iris.* See 154].

Han Pao-sheng [10th cent.]:—The *she kan* grows from 2 to 3 feet high, has yellow flowers and black fruits. The root shows many radical fibres, has a yellowish black skin and yellowish red flesh.

Ch'en Ts'ang-ki [8th cent.]:—The *she kan* resembles the *yüan wei.* It is also called 鳳翼 *feng i* (phœnix' wing), from the shape of the leaves. It blossoms in autumn. Flowers red and dotted. The *yüan wei* has blue flowers.

Sung Sung [11th cent.]:—The *she kan* is much cultivated. It grows from one to two feet high and has long narrow leaves which spread out like the wings of a bird. The stem rises from the midst of the leaves like that of the *süan ts'ao* (*Ilemerocallis*). The flowers, which appear in the 6th month, are orange coloured and with small spots on the corolla. The fruit is a capsule which contains black seeds.

Li Shi-chen :—The *she kan* is also called 扁竹 *pien chu* (flat bamboo) and is much cultivated.

Ch., XXIV, 37 :—*She kan.* The drawing represents *Pardanthus chinensis,* Ait. The genus name *Pardanthus* (leopard flower) refers to the spotted flowers. The *pai* (white) *she kan* [*Ch.,* XXIV, 40] is *Pardanthus dichotomus,* Ledeb. (*Iris dichotomus,* Pall.). The first is much cultivated at Peking, the second grows wild there in the mountains. The above Chinese descriptions agree in a general way.

LOUR., *Fl. cochin.*, 46 :—*Ixia chinensis* (same as *Pardanthus chinensis*). Sinice : *xe can* [*she kan*]. Radix bulbis teretibus croceis. Flos magnus, terminalis, aureus, rubro punctatus.

TATAR., *Cat.*, 52 :—*She kan.* Radix *Pardanthi chinensis.*—Gauger [35] describes and figures the root *she kan*, which, he thinks, belongs to an *Iridacea.*—P. SMITH, 167.

Cust. Med., p. 280 (108):—*She kan* exported 1885 from Amoy 3.14 piculs.—According to the *Hank. Med.* [p. 36], exported also from Hankow.

Amœn. exot., 872 :—射 干 *jakan*, vulgo *karasu oogi* et *fi oogi.* Iris flore liliaceo parvo, puniceo, punctis sanguineis intus asperso. This, as well as the figure in the *So moku* [II, 12] sub 射 干, is *Pardanthus chinensis.*

154.—鳶 尾 *yüan wei.*   *P.*, XVII*b*, 35.   *T.*, CLXVIII.

Pen king :—*Yüan wei* (kite's tail), 烏 圜 *wu yüan.* The root is officinal. Taste bitter. Nature uniform. Poisonous.

Pie lu :—The *wu yüan* grows in Kiu i [in Kiang si, App. 155] in mountain-valleys. It is gathered [dug up] in the 5th month.

T'AO HUNG-KING :—The root of this plant is called 鳶 頭 *yüan t'ou* (kite's head).

SU KUNG [7th cent.]:—This plant is cultivated. Its leaves resemble those of the *she kan* [see 153] but are broader and shorter. The stem is also not long. The flowers are of a violet colour. The root resembles that of the *kao liang kiang* [*Alpinia.* See 57], it has a yellow skin and white flesh. When chewed it causes scratching in the throat.

Ch., XXIV, 40 :—*Yüan wei.* The figure represents a large *Iris.*

Amœn. exot., 873 :—鳶 尾 *ssibi*, vulgo *itz fatz.* Iris pumila flore magno pleno,

So moku, II, 3 :—Same Chinese name, *Iris tectorum*, Maxim.

This plant is much cultivated at Peking as an ornamental plant. Its popular name there is 草玉蘭

155.—羊鸔圚 *yang chi chu.* P., XVII*b*, 40. *T.*, CLXXIX.

Pen king :—Yang chi chu. The flowers are officinal. Taste acrid. Nature warm. Very poisonous.

Pie lu :—Other name : 玉枝 *yü chi.* The *yang chi chu* grows in the valleys of the T'ai hing shan mountains [in N. China, App. 323], also in the mountains of Huai nan [An hui, Kiang su, App. 90]. The flowers are gathered and dried in the shade.

T'AO HUNG-KING explains the above name [*yang* = sheep, *chi chu* = to reel] by the fact that sheep when eating the leaves of this plant begin to reel and die. It grows in Mid China on mountain-slopes and has yellow flowers resembling those of the *lu ts'ung* (*Hemerocallis fulva*). It cannot be approached to the eye.[38]

Su KUNG [7th cent.] observes that the flowers of this plant do not resemble the *lu ts'ung* flowers but rather the *süan hua* [*Calystegia.* See 169].

HAN PAO-SHENG [10th cent.]:—It is a shrub, 2 feet high. Leaves like peach-leaves. Yellow flowers resembling melon-flowers. There are gathered in the 3rd and 4th months and dried in the sun.

Su SUNG [11th cent.] compares its flowers to those of *ling siao* [*Tecoma, Bignonia.* See 170] and the *shan shi liu* [mountain pomegranate. See further on]. The flowers are yellow. The plant grows from 3 to 4 feet high. It is poisonous to sheep. In the mountains of Ling nan [South China, App, 197] and Shu (Sz ch'uan) grows a sort with deep red flowers. This is not used in medicine.

[38] 不可近眼

LI SHI-CHEN:—The corolla of the flower is five-lobed and of a yellow colour, as also the stamina. It has an unpleasant taste and smell. The red flowered species mentioned by SU SUNG is the 紅 | | *hung* (red) *chi chu*, which is also called 山 | | *shan* (mountain) *chi chu* and 山 石 榴 *shan shi liu* (mountain pomegranate). It is not poisonous. Other names for the *shan chi chu*, which is common in the mountains, are 杜 鵑 花 *tu küan hua* (cuckoo flower) and 映 山 紅 *ying shan hung*.

Other names for the common (yellow) *yang chi chu* are : 鬧 羊 花 *nao yang hua* (flower which makes sheep giddy), 驚 羊 花 *king yang hua* [similar meaning], 黃 杜 鵑 花 *huang* (yellow) *tu küan hua*.

Ch., XXIV, 19 :—*Yang chi chu*, a *Rhododendron* figured.

TATAR., *Cat.*, 29 :—*Yang chi chu*, *Hyoscyamus niger*. *Ibidem*, 41 :—*Nao yang hua*, *Hyoscyamus*.

HANBURY, *Sc. pap.*, 266 :—*Nao yang hua*. Flowers of *Rhododendron?*—P. SMITH, 29, 84, 115.—PARKER, *Canton pl.*, 85 :—*Nao yang hua*, *Datura alba*, L.

HENRY, *Chin. pl.*, 218 :—老 虎 花 *lao hu hua* (tiger flower), *Rhododendron* (*Azalea*) *sinense*, Sw. An *Azalea* with yellow flowers, reputed dangerous to cattle that browse on it. This is the *yang chih chu* of *Ch.*, XXIV, 19, where the local name is given as a synonym.

HENRY, *l.c.*, 558 :—映山紅 *ying shan hung*, *Rhododendron* (*Azalea*) *indicum.*, Sw., var. This red flowering species is also known as *hung chi chu* and 紅杜鵑 *hung tu küan.*—*Huang* (yellow) *tu küan* is a name for the [yellow] *Azalea sinensis* [*v. supra*].

At Peking 杜 鵑 花 *tu küan hua* is a general name for *Rhododendron* (*Azalea*). The same at Canton. See BRIDGM., *Chin. Chrest.*, 472. At Peking *hung* (red) *tu kuan hua* is *Azalea indica*, L., var. *macrantha*. Large crimson flowers.

Cultivated.—*Pai* (white) *tu küan hua* is *Rhod. leucanthum*, Bge. Cultivated. White flowers.—*Ye* (wild) *tu küan hua* is *Rhod. dauricum*, L. Wild-growing. rose-coloured flowers.

China is very rich in *Rhododendrons*. The *Ind. Fl. sin.* [II, 19-32] enumerates 65 Chinese species, including the sub genus *Azalea*. The flowers are for the greater part red, crimson, rose-coloured and white. Only a few Chinese species have yellow flowers, viz. *Rhod.* [*Azalea*] *sinense*, Sw. [*A. pontica*, L. var. *sinensis*, Lindl., *Bot. reg. t.* 1253],—*Rhod. sulfureum*, Franchet, from Yünnan,—*Rhod. lutescens*, Franch., Father DAVID, from Mupin.—*Rhod. sinense* is common in the mountains of Mid China, it is also much cultivated in Chinese gardens. European observers do not mention its poisonous properties. We know that the nearly allied yellow flowered *Azalea pontica*, L., possesses dangerous narcotic qualities.—

According to Abbé DESGODINS, in E. Tibet sheep and goats are poisoned by *Rhododendron* leaves [*Bull. Soc. Géogr.*, 1873, 1, 333]. AITCHISON, in his "Flora of the Kuram Valley" [*J. Linn. Soc.*, XVIII], states that *Rhododendron afghanicum* [whitish green flowers] is poisonous to goats and sheep.

Amœn. exot., 845:—鬫躅 *tecki tsjoku*, vulgo *tsutsusi*, cum icone, p. 846. This is, according to MAXIMOWICZ [*Rhodod. Asiæ orient*, p. 37], *Rhododendron* [*Azalea*] *indicum*, Sw.

Amœn. exot., 849 :—杜鵑 *to ken*, vulgo *satsuki*. Cytisus Liliifer autumnalis, etc. According to MAXIMOWICZ [*Rhod. Asiæ orient*, p. 39], this is probably *Rhododendron* [*Azalea*] *indicum*, var. *macranthum*.

Phon zo, XXIII, 21-24 —羊鬫躅. Various species of *Rhododendron* are figured under this Chinese name, *viz.* 21, 22, yellow and rose-coloured flowers, *Rhod. Keiskii*, Miq.,— 22, *Rh. Schlippenbachii*, Maxim., rose-coloured,—23, *Rhod. Albrechti*, Max., rose-coloured,—23, *Rhod. rhombicum*, Miq., violet,—24, undetermined, white.—

Ibid., 24-26 :—山蹢躅, *viz.* 24, *Rhod. sublanceolatum*,
Miq., rose-coloured,—25, *Rhod.*, undetermined, violet, white,
rose-coloured,—26, *Rhod. linearifolium*, S. & Z., and *Rhod.
macrosepalum*, Max. ·

Ibid., 27, r.:—杜鵑花. Two species of *Rhododendron*,
undetermined.

SIEB., *Icon. ined.*, V :—杜鵑花. *Rhod. satsuki* [=*Rhod.
indicum* var. *macranthum* [*v. supra*].

Ibid.:—羊蹢躅. *Azalea pontica?* *Azalea chinensis*,
SIEB.—紅蹢躅. *Rhod. japonicum*, SIEB.

156.—芫花 *yüan hua.* *P.*, XVIIb, 42. *T.*, CX.
Comp. *Bot. sin.*, II, 258, 465.

Pen king :—Yüan hua, 去水 *kü shui*. The flowers are
officinal. Taste bitter and acrid. Nature warm. Poisonous.

Pie lu:—Other names 杜芫 *tu yüan*, 毒魚 *tu yü*
(poisoning fish), 蜀桑 *Shu sang* [Sz ch'uan mulberry]. The
yüan hua grows in Huai yüan [in Ho nan, App. 92] in
river-valleys. The flowers are gathered on the 3rd day of
the 3rd month and dried in the shade.

WU P'U [3rd cent.]:—Other names : 赤芫 *ch'i* (red)
yüan, 兒草 *rh ts'ao*, 敗華 *pai hua* (injurious flower). The
root is called 黃火戟 *huang* (yellow) *ta ki* [comp. 136].
The 芫根 *yüan ken* [root] grows in Han tan [in S. Chih li,
App. 56]. Its leaves appear in the 2nd month, are green at
first, but when growing thicker they become black. The
flowers are purple, red or white. Flowers, leaves and root are
officinal. The flowers are gathered in the 3rd month, the
leaves in the 5th, and the root is taken up in the 8th or 9th
month.

T'AO HUNG-KING :—It (the flowers?) is boiled for medical
use, and cannot be approached to the eyes.

HAN PAO-SHENG [10th cent.]:—It is a common plant in Mid China, and grows from 2 to 3 feet high. The leaves resemble those of the *pai ts'ien* [*Vincetoxicum*. See 45] or willow-leaves. The root has a yellow skin resembling the root of the mulberry tree. The flowers are violet; they appear in the first or second month and fall off when the leaves begin to expand.

SU SUNG [11th cent.]:—The plant has a perennial root. The old branches and the stem are of a purple colour It grows from 2 to 3 feet high. The root penetrates from 3 to 5 inches into the ground, is of a white colour and resembles the root of the elm. The leaves are small and narrow like willow-leaves. It blossoms in the second month. Purple flowers in spikes resembling those of the *tsz' king* [*Cercis*]. There is one sort in Kiang chou [in Shan si, App. 123], with yellow flowers, which is called 黃芫花 *huang* (yellow) *yüan hua*.

LI SHI-CHEN:—The name is also written 杬 *yüan*. Thrown into the water it [flowers or leaves] poisons fish. Its smell causes head-ache, whence it is popularly called 頭痛花 *t'ou t'eng hua* (head-ache flower) The dictionary *Yüe pien* [6th cent.] states that the *yüan* tree grows in Yü chang [Kiang si. See *Bot. sin.*, II, sub 513]. Its juice when boiled preserves fruit and eggs from spoiling. The *Yung chui sui pi* [12th cent.] says that it is common in Jao chou [in Kiang si, App. 109]. From the stem it cannot be decided whether it is a tree. People sometimes rub the skin with this drug [it is not said what part of the plant, probably the bark] to produce an inflammatory swelling, in order to simulate wounds. Eggs when rubbed with a mixture of this drug with salt assume an ochre colour.

TATAR., *Cat.*, 31 :—*Yüan hua, Passerina Chamædaphne*, Bunge [*Wickstræmia Chamædaphne*, Meissn.]. This plant, common near Peking, has small, yellow, fragrant flowers. It is poisonous, and belongs to the order *Thymelaceæ,—*

P. Smith, 168.

Henry, *Chin pl.*, 281 :—悶頭花 *men t'ou hua* (plant which stupefies the head). *Daphne genkwa*, S & Z. This is the 芫花 *yüan hua* in *Ch.*, XXIV, 44. The figure on p. 46 probably represents the same, and the name given there, 金腰帶 *kin yao tai*, is also used at I chang.

Cust. Med., p. 128 (110):— *Yüan hua* flowers exported 1885 from Chin kiang 3.25 piculs.—According to *Hank. Med.*, 53, exported also from Hankow.

Sieb. & Zucc., *Fl. jap.*, I, 137, tab., 75 :—*Daphne genkwa* (order *Thymelaceæ*), sinice 芫花, said to have been introduced from China into Japan. Lilac flowers. The flowers and the bark are officinal in Japan, and the latter is used as a vesicatory.

Phon zo, XXIV, 2 :—Same Chinese name, *Daphne genkwa*.

157.—蕘花 *jao hua*. P., XVII*b*, 45. T., CX.

Pen king :—*Jao hua*. The flowers are officinal. Taste bitter. Nature cold. Poisonous.

Pie lu :—The *jao hua* grows in Hien yang [in Shen si, App. 65] in river-valleys, also in Ho nei [S.E. Shan si, App. 77] and Chung mou [in Ho nan, App. 30]. The flowers are gathered in the 6th month and dried in the shade.

T'ao Hung-king :—The *Chung mou* drug now comes from the Yellow River. It (the flowers?) resembles the *yüan hua* [see 156] but is very small and white.

Su Kung [7th cent.]:—The plant resembles the *hu sui* (*Coriandrum*), has no spines, small yellow flowers. It has no resemblance to the *yüan hua*.

Han Pao-sheng [10th cent.]:—The best sort comes from Yung chou [in Shen si, App. 424]. The plant grows in the mountains, and is about 2 feet high.

Li Shi-chen suggests that this plant (*jao hua*) is the *yüan hua* with yellow flowers noticed by Su Sung as growing in Kiang chou [comp. 156]. The fresh flowers of the *jao hua* are yellow, but when dried they become white, wherefore T'ao Hung-king states that the plant has white flowers.

Ch., XXIV, 48 :—*Jao hua*. Rude drawing. Only leaves.

Phon zo, XXIV, 2, 3 :—芫花. Two plants represented, one with yellow, the other with white flowers. Not identified by Franchet.

Sieb., *Icon. ined.*, VI :—*Passerina japonica*, S. & Z. [*Wickstræmia japonica*, Miq.]. Sinice 芫花.

Sieb., *Œcon.*, 132:—Same Chinese name, *Stellera ganpi*. E cortice conficitur charta ob firmitatem laudata. *Ibidem*, 131, Siebold says the same with respect to *Stellera japonica*. This latter is the same as *Wickstræmia japonica*. Siebold's *Stellera* (*Passerina*) *ganpi* is *Wickstræmia canescens*, Meisn.

In Hoffm. & Sch., 415, 眼皮花 [Chinese pronunciation *yen p'i*] is given as the Chinese name for *Passerina ganpi*. This name is not found in the *P.*

158.—莽草 *mang ts'ao*. *P.*, XVII*b*, 47. *T.*, CX.

Comp. *Rh ya*, 147, *Classics*, 464.

Pen king :—*Mang ts'ao*. Leaves used in medicine. Taste pungent. Nature warm. Poisonous.

Pie lu :—Other names :—䓚 *mi*, 春草 *ch'un ts'ao*. The *mang ts'ao* grows in Shang ku [in N. Chi li, App. 272] in mountain-valleys, and in Yüan kü [in Shan tung, App. 415]. The leaves are gathered in the 5th month and dried in the shade.

T‘AO HUNG-KING:—The name *mang* was originally written 罔 *wang*. [Some editions of the *P.* write 芒 *wang*. The *K.D.* does not say that these two characters *wang* are identical.] This is now a common plant in the eastern provinces. The people bruise the leaves, mix the powder with old millet-flour and throw it into the water to stupefy fish, which are then easily caught and can be eaten without any danger.

SU SUNG [11th cent.]:—It is a common plant in the southern provinces and in Shu [Sz ch‘uan]. It is a tree which resembles the *shi nan* [see 347], but the leaves are sparely produced. It bears neither flowers nor fruit. The leaves are gathered in the 5th and 7th months. Others say that it is an herbaceous plant which climbs on trees.

K‘OU TSUNG-SHI [12th cent.]:—It is a tree with leaves like the *shi nan* [*v. supra*] which have the smell of the *tsiao* [*Zanthoxylon.* See 288].

LI SHI-CHEN:—This plant is poisonous. When eaten it causes man to lose his senses. The mountain people use it for killing rats, whence the name 鼠莽 *shu* (rat) *mang*. The name *ch‘un ts‘ao* given in the *Pie lu* as a synonym of *mang ts‘ao* is also applied to the plant *pai wei* [*Vincetoxicum.* See 44].

For further particulars regarding this poisonous plant see *Bot. sin.*, II, 464. In Japan the name *mang ts‘ao* is applied to *Illicium religiosum*, but in China *mang ts‘ao* is a different plant.

159.—茵芋 *yin yü.*   *P.*, XVII*b*, 49.   *T.*, CXLV.

Pen king:—*Yin yü.* The leaves are officinal. Taste bitter. Nature warm. Poisonous.

Pie lu:—Other names: 莞草 *kuan ts‘ao*, 畢共 *pi kung.* The 茵蕷 *yin yü* grows in T‘ai shan [in Shan tung, App.

322] in mountain-valleys. The leaves are gathered on the 3rd day of the 3rd month.

T'AO HUNG-KING :—The best comes from P'eng ch'eng [in Kiang su, App. 247]. The plant grows in Mid China. In its stem and leaves it resembles the *mang ts'ao* [see 158] but is more slender and weak. It is seldom used in medical prescriptions.

TA MING [10th cent.]:—This drug comes from Hai yen [in Che kiang, App. 51]. The plant resembles the *shi nan* tree [see 347]. Its leaves are thick. They are gathered from the 5th to the 7th month.

SU SUNG [11th cent.]:—The plant is now produced in Yung chou [in Shen si, App. 424], Kiang chou [in Shan si, App. 123], Hua chou [in Shen si, App. 85] and in Hang chou [in Che kiang, App. 58]. It grows from 3 to 4 feet high and has a red stem. Leaves like those of the pomegranate tree but shorter and thicker ; they resemble also the leaves of the *shi nan* [*v. supra*]. In the 4th month it produces small white flowers, and in the 5th month fruit. Stem and leaves are officinal.

LI SHI-CHEN observes that the name *kuan ts'ao*, given in the *Pie lu* as a synonym of *yin yü*, is properly applied to another plant, a kind of sedge. [Comp. *Bot. sin.*, II, 455.]

Ch., XXIV, 42 :—*Yin yü.* Rude drawing, too indistinct to permit of identification.

SIEB. & ZUCC., *Fl. jap.*, I, 127, tab. 68 :—*Skimmia japonica*, Thbg. Order of *Rutaceæ*. Sinice 茵 芋. See also SIEB., *Icon. ined.*, II. This is an evergreen shrub, from 3 to 4 feet high. Flowers fragrant, white, with a tinge of red. The Japanese and the Chinese consider it to be poisonous, as the Japanese name *mijama sikimi*, or mountains *sikimi*, indicates. *Sikimi*, a name applied to *Illicium religiosum*, means "malignant fruit."

Phon zo, XXIV, 5 :—Same Chinese name. *Skimmia japonica.*—*Kwa wi,* 113.—

The *Skimmia japonica* was first described by KÆMPFER in his *Amœn. exot.,* 779 [*Sin san,* vulgo *mijama skimmi*], but the Chinese name affixed there is wrong. One of the characters indecipherable.

The descriptive details regarding the *yin yü* as given in the *P.* are too vague to permit of deciding whether *Skimmia* is meant. R. FORTUNE [*Tea Countries,* 329] notices *Sk. Reevesiana* as cultivated in Shanghai gardens. No *Skimmia* is mentioned in the *Ind. Fl. sin.*

160.—石龍芮 *shi lung jui.* P., XVII*b*, 50. T., LIX.

Pen king :—*Shi lung jui,* 地椹 *ti shen* (ground mulberry). The seeds and the skin of the root are officinal. Taste bitter. Nature uniform. Non-poisonous.

Some of the ancient authors identify the *shi lung jui* with *Rh ya,* 128, others with 134.

Pie lu :—Other names : 天豆 *t'ien tou,* 石能 *shi neng,* 魯果能 *lu kuo neng,* 彭根 *p'eng ken.* The *shi lung jui* grows in T'ai shan [in Shan tung, App. 322] among stones in rivers and marshes On the 5th day of the 5th month the seed is gathered, and in the 8th month the skin [of the root] and dried in the shade.

WU P'U [3rd cent.] calls it 水菫 *shui k'in* [a name properly applied to *Œnanthe,* an umbelliferous plant. See 250].

T'AO HUNG-KING :—The *shi lung jui* which is found in Mid China has seeds resembling those of the *she ch'uang* [*Cnidium.* See 49], but they are flat. This, which is not the true *shi lung jui,* is called 菌荣子 *tsz' ts'ai tsz'.* The genuine plant grows in the eastern mountains on rocks. It has soft, small and short leaves. Its seeds resemble those of the *t'ing li* [*Sisymbrium.* See 114], are of a yellow colour and somewhat pungent taste.

Su Kung [7th cent.]:—The plant which is now generally used is popularly called *shui k'in* [*v. supra*]. Its leaves resemble those of the *fu tsz'* [*Aconitum.* See 143]. Its fruit resembles a mulberry, whence the name *ti shen* [*v. supra*]. It grows in low marshes. The leaves and the seeds are of a pungent taste. The plant produced in Shan nan [S. Shen si, App. 268] has seeds as large as mallow-seeds, but that from Kuan chung [Mid Shen si, App. 158] and Ho pei [S. Chi li, App. 78] has small seeds like the *t'ing li* [*v. supra*], and they are less potent than those from Shan nan. It is not clear why T'AO HUNG KING asserts that the small-seeded is the true *shi lung jui.*

Su Sung [11th cent.]:—This plant now grows in Yen chou [in Shan tung, App. 404]. It grows in a bushy manner, producing many stems of a greenish purple colour. Three leaves at the top of one stalk. These leaves are small, short, and incised. Seeds like those of the *t'ing li* [*v. supra*] and yellow. Su Kung states that this plant is also called *shui k'in*. But this latter is a quite different plant. The plant from Yen chou is that spoken of in the *Pen ts'ao king*, and which T'AO HUNG-KING calls the true *shi lung jui.*

K'ou Tsung-shi [12th cent.]:—There are two kinds of the *shi lung jui.* That which grows in water has glabrous leaves and round seeds, that growing on dry land has its leaves covered with hair, and pointed seeds. Only the first is used in medicine.

Li Shi-chen:—The ancient authors have erroneously identified the *shi lung jui* with the *shui k'in*, which is a different plant and is used as a vegetable. The hairy *shi lung jui*, growing on dry land, spoken of by K'ou Tsung-shi, is very

36

poisonous. But that growing in water, when boiled can be used for food. It is also called 胡椒菜 *hu tsiao ts'ai* (pepper vegetable). It is a plant about one foot high. The root resembles that of the *tsi* [*Capsella*. See 251]. Ternate, dissected, glabrous leaves, small yellow flowers. The fruit is green, as large as a bean, and looks like an unripe mulberry. When rubbed between the fingers the small seeds, like *t'ing li* seeds, fall out. This is the *shi lung jui*. The people in Kiang and Huai [Kiang su and An hui] gather the leaves of this plant in the 3rd and 4th month and eat them boiled.

Ch., **XXIV**, 41 :—*Shi lung jui*. The drawing represents a *Ranunculus*, probably *R. sceleratus*, L. LI SHI-CHEN'S description agrees.

Phon zo, **XXIV**, 6 :—石龍芮, *Ranunculus sceleratus.*— *So moku*, X, 55 :—Same Chinese name, *R. sceleratus*, and [52] *R. ternatus*, Thbg.

SIEB., *Icon. ined.*, I :—Same Chinese name, *R. sceleratus*.

161.—牛扁 *niu pien*. *P.*, XVII*b*, 53. *T.*, CLXVI.

Pen king :—*Niu pien*. The root is official. Taste bitter. Nature slightly cold. Non-poisonous. It is used to cure ulcers. A decoction of it is good for killing the lice of cattle and other parasites. It is used in the treatment of diseases of cattle, whence the name (*niu* = cattle).

Pie lu:—The *niu pien* grows in Kui yang [S.E. Hu nan, App. 167] in river-valleys.

T'AO HUNG-KING :—This drug is little known now.

SU KUNG [7th cent.] :—This drug resembles the *shi lung jui*. The root is like the *ts'in kiao* [see 28] but is more slender. It grows in low marshy places. Its common name is *niu pien*, but in the *Court of Sacrifices*[39] it is called 扁特 *pien te* or 扁毒 *pien tu* (poison). It is very efficacious in destroying lice of cattle.

HAN PAO-SHENG [10th cent.]:—This plant is now produced in Ning chou [in Kan su, App. 234]. Its leaves resemble those of the *shi lung jui* [*Ranunculus.* See 160] and the *fu tsz'* [*Aconitum*, App. 143]. The root is gathered in the 2nd and 8th months.

SU SUNG [11th cent.]:—There is a peculiar sort of *pien te* in Lu chou [in Shan si, App. 204]. It flowers in the 6th month and bears fruit in the 8th. The leaves and the root are useful in destroying lice.

Ch., XXIV, 47 :—*Niu pien.* Rude drawing of a plant with palmate leaves.

So moku, X, 26 :—牛扁 *Aconitum Lycoctonum*, L.— See also *Phon zo*, XXIV, 11, 12 :—*Kwa wi*, 17.[40]

162.—鉤吻 *kou wen.* *P.*, XVIIb, 55. *T.*, CXXXV.

Pen king :—*Kou wen*, 野葛 *ye ko.* The root is officinal. Taste acrid. Nature cold. Very poisonous.

Pie lu :—The *kou wen* grows in Fu kao [App. 45] in mountain-valleys, also in Hui ki [in Che kiang, App. 98] and Tung ye[41] [App. 377]. One sort is called 固活 *ku huo.* When broken it emits a blue vapor.[42] It is taken up in the 2nd and 8th months.

[39] 太常.

[40] HOFFM. & SCHLT. give the above Chinese characters as the Chinese name for *Geranium Thunbergii.* But this seems to be a mistake. According to the *So moku* [XII, 46] the Chinese name of this plant is 犢牛兒, and in the *Ch.* [XII, 41], under the same Chinese name, *mang niu rh*, a *Geranium* is figured. See also *Kiu huang*, L, 19.

[41] Or perhaps Tung ye in the prefecture of Hui ki.

[42] 折之青烟出.

Wu P'u [3rd cent.]:—Other names : 除辛 *ch'u sin*, 毒根 *tu ken* (poisonous root). The *kou wen* grows in the mountains of Nan Yüe [S. China, App. 233], also in Han shi shan [App. 57] and in I chou [Sz ch'uan, App. 102]. Its leaves resemble those of the *ko* [*Pachyrhizus*. See 194]. Red, square stem, resembling an arrow. Yellow root, which is dug up in the 1st month.

T'ao Hung-king :—This drug when introduced into the mouth causes pain in the throat. There is one kind of the *kou wen* which in its leaves resembles the *huang tsing* [*Polygonatum*. See 7]. It has a purple stem and yellow flowers. The young plant has a strong resemblance to the *huang tsing*, and the plants are frequently confounded. As the first-mentioned plant is poisonous it causes death when taken.—[Comp. sub 7 the legend regarding these two plants.]

Su Kung [7th cent.]:—The *ye ko* grows in Kui chou [in Kuang si, App. 164]. In the south it is a common plant near villages, on burial wastes and on roads. The people there call it *kou wen*, but properly *kou wen* is the name for the herbaceous part of the plant, whilst the root is called *ye ko*. It is a climber. Leaves like those of the *shi* [*Diospyros*. See 279]. The fresh root, when just taken out, has a white skin and a yellow bone (centre). The old root resembles the *ti ku* [*Lycium*. See 345], the young root the *Han fang ki* [*Cocculus*. See 183]. It resembles also the *pai hua t'eng* [white-flowered *Liana*. Unknown to me]. When the fresh root is broken it does not exhale any vapor (odour), but when one year old a vapor is emitted from the small pores of the substance of the root. The root of the *kou k'i* [*Lycium*. See 345] shows the same peculiarity. That which the *Pie lu* states regarding the blue vapors [*v. supra*] is not intelligible. When people by mistake happen to eat the leaf of this plant they die, whilst sheep browse on it without being injured and even grow fat.

LI SHI-CHEN :—The *Nan fang ts'ao mu chuang* [see *Bot. sin.*, I, 38] says :—The *ye ko* is a climber. Its leaves resemble those of the *lo le* (*Ocimum basilicum*), are glabrous and thick. It is also called 胡蔓草 *hu man ts'ao*. When people happen by mistake to eat the leaves mixed with vegetables, they die in the course of half-a-day. The *Yu yang tso tsu* [8th cent.] states :—The *hu man ts'ao* grows in Yung chou and Yung chou [differently written in Chinese. Both in Kuang si, App. 429, 428]. Flowers flat (?), resembling the flowers of the *chi tsz'* (*Gardenia*). [See 335] but are slightly larger and of a yellowish white colour. Leaves of a blackish colour. When eaten by mistake they cause death in a few days. Counter-poison : the blood of white goose and duck.—The *Ling nan wei sheng fang* (Prescriptions for Preserving Life in Southern China) says :—The leaves of the *hu man ts'ao* resemble tea-leaves. Small yellow flowers. One leaf when introduced into the mouth causes hemorrhagies from all orifices (pores), followed by death. LI SHI-CHEN adds that this plant is also called 斷腸草 *tuan ch'ang ts'ao* (herb which cuts the bowels) and 爛腸草 *lan ch'ang ts'ao*. When it comes in contact with the bowels of man or beast they become black and gangrenous in half-a-day. The leaves are round and glabrous. The younger leaves in spring and summer are especially dangerous, the old leaves in autumn and winter are less injurious. The plant flowers in the 5th or 6th month. The flowers resemble those of the *kü liu* [see *Bot. sin.*, II, 238] and appear in dense clusters forming a spike. The kind which grows in Ling nan [South China] has yellow flowers, that of Tien nan [Yün nan, App. 338] has red flowers, and is also called 火把花 *huo pa hua*.

Ch., XXIV, 54, 55 :—*Kou wen.* The drawing represents only leaves. It cannot be decided what plant is meant. —*Ibid.*, 55 :—Verso sub 滇 | | *Tien* (Yün nan) *kou wen*, a plant is figured which resembles a *Polygonatum*.

In an interesting article on Chinese drugs published by FORD, CROW, etc., in the *China Review* [XV, 214], it is proved that the plant *kou wen* of the Chinese herbal is a poisonous twining plant, *Gelsemium elegans*, Benth., Order *Loganiaceæ*, yellow flowers, the root used in medicine. It is known at Hong kong under the name of 胡蔓強 *hu man kiang*, also 斷腸草 *tuan ch'ang ts'ao* and 大茶藥藤 *ta ch'a ye t'eng*.

Cust. Med., p. 152 (197):—*Kou wen* imported 1885 to Shang hai 68 piculs. It is stated to come from Han kow, but is not mentioned as exported from Hankow.

Phon zo, XXIV, 24:—鉤吻 or 野葛, *Rhus toxicodendron*, L., var. *radicans.*—*Ibid.*, 24, 25:—黃精葉鉤吻 [*kou wen* with *Polygonatum* leaves]. This drawing seems also to represent a *Rhus*.

But the *So moku* [II, 42] figures under the latter Chinese name *Croomia japonica*, Miq., Order *Stemonaceæ*.

SIEB., *Icon. ined.*, VIII:—Same Chinese name [*Polygonatum*-leaved *kou wen*], *Croomia pauciflora*, Torr.

163.—菟絲子 *t'u sz' tsz'.*[48] P., XVIIIa, 1. T., CLXIX.

Comp. *Rh ya*, 131, *Classics*, 450, 451.

Pen king:—*T'u sz' tsz'* (hare's silk), 菟蘆 *t'u lu*. The seeds are officinal. Taste pungent sweet. Nature uniform. Non-poisonous. The whole plant is also used in medicine.

Pie lu:—The *t'u sz' tsz'* grows in Chao sien [Corea, App. 9] in marshes and fields. It is a twining plant which fastens itself to herbaceous plants and trees (shrubs). The fruit is gathered in the 9th month and dried in the sun. That of a yellow colour, which is slender, is called 赤綱

---

[48] The first character is more generally written 菟 hare.

ch'i kang (red string). The larger (coarser) kind, which is of a light colour, is called 菟蔂 *t'u lei*. The medical virtues in both are the same.

In the *Kuang ya* [3rd cent.] this plant is termed 菟丘 *t'u k'iu*.

T'AO HUNG-KING :—This plant grows abundantly in the fields. It twines about herbaceous plants as the *lan* [*Polygonum*. See 123], the *ch'u ma* [*Bœhmeria*. See 88] and the *hao* (*Artemisia*). Seeds used in medicine.

TA MING [10th cent.]:—The plant looks like yellow floss-silk. It has neither leaves nor stem, twines about herbaceous plants of the fields, and continues growing even when these plants die. It flowers and bears fruit at the same time. The seeds are small like millet.

CHANG YÜ-HI [11th cent.] quotes the *Lü shi ch'un tsiu* [3rd cent. B.C.], which states that the *t'u sz'* has no root. The *Pao p'o tsz'* [13th cent.] states that where the *t'u sz'* grows the *fu t'u* [or *fu ling*, *Pachyma cocos*. See 350] is found in the ground, but it is not attached to the former.

SU SUNG [11th cent.] says that the statement of the *Pao p'o tsz'* has not been confirmed, and that the *t'u sz'* has nothing to do with the *fu ling*. The *t'u sz'* grows in Mid China. The best is found in Yüan kü [in Shan tung, App. 415]. The plant consists of fine filaments which twine about other plants. Its root soon abandons the ground and attaches itself to the pores of the foster plant. Some say that it has no root at all and lives upon the air.

LI SHI-CHEN :—The *t'u sz'* is a common plant in neglected places, gardens, and on old roads. Its seeds germinate in the ground, but the plant thus sent up [after having reached the foster plant] becomes detached from the root. It has no leaves. Flowers white or of a delicate red colour, and fragrant. The fruit, like a small bean and of a yellow colour, is produced on the stem of the plant. The plant is common in the forests of Meng chou in the prefecture of of Huai k'ing fu [in Ho nan, App. 220, 94] and is a much-valued medicine. Other names for it are 火燄草 *huo yen tsao* (blaze plant), 野狐絲 *ye hu sz'* (wild fox silk), 金線草 *kin sien ts'ao* (gold thread plant).

Cust. Med., p. 16 (137):—*T'u sz' tsz'* seeds exported 1885 from New chwang 84 piculs,—p. 36 (181), from Tien tsin 111 piculs,—p. 200 (249), from Ning po 29 piculs,—p. 80 (189), from Han kow 26 piculs.

Ch., XXII, 1:—*T'u sz' tsz'*. The drawing represents *Cuscuta*, Dodder.

TATAR., *Cat.*, 61:—*T'u sz' tsz'*. Semina *Cuscutæ europeæ*.

HANB., *Sc. pap.*, 240:—*T'u sz' tsz'*. *Cuscuta chinensis*, Lam.—P. SMITH, 87.

Phon zo, XXV, 2:—*Cuscuta japonica*, Chois., 兎絲 or 金線藤.

For further particulars see *Bot. sin.*, II, 450.

164.—五味子 *wu wei tsz'*. *P.*, XVIIIa, 4. *T.*, CLXX.
Comp. also *Rh ya*, 149, 240.

Pen king:—*Wu wei tsz'* (five tasted). The fruit is officinal. Taste sour. Nature warm. Non-poisonous.

Pie lu:—Other name: 玄及 *hüan ki*. The *wu wei tsz'* grows in Ts'i shan [in An hui, App. 350] in mountain-valleys, also in 代 Tai [in N. Shan si, App. 321]. The fruit is gathered in the 8th month and dried in the shade.

T'AO HUNG-KING:—Now the best comes from Kao li [N. Corea, App. 116]. The berry has much flesh (pulp), is sour and sweet. The next in quality is brought from Ts'ing chou [in Shan tung, App. 363] and Ki chou [in Chi li,

App. 119]. This is very sour. Its seeds are kidney-shaped. There is also the drug from Kien p'ing [in Hu pei, App. 139]. It has but little flesh, and the seeds are different. It is of a bitter taste.

Su Kung [7th cent.] in explaining the name *wu wei* (five tasted) says :—The skin and the pulp [of the fruit] are sweet and sour, the kernels are pungent and bitter, and the whole drug has a salty taste. The *wu wei tsz'*, which is also called 會 及 *hui ki*, is a plant which climbs on trees. Leaves like apricot-leaves but larger. The fruit is a berry like the *lo k'ui* [*Basella*. See 258] or the *ying tsz'* (wild grape). It is produced in P'u chou [in Shan si, App. 264] and Lan t'ien [in Shen si, App. 175]. It is sent as an annual tribute from Ho chung fu [in Shan si, App. 74].

Han Pao-sheng [10th cent.]:—It is a climber. Yellowish white flowers. Fruit purple when ripe and of a sweet taste.

Su Sung [11th cent.]:—It is very common in Ho tung and Shen si [*i.e.* in North China. See App. 80, 284] and is also found in Hang and Yüe [both in Che kiang, App. 58, 418]. The young plant is of a red colour. It climbs on high trees and is 5 or 6 feet long. The leaves are round and pointed, resembling apricot-leaves. It blossoms in the 3rd or 4th month. Flowers yellowish white, resembling in shape a *Nelumbium* flower. The fruit ripens in the 7th month. Many of these berries are collected together at the end of a common stalk. They are of a reddish purple colour and are used in medicine, for which purpose they are dried in the sun with the seeds.

Li Shi-chen :—There are two sorts of *wu wei tsz'*—the northern and the southern. The southern is red and the northern black. The plant is also cultivated. The people plant the root or raise it from seed.

Ch., XXII, 5 :—*Wu wei tsz'*. Rude drawing. It seems

that *Schizandra chinensis*, Baill., is intended, for this plant, very common in the Peking mountains, is known there under the above Chinese name. It produces red berries, each containing two kidney-shaped seeds.

TATAR., *Cat.*, 64 :— *Wu wei tsz'*, Baccæ *Kadsuræ chinensis*, Turcz. [same as *Schizandra chinensis*].—P. SMITH, 126.—

The *Kadsura chinensis*, Hance, is a different plant, from Southern China. *See* BENTH., *Fl. hongk.*, 8. The genus *Kadsura* is nearly allied to *Schizandra*, and in Japan the Chinese name 五昧子 is applied to *Kadsura japonica* and *Schizandra chinensis*. [Comp. *Bot. sin.*, II, 149.] The berries of both these species are red. *Schizandra nigra*, Max., from Japan, has black berries.

Cust. Med., p. 16 (138):—*Wu wei tsz'* exported 1885 from New chwang 806 piculs,—p. 202 (255), from Ningpo 199 piculs,—p. 36 (183), from Tien tsin, 2.3 piculs.— According to the *Hank. Med.*, exported also from Hankow.

165.—蓬蘽 *p'eng lei*. *P.*, XVIIIa, 7. *T.*, CXLIII.

Comp., *Classics*, 436.

Pen king :—*P'eng lei*. The fruit is officinal. Taste sour and pungent. Non-poisonous.

Pie lu :—Other names : 覆盆 *fu p'en*, 陵 | *ling lei*, 陰 | *yin lei*. The *p'eng lei* grows in King shan [in Hu pei, App. 149], in marshes, also in Yüan kü [in Shan tung, App. 415].

T'AO HUNG-KING :—*P'eng lei* is the name of the root. It is not used in medicine. *Fu p'en* is the name of the fruit. LI TANG-CHI [3rd cent.] says it is the edible fruit 莓子 *mei tsz'*, the juice of which has a pleasant taste, and which contains small kernels. But it seems to T'AO HUNG-KING that the *fu p'en* then used in medicine was somewhat different.

MA CHI and SU SUNG [10th and 11th cent.] say that the name *p'eng lei* refers to the whole plant, whilst *fu p'en* is the fruit of it. SU SUNG says it is a common plant, abounding in Ts'in [Shen si, App. 358] and Wu [Kiang su, App. 389]. It is only one foot high, and the whole plant is covered with spines. White flowers, reddish yellow fruit, like one half of a small ball with a pedicle below. Children like to eat it. The fruit is gathered in the 5th month, the leaves and the stem at all times of the year. In Kiang nan [Kiang su, An hui, etc., App. 124] it is called 苺 *mei*. But it is tardy in that soil, for it does not flower before the 8th or 9th month, and the fruit is gathered in the 10th.

TA MING [10th cent.] says *mei tsz'* is the fruit of the *p'eng lei*. The *fu p'en tsz'* [which is a different plant] is also called 樹 | *shu* (tree) *mei* [see 166].

WANG KI [16th cent.] states that in Hui chou [in An hui, App. 97] the *p'eng lei* is called 寒 | *han* (cool) *mei*.

LI SHI-CHEN :—There are five distinct species of the same genus [*Rubus, v. infra.*]. One is a climbing plant. The stem is covered with spines which are bent downward. The leaves which spring from the joints are as large as the palm of the hand, resemble mallow-leaves, are white underneath, thick, and covered with hair. It blossoms in the 6th or 7th month. Small white flowers. The fruit is on a pedicle, from 30 to 40 berries together in a cyme, at first greenish yellow, but dark red when ripe and covered with sparse black hair. The berry resembles a mulberry but is flat. The leaves do not fall off in winter. The common name of the fruit is 割田藨 *ko t'ien pao*.[44] This is the *p'eng lei* of the *Pen king.*

Another species, likewise a climber and smaller than the *p'eng lei*, has hooked spines. Small five foliolate leaves. They

[44] LI SHI-CHEN pronounces *pao*, but *W.D.* [682] *piao.*

are not white underneath, but glabrous, thin and without hairs.
It blossoms in the 4th or 5th month. White flowers. The
berries are smaller than those of the *p'eng lei*, less numerous,
at first greenish yellow, and when ripe they are of a dark red
colour. The leaves are deciduous. The popular name of this
is 插田藨 *ch'a t'ien pao*. This is the *fu p'en tsz'* of the
books [see 166] and the *kui* or *küe p'en* of the *Rh ya* [133].
This, as well as the *p'eng lei*, is officinal.

There is one species, also climbing, smaller than the
p'eng lei. It has trifoliolate leaves which are white underneath and covered with sparse hair. Small white flowers.
Red fruit like a cherry, appearing in the 4th month. This is
commonly called 耨田藨 *nou t'ien pao*. It is the *pao* of the
Rh ya [190]. This is not officinal.

One kind is like a tree, from 4 to 5 feet high. Leaves
like cherry-leaves but narrower and longer. It blossoms in
the 4th month. Small, white flowers. The fruit resembles
that of the *fu p'en tsz'* and is of a plain red colour. This is the
shan mei of the *Rh ya* [127] and the 懸鉤子 *hüan kou tsz'*
described by CH'EN TS'ANG-KI [8th cent]. The *P.* [XVIIIa,
11] devotes a special article to this plant. The name means
"hook hanging upside down," and refers to the recurved
spines. It is also called 沿鉤子 *yen kou tsz'*, 木 | *mu mei*,
樹 | *shu* (tree) *mei*.

One kind creeps on the ground. It is a small plant with
yellow flowers and fruit of a bright red colour resembling
those of the *fu p'en*. Not edible. This is the *she mei*
[*Fragaria indica*. See 167].

For the botanical identification of the names *p'eng lei*,
fu p'en tsz' and *hüan kou tsz'*, in China and Japan, see *Bot.
sin.*, II, 436, 133 and 127. Comp. also P. SMITH, 188.—
HENRY, *Chin. pl.*, 344-349.

According to PARKER [*in litteris*] 蒿央藨 *hao ying pao*

in Sz ch'uan is *Rubus trifidus*, Thbg.—HENRY, 345, has 栽秧 | | *ts'ai yang pao tsz'*, *Rubus parvifolius*, L. He means that this is the *nou t'ien pao* of the *Pen ts'ao*. He identifies [346] the *ch'a t'ien pao* of the *Pen ts'ao* with *Rubus coreanus*, Miq., *Ibidem*, 349 :—*T'ung* (winter) *pao tsz'*. R. *Lambertianus*, Ser., and *R. ichangensis*, Hemsl., are known by this name. These species have white flowers early in autumn and bear fruit at the beginning of winter.

Phon zo, XXV, 10 :—耨田藨, *Rubus parvifolius*, L.— *Ibid.*, 11 :—Same Chinese name, *R. phœnicolasius*, Maxim.

SIEB., *Œcon.*, 340 :—Same Chinese name, *Rubus trifidus*, Thbg. E. China.

Amœn. exot., 787 :—苺 *foo*, it. *moo*, vulgo *itzingo*. *Rubus vulgaris* fructu nigro. According to THBG. [*Fl. jap.*, 216] this is *Rubus cæsius*, L. [=*R. triflorus*, Rich.].

SIEB., *Œcon.*, 342 :—*Rubus moluccanus*, Thbg. [= *R. Buergeri*, Miq.]. Japonice : *fuju itsigo*. Sinice 寒苺.

166.—覆盆子 *fu p'en tsz'*. P., XVIIIa, 9. T., CXIV.

Comp. *Rh ya*, 133.

The *Pie lu*, as we have seen, gives *fu p'en tsz'* as a synonym of *p'eng lei* [see 165]. But subsequent writers agree in keeping it apart.

LI TANG-CHI [3rd cent.] in explaining the name says that *fu p'en* means a turned-over bowl, and refers to the shape of the fruit [*Rubus*].

CHEN KUAN [6th cent.] gives as synonyms 馬瘰 *ma lou* and 陸荆 *lu king*.

CH'EN TS'ANG-K'I [8th cent.] identifies the *fu p'en tsz'* [arbitrarily] with the 蘇蜜那 *su mi na* of Sanscrit books [*soma*].

The *T'u king Pen ts'ao* [11th cent.] gives as synonyms 西果草 *si kuo ts'ao* and 畢楞伽 *pi leng kia* [apparently a foreign name].

K'ou Tsung-shi [12th cent.]:—It is a common plant, especially in Ts'in chou [in Kan su, App. 358], in Yung chou [in Hu nan, App. 425] and in Hing chou and Hua chou [both in Shen si, App. 66, 85]. The plant has long branches (rods). The fruit ripens in the 4th or 5th month and is of a red colour. The mountain people then gather it and offer it for sale. It has a pleasant sourish sweet taste. The fruit is dried in the sun or boiled into jam.

Li Shi-chen [see above sub 165]:—Other names 烏薦子 *wu* (black) *pao tsz'*, 大麥莓 *ta mai mei*, 栽秧 | *ts'ai yang pao*. It is also called 插田 | *ch'a t'ien pao*, because it blossoms in the 4th or 5th month [when rice is transplanted. *Ch'a* = to transplant, *t'ien* = field], whilst the *p'eng lei* [see 165] is called 割田 | *ko t'ien pao*, because it blossoms in the 6th or 7th month [when corn is reaped. *Ko* = to cut].

The *fu p'en tsz'* is a raspberry, *Rubus*. For further particulars see *Bot. sin.*, II, 133.

In Tatar., *Cat.* [24], *fu p'en tsz'* is erroneously identified with Fructus *Humuli Lupuli*, and [*ibid.*, 24] *fou p'ing tsz'*, which is a common name for *Lemna* [see below, 198], is identified with Fructus *Rubi parvifolii*.—P. Smith, 115, 188.—

Cust. Med., p. 198 (203):—*Fu p'en* exported 1885 from Ning po 100 piculs,—p. 292 (284), from Amoy, 0.66.— According to *Hank. Med.*, [p. 13] exported also from Hankow.—

167.—蛇莓 *she mei*. *P.*, XVIIIa, 12. *T.*, CLXVII.

Pie lu:—*She mei* (snake berry). Taste of the juice of the berries sweet and sour. Nature very cold. Poisonous.

T'AO HUNG-KING :—The *she mei* is common in neglected gardens. The fruit is red, resembling the *mei* [raspberry. See 166] but is not good to eat, also not used in medicine.

HAN PAO-SHENG [10th cent.]:—It grows in damp places, has trifoliolate leaves, yellow flowers, and a red fruit resembling the *fu p'en tsz'* [*Rubus*. See 166]. The root is similar to that of the *pai tsiang* [*Patrinia*. See 108]. The fruit is gathered in the 4th and 5th, the root in the 2nd and 8th months.

K'OU TSUNG-SHI [12th cent.]:—It is a common plant in fields and by roadsides. It spreads along the ground. The leaves resemble those of the *fu p'en tsz'* but are smaller, glabrous and slightly wrinkled. Yellow flowers larger than those of the *tsi li* [*Tribulus*. See 128]. At the end of spring and in summer it produces red fruit resembling [in colour] the *li chi* (*Nephelium litchi*).

WU SHUI [Mongol period]:—It (the fruit) ripens and becomes red at the same time as the silkworm (*ts'an*) is full grown. That with a hollow centre is called 蠶 | *ts'an mei*, and that with a solid centre is 蛇殘 | *she ts'an mei* (berry destroyed by snakes).

WANG KI [16th cent.]:—It spreads along the ground, and is therefore called 地 | *ti* (ground) *mei*.

LI SHI-CHEN :—It is a slender plant which creeps on the ground and sends out roots from every joint (runner). Trifoliolate leaves. Leaflets serrated. It blossoms in the 4th or 5th month. Small yellow flowers with five petals. The fruit is of a bright red colour, resembling the *fu p'en tsz'*, but the insertion of the fruit-stalk is different. The root is fibrous. The juice of the fruit and the leaves together with the root are used in medicine.

Ch., XXII, 59 :—*She mei*. The drawing represents a *Fragaria*, probably *Fr. indica*, L., for the text says that the flowers are yellow. The description in the *P*. agrees well.

P. SMITH, 99 :—*She mei. Fragaria vesca.*—According to PARKER, *she pao* in Sz ch'uan is *Fragaria indica.*

HENRY, *Chin. pl.*, 350 :—*She pao tsz'. Fragaria indica.* A strawberry, with yellow flowers and beautiful red fruit, which has not the slightest flavour and is believed by the Chinese to be poisonous.

Cust. Med., p. 360 (272):—*She pao li* exported 1885 from Canton 1.06 picul.

So moku, IX, 31 :—蛇苺, *Fragaria indica*, and [28], same Chinese name, *Fragaria chilensis*, Ehrh.

Phon zo, XXV, 19-20 :—Same Chinese name, *Fragaria indica* and other species.

168.—牽牛子 *k'ien niu tsz'.* P., XVIIIa, 21. T., CLXX.

Pie lu :—K'ien niu tsz'. The seeds are officinal. Taste bitter. Nature cold. Poisonous.

T'AO HUNG-KING :—The *k'ien niu* is a climbing plant. The flowers resemble those of the *pien tou* [*Lablab vulgaris*] and are of a yellow colour. The seeds are produced in a small capsule. They are black, resembling those of the *k'iu tsz'* (*Cratægus*).

The *P'ao chi lun* [5th cent.] calls this plant 草金鈴 *ts'ao* (herbaceous) *kin ling* (golden bell, a name applied to the fruit of *Melia*).

SU KUNG [7th cent.]—The flowers of this plant resemble those of the *siïan hua* [*Calystegia* or *Convolvulus*. See 169], are of a blue colour, not yellow and also not like the *pien tou* [as T'AO HUNG-KING asserts].

SU SUNG [11th cent.]:—It is a common cultivated plant raised from seed which climbs on fences and walls. It grows from 20 to 30 feet long. Trilobed leaves. It blossoms in the 7th month. The flowers are reddish blue, resembling the

ku tsz' flowers [*Convolvulus*. See 169] but are larger. The
fruit (capsule) is produced in the 8th month. It has an
exterior white skin which encloses several small balls (cells)
each containing from 4 to 5 seeds as large as buck-wheat and
three-edged. There are two kinds, the black (seeded) and the
white.

LI SHI-CHEN :—There are two sorts of *k'ien niu*, the
black and the white. The first is a common wild plant. It is
covered with white hair [*tomentose*]. The stem when broken
discharges a white juice. Leaves trilobed like those of the
feng tree [*Liquidambar*. *Bot. sin.*, II, 261]. The corolla is
monopetalous and resembles that of the *sñan hua* but is larger.
The covering of the fruit (capsule) is at first green but when
dried it becomes white. The kernels (seeds) are like those of
the *t'ang k'iu tsz'* [*Cratægus*] but of a deep black colour. The
other kind, the white sort, is much cultivated. Its stem is
slightly red, not covered with hair, and has weak spines. When
broken it discharges a thick juice. The leaves are oblique,
round and pointed, and resemble the leaves of the *shan yao*
[*Dioscorea*. See 262]. The flowers of this species are smaller
than those of the black kind, and are of a pale blue colour,
tinged with red. The fruit (capsule) has a peduncle more than
an inch long. It (the peduncle) is at first green, but when
dried becomes white. The kernels (seeds) are white and coarse.
The people gather the unripe fruit and roast it with honey to
prepare a sweet meat. They call it 天茄 *t'ien k'ie*, because
its fruit-stalk is like that of the *k'ie* or brinjal. The black
and the white *k'ien niu* are also called 黑丑 *hei ch'ou* and
白丑 *pai ch'ou* (black and white *ch'ou*. I suspect, by the
second character (鈕 *niu*) a button was originally intended,
for the capsule of the plant resembles a Chinese button).
Other names 盆甑草 *p'an tseng ts'ao* [from the *Yu yang tsa
tsu*, 10th cent.], 狗耳草 *kou rh ts'ao* (dog's ear).

38

Ch., XXII, 61 :—*K'ien niu tsz'*. Good drawing. *Ipomœa* [*Pharbitis*] *Nil*, Roth.

LOUR., *Fl. cochin.*, 133 :—*Convolvulus tomentosus*, L. [*Pharbitis tomentosa*, Chois.]. Sinice *khien nieu*. Seminibus nigris. The seeds officinal.

TATAR., *Cat.*, 1, 7 :—*Pai* (white) *ch'ou* and *hei* (black) *ch'ou*. Semina *Pharbitidis Nil*.—P. SMITH, 170.

At Peking *pai ch'ou* are the white seeds of *Ipomœa* [*Pharbitis*] *Nil*, cordiform, not lobed, tomentose leaves, frequently cultivated,—and *hei ch'ou*, the black seeds of the wild-growing *Ipomœa hederacea*, L. [*I. triloba*, Thbg., *Pharbitis triloba*, Miq.], with trilobed leaves. Both species have beautiful blue flowers. Comp. MAXIMOWICZ, *Diagn. pl. asiat.*, fasc. VI (1886), p. 482 : —*Ipomœa Nil*, seminibus albis pedunculis crassis, elongatis;—*I. hederacea* [*triloba*] seminibus nigris pedunculis brevibus, tenuibus.

HENRY, *Chin. pl.*, 60 :—*K'ien niu*. *Pharbitis hederacea*. Occurs at Ichang as a weed in gardens.

Cust. Med., p. 128 (123) and 130 (140):—Exported from Chin kiang in 1885 : *hei ch'ou*, 60 piculs, *pai ch'ou*, 11 piculs.—Both sorts also exported from Hankow [*Hank. Med.*, 12].

Amœn. exot., 856 : — 牽 牛 *kingo*, vulgo *asagawo*, *i.e.* mane aperiens. *Convolvulus vulgaris*, flore majore albo matutino.—THBG. [*Fl. jap.*, 86] identifies this with *Ipomœa triloba*.

So moku, IV, 18 :—Same Chinese name. *Pharbitis triloba*, Miq.

169.—旋花 *süan hua*. *P.*, XVIIIa, 27. *T.*, CXXI.

Comp. *Rh ya*, 51. *Classics*, 442.

Pen king:—*Silan hua* [this name has the same meaning as *Convolvulus*], 筋 根 *kin ken* (tendon root). The flowers and the root are officinal. Taste of the flowers sweet, and of the root pungent. Nature warm. Non-poisonous.

Pie lu :—Other name 美 草 *mei ts'ao* (beautiful plant). The *silan hua* grows in Yü chou [An hui, App. 413] in marshes. Gathered in the 5th month and dried in the shade. The flowers are called 金 沸 *kin fu.* The root heals broken tendons (whence the above name).

Su Kung [7th cent.]:—The *silan hua* is more properly called 旋 葍 *silan fu.* Its root resembles a tendon, whence the name *kin ken* [*v. supra*]. The *Pie lu* confounds this plant with another of a homophonous name, *viz.* the 旋 覆 *silan fu* [*Inula.* See 81], the flowers of which are officinal, and the *kin fu* belongs to the latter.

Han Pao-sheng [10th cent.]:—The *silan fu hua* grows in marshes. It is a twining plant. Its leaves resemble those of the *shu yü* [*Dioscorea.* See 262] but are narrower and longer. It has red flowers. The root has neither hairs (radical fibres) nor joints. It is eaten when steamed or cooked, and has a pleasant sweet taste. It is also known under the name of *kin ken* (tendon root) and is gathered [for medical use] in the 2nd and 8th months and dried in the sun.

Su Sung [11th cent.]:—Owing to the property of the root to heal broken tendons, the southern people call it 續 筋 根 *su kin ken.* Another name, which refers to its shape, is 独 腸 草 *tun ch'ang ts'ao* (sucking-pig's bowels).

K'ou Tsung-shi [12th cent.]:—It is a common plant in Ho pei [S. Chi li, App. 78], Pien si [in Ho nan, App. 249] and in Kuan and Shen [Shan si and Shen si, App. 158, 284], where it grows in fields. It blossoms in the 4th or 5th month. A small cutting of the root, when placed in the ground and properly watered, will produce a young plant

after ten days. A popular name for this plant is 鼓子花
ku tsz' hua (drum flower ; but, as LI SHI-CHEN explains,
ku tsz' means also a war trumpet).

LI SHI-CHEN :—The *süin hua* is a twining plant which
grows in fields, on walls, etc. Leaves like those of the
po ts'ai [Spinage] but smaller. It blossoms until autumn.
The flowers are of a pale red colour, resembling in shape those
of the white *k'ien niu* [*Pharbitis.* Sec 168] ; sometimes they
are double. The root is white and resembles a tendon. It does
not produce seeds (不結子,—but the last character sometimes
also means tubers). Another name for this plant is 天劍草
t'ien kien ts'ao (heavenly sword plant). There is also one
kind with double flowers, which is called 纒枝牡丹 *ch'en chi
mou tan.*

Ch., XXII, 13 :—*Süan hua.* Figures of two different
convolvulaceous plants. One of them is probably *Calystegia
sepium,* R. Br. The same plant seems also to be represented
in the *Kiu huang* [LI, 10] sub 菖子棍 *fu tsz' ken,* also called
燕 | *yen fu,* 打碗花 *ta wan hua,* and 兎兒茜 *t'u rh miao*
(hare leaf).—According to HENRY [*Chin. pl.,* 479] the last
name is used in Hu pei for *Calystegia sepium.*—*Ta wan hua*
at Peking is the common name for *Convolvulus arvensis,* L.,
var. *sagittatus.* The name *yen fu* is also used for it.

TATAR., *Cat.,* 48 :—*Suan hua. Calystegia sepium?*—
P. SMITH, 47.

Amœn. exot., 856 :—鼓子 *kos* et *kudsi,* vulgo *firagawo, i.e.*
meridie aperiens. *Convolvulus vulgaris* flore majori albo
meridiano. In hortulis ob ornatum colitur.—According to
FRANCHET & SAV., *firagawo* is the Japanese name for *Calystegia
japonica,* Miq.

So moku, IV, 22 :—旋花 *Calystegia japonica,* Miq.—
Phon zo, XXVI, 15 :—Same Chinese name and same
identification.—*Ibid.,* 16 :—纒枝牡丹 *Convolvulacea?* not
determined by FRANCHET.

170.—紫葳 *tsz' wei.* *P.,* XVIIIa, 29. *T.,* CVI.
Rh ya, 164, 165.—*Classics,* 448.

Pen king:—*Tsz' wei,* 陵苕 *ling t'iao,* 茇華 *po hua.*
The flowers and the root are officinal. Taste sour. Nature
slightly cold. Non-poisonous.

Pie lu:—The *tsz' wei* grows in Si hai [in Shan tung,
App. 298] in river-valleys, also in Shan yang [in Shan tung,
App. 270].

WU P'U [3rd cent] gives the synonyms 武威 *wu wei,*
瞿陵 *k'u ling* and 鬼目 *kui mu.*

T'AO HUNG-KING takes the *tsz' wei* erroneously to be
the root of the 瞿麥 *k'u mai* [*Dianthus.* See 112], and is
refuted by subsequent writers.

CHEN KUAN [7th cent.] gives the synonym 女] *nü wei.*
SU KUNG [7th cent.] calls it 凌霄花 *ling siao hua.*

SU SUNG [11th cent.]:—It is a common plant in the
mountains and also much cultivated in gardens. It climbs on
high trees up to the summit. It blossoms in the summer.
Yellowish red (orange-coloured) flowers. They are much
collected for medical use, especially in diseases of women.

LI SHI-CHEN:—The *ling siao* grows wild. It first creeps,
but having reached a tree it climbs up it. It has a very long
stem. Several leaves (leaflets) are on the same [common]
stalk (pinnate leaves). The leaves are of a dark green colour
and serrated. It blossoms from the beginning of summer till
autumn. Racemes composed of numerous large flowers like
those of the *k'ien niu* [*Pharbitis.* See 168]. The corolla is five-
cleft, orange-coloured, with small spots. In autumn the
colour becomes darker. The fruit, which appears in the 8th
month, is a pod 3 inches and more long. The seeds are light,
thin, and resemble those of the elm or the *ma tou ling*
[*Aristolochia.* See 54]. The root is long and resembles that

of the *ma tou ling*. Flowers and root officinal. LI SHI-CHEN says that the name *tsz' wei* means "beautiful red" [but *tsz'* is purple, and the plant has orange-coloured flowers]. *Ling siao* means "to strive skyward." This name is given to this *liana* on account of its climbing to the summit of high trees.

Ch., XXII, 25 :— *Tsz' wei*. *Tecoma (Bignonia) grandiflora*, Delaun. Good drawing.

This beautiful climber is frequently cultivated at Peking under the name of *ling siao hua*. The flowers are gathered for medical use. The description in the *P.* agrees well.

LOUR., 458 :—*Campsis adrepens* [same as *Tecoma grandiflora*]. Sinice, *lien sieu (ling siao)*.

TATAR., *Cat.*, 36, 61 :—*Ling siao, tsz' wei hua, Bignonia.* —P. SMITH, 38.

PARKER, *Canton pl.*, 261 :—*Tecoma grandiflora, ling siao hua*.

Cust. Med., p. 154 (234):—*Ling siao hua*, exported 1885 from Shang hai, 0.15 picul,—p. 364 (310) from Canton 0.12 picul.

For Chinese names applied to this plant in Japan, see *Bot. sin.*, II, 164, 165.

171.—營實 *ying shi*. *P.*, XVIIIa, 31. *T.*, CXVIII.

Comp. *Rh ya*, 295.

Pen king :—*Ying shi*, name of the fruit, 墻蘼 *ts'iang mi* (wall rose, name of the whole plant), 牛棘 *niu ki* (ox thorn). The fruit and the root are officinal. Taste of the fruit sour. Nature warm. Non-poisonous. The root is bitter, harsh and cold.

Pie lu:—Other names : 薔薇 *ts'iang wei*, 山棘 *shan ki* (mountain thorn), and 牛勒 *niu le*. The *ying shi* grows in

Ling ling [in Hu nan, App. 196], in river-valleys, also in Shu (Sz ch'uan). It is gathered in the 8th and 9th months and dried in the shade.

T'AO HUNG-KING :— *Ying shi* is the name for the fruit of the *ts'iang wei* plant. The best sort is that with white flowers. By boiling the stem and the leaves a beverage (decoction) can be made. The root is used for the preparation of a wine.

HAN PAO-SHENG [10th cent.]:—It is a twining plant. The stem is covered with spines. The flower has 6, 8 or also numerous petals. It is red or white. The fruit resembles that of the *tu t'ang* [*Pyrus*. See *Bot. sin.*, II, 482].

LI SHI-CHEN :— The *ts'iang wei*, which is also called 刺花 *ts'z' hua* (spiny flower), grows wild in the forests and on the borders of ditches. In spring children eat the young shoots after having stripped off the skin with the spines. It is a twining plant. The stem becomes hard (ligneous) and is covered with numerous spines. The leaves (leaflets) are small, pointed, thin and serrated. It blossoms in the 4th or 5th month. The flower has four petals. Its heart (anthers) is yellow. The flowers are white or pale red. The fruit when ripe is of a red colour. The kernels (seeds) are covered with white hair like those of the *kin ying tsz'* (*Rosa*). It is gathered in the 8th month. The root can be taken up at any season of the year. This plant is also cultivated in gardens. Then the stem becomes coarser, and the leaves and flowers also become larger. There are white, yellow, red and purple flowered varieties. One kind, with large flowers, is called 佛見笑 *Fo kien siao* (Buddha sees it and smiles); another, with small flowers and very fragrant, is 木香 *mu hiang*. The people in the south prepare a fragrant water from the petals of the *ts'iang wei* flowers.

Ch., XXII, 16 :— *Ying shi* or *ts'iang mi*. The figure represents a Rose, as also the drawing sub *ts'iang mi* in the

Kiu huang [XLIX, 24].—*Ch.*, XXIX, 24:—*Pai ts'iang wei*, a white Rose from Yünnan.—*Ibid.*, XXI, 43:—*Fo kien siao*, a large double-flowered Rose, and 47 :—*Mu hiang*. Seems to represent *Rosa Banksiæ*, which is cultivated at Peking under the name of *mu hiang*.

LOUR., *Fl. cochin.*, 396 :—*Rosa indica*, L. Sinice *tsiam hoa.*—

P. SMITH, 187.

Amœn. exot., 862 :—薔薇 *foo sen*, it. *kin fo qua*, vulgo *ibara*, it. *iyi*, *i.e.* spina, *igino fanna*, *i.e.* flos spinæ, vel mutuato a Lusitanis vocabulo : *Rosa*. *Rosa* frutex spinosus nostras.— THBG. [*Fl. jap.*, 214] means that this is *Rosa canina*, L.

Phon zo, XXVII, 2 :—營 實 Rose, not determined by FRANCHET.—*Ibid.*, 2 :—薔 薇 *Rosa multiflora*, Thbg., and 3-6 [under the same Chinese name] several varieties of the same.—*Ibid.*, 7 :—黄 薔 薇, a yellow Rose.—*Ibid.*, 8 :— 木 香 *Rosa Banksiæ*.

172.—栝 樓 *kua lou*.   *P.*, XVIIIa, 34.   *T.*, CLI.

Comp. *Rh ya*, 23, *Classics*, 385.

Pen king :—*Kua lou*, 地 樓 *ti lou*. The fruit and the root are officinal. Taste of the fruit bitter. Nature cold. Non-poisonous.

Pie lu :—Other names : 天瓜 *t'ien kua* (heavenly gourd), 黄瓜 *huang* (yellow) *kua*, and 澤姑 *tse ku*. The *kua lou* grows in Hung nung [in Ho nan, App. 99] in river-valleys and shaded places in the mountains. The best sort is that with a root penetrating deep into the ground. That which grows in a saliferous soil is poisonous. The root is gathered in the 2nd and 8th months and dried in the sun during 30 days.

T'AO HUNG-KING :—It is a climbing plant of Mid China. It resembles the *t'u kua* [see the next], but its leaves are lobed. The root penetrates from 6 to 7 feet into the ground, grows very large, and is used for food. The fruit is employed in the preparation of ointments.

The *P'ao chi lun* [5th cent.] states that the round sort [apparently the root is referred to] is called 括 *kua*, the long one is 樓 *lou*. They are also distinguished as female and male [root].

Su Kung [7th cent.]:—That which is produced in Shen chou [in Ho nan, App. 283] and has a white fruit is considered the best.

Su Sung [11th cent.]:—It is a climbing plant. Leaves like those of the *t'ien kua* (melon), but narrower, lobed, and covered with fine hair. In the 7th month it blossoms. The flowers resemble the *hu lu* flowers [*Lagenaria*] and are of a pale yellow colour. The fruit is produced beneath the flower. It is as large as a fist, at first green but in the 9th month it ripens and then assumes an orange colour. It is globular in shape, sometimes also pointed or oblong. The medical virtues in all sorts are the same. The root is known under the name of 白藥 *pai yao* (white drug). It has a yellow skin and white flesh.

Li Shi-chen :—The name is also written 瓜蔞 *kua lou.* The root penetrates perpendicularly into the ground, and after some years it is several feet long. When dug up after autumn, when the plant bears fruit, the root contains a white flour like snow. This is called 天花粉 *t'ien hua fen* (heavenly flower starch), also 瑞雪 *shui süe*. In the summer no starch is found in the root, it then shows only tendons (coarse fibres). The fruit is globular or oblong, resembles a gourd or a persimmon, and is of a yellow colour. In the mountains children eat it. It contains flat seeds as large as the seeds of the *s:' kua* (*Luffa*). The outer skin of these seeds is of a grayish colour ; the kernel is green and contains much oil, which is expressed and can be used as lamp-oil.

39

Ch., **XXII**, 27 :—*Kua lou*. Two figures. Both seem to represent *Trichosanthes*. See also *Kiu huang*, LIII, 18.—*Ch.*, XX, 59 :—*T'ien hua fen*. Rude drawing of a plant said to grow in Miug chou [in Che kiang, App. 224]. It is bitter and poisonous, and not the same as the *kuo lou*.

Lour., *Fl. cochin.*, 629 :—*Solena heterophylla* [*Bryonia*, see *D.C. Prodr.*, III, 306]. Sinice : *khu leu, tien hoa fuen*. Bacca coccinea, seminibus nigricantibus. Radix tuberosa, fasciculata, tuberibus farinaceis, albicantibus, edulibus.

Tatar., *Cat.*, 27 :—*Kua lou*, Fructus globosus *Trichosanthes palmatæ*, and [p. 56] *t'ien hua fen*, Radix *Trichosanthes palmatæ*, and [p. 9] *hua fen*, Radix *Bryoniæ*.—P. Smith, 43, 221.

Kua lou at Peking is the common name for *Trichosanthes Kirilowii*, Maxim. [See *Bot. sin.*, II, 385.] But the drug *t'ien hua fen* obtained from a Chinese apothecary's shop there had cylindrical roots, four inches long, one inch in diameter, and contained a white flour ; it did not seem to be the root of the above plant.

Henry, *Chin. pl*, 193 :—*Hua k'u kua, t'ien hua fen, kua lou t'eng* in Hu pei names for *Trichosanthes multiloba*, Miq., and *T. Kirilowii*.

Cust. Med., p. 72 (96):—Hankow exported 1885 *kua lou p'i* (skin of the fruit) 220 piculs,—p. 76 (157), *kua lou jen* (seeds) 65 piculs,—p. 66 (24), *t'ien hua fen* 54 piculs.

Ibid., p. 126 (77):—Chin kiang exported *kua lou p'i*, 71 piculs,—p. 130 (132), *kua lou jen*, 38 piculs,—p. 124 (59), *t'ien hua fen*, 21 piculs.

Ibid., p. 368 (378):—Canton *kua lou* 23 piculs,—p. 368 (377), *kua lou jen* 40 piculs.

So moku, XX, 35 :—括樓 *Trichosanthes japonica*, Regel.

Sieb., *Œcon.* 235 :— *Trichosanthes cucumerina*, Miq. Japon.: *tenk ha fun*. Sinice : 天花粉, Adhibetur radix ab agricolis in exanthematibus.   In a MS. note [St. Petersburg] SIEBOLD says that the root of this plant sometimes weighs several pounds, and that starch is extracted from it.

173.—王瓜 *wang kua*.   P., XVIIIa, 40.   T., XLV.

Comp. *Rh ya*, 34, 152,—*Classics*, 386.

Pen king:—*Wang kua* (royal gourd), 土瓜 *t'u kua*. Root and fruit officinal.   Taste of the root bitter.   Nature cold.   Non-poisonous.   The fruit is sour and bitter.

Pie lu:—It grows in the country of Lu [Shan tung, App. 202] in marshes and fields, near dwellings and on walls. The root is dug up in the 3rd month and dried in the shade.

T'AO HUNG-KING:—The *t'u kua* grows on fences and walls.   The fruit when ripe represents a red ball.

SU KUNG [7th cent.]:—It is a climbing plant.   The leaves resemble those of the *kua lou* [*Trichosanthes*.   See 172] but are not lobed.   They are hispid.   It blossoms in the 5th month.   Yellow flowers.   The fruit is produced beneath the flower.   It is of a red colour when ripe and of a globular shape.   The root resembles that of the *ko* [*Pachyrhizus*.   See 174] but is smaller and contains much starch.   It is called *t'u kua* root.   In that sort which grows in the north the root has tubers (子) as large as a jujube with a yellow skin and white flesh.   The leaves and the fruit are similar [to the southern plant] but the root is different.   For medical use the southern drug is preferable.

SU SUNG [11th cent.] states that in Kün and in Fang [both in Hu pei, App. 172, 35] the plant is called 老鸦瓜 *lao ya kua* (crow melon), for crows are fond of the fruit. Another name is 菟瓜 *t'u kua*.

K'OU TSUNG-SHI [12th cent.]:—The fruit of the *wang kua* is one inch thick and two inches long, the upper end is round the lower end pointed.   It ripens in the 7th and 8th months,

and is then of a scarlet colour. The seeds within resemble
the head of a *Mantis*. Its common name is 赤雹 *ch'i pao*
(red hailstone). The root is called 土 瓜 根 *t'u kua ken*.
It is fibrous, and produces from three to five pale yellow tubers
of the size of a finger. The seeds as well as the fruit are used
in medicine.

LI SHI-CHEN :—The *wang kua* is a climbing plant. It is
hispid. The young plant is eaten. The leaves are roundish
and resemble a horse's hoof in shape, but are pointed, green on
the upper side, and paler and rough underneath. It blossoms
in the 6th or 7th month. Small yellow flowers with a 5-cleft
corolla, in racemes. The ripe fruit is red or yellow with a
rough skin. The root does not resemble the *ko* root [as SU
KUNG asserts] but is rather like a small *kua lou* root. The
flour contained in the root is very white and fat. To obtain
the real root it must be dug up to the depth of from 2 to 3 feet.
The people of Kiang si cultivate it in a rich soil, and use the
root for food, as they use the *shan yao* (*Dioscorea*).—Other
names of the plant: 馬㼋瓜 *ma pao kua*, 野甜瓜 *ye t'ien
kua*, 師姑草 *shi ku ts'ao*, 公公鬚 *kung kung sü.*

Ch., XXII, 30 :—*Wang kua* or *ch'i pao*. *Thladiantha
dubia*, Bge. [order *Cucurbitaceæ*].—See also *Kiu huang*, LII,
15, *ma pao rh.*

TATAR., *Cat.*, 15 :—*Ch'i pao*. Fructus *Thladianthæ
dubiæ*.—This plant is much cultivated at Peking under the
name of *ch'i pao rh*. The description in the *P.* agrees in a
general way.

In Japan, where no species of *Thladiantha* is met with,
the Chinese name 王瓜 is applied to *Trichosanthes cucumeroïdes*,
Ser. See *So moku*, XX, 34.

174.—葛 *ko*. *P.*, XVIIIa, 42. *T.*, CXI.
Comp. *Classics*, 390.
Pen king :—葛根 *ko ken*, name of the root, 葛穀 *ko ku*,

name of the fruit, 鷄齊 *ki ts'i*. The root, fruit (seeds) and the flowers are officinal. Taste of the root sweet and pungent. Nature uniform. Non-poisonous. Taste of the seeds sweet. Non-poisonous.

Pie lu:—Other names 鹿藿 *lu huo,*[45] 黃斤 *huang kin.* The *ko ken* grows in Wen shan [in Sz ch'uan, App. 388] in mountain-valleys. The root is dug up in the 5th month and dried in the sun.

T'AO HUNG-KING:—Now the people use the *ko ken* root for food, and eat it steamed. It is a large root which penetrates deep into the earth. It is broken to pieces and dried in the sun. The best comes from Nan k'ang and Lü ling [both in Kiang si, App. 229, 208]. It has much flesh and but few fibres, is of a sweet pleasant taste, but as a medicine it is unimportant.

SU KUNG [7th cent.]:—The *ko* root penetrates into the ground from 5 to 6 inches. The upper part is called 葛脰 *ko tou* (neck of the *ko*). It has emetic power and is somewhat poisonous.

For further particulars regarding the *ko*, which is the *Pachyrhizus Thunbergianus*, S & Z. (*Pueraria Thunbergiana*, Benth.), a plant much cultivated in China and Japan for its textile fibres and for its edible root [see *Bot. sin.*, II, 390] I defer a more detailed account of this plant to another part of my *Botanicon sinicum.*

TATAR., *Cat.*, 26 :—葛根 *ko ken.* Lignum griseum, and 葛條花 *ko t'iao hua,* Flores *Pachyrhizi trilobi.*—*Ko t'iao* at Peking is a common name for the wild-growing creeper *Pachyrhizus Thunbergianus.*— P. SMITH, 88, sub *Dolichos trilobus.*

HENRY, *Chin. pl.*, 176 :—葛藤 *ko t'eng, Pueraria Thunbergiana.* The root is made into *ko fen*, an arrowroot-like preparation.

[45] This name is properly applied to another plant. See 260.

Cust. Med., No. 601 :—*Ko ken, Pachyrhizus angulatus,* Rich.

Cust. Med., p. 342 (58) :—*Kan ko* (dried *ko* root) exported 1885 from Canton 215 piculs,—p. 276 (53), from Amoy 3.34 piculs,—p. 142 (51), from Shanghai [*ko ken*] 1.71 picul.—*Hank. Med.*, 13 :—*Fen ko* (*ko* starch) exported from Hankow.

Cust. Med., p. 344 (65):—From Canton exported 葛頭 *ko t'ou* (head) 64 piculs.—*Ibid.*, p. 362 (304):—*Ko hua* (flowers) from Canton 5 piculs,—p. 196 (182):—From Ning po 4.30 piculs,—p. 290 (241):—From Amoy 0.10.—

175.—黃環 *huang huan. P.*, XVIIIa, 46. *T.*, CLIV.

Pen king :—*Huang huan* (yellow ring or circle), 大就 *ta tsiu,* 凌泉 *ling ts'üan.* The root is officinal. Taste bitter. Nature uniform. Poisonous.

Pie lu :—The *huang huan* grows in Shu [Sz ch'uan] in mountain-valleys. The root is dug up in the 3rd month and dried in the shade. The fruit (seed) of the *huang huan* is called 狼跋子 *lang po tsz'* (wolf step. LI SHI-CHEN explains that the pod resembles a wolf's foot). It is of a bitter taste and slightly poisonous.

WU P'U [3rd cent.]:—The *huang huan* of Shu is also called 生芻 *sheng ch'u* and 根韭 *ken kiu.* The plant begins to grow in the 2nd month and then is of a red colour. It attains a height of 2 feet. The leaves are round and large. The *King* [*Pen king*] states that the leaves contain a yellowish white juice. In the 5th month the plant produces round fruits. The root is dug up in the 3rd month. It has veins like the radiating spoke of a wheel [probably seen on a transverse section].

T'AO HUNG-KING :—It resembles the *fang ki* [*Cocculus?* See 183]. It (the root) shows veins like a radiating spoke of a wheel, is seldom used in medicine and little known.—The *lang po tsz'* is produced in Kiao and Kuang [Kuang tung and Kuang si, App. 112, 160]. The seeds are flat. They are bruised and thrown into the water to kill fish.

SU KUNG [7th cent]:—The *huang huan* abounds in Siang yang [in Hu pei, App. 306]. The people of Pu si [N. Sz ch'uan, App. 236] call it 就葛 *tsiu ko*. It is cultivated in gardens. It is a climber. Large specimens (the stem) measure from 6 to 7 inches in diameter. The root resembles the *ko* root [see 174], but when eaten it provokes violent vomiting. Rice-water is used to stop it. This is the true *huang huan*. Now that produced in Kien nan [Sz ch'uan, App. 136], which is kept in store in the Sacrificial Court,[46] and known also under the name of 鷄屎 | | *ki shi ko ken*, is not the true *huang huan*. The latter has purple flowers. Its fruit, which is called *lang po tsz'*, is a pod like that of the *tsao kia* [*Gleditschia*. See 325]. Now a drug from Kiao and Kuang [*v. supra*] is also received in the Sacrificial Court. This is the true fruit of the *huang huan*. Another name for the *huang huan* is 度穀 *tu ku*.

LI SHI-CHEN says that WU P'U's account of the *huang huan* is correct, but he does not understand what plant SU KUNG means.

Ch., XXII, 40 :—*Huang huan*. The figure seems to be intended for a leguminous plant.

Phon zo, XXVII, 17, 18 :—黃環. 狼跋子. The figure represents a leguminous plant. Not identified by FRANCHET.

176.—天門冬 *t'ien men tung*. P., XVIIIa, 47. T., CLXXIII.

"太常

Comp. *Rh ya*, 92, 108, and above, 104.

Pen king :—T'ien men tung, 顛 勒 *tien le*. The root
(tubers) is official. Taste bitter. Nature uniform. Non-
poisonous.

Pie lu :—The *t'ien men tung* grows in Feng kao [in Shan
tung, App. 41] in mountain-valleys. The root is gathered in
the 2nd, 3rd, 7th and 8th months, and dried in the sun.

T'AO HUNG-KING :—Feng kao is the name of a *hien*
(district) near the T'ai shan mountain [in Shan tung]. Now
the *t'ien men tung* is found everywhere on elevated places.
That with a large root and of a sweet taste is the best. The
ancient *Ts'ai yao lu* says that it is a creeping plant with
prickly leaves. It blossoms in the 5th month. White flowers.
In the 10th month it bears black fruits. The root consists of
twenty and more pieces (tubers). The *Po wu chi* [3rd cent.]
says :—The *t'ien men tung* has prickles on its stem, and smooth
leaves. It is also called 絺 體 *ch'i t'i* and 顛 棘 *tien ki*. The
root, when steeped in hot water, yields very white textile
fibres like those of the *ch'u* (*Bœhmeria*). The people of Yüe
[Che kiang, App. 418] call it 浣 草 *huan ts'ao* (clean herb).
But it is dubious whether this is the *t'ien men tung.*

SU KUNG [7th cent.]:—There are two sorts of *t'ien men
tung.* One is prickly and rough, the other has no prickles and
is smooth (glabrous). The plant has many names. *Huan
ts'ao* is the name for the root when cleaned [for the use of the
textile fibres].

SU SUNG [11th cent.]:—It is a common creeping plant
more than 10 feet high (long). The stem is as thick as a hair-
pin. The leaves resemble those of the *hui hiang* [*Fœniculum*],
are linear and glabrous. Sometimes the plant is provided with
prickles, or it has no prickles but is rough. It blossoms in
the summer. Small white flowers, in some varieties they are
yellow or purple coloured. In the autumn it produces black

fruits. After summer, when the plant has ceased flowering, it produces tubers which are attached to the fibrous root. These tubers are white or yellow or of a purplish colour, of the size of a finger, oblong, 2 or 3 inches long and from 10 to 20 in number. They somewhat resemble the tubers of the *po pu* [see 177]. That kind of *t'ien men tung* which is produced in Lo [Lo yang, in Ho nan] has large leaves, a coarse root, and is different. Another kind in Ling nan [S. China, App. 197] does not produce flowers.

CHANG YÜ-HI [11th cent.]:—The *t'ien men tung* is the *ts'iang mei* or *men tung* of the *Rh ya* [108]. PAO P'O-TSZ' [3rd or 4th cent.] calls it *tien ki* [*v. supra*] also 地門冬 *ti men tung*, 筵門冬 *yen men tung*. He states also that on each of the five sacred mountains [wu Yo. See *Bot. sin.*, I, p. 223] it has a different name, *viz.*: On the Eastern Yo [T'ai shan in Shan tung] it is called 淫羊藿 *yin yang huo*,—on the Central Yo [Sung shan in Ho nan] it is *t'ien men tung*,—on the Western Yo [Hua shan in Shen si] it is 菅松 *kien sung*,—on the Northern Yo [Heng shan in Shan si] it is 無不愈 *wu pu yü*,—on the Southern Yo [Heng shan in Hu nan] it is 百部 *po pu*. CHENG YÜ-HI observes that the *po pu* is not identical with the *t'ien men tung* as PAO P'O-TSZ' asserts [see the next]. The *yin yang huo* is likewise another plant [*Aceranthus*. See 18].

LI SHI-CHEN identifies the *t'ien men tung* with *Rh ya*, 92, 髦 *mao* or 顛蕀 *tien ki*, which according to the commentator is a plant with fine (linear) leaves and prickles. But other authors refer *Rh ya*, 108, to *t'ien men tung*. According to LI SHI-CHEN it is much cultivated.

Ch., XXII, 9 :—*T'ien men tung.* The figure represents a plant with linear leaves and tuberous roots, probably *Asparagus lucidus*, Lindl., for at Peking the latter plant is

40

cultivated under the above Chinese name.—See also *Kiu huang* [LI, 4], *t'ien men tung.* Synonyms : 萬歲藤 *wan sui t'eng* and other names.

Lour., *Fl. cochin.,* 268 :—*Melanthium cochinchinense.* Sin. : *lien muen tum.* Caulis tenuis, procumbens, aculeatus, aculeis brevibus, sparses. Folia linearia, triquetra, minuscula. Radix fasciculata, tuberibus oblongis, carnosis, fuscorufis. Flos albus. Tuber humectans, expectorans. Prodest in phthysi, siti et calore febrili. In usum etiam venit et jucunde sapit saccharo conditum. Hance suggests that Loureiro's plant may be *Asparagus lucidus.* [*Flora hongk.,* 371.]

Tatar., *Cat.,* 56 :—*T'ien men tung.* Radix *Asparageæ, Melanthii cochinchinensis,* Lour.—Gauger [43] describes and depicts the tubers of the *t'ien men tung.*—Hanbury, *Sc. pap.,* 257.—P. Smith, 145.

Henry, *Chin. pl.,* 463 :—*T'ien men tung* in Hu pei is *Asparagus lucidus.*

Cust. Med., p. 78 (184):—*T'ien men tung* exported 1885 from Han kow 920 piculs,—p. 60 (26), from I chang 139 piculs,—p. 212 (37), from Wen chow 52 piculs,—p. 190 (96), from Ning po 13 piculs,—p. 226 (92), from Fu chow 13 piculs.—

So moku, VII, 7 :—天門冬, *Asparagus lucidus.*

Sieb., *Œcon.,* 80 :—*Asparagus japonicus. Ten mondoo* 天門冬. Radices non vero caules juniores inter fercula ponuntur.

Miquel, *Prol. Fl. jap.,* 315 :—*Asparagus lucidus.* Radix (haud soboles) edulis statuitur, et prostant specimina juvenilia humilia radicis fibris in tubera oblonga stipitata tumentibus.

177.—百部 *po pu.* P., XVIIIa, 52. T., CXXXII.

Pie lu :—*Po pu.* The root is officinal. Taste sweet. Nature slightly warm. Non-poisonous.

T'AO HUNG-KING :—It is a common wild plant in the mountains. Its root has twenty and more tubers attached to the root, similar to those of the *t'ien men tung* [see 176], and is of a bitter taste.

CH'EN TS'ANG-K'I [8th cent.] :—The *t'ien men tung* [see 176] has ten and more short, roundish, fleshy tubers, of a sweet taste, attached to its root. The tubers of the *po pu* are 50 or 60 in number, stem-like, long, pointed, hollow in the centre, bitter.

The *Ji hua Pen ts'ao* [10th cent.] calls this plant 婆婦草 *po fu ts'ao*.

SU SUNG [11th cent.]:—It is a common plant in Kiang [Kiang su, An hui etc., App. 124], Hu [Hu kuang, App. 83], Shen [Shen si, App. 284], Ts'i and Lu [Shan tung, App. 348, 202]. Twining plant. Large, long, pointed, glabrous leaves, somewhat resembling bamboo-leaves. Beneath the root about 15 or 16 tubers are produced, of a white colour. They are used in medicine.

The *Cheng T'siao tung chi* [12th cent.] states:—The *po pu*, also called *po fu ts'ao*, is used for destroying worms, insects and flies. Its leaves resemble those of the *shu yü* [*Dioscorea.* See 262]. The root is similar to that of the *t'ien men tung*.

LI SHI-CHEN :—The *po pu* has fine leaves like the *hui hiang* (Fennel). It has a green, fat stem, which when young is eaten boiled. The root is nearly a foot long. The fresh root is fleshy and succulent, when dried it becomes hollow and poor within. It is also called *ye* (wild) *t'ien men tung.*

Ch., XXII, 32 :—*Po pu.* The figure represents a plant with oblong tubers. It seems *Roxburghia* is intended. The descriptions of the ancient Chinese authors agree, with the exception of LI SHI-CHEN's statement regarding the leaves.

LOUR., *Fl. cochin.,* 490 :—*Stemona tuberosa* [*Roxburghia gloriosoides,* Kth.]. Sinice : *pe pu tsao.* Caule scandente, foliis ovatis, septemnerviis, radice tuberosa, fasciculata, tuberibus longis, teretibus, utrinque attenuatis, albicantibus, edulibus.

TATAR., *Cat.*, 4 :—*Po pu.* Radix?—P. SMITH, 145. *Melanthium?*

The drug which I obtained, under the name of *po pu*, from a Peking drug-shop, were white cylindrical tubers, somewhat resembling stalks, hollow inside, and of a sweetish taste.

HENRY, *Chin. pl.*, 860 :—*Po pu ken* in Hu pei, the roots of *Stemona tuberosa*, Lour., which are used in medicine.

Cust. Med., p. 122 (43):—*Po pu* exported 1885 from Chin kiang 64 piculs,—p. 348 (122), from Canton 37 piculs,—p. 188 (66), from Ning po 3.38 piculs,—p. 280 (95), from Amoy 1.65 piculs.—Exported also from Han kow. See *Hank. Med.*, 32.

The plant described by KÆMPFER [*Amœn. exot.*, 784, sub 百部根 *fuckubukon, vulgo fekuso kadsura*] is *Pæderia fœtida*, L. See also BANKS, *Icon. sel. Kæmph.*, tab. 9. But according to Japanese botanists the above Chinese and the first Japanese name belong to *Stemona*, only the second Japanese name is applied to *Pæderia*.

So moku, II, 56 :—百 部 根. *Stemona (Roxburghia) sessilifolia*, Miq.—*Ibid.*, 57 :—蔓生百部根 [the first two characters mean "climbing"], *Stemona japonica*, Miq. The same Chinese name applied in SIEB., *Icon. ined.*, VIII, to *Roxb. phyllantha.*—Same identifications in *Fhon zo*, XXVIII, 3-5.—*Kwa wi*, 61.—

178.—萆 薢 *pei (pi) hiai.* P., XVIII*b*, 1. T., CXLVI.

Pen king:—*Pei hiai.* Root officinal. Taste bitter. Non-poisonous.

Pie lu :—Pei hiai, 赤 節 *ch'i* (red) *tsie* (joint). The *pei hiai* grows in Chen ting [iu Chi li, App. 11] in mountain-valleys. The root is gathered in the 2nd and 8th months and dried in the sun. Taste sweet.

Wu P'u [3rd cent.] calls it 百 枝 *po chi* [a name likewise applied to the *kou tsi, v.* 13]; in the *P'ao chi lun* [5th cent.] it is termed 竹 木 *chu mu.*

T'ao Hung-king :—It is a common plant. The root resembles that of the *pa k'ia* [*Smilax.* See 179] but it is larger, has but few excrescences and joints, and the colour is paler.

Su Kung [7th cent.]:—There are two kinds. One has a prickly stem and a white fleshy root; the other, which has no prickles, has a hollow, weak root. Climbing plant. Leaves like those of the *shu yü* [*Dioscorea.* See 262].

The *Ji hia Pen ts'ao* [10th cent.] says that the popular name of this plant is 白 菝 葜 *pai* (white) *pa kia* [comp. 179].

Su Sung [11th cent.]:—This plant is common in all the prefectures of Ho [Shan si and Chih li, App. 72], Shen [Shen si, App. 284], Pien tung [in Ho nan, App. 250], King [Hu pei, App. 145] and Shu [Sz ch'uan]. Climbing plant, trilobed leaves like those of the *shan shu* (*Dioscorea*), or the *lü tou* (*Phaseolus Mungo*). The flowers are yellow, red or white. Sometimes the plant does not flower, but produces white seeds or fruits [axillar bulbils?]. The root is yellowish white, as thick as three fingers, and has many joints. It is dug up in spring and in autumn and dried in the sun. The sort produced in Ch'eng te kün [not ascertained. App. 19] has a hard root like the *shan shu.* It climbs. The leaves are like those of buckwheat. Seeds three-edged.

Li Shi-chen :—The *pei hiai* is a climbing plant. Large leaves resembling those of the *pa k'ia* [*v.* 179]. The root is long and hard, resembling in size the *shang lu* root [*Phytolacca.* See 131]. The people confound it sometimes with the *t'u fu ling* [*Smilax pseudo-china.* See 179], but that is a quite different plant. WU P‘U confounds it with the *lou tsi* [*v. supra*]. In the Sung period the *pei hia* was brought as a tribute from Huai k‘ing [in Ho nan, App. 94].

Ch., XXII, 52 :—*Pei hiai.* The figure seems to represent a *Smilax.* Another name given there, 硬飯圍 *ying fan t‘uan* (hardened lumps of cooked rice).

Cust. Med., p. 68 (51):—*Pei hiai* exported 1885 from Han kow 185 piculs,—p. 58 (20), from I chang 146 piculs,—p. 212 (30), from Wen chow 64 piculs,—p. 188 (74), from Ning po 8 piculs,—p. 280 (92), from Amoy 0.25 picul.—

Amœn. exot., 827 :—蘋 [the first character of the name is omitted] *kai*, vulgo *tokoro.* Herba sylv. scandens, Bryoniæ albæ affinis, radice Zingiberis facie, eduli, etc. The plant is figured in BANKS, *Icon. Kumpf. sel.* [15]. It is *Dioscorea quinqueloba*, Thbg.

Phon zo, XXVIII, 9, 10 :—萆薢, *Dioscorea.*

So moku, XX, 55, 57 :—山萆薢, *Dioscorea sativa*, L., and *D. quinqueloba*, Thbg.

179.—菝葜 *pa k‘ia.* P., XVIIIh, 3. T., CXLVI.

Pie lu :—*Pa k‘ia.* The name is also written 菝葜 *pa k‘ia.* The root is officinal. Taste sweet and sour. Nature uniform, warm. Non-poisonous. It grows in the mountains. The root is gathered in the 2nd and 8th months and dried in the sun.

T‘AO HUNG-KING :—The *pa k‘ia* has a short purple stem covered with prickles. It (the root) is smaller than that of the *pei hiai* [*v.* 178] and of a darker colour. The people use it for making a beverage.

SU KUNG [7th cent.]:—The *pei hiai* has a thin, long, white root, but the root of the *pa k‘ia* consists of nodular masses of a yellowish red colour.

Su Sung [11th cent.]:—This plant is common in Mid China, in Kiang [Kiang su, An hui, etc., App. 124] and Che [Che kiang, App. 10]. It is a prickly climbing plant, from 3 to 4 feet long. The leaves resemble those of the *tung ts'ing* [*Ligustrum*. See 342] and the *wu yao* (*Daphnidium*) but are larger. It flowers in the autumn. Yellow flowers and black fruits of the size of a cherry. The root is massive. The people call it 金剛根 *kin kang ken* (diamond root).

Li Shi-chen :—The *pa k'ia* is a common mountain plant. It sometimes climbs, but the stem is strong, hard and covered with prickles. The leaves are large, round, like a horse's hoof, shining, and resemble the leaves of the *shi* (*Diospyros*), but not those of the *tung ts'ing* [*v. supra*]. In autumn it produces yellow flowers, followed by red fruits. The root is very hard and is covered with strong hairs like prickles. A decoction is made of the leaves. It is sour and harsh. The savages gather the leaves and the root and use them as a dye. It is also called 鐵菱角 *t'ie ling küe* (iron-water caltrop). Wu P'u [3rd cent.] erroneously identifies the *pa k'ia* with the *kou tsi* [*v. supra*].

Ch., XXII, 55 :—*Pa k'ia.* The figure represents probably *Smilax China*, L. The descriptions in the *P.* agree. The stem is prickly.—Comp. Henry, *Chin. pl.*, 65, 478. The plant from which China-root is obtained has been supposed to be *Smilax China*. But this is certainly not the source of the drug.

Amœn. exot., 782 :—拔葜. The drawing under this Chinese name is *Smilax China*. In the description [p. 781, *sankira*, etc.] another Chinese name is given, which is probably erroneous.

Phon zo, XXVIII, 14 :—Chinese name as above. *Smilax China.*

Sieb., *Œcon.*, 71 :—*Smilax China, kakubara* 拔葜. Radix antisyphilitica, agricolis laudata, folia hinc ac inde pro tabaco fumantur.

土茯苓 *t'u fu ling*.　P., XVIII*b*, 4.　T'., CLXXVI.

T'u [= native] *fu ling* = *Pachyma Cocos*.　See 350].　This
is a more modern name for another sort of China-root which
is known also by many other names, *viz*. 土萆薢 *t'u* (native)
pei hiai, 刺猪苓 *ts'z'* (prickly) *chu ling* [pig tubers.　See
352], 山猪糞 *shan chu fen* (mountain-pig's dung), 冷飯圑
leng fan t'uan [comp. WILLIAMS' *Chinese Commercial Guide*,
114], 仙遺糧 *sien i liang* (food left by the immortals), 硬飯
ying fan (hard food), and 山地栗 *shan ti li* (mountain
ground chestnut).

Lɪ SHI-CHEN says that this is the plant noticed by
T'ao Hung-king under the name of 禹餘糧 *Yü yü liang* [47] in

---

[47] This name (*Yü yü liang*) is properly that of a mineral. [See *P*., X, 9.]
As my friend Professor A. BILLEQUIN informed me, it is the so-called eagle-
stone, a kind of argillaceous oxide of iron.　But the same name is applied
in the *P*. to three different plants, viz. *Smilax*, *Ophiopogon* [see 104]
and the plant 蒒草 *shi ts'ao*. *P*., XXIII, 17.—*T*., CV [without figure].

The *shi ts'ao* is first noticed in the *Po wu chi* [3rd cent.], where it is
stated that in the islands of the Eastern Sea there grows a plant which is
called *shi*.　It has an edible grain-fruit, like barley, which ripens in the
7th month, and is gathered by the people till the beginning of winter.　It
is also called 自然穀 *tsz' jan ku* (spontaneous grain) and *Yü yü liang*,
not to be confounded with the mineral of the same name.

Lɪ SUN [8th cent.] says:—The fruit of the *shi* is like a ball.　It is
gathered in the 8th month and eaten by those [eastern] people, but in China
it is unknown.

WILLIAMS [*Dict*., 758] understands that the *shi* is a floating plant,
probably *Zostera*, but the Chinese text above translated does not say that
it floats on the sea.

Amœn. exot., 900 :—蒒 *setz*, vulgo *suge*.　Herba palustris, foliis aruodi-
naceis brevioribus tensis, ex quibus ad albedinem redactis construuntur
e'egantissimi pilei, quibus teguntur deambulantes fœminæ.

Phon zo, XLII, 3 :—Same Chinese name, *Carex pumiia*, Thbg. Japonice : *gobo mugi*.

SIEB., *Œcon*., 9 :—Same Chinese name.　Japonice : *futegusa, Curicea?*
Radices fibrosæ adhibentur pro penicellis scriptoriis.

MATSUMURA, 41 :—Same Chinese name.　Japonice : *kobo-mugi, Carex
macrocephala*, Willd.

the following terms :—In Southern China, in marshes, there grows a climbing plant with leaves resembling those of the *pa k'ia* [*r. supra*]. The root forms nodular, jointed masses like the *pa k'ia* root, of a red colour. It tastes like the *shu yü* [*Dioscorea*. See 262]. It is called *Yü yü liang* (Yü's extra food). When Emperor Yü once travelled in the mountains it occurred that there was no food. Then this root was gathered, whence the name.

CH'EN TS'ANG-K'I [8th cent.]:—The *ts'ao* (herbaceous) *Yü yü liang* grows near the seashore and in mountain-valleys. The root is like a cup, a conglomerate of nodular masses, one-half of it above the ground. The skin of the root is like that of the *fu ling* [*Pachyma Cocos*. See 350]. The flesh is red and of a harsh taste. The people use it for food as a substitute for grain in times of scarcity.

SU SUNG [11th cent.]:—In Shi chou [in Hu pei, App. 288] there grows one sort of this plant which is called *ts'z' chu ling* [*r. supra*]. It is a climber. In spring and in summer the people dig up the root and, having taken off the skin, dry it by artificial heat. It is used for curing contagious ulcers.

LI SHI-CHEN :—The *t'u fu ling* grows plentifully in the mountain-forests of Ch'u [Hu kuang, App. 24] and Shu (Sz ch'uan). It is a climber, resembles the *shun*,[48] and has a spotted stem. The leaves are not opposite, somewhat resemble large bamboo-leaves, but are thicker, shining and from 5 to 6 inches long. The root resembles that of the *pa k'ia*, is roundish, consists of a conglomeration of tubers of the size of a hen's or duck's egg, more or less deep in the ground, one foot or but a few inches. The flesh is tender and can be eaten raw. There are two sorts, a red and a white. The latter is used in medicine.

---

* 蓴 probably an erroneous letter, for this is a water-plant. See 199.

41

T'u fu ling is the Chinese name of the drug which we call China-root. As has been stated above, this is not yielded by *Smilax China*, as has been supposed in former times, but by one or several other species of *Smilax*. One of them is *Smilax glabra*, Roxb. See HANCE, on the source of the China-root of commerce [*Journ. Bot.*, 1872, p. 102]. This seems to be the plant represented in the *Ch.* [XX, 1] sub *t'u fu ling*.

A. CLEYER, *Specimen Medicinæ sinicæ*, 1682, 137 :—*Tu fo lim, Pao de China* (China-wood), uti vocant Lusitani, rubei coloris fere est. Est et aliud præstantius, coloris albi, *pe fo lim* dictum.—Comp. *infra*, 350.

LOUR., *Fl. cochin.*, 763, 710 :—*Smilax China*, L. Sinice : *thu fu lin.* Radix *sinensis rubra*. Radix horizontalis, longissima serpens : tuberibus oblongis, nodoso-verrucosis, interdum ramosis, intus pallidis, vel rubescentibus, semipedalibus, sub teretibus, sparsis. — MAXIMOWICZ [*Dec.*, X, 410] doubts whether LOUREIRO'S plant is really *Smilax China*.

DU HALDE, *la Chine*, I, 30, III, 647.—GROSIER, *la Chine*, III, 324, 328.

TATAR., *Cat.*, 61 :—*T'u fu ling*. Radix *Smilacis*.—GAUGER [44] describes and figures the same drug.—P. SMITH, 198 :—*T'u fu ling, Smilax chinensis*.

HENRY, *Chin. pl.*, 478 :—*T'u fu ling, Smilax* sp., from which China-root is obtained. It has often been supposed that China-root is obtained from *Smilax China*, but this is very common at I chang and Pa tung, and certainly is not the source of the drug. It is to be noted that the drug exported from Sz ch'uan as China-root is quite a different substance, viz. *Pachyma Cocos* [see 350]. Both this and *Smilax* root pass through the Customs under the same name. In Chinese books the *Pachyma* is *fu ling* or *pai* (white) *fu ling*, while the *Smilax* is distinguished as *t'u fu ling*.

Cust. Med., p. 86 (8):—*T'u fu ling* exported 1885 from Kiu kiang 4,831 piculs,—p. 60 (28), from I chang 94 piculs,—p. 416 (50), from Pakhoi 61 piculs.

HOFFM. & SCHLT., 553 :—土茯苓 *Smilax pseudo-China*, Thbg.—*Phon zo*, XXVIII, 17, 18 :—Same Chinese name, *Smilax*, not determined by FRANCHET.

180.—白蘞 *pai lien.* *P.*, XVIII*b*, 6. *T.*, CLVIII.

Comp. Classics, 453.

Pen king :—*Pai* (white) *lien,* 白草 *pai ts'ao.* The root is officinal. Taste bitter. Nature uniform. Non-poisonous.

Pie lu :—Other names : 白根 *pai ken* (white root), 兔核 *t'u ho* and 崑崙 *k'un lun.* The *pai lien* grows in Heng shan [in Hu nan, App. 61] in mountain-valleys. The root is dug up in the 2nd and 8th months and dried in the sun. Taste sweet.

T'AO HUNG-KING :—It is a common plant in Mid China, a climber, with a root like that of the *pai chi* [*Angelica.* See 51].

SU KUNG [7th cent.]:—The root resembles the *t'ien men tung* [*Asparagus.* See 176]. It consists of more than 10 tubers. Its skin is reddish black, and the flesh white, similar to that of the *shao yo* [*Pœonia.* See 52]. It does not resemble the *pai chi.* It is a climber with digitate leaves.

SU SUNG [11th cent.]:—This plant now grows in King chou, Siang chou [both in Hu pei, App. 146, 305], Huai chou, Meng chou [both in Ho nan, App. 93, 220], Shang chou [in Shen si, App. 278] and Ts'i chou [in Shan tung, App. 348]. A common climber in forests. Red stem. Leaves like small mulberry-leaves. It blossoms in the 5th month and afterwards produces fruit. The root consists of from 3 to 5 oblong tubers, like duck's eggs, collected together.

The skin is black and the flesh white. One sort is called 赤 |
ch'i (red) *lien*, for it (the root) has a red skin. But the
flowers, the fruit and the medical virtues are the same.

LI SHI-CHEN :—Another name for this drug is 猫兒卵
mao rh luan (cat's testicles).

Ch., XXII, 46 :—*Pai lien*. Rude drawing which does
not permit of identification.

HENRY, *Chin. pl.*, 275 :—*Mao rh luan, Vitis serianæfolia*,
Bge.—This plant is figured in the *Ch.* [XIX, 47] under the
name of 鵝抱蛋 *ó pao tan* (goose sitting on eggs).

Cust. Med., p. 346 (108):—*Pai lien* exported 1885 from
Canton, 3.75 piculs,—p. 280 (88), from Amoy 0.45 picul.

Vitis serianæfolia is a common plant in North and Mid
China and Japan. It has palmately 5-parted leaves and
tuberous roots.

So moku, IV, 17 :—白蘞, *Vitis serianæfolia.*—Comp.
also *Kwa wi*, 45.

181.—蘮魁 *che k'ui*. P., XVIIIb, 8. T., CLVI.

Pen king :—*Che k'ui*. The root is officinal. Taste sweet.
Nature uniform. Non-poisonous.

Pie lu :—It is a mountain plant. The root is dug up in
the 2nd month.

T'AO HUNG-KING :—The root resembles a small *yü*
[*Colocasia*. See 261]. The flesh is white and the skin
yellow. It grows in Mid China.

SU KUNG [7th cent.]:—This drug is slightly poisonous.
Large specimens of the root are as large as a *tou*, the smaller
as large as a *sheng*.[49] The plant climbs on herbaceous plants
and trees. Leaves like those of the *tu heng* [*Asarum*,

---

[49] Chinese dry measures. See *W.D.*, p. 771, 874.

See 41]. The plant noticed by T'AO HUNG-KING is not the che k'ui but rather the 土卵 t'u luan (ground testicle), which is also called 黃獨 huang tu. It is seldom used in medicine, but the people of Liang Han [probably Sz ch'uan, App. 188] eat it (the root) steamed.

HAN PAO-SHENG [10th cent.]:—It is a climbing plant. The leaves resemble those of the lo mo (Metaplexis) and the root is like the pa k'ia root [Smilax. See 179], with a purplish black skin and orange-coloured flesh. Larger specimens are contorted and as large as a sheng, the smaller are of the size of a fist.

LI SHI-CHEN:—The name che k'ui means "brown wine vessel." It refers to the shape of the root, which contains a brown juice. The people of Min (Fu kien) use the plant for giving the inner surface of jars a blue colour.[60]

Ch., XXII, 47:—Che k'ui. Figure of a plant unknown to me.

182.—茜草 ts'ien ts'ao. P., XVIIIb, 19. T., CIV.

Comp. Rh ya, 22, Classics, 393.

Pen king:—Ts'ien ts'ao, 茜根 ts'ien ken [root]. The root is officinal. Taste bitter. Nature cold. Non-poisonous.

Pie lu:—Other name: 地血 ti hüe (earth blood). The ts'ien ken grows in K'iao shan [in Shen si, App. 134] in mountain-valleys. The root is taken up in the 2nd and 3rd months and dried in the sun. The plant grows in shady mountain-valleys. It twines around herbaceous plants and trees. The stem is prickly. The fruit is like that of the tsiao (Zanthoxylon).

T'AO HUNG-KING:—The ts'ien ts'ao is used for dyeing a dark red colour. It is more common in the western

[60] 諸魁閩人用入染靑頗中云易上色. Not quite intelligible.

provinces than in Eastern China. The Chinese character applied to the plant is derived from 西 (west).

HAN PAO-SHENG [10th cent.]:—This plant is also called 染緋草 *jang fei ts'ao* (plant which dyes a dark red colour). Its leaves resemble jujube-leaves, are pointed and have a broad base (heart-shaped). Stem and leaves scabrous. The leaves issue from the joints, 5 together. It climbs on herbaceous plants and trees. The root is purplish red. It is taken up in the 8th month.

SU SUNG [11th cent.]:—It is now much cultivated in gardens. The *Shi ki* (Historical Records) state, that a man who plants a thousand *mou* (acres) with the *ts'ien* and the *chi* [51] plants, is considered to equal in wealth a nobleman who possesses thousand families. This proves the great importance of these plants.

LI SHI-CHEN:—The *ts'ien ts'ao* begins to grow in the 12th month [in January]. It is a twining plant several feet long, with a square stem, hollow inside, covered with small prickles. It has joints several inches apart. Five leaves issue from every joint. The leaves, which resemble the *wu yao* leaves (*Daphnidium myrrha*), are scabrous, and darker on the upper side. It blossoms in the 7th or 8th month. The fruit is like that of the *tsiao* (*Zanthoxylon*), with small seeds within. The plant has many names, viz. 血見愁 *hüe kien ts'ou*, 風車草 *feng ch'e ts'ao*, 四天王章 *sz' t'ien wang chang*, 四嶽近陽草 *sz' yo kin yang ts'ao*, 四補草 *sz' pu ts'ao*, 鐵塔草 *t'ie t'a ts'ao* and 過山龍 *kuo shan lung*. These names are from Taoist books.

Ch., XXII, 20:—*Ts'ien ts'ao.* The figure represents *Rubia cordifolia*, L.—See also *Kiu huang*, LII, 24, sub 土茜苗 *t'u ts'ien miao.*

---

[51] Another tinctorial plant, *Gardenia.* See 335.

TATAR., *Cat.*, 12 :—*Ts'ien ts'ao ken.* Radix *Rubiæ.*—
P. SMITH, 188.

HENRY., *Chin. pl.*, 57 :—*Ts'ien ts'ao, Rubia cordifolia*, L.
Cust. Med., p. 80 (199):—*Ts'ien ts'ao* exported 1885
from Han kow 134 piculs,—p. 132 (170), from Chin kiang
55 piculs,—p. 168 (412), from Shang hai, *hūe kien ts'ou*
[*v. supra*] 0.01 picul.

Amœn. exot., 912 :—茜 *sen*, vulgo *akanni.* Herba spitha-
malis vel pedalis, ramosa, procumbens, radice fibrosa densa,
foliis Nummulariæ, infectoribus pro colore serviens.

So moku, II, 64, 65 :—Same Chinese name, in Japanese
akane, *Rubia cordifolia.*

SIEB., *Œcon.*, 332 :—*Queria trichotoma*, Thbg.[52] Japonice:
akane; Sinice: 茜. Adhibetur Rubiæ nostratis ad instar tinctoria.

183.—防己 *fang ki.* P., XVIII*b*, 23. T., CXXX.

Pen king :—*Fang ki*, 解離 *kiai li.* Root officinal.
Taste pungent. Nature uniform. Non-poisonous.

Pie lu :—The *fang ki* grows in Han chung [S. Shen si,
App. 54] in river-valleys. The root is dug up in the 2nd
and 8th months and dried in the shade.

LI TANG-CHI [3rd cent.]:—It has a twining stem like the
ko [*Pachyrhizus.* See 174]. The root is white externally and
yellow internally, like the *kie keng* [*Platycodon.* See 6]. It
shows black veins in the substance, radiating like the spokes
of a wheel.

T'AO HUNG-KING :—Now the drug which is produced in
I tu and Kien p'ing [both in Hu pei, App. 104, 139] is
large, of a greenish white colour, empty and soft. This is

---

[x] SIEBOLD is evidently mistaken. *Queria trichotoma*, same as THUN-
BERG'S *Rubia* spicis ternis, is *Wickstræmia japonica*, Miq., which is not a
tinctorial plant.

considered the best. Another sort, with black spots and of a ligneous structure, is not valued. The root is also used for food.

CH'EN TS'ANG-KI [8th cent.]:—The two sorts spoken of by T'AO HUNG-KING are the 漢防己 *Han fang ki* and the 木丨丨 *mu* (ligneous) *fang ki*.

SU SUNG [11th cent.]:—Now this drug is produced in K'ien [N. Kui chou, App. 141]. That from Han chung [S. Shen si, App. 54] when broken (or rather cut transversely) shows radiating veins. It is yellow, solid and fragrant. Slender, weak stem, small leaves having some resemblance to the *k'ien niu* [*Pharbitis*. See 168]. If a section of the stem be taken and air be blown at one end it passes through it, the same as in the *mu t'ung* [see the next]. This is the drug of a greenish white colour, empty and soft [noticed by T'AO HUNG-KING]. There is another kind, which has an unpleasant odour. The skin of the drug is wrinkled and covered with warts.[53] This is called 木防己 *mu fang ki.* SU KUNG says that it is not used in medicine. This name, however, is met with in the prescriptions of CHANG CHUNG-KING [2nd cent.] and others.

Ch., XXII, 38 :—*Fang ki.* Rude drawing. Plant with berries.

TATAR., *Cat.*, 23 : — *Fang ki.* Radix *Convolvuli.* — P. SMITH, 70.

Cust. Med., 340 (28):—*Fang ki* exported 1885 from Canton 268 piculs.—Exported also from Han kow. See *Hank. Med.*, 12.

Amœn. exot., 909 :—防己 *boi*, vulgo *awu kadsura.* Hedera major sterilis. C. Pauh. P.

HOFFM. & SCHLT., 160 :—漢防己. *Cocculus japonicus,* DC. *Ibid.*, 161 :—木防己. *C. Thunbergii*, DC.

---

[53] 上有丁足了 literally nail-feet.

Phon zo, XXX, 4 :—防 己 *Cocculus Thunbergii*, DC.—
4, 5 :—滇防己, *Cocculus Thunbergii*. Japonice: *awa kadsura*,
female plant; *kampœ*, male plant.

Phon zo, 6, 7 :—防 己. This Chinese name applied to
Cocculus diversifolius, *Menispermum dauricum* and *Stephania
hernandifolia*. All these plants belong to the order *Meni-
spermaceæ*.

From the Chinese descriptions of the *fang ki* [the authors
describe only the root] it cannot be decided what plant is
meant. According to HENRY [*Chin. pl.*, 71, 186], *Cocculus
Thunbergii* is known by other Chinese names in Hu pei.

184.—通 草 *t'ung ts'ao*. *P.*, XVIII*b*, 25. *T.*, CXI.

Pen king :—T'*ung ts'ao*, 附支 *fu chi*. The twigs and
the fruit are officinal. Taste pungent. Nature uniform.
Non-poisonous.

Pie lu :—The *t'ung ts'ao* grows in Shi ch'eng [in An hui,
App. 285] in mountain-valleys, also in Shan yang [in Shan
tung, App. 270]. The twigs are gathered in the 1st and
2nd months and dried in the shade.

T'AO HUNG-KING :—It is now produced in Mid China.
The plant climbs on trees and contains a white juice. The
stem shows [on a transverse section] small holes (or rather
longitudinal tubes). If air be blown at one end of a section
of it, it passes through it, whence the name [*t'ung* means
"permeable"]. Some say that it is the 菖藤莖 *fu t'eng heng*.

WU P'U [3rd cent.] calls it 丁翁 *ting weng*.

SU KUNG [7th cent.] :—It is, when full grown, 3 inches
in diameter. 3 or 4 twigs (stalks) spring from each joint,
each with 5 leaves (leaflets) at the end. The fruit is from 3
to 4 inches long. It has a white pulp with black kernels, is
edible and of an agreeable, sweet taste. The southern people

42

call this fruit 燕覆子 *yen fu tsz'* also 烏 | | *wu* (black) *fu tsz'*. It is gathered in the 7th and 8th months.

CHEN KUAN [7th cent.] calls it 萬年藤 *wan nien t'eng*, and says that the root is used in medicine.

CH'EN TS'ANG-K'I [8th cent.]:—The people of Kiang tung [Kiang su, An hui, etc., App. 124] call it 畜菖子 *ch'u fu tsz'*, in Kiang si [App. 124] it is 挐子 *na tsz'*. It looks like an abacus-bag (?).[51] The pulp is yellow and the seeds black. It is eaten after the skin has been removed. SU KUNG, in stating that the pulp is white, means the 猴菖 *hou fu* (fruit).

SU SUNG [11th cent.]:—The plant is now found in Tse chou, Lu chou [both in Shan si, App. 345, 204], in Han chung [S. Shen si, App 54], in Kiang and Huai [Kiang su and An hui, App. 124, 89] and in Hu nan. It is a climber [the twigs] as thick as a finger, but the stem is sometimes 3 inches in diameter. The leaves are five at the end of a common petiole, somewhat resembling the leaves of the *shi wei* [*Niphobolus lingua*. See 205]. They have also some resemblance to *Pæonia* leaves. They stand in pairs opposite. It blossoms in summer and autumn. The flowers are purple or white. The fruit resembles a small quince, is edible and of an agreeable taste. CHEN SHI-LIANG [10th cent.] calls it 桴棪子 *fou yen tsz'*. Its twigs are now known under the name of 木通 *mu t'ung* also 通草 *t'ung ts'ao*, which latter name is also applied to the 通脱木 *t'ung t'o mu* [*Aralia papyrifera*. See *Bot. sin.*, II, 82].

LI SHI-CHEN:—There are two sorts of *mu t'ung*, the purple and the white. The [fruit of the] purple has a thick skin and is of a pungent taste, the white has a thin skin and is insipid.

Ch., XXII, 37:—Rude drawing. *Convolvulacea?*

[51] 算袋.

LOUR., *Fl. cochin.*, 422 :—*Clematis sinensis.* Sinice : *mu tum.*

TATAR., *Cat.*, 40 :— *Mu t'ung.* Radix *Clematidis?*—
P. SMITH, 66.—

The drug *mu t'ung* obtained from a Peking drug shop was in thin slices, transverse sections of a ligneous stem, half-an-inch in diameter, the marrow showing small holes like a sieve (longitudinal canals) and was examined at Kew. It proved to belong to *Akebia quinata.* The description in the *P.* agrees. The *Ind. Fl. sin.* [I, 30] notices for China *Akebia quinata*, Dene., and *A. lobata*, Dene.

The name *t'ung ts'ao* is nowadays, it seems, more generally applied to *Aralia papyrifera*, in China. HENRY, *Chin. pl.*, 299 :—*Mu t'ung* at Pa tung is the name for several species of *Clematis*, and [488], *t'ung ts'ao*, *Fatsia* (*Aralia*) *papyrifera.*

Cust. Med., p. 12 (92):—*Mu t'ung* exported 1885 from New chwang 688 piculs,—p. 74 (109), from Han kow 291 piculs, p. 194 (162),—from Ning po 112 piculs.

SIEB., *Icon. ined.*, I, and SIEB. & ZUCC., *Flora japon.*, 1, 143, 145, tab., 77, 78, the Chinese names 木通 and 通草 applied both to *Akebia quinata* and *A. lobata.*—Same in *Phon zo*, XXX, 8, 9, and HOFFM. & SCHLT., 33.

185.—釣藤 *kou t'eng.* *P.*, XVIII*b*, 29. *T.*, CXII.

Pie lu :—*Kou t'eng* (hooky climber). Only the name given and medical virtues explained. It seems the thorns of the stem used in medicine.

T'AO HUNG-KING :—It is produced in Kien p'ing [in Hu pei, App. 139]. It is also called 弔藤 *tiao t'eng* (suspended climber) and employed in the treatment of diseases of children.

Su Kung [7th cent.]:—The *kou t'eng* grows in Liang chou [Sz ch'uan, etc., App. 187]. Small, long leaves. The stem is provided with hooked thorns.

Su Sung [11th cent.]:—It is found in Hing yüan fu in Ts'in [in Shen si, App. 68, 358]. The drug is gathered in the 3rd month.

K'ou Tsung-shi [12th cent.]:—It is common in the mountains of Hu nan, Hu pei and Kiang si. It is a climber from 8 to 20 feet long, of the thickness of a finger and [the stem] hollow. Thieves employ this hollow stem to suck out wine from a closed jar without damaging the latter.

Li Shi-chen :—It is a climber, like the vine, with purple coloured recurved thorns like hooks. In ancient times at first the bark was used in medicine, afterwards the hooks became officinal.

Ch., XXII, 57 :—*Kou t'eng*. The figure represents a plant with hooked spines, probably an *Uncaria*. The text says that it grows in Yün nan, Chen hiung chou.

Tatar., *Cat.*, 27 :—*Kou t'eng*. Rami scissi *Uncariœ Gambir*.—P. Smith, 224.—The drug *kou t'eng* received from Wen chow consisted of recurved spines.

Cust. Med., p. 358 (248):—*Kou t'eng* exported 1885 from Canton 58 piculs,—p. 214 (59), from Wen chow 35 piculs,—p. 228 (128), from Fu chow 1.80 picul,—p. 74 (108), 米鈎 *mi kou* [stated in *Hank. Med.* (27) to be the same as *kou t'eng*] from Han kow 340 piculs.

In the *Kwa wi* [111] 鈎藤 or 天弔藤 is *Uncaria rhynchophylla*, Miq. Comp. also the drawing in *Phon zo* [XXX, 13, 14].

The genus *Uncaria* (order *Rubiaceœ*, tribus *Naucleœ*) is characterised by its climbing habit and by the old or barren flower-stalks being converted into hard, woody spines, directed downwards so as to form hooks. No Chinese *Uncaria* appears in the *Index Fl. sin.* TATARINOV'S identification of the *kou t'eng* with *Uncaria Gambir* is very doubtful.

In Hu peh the drug *kou t'eng* is yielded by *Nauclea sinensis*, Oliv. *See* Dr. HENRY'S Note in *Hook Icon. plant.*, tab., 1956.

Uncaria Gambir, Roxb. (*Nauclea Gambir*, Hunt.), is the plant which affords the adstringent, earthy-looking, masticatory and tanning substance called *Gambir* or *Terra japonica*. This is the 檳榔膏 *pin lang kao* (betel-nut extract) of the Chinese. *See* RONDOT, *Commerce d'Export de la Chine*, p. 198; WILLIAMS, *Chin. Comm. Guide*, p. 92; P. SMITH, 100 [Pale Catechu of Commerce]. This drug has frequently been confounded with another drug—very similar in composition but widely diverse in botanical origin—the *Cutch* or *Catechu*, the extract of the wood of *Acacia Catechu*, Willd., which in Chinese is 兒茶 *rh ch'a*. *See* RONDOT, *l.c.*, and P. SMITH, 55, sub *Catechu*. The *P.* [VII, 15], which includes this foreign drug among the earths, calls it 烏爹泥 *wu tie ni* or 孩兒茶 *hai rh ch'a* [meaning "infants' tea." But as the Bengal name of the drug is *khaiar*, this name may perhaps be rendered by the above Chinese characters]. The Chinese also, it seems, sometimes confound *Gambir* and *Catechu*.

186.—白兔藿 *pai t'u huo*. *P.*, XVIII*b*, 30. *T.*, CLXXIX.

Pen king:—Pai t'u huo. The root is officinal. Taste bitter. Nature uniform. Non-poisonous.

Pie lu:—It grows in Kiao chou [Kuang tung, App. 132] in mountain-valleys.

Wu P'u [3rd cent.] calls it 白葛 *pai ko*.

T'ao Hung-king says that it is a counter-poison.

Su Kung [7th cent.]:—It grows in the mountains of King and Siang [both in Hu pei, App. 145, 305], and is a climber. The people of Shan nan [S. Shen si, App. 268] call it *pai ko*. It resembles the *lo ma* (*Metaplexis*), and has round, thick leaves. The stem is covered with white hair.

Ch., XXII, 23 :—*Pai t'u huo.*

Phon zo, XXX, 15, 16 :—白兔藿, *Asclepiadea*.

187.—白英 *pai ying*. P., XVIII*b*, 31.   *T.*, CLVIII.
Comp. *Rh ya*, 79.

Pen king:—*Pai* (white) *ying*. Root, leaves and fruit officinal. Non-poisonous. The root is sweet and the fruit sour.

Pie lu:—Other names 穀菜 *ku ts'ai* and 白草 *pai ts'ao*. The *pai ying* grows in I chou [Yün nan or Sz ch'uan, App. 102] in mountain-valleys. The leaves are gathered in spring, the stem in summer, the flowers in autumn and the root in winter. The same *Pie lu* says :—The 鬼目 *kui mu* (devil's eye) or 來甘 *lai kan* has a red fruit like the *wu wei tsz'* [*Schizandra*. See 164]. It is gathered in the 10th month.

T'ao Hung-king :—*Kui mu* is the popular name for the fruit of the *pai ts'ao*.

Su Kung [7th cent.]:—The *pai ying* or *kui mu* is a climbing plant. Its leaves resemble those of the *wang kua* [*Thladiantha*. See 173] but are smaller, longer, and five-lobed. The fruit is globular, like that of the *lung k'ui tsz'* (*Solanum nigrum*), and is at first green but purplish black when ripe. The people of Eastern China call it *pai ts'ao*.

Ch'en Ts'ang-k'i :—The *pai ying* or *kui mu* is a vege-table mentioned in the *Rh ya* [79]. In Kiang tung [Kiang su, An hui, etc., App. 124] the people in summer gather the leaves and the stem and boil them with rice for food. It is a counter-poison. Other names for the plant: 白幕 *pai mu*, 排風子 *p'ai feng tsz'*.

LI SHI-CHEN:—The young leaves are white and can be eaten. It blossoms in autumn. Small, white flowers. Fruit like the *lung k'ui tsz'* [*v. supra*]. It is dark red when ripe. The name *kui mu* is also applied to several other plants.

Ch., XXII, 18:—*Pai ying.* The figure represents a *Solanum* with hastate or lobed leaves, flowers and berries.

So moku, III, 52:—白英, *Solanum dulcamara*, L.

Phon zo, XXX, 17, 18:—Same Chinese name, *Solanum lyratum*, Thbg. This plant has downy leaves.

188.—羊桃 *yang t'ao.* *P.*, XVIIIb, 37. *T.*, CLI.

Comp. *Rh ya*, 198, *Classics*, 493.

Pen king:—*Yang t'ao* (goat's peach), 鬼桃 *kui t'ao* (devil's peach), 羊腸 *yang ch'ang* (goat's bowels). The stem and the root are officinal. Taste bitter. Nature cold. Poisonous. [Subsequent writers say "non-poisonous."]

Pie lu:—The *yang t'ao* grows in mountain-forests, in river-valleys and in fields. It is gathered in the 2nd month and dried in the shade.

T'AO HUNG-KING:—It is a common mountain-plant. It resembles the cultivated peach but is not what is called the mountain-peach. The flowers are deep red. The fruit is small, bitter, and not much eaten. Not used in medicine now. In the *Shi king* it is called *ch'ang ch'u.*

HAN PAO-SHENG [10th cent.]:—It is a common plant in marshes. The stem is long and weak, never tree-like (woody). Leaves and flowers resemble those of the peach.

Small fruit of the size of a jujube-stone. It is commonly called 細子 *si tsz'* (small fruit). .he root resembles that of *Pæonia moutan.* The lower portio of the stem is placed in hot ashes, when the bark becomes loose and can be worked into pencil-holders.

Ch., XXII, 42, 43 :— *Yang t'ao.* Two figures representing herbaceous plants.

LI SHI-CHEN :—The *yang t'ao* has a stem of the thickness of a finger, is tree-like but weak and creeping. Leaves as large as the palm of the hand, green on the upper side, white and downy underneath, resembling the *ch'u ma* (*Bœhmeria nivea*) leaves but round. The branches when steeped in water become viscid.

Cust. Med., p. 364 (320):— *Yang t'ao hua* (flowers) exported 1885 from Canton 9 piculs,—p. 326 (159), from Swatow 4 piculs,—p. 362 (286), from Canton *yang t'ao ye* (leaves) 0.10 picul.—The compilers of the *Cust. Med.* identify *yang t'ao* with *Averrhoa Carambola,* but this is a mistake, which has already been pointed out in *Bot. sin.,* II, 493.

Phon zo, XXXI, 8 :—羊桃. The drawing represents a creeper with large red flowers. Not identified by FRANCHET.

189.—絡石 *lo shi. P.,* XVIII*b*, 38. *T.,* CXXXIX.

Pen king :—Lo shi. The stem and the leaves are officinal. Taste bitter. Nature warm. Non-poisonous.

Pie lu :—Other names: 石龍藤 *shi lung t'eng,* 懸石 *hüan shi,* 雲珠 *yün chu,* 略石 *lio shi,* 領石 *ling shi,* 明石 *ming shi,* 石鯪 *shi ts'o.* The *lo shi* grows in T'ai shan [in Shan tung, App. 322] in river-valleys, also on rocks in the high mountains, in shady places, and near dwellings (cultivated). It is gathered in the 5th month.

Wu P'u [3rd cent.] gives the following synonyms: 石鯪 *shi ling,* 雲英 *yün ying,* 雲花 *yün hua* and 雲丹 *yün tan.*

T'ao Hung-king considers it a dubious plant. Some say it is a stone.

Su Kung [7th cent.]:—This plant grows in shady, damp places. It is an evergreen with a round black fruit. It climbs upon trees and rocks. That found on rocks has small, thick, round leaves, and that which climbs on trees has large thin leaves. The people cultivate it also as an ornamental plant. Its popular name is 耐冬 *nai tung* (enduring the winter). The name *lo shi* (entangling rocks) refers to its climbing on rocks. The people of Shan nan [S. Shen si, App. 268] call it 石血 *shi hüe* (stone blood). It is useful in hemorrhage after childbirth.

Han Pao-sheng [10th cent.]:—It is an evergreen climber on trees and rocks. Its leaves, which proceed from the joints of the stem, resemble small orange-leaves. It clings to the rocks by the fibres of its root. White flowers and large black fruit.

Li Shi-chen:—The *lo shi* climbs on rocks. Its stem contains a white juice. The leaves are small, of the size of a finger-nail, thick, firm, green on the upper side, and paler, scabrous, not shining, underneath. There are two kinds, one with pointed and the other with round leaves. The medical virtues in both are the same. Su Kung's account is not incorrect but it is incomplete.

Ch., XXII, 22:—*Lo shi.* Rude drawing. Twining plant. Only leaves.

Phon zo, XXXI, 11:—絡石 *Rhynchospermum jasminoides,* Lindl. (*Nerium divaricatum,* Thbg., *Malouetia asiatica,* S. & Z.). Order *Apocyneæ.*—*Ibid.,* 8-10, four other plants figured with the above Chinese name.

43

Comp. *Classics*, 452.

Pie lu :—*Ts'ien sui lei* (thousand years' creeper), 蒘 蕪
lei wu. The *ts'ien sui lei* grows in T'ai shan [in Shan tung,
App. 322] in mountain-valleys. The root is officinal. Taste
sweet. Nature uniform. Non-poisonous.

T'AO HUNG-KING :—It is a climber which grows like
the vine. The leaves resemble those of the *kui t'ao* [see 188].
It climbs on trees, and contains a white juice.

CH'EN TS'ANG-K'I [8th cent.]:—It is a twining plant
similar to the *ko* [*Pachyrhizus*. See 174]. The leaves are
white underneath. It has a red fruit. The branches contain
a white juice. It is mentioned in the *Shi king*.

SU SUNG [11th cent.]:—It is a common plant which
climbs on trees. Leaves like those of the vine but smaller.
The stem when broken discharges a white juice of a sweet
taste. It flowers in the 5th and produces fruit in the
7th month. The fruit is greenish black with a tinge of red.

Ch., XXII, 50 :—*Ts'ien sui lei*. The figure represents
a vine or *Vitis*.

Phon zo, XXXII, 3 :—千歲蘽 *Vitis flexuosa*, Thbg.

191.—忍冬 *jen tung*. P., XVIII*b*, 43. *T.*, CXIX.

Pie lu :—*Jen tung* (enduring the winter). It [not said
what part of the plant] is gathered in the 12th month and
dried in the shade. Taste sweet. Nature warm. Non-
poisonous.

T'AO HUNG-KING :—The *jen tung* is a common climbing
plant. It does not wither in winter, whence the name.

Su Kung [7th cent.]:—It twines around herbaceous plants and trees. The stem and the leaves are of a purplish red colour. The old stem has a thin epidermis, and the young stem is covered with hair. The leaves resemble those of the *hu tou* (pea), tomentose on both sides. Flowers white with purple filaments (anthers). This plant is sometimes confounded with the *lo shi* [see 189], but that is another plant.

Li Shi-chen:—The *jen tung* is a climber. Its stem has a tinge of purple. The leaves proceed in pairs from the joints. They resemble the leaves of the *pi li* [see *Bot. sin.*, II, 415] but are hirsute. It blossoms in the 4th month. The flower is more than an inch long. One peduncle produces two flowers. The corolla has two lips, one large and one small. It looks like one-half of a flower. Long filaments. The flowers when they first open are all white, but after two or three days they become yellow. Owing to the plants producing yellow and white flowers at the same time it is also called 金銀花 *kin yin hua* (gold and silver flowers). The flowers are very fragrant. They are gathered for medical use in the 4th month and dried in the shade. The leaves are gathered at all times of the year. Other names for the plant are 鴛鴦藤 *yüan yang t'eng*, 鷺鷥 | *lu ts'z' t'eng*, 老翁鬚 *lao weng sü*, 纒 | *ch'an t'eng*, 金釵股 *kin ch'a ku*. In Taoist books it is 蜜桶 | *mi t'ung t'eng*, 陰草 *yin ts'ao* and 通靈草 *t'ung ling ts'ao*.

Ch., XXII, 48 :—*Jen tung*. Good drawing. *Lonicera japonica*, Thbg. (*L. chinensis*, Wats.). The description in the *P.* agrees well. See also *Kiu huang*, LIII, 21.

Loub., *Fl. cochin.*, 185 :—*Lonicera periclymenum* [probably *L. japonica* is meant]. Sinice : *gin tum*. Flos albo luteus. Flores ad usum medicum adhibentur.

TATAR., *Cat.*, 58 :—*Kin yin hua*. Flores *Lonicerœ chinensis*.—P. SMITH, 138, 50, 114.—*Kin yin hua* is the common name for the plant at Peking, where it is much cultivated. *See also* HENRY, *Chin. pl.*, 66.

Cust. Med., p. 128 (100):—*Kin yin* flowers exported 1885 from Chin kiang 153 piculs,—p. 30 (117), from Tien tsin 40 piculs,—p. 356 (227, 228), from Canton flowers and leaves 0.7 picul,—p. 276 (48), *jen tung* from Amoy 0.45 picul.—Exported also from Han kow. See *Hank. Med.*, 5.

Amœn. exot., 785 :—忍冬 *nin too*, it. *sin too*, vulgo *sui kadsura* et ex floris colore *kin gin qua*, i.e. auri et argenti flos appellata. Periclymenum vulgare, sive Caprifolium non perfoliatum, baccis atro-purpureis vel nigris.

Phon zo, XXXII, 3, 4 :—忍冬, *Lonicera japonica* and other species.

SIEB., *Œcon.*, 238 :—*Lonicera japonica*. Japonice : *nindœ*. Sinice : 忍冬. Stipites ab agricolis in syphile inveterata optimum prædicantur remedium.

192.—澤瀉 *tse sie*. P., XIX, 1. T., CXL.

Comp. *Rh ya*, 95, *Classics*, 437.

Pen king :—*Tse sie*, 水瀉 *shui sie*, 鵠 | *ku sie*. The leaves, the root and the fruit are officinal. Taste of the root sweet. Nature cold. Non-poisonous. Taste of the leaves saltish and of the fruit sweet.

Pie lu :—Other names : 及瀉 *ki sie* and 芒草 *mang ts'ao*. The *tse sie* grows in Ju nan [in Ho nan, App. 110] in ponds and swamps. The leaves are gathered in the 5th, the root in the 8th, the fruit in the 9th month, and dried in the shade.

T'AO HUNG-KING :—Ju nan is in the province of Yü chou [Ho nan. See App. 413]. This plant occurs in Mid

China but is not much used. The drug from Han chung [S. Shen si, App. 54], Nan cheng [in Shen si, App. 226], Ts'ing chou [in Shan tung, App. 363] and Tai chou [in N. Shan si, App. 321] is more generally employed. It (the root) is large, with a long tail, and has two protuberances [perhaps branches]. The sort which grows in shallow water has long, narrow leaves.

Su Kung [7th cent.]:—Now the drug is not gathered in Ju nan. That from King chou [in Kan su, App. 153] and Hua chou [in Shen si, App. 85] is considered the best.

Su Sung [11th cent.]:—This plant is now found in Shan tung and Ho [provinces near the Yellow River, App. 72], Shen [Shen si, App. 284], Kiang and Huai [Kiang su and An hui, App. 124, 89], but the best drug comes from Han chung. It is a common plant in shallow water. The leaves resemble an ox-tongue. It has a long single stem. It blossoms in autumn. Numerous white flowers like the *ku tsing ts'ao* (*Androsace*). At the end of autumn the root is dug up and dried in the sun.

Li Shi-chen :—The plant is also called 禹 孫 *Yü sun.*

Ch., XVIII, 1, and *Kiu huang,* XLVII, 5, sub *tse sie,* good drawings of *Alisma plantago,* L.

Tatar., *Cat.,* 57 :—*Tse sie.* Radix *Alismæ plantaginis.* —P. Smith, 7.—The drug obtained under the name *tse sie* from a Peking drug-shop consisted of hard, fragrant, white tubers, one inch in diameter.

Cust. Med., p. 80 (188):—*Tse sie* exported 1885 from Han kow 1,207 piculs,—p. 226 (96), from Foo chow 640 piculs,—p. 320 (91), from Swatow 235 piculs,—p. 60 (27), from I chang 56 piculs,—p. 190 (100), from Ning po 28 piculs,—p. 92 (77), from Kiu kiang 26 piculs.

So moku, VII, 35 :—澤 瀉, *Alisma plantago.*—*Ibid.,* 35, 36 :—水 | |, the same, with lanceolate leaves.

193.—羊蹄 *yang t'i*.   *P.*, XIX, 4.   *T.*, LXIV.

Comp. *Rh ya*, 117, *Classics*, 440.

Pen king:—*Yang t'i* (sheep's hoof), 蓄 *ch'u*, 鬼目 *kui mu*, 東方宿 *tung fang su*, 連蟲陸 *lien ch'ung lu*. The root is officinal. Taste bitter. Nature cold. Non-poisonous.

Pie lu:—The *yang t'i* grows in Ch'en liu [in Ho nan, App. 13] in river-valleys.

T'AO HUNG-KING:—It is now also called 禿菜 *t'u ts'ai*.

HAN PAO-SHENG [10th cent.]:—It grows in low, damp places. The plant is from three to four feet high. The leaves are narrow and long, somewhat resembling those of the *wo kü* (*Lactuca*), but they are of a darker green colour. The stem and the joints are of a purplish red colour. Greenish white flowers in racemes. The seeds are three-edged. The plant begins to wither in summer. The root resembles that of the *niu p'ang* [burdock. See 91]. It is hard and solid.

K'OU TSUNG-SHI [12th cent.]:—The leaves resemble those of the *po ling* (spinage) but are not hastate and are thicker. The flowers resemble the seeds. The leaves are used for polishing (or cleaning) certain stones. The fruit is called 金蕎麥 *kin k'iao mai* (golden buckwheat). Artificers use it in the working of lead.

LI SHI-CHEN:—This plant grows abundantly by river-sides and in moist places. The leaves are more than a foot long and resemble an ox-tongue but have no resemblance to spinage-leaves. The stem rises in the beginning of summer, and the plant produces flowers and seeds. The flowers are of the same colour as the leaves. At the end of summer it withers, but thrives again late in autumn and then does not wither in winter. The root is nearly a foot long, of a reddish yellow colour and resembles the Rhubarb root and also a carrot. Other names: 羊蹄大黃 *yang t'i ta huang* (sheep's-hoof Rhubarb), 敗毒菜 *pai tu ts'ai*, 牛舌菜 *niu she ts'ai* (ox-tongue vegetable), and 水黃芹 *shui huang k'in*.

Ch., XVIII, 8:—*Yang t'i*. Rude drawing. Probably *Rumex* is intended. Also *Kiu huang*, LII, 21.

At Peking *yang t'i* or *niu she* is *Rumex crispus*, L., and other species. A *Rumex* in the Peking mountains is called 土大黃 *t'u* (native) *ta huang* (Rhubarb). Its root resemble a cloven hoof.

P. SMITH, 87:—Dock, *yang t'i*.—HENRY, *Chin. pl.*, 308:—牛舌頭 *niu she t'ou*, in Hu pei, *Rumex acetosa*, L.

Amœn. exot., 911:—羊蹄 *jotei*, communiter *si*. Thlaspi majus, foliis Lapathi, caulibus in spicas excurrentibus, capsulis Bursæ pastoris, intermixtis foliolis, confertas. [It does not seem that a *Rumex* is meant.]

So moku, VII, 27:—羊蹄 *Rumex japonicus*, Meisn.

SIEB., *Œcon.*, 108:—*Rumex crispus*. Japonice: *gisi gisi*. Sinice: 羊蹄. Remedium agricolis in exanthematibus.

194.—菖蒲 *ch'ang p'u*.　*P.*, XIX, 8.　*T.*, LXVIII.

Comp. *Classics*, 376.

Pen king:—*Ch'ang p'u*. The root is officinal. Taste pungent. Nature warm. Non-poisonous.

Pie lu:—Other names: 昌陽 *ch'ang yang*. The *ch'ang p'u* grows in Shang lo [in Shen si, App. 274] in ponds and swamps, also in Yen tao, belonging to Shu [Sz ch'uan, App. 406, 292]. The best drug is that which contains nine joints within a space of an inch of the root. The bedewed root (?)[55] cannot be used. The root is dug up in the 5th and 12th months and dried in the shade.

"露根.

Wu P'u [3rd cent.] calls it 堯韭 *yao kiu.*

T'ao Hung-king:—One kind, which is a common plant, grows on stones. That with numerous joints is the best. That with a large root, which grows in low, damp places, is called 昌陽 *ch'ang yang.* It is not much used for food. The true *ch'ang p'u* has a leaf which shows a ridge (an elevated line) like a sword. In the 4th and 5th months it produces minute flowers. There is one kind found in Eastern China, in rivulets and swamps, which is called 溪蓀 *k'i sun.* In odour and colour its root is much like the *ch'ang p'u* which grows on stones, but its leaves are very like the leaves of the *p'u* (*Typha*) and have no ridge (elevated line). It is frequently confounded with the stone *ch'ang p'u.* It is not eaten, but is employed as an expectorant, and is also useful in destroying fleas and lice.

Ta Ming [10th cent.]:—The root of the *ch'ang p'u* which grows in stony rivulets is small and hard. The best is that with nine joints within one inch of the root. It is produced in Süan chou [in An hui, App. 315].

Su Sung [11th cent.]:—It is a common plant. The best sort is produced in Ch'i chou [in An hui, App. 21] and in Jung chou [in Sz ch'uan, App. 112]. The leaf is from one foot to two feet long, has an elevated middle ridge like a sword. It has neither flowers nor fruit. Its root is contorted, creeping, has joints, and resembles a switch. The principal root sends out three or four lateral roots with joints close together, 12 within an inch. The fresh root is soft and hollow, but after having been dried in the sun it becomes hard and solid. When broken the heart shows a reddish tinge. Its taste is acrid and aromatic. The plant is much cultivated for medical use, but the best is that brought by the savages (Man) of K'ieh [N. Kui chow, App. 141] and Shu (Sz ch'uan). This is the *shi* (stony) *ch'ang p'u.* Another sort is called 水 | | *shui* (water) *ch'ang p'u.* It grows in rivulets and swamps, and is seldom used in medicine. These two sorts are frequently mixed together by the druggists and are not easily distinguished.

Li Shi-chen:—There are five kinds of *ch'ang p'u*:—

1.—That which grows in ponds and marshes and has a leaf like the *p'u*, a fat (succulent) root, and is from two to three feet high, is the 泥 | | *ni ch'ang p'u* (*ni* = mud) or 白 菖 *pai ch'ang* [see 195].

2.—That which grows in rivulets, has a *p'u* leaf, a meagre root, and is from two to three feet high, is the 水 | | *shui* (water) *ch'ang p'u* or *k'i sun* [*v. supra*]. It is also called 水劍草 *shui tsien ts'ao* (water sword-plant).

3.—That which grows in the water among stones, and the leaves of which show an elevated ridge like that of a sword, root meagre, joints close together, and the plant about one foot high, is the 石 | | *shi* (stone) *ch'ang p'u.*

4.—One kind of *shi ch'ang p'u* is cultivated by the people in coarse sand. It is only from four to five inches high, and has fine leaves like the *kiu* (*Allium odorum*). Root like the handle of a spoon.

5.—The smallest kind, the root of which measures but two or three tenths of an inch, with leaves one inch long, is called 錢蒲 *ts'ien p'u.*

For food and for medical use only the two sorts of *shi ch'ang p'u* can be employed. The other sorts are worthless.

Su Sung is erroneous in stating that the *ch'ang p'u* has neither flowers nor fruit. It produces in the 2nd or 3rd month a spike of small yellow flowers.

For botanical identification see the next.

195.—白昌 *pai ch'ang.*　*P.,* XIX, 13.　*T.,* LXVIII.

Pie lu:—Other names : 水菖蒲 *shui ch'ang p'u,* 水宿 *shui su* and 莖 | *heng p'u.* The *pai* .(white) *ch'ang* is dug up in the 10th month. The root is officinal. Taste sweet and pungent. Non-poisonous.

T'AO HUNG-KING gives the synonym 蘭蓀 *lan sun.*

CH'EN TS'ANG-K'I [8th cent.]:—This is the plant which is now called *k'i sun* [see 194] and also 昌陽 *ch'ang yang.* It grows by the sides of water. It is different from the *ch'ang p'u* which grows among stones [see 194]. It has a large, very white root, of a disagreeable smell.

SU SUNG [11th cent.]:—The *shui ch'ang p'u* grows abundantly in rivulets and marshes. It perishes when the water dries up. The leaves resemble those of the *shi* (stone) *ch'ang p'u* but they have not the elevated rib [*v. supra,* 194]. The root after drying becomes light, shrunken, and of a dirty appearance. It is not fit for medical use.

LI SHI-CHEN :—There are two sorts. One has a large, white succulent root with the joints wide apart. This is the *pai* (white) *ch'ang,* commonly called 泥 | | *ni* (mud) *ch'ang p'u* [comp. 194]. The other kind has a meagre root of a red colour, and the joints close together. This is the *k'i sun,* commonly called *shui* (water) *ch'ang p'u.* Both these sorts do not show that [above-mentioned] elevated ridge on their leaves. The taste and the smell of the *k'i sun* are superior. Both drugs are useful in destroying vermin, but are not fit for being eaten.

Ch., XVIII, 2 :—*Ch'ang p'u* or *shi ch'ang p'u.* The figure seems to represent an *Acorus.* Linear leaves. Root not represented.—*Kiu huang,* LI, 9 :—*Ch'ang p'u.* The drawing shows only young leaves and a ringed rhizome.

LOUR., *Fl. cochin.*, 259:—*Acorus Calamus*, L. [LOUREIRO'S plant is, according to KUNTH, *Enum. pl.*, III, 87, not the common sweet flag but the *Acorus terrestris* of RUMPHIUS, *Amb.*, V, tab. 72]. Sinice : *re cham pu* (*shi ch'ang p'u*). In montibus et locis petrosis Cochinchinæ et Chinæ. Radix utitur in medicina.

TATAR., *Cat.*, 14, 53 :—*Ch'ang p'u* and *shi ch'ang p'u*. Rad. *Acori terrestris.*—P. SMITH, 4.

LOUR., *l.c.*, 258 :—*Orontium cochinchinense* [this is *Acorus cochinchinensis*, Schott. *See* KUNTH, *Enum. pl.*, III, 87]. Sinice : *rui cham pu* (*shui ch'ang p'u*). Habitat in paludibus et locis aquosis Cochinchinæ et Chinæ.

At Peking the common *Acorus Calamus* is called *ch'ang p'u*. It has the same Chinese name in Hu pei. *See* HENRY, *Chin. pl.*, 18.

The *pai ch'ang* or *shui ch'ang p'u* of the *Pen ts'ao* is probably *Acorus Calamus*.

Cust. Med., p. 338 (2):—*Ch'ang p'u* exported 1885 from Canton 47 piculs.—Exported also from Han kow. See *Hank. Med.*, p. 1.—The *shi ch'ang p'u* is mentioned in the *Cust. Med.* as imported to New chwang, Shang hai and Tamsui, and said to come from Ning po and Amoy, but it is not noticed in these ports among the exports.

Amœn. exot., 900 :—莒 *sju*, vulgo *soobu*. Herba arundinacea palustris, foliis liliaceis, ob pulchritudinem in hortis et cisternis recepta ; cujus tres habentur species foliorum magnitudine differentes : *soo* foliis longissimis,—*ajami* mediocribus,—et *sikisoba* parvulis, quæ in fictilibus asservatur.— THUNBERG identifies this with his *Iris versicolor* [*Flora jap.*, 34], but from the Japanese names it would seem that KÆMPFER's description refers to *Acorus*.

Phon zo, XXXIII, 16, 17, and *So moku*, VII, 10 :— 白莒, japonice *sjobu*, is *Acorus spurius*, Schott. [=*A. Calamus*, Thbg., *Fl., jap.*, 144]. In aquosis Japoniæ (Buerger).

So moku, VII, 11 :—石菖, *Acorus gramineus*, Ait. Comp. also *Phon zo*, XXXIII, 13-14, s. nom. sin. *Acorus*, only leaves figured.

SIEB., *Icon. ined.*, VIII, and SIEB., *Œcon.*, 4 :—*Acorus Calamus*, L., var. *asiatica*. Japonice : *sjoobu*; sinice : 泥菖. Usus pro ceremoniis.

SIEB., *Icon. ined.*, VIII, and SIEB., *Œcon.*, 5 :—*Acorus gramineus*. Japonice : *seki sjoo* ; sinice : 石菖蒲.

SIEB., *Icon. ined.*, VIII, and SIEB., *Œcon.*, 6 :—*Acorus pusillus*. Japonice : *biroodo seki sjo* ; sinice : 錢蒲. Utraque planta [i.e. *A. gramineus* et *pusillus*] pro septis lacuum artificialium in hortis, nec non ob virtutem in contusionibus colitur.

MIQUEL, *Prol. Fl. jap.*, 135, 356 :—*Acorus pusillus*. Acoro gramineo valde affinis, omnibus partibus minor ; folia angustissima, etc.

MATSUMURA, 4 :—*Acorus Calamus*, 水菖蒲.

196.—香蒲 *hiang p'u*. P., XIX, 13. T., LXVIII.

Comp. *Classics*, 375.

Pen king :—*Hiang* (fragrant) *p'u*, 蒲黃 *p'u huang* (yellow), name of the yellow pollen of the flowers, which is used in medicine. Taste sweet. Nature uniform. Non-poisonous.

Pie lu :—The *hiang p'u* is produced in Nan hai [Kuang tung, App. 228] in pools and swamps, and the *p'u huang* in Ho tung [in Shan si, App. 80] in pools and swamps. It is gathered in the 4th month.

WU P'U [3rd cent.] gives the synonym 醮石 *tsiao shi*.

SU KUNG [7th cent.]:—The *hiang p'u* is also called 甘｜ *kan* (sweet) *p'u*. It is fit for making mats. In the spring the white sprouts of this plant are collected and pickled. They may also be eaten when steamed. The people of Shan nan [S. Shen si, App. 268] call it *hiang* (fragrant) *p'u*, whilst by 臭 | *ch'ou* (stinking) *p'u* they understand the *ch'ang p'u* [Acorus. See 195]. *P'u huang* is the name of the flowers of the *hiang p'u*.

Su Sung [11th cent.]:—The *hiang p'u* is the name of the plant which produces the *p'u huang*. It is a common plant. The best comes from Ts'in chou [in Kan su, App. 358]. The young plant, in the spring, when rising from the bottom of the water is of a reddish white colour. The heart of the tender plant which enters the mud, and which is called 白蒻 *pai jo*, is of the size of the handle of a spoon, and can be eaten raw. It is sweet and delicate. It is also very palatable when steeped in vinegar, like bamboo-sprouts. This is mentioned in the *Chou li*. Nowadays it is rarely used for food. In the beginning of summer the stem shoots upwards from the midst of the leaves. It bears at the top a kind of mace which contains the flowers. It is called 蒲槌 *p'u ch'ui* (club, mace) and | 蕚 *p'u o* (receptacle). *P'u huang* is the name for the powder of the filaments of the flowers (pollen). It is fine, like golden dust. The people collect it at the proper time, mix it with honey and sell it as a sweetmeat.

Li Shi-chen :—The *p'u* grows in a bushy manner by the sides of the water, like the *kuan* [a rush. See *Bot. sin.*, II, 455], but it is smaller and [the leaf] has a ridge. In the 2nd and 3rd months the young roots are taken out and prepared with a condiment of fish. The old root is also edible when boiled in fat or steamed. Or it is dried in the sun and reduced to powder, of which cakes are made. In the 8th or 9th month the leaves are gathered and woven into mats. Fans can also be made of the leaves. They are pliable, smooth and keep warm.

Ch., XVIII, 4 :—*Hiang p'u.* Rude drawing. *Typha* is evidently intended. The figure in the *Kiu huang* [LIII, 12, sub *p'u sun* (sprouts)] is also *Typha.*

LOUR., *Fl. cochin.*, 675 :—*Typha latifolia*, L. In paludibus Chinæ et Cochinchinæ. Sinice : *pu hoam.*

TATAR., *Cat.*, 46 :—*Hiang p'u, Typha Bungeana.*— P. SMITH, 223.

Typha angustifolia, L., is a common plant in the marshes near Peking. Popular name *p'u tsz'.*

Cust. Med., p. 128 (109) :—*P'u huang* exported 1885 from Chin kiang 4 piculs,—p. 204 (282), from Ning po 0.47 picul,—p. 374 (466), from Canton *hiang p'u ts'ao* 0.05 picul.—*P'u huang* is also exported from Han kow. See *Hank. Med.*, 34.

Amœn exot., 900 :—蒲 *fo*, vulgo *kamma*, gramen cyperinum palustre.—It does not seem that KÆMPFER means *Typha*, although *gamma* is the Japanese name for *Typha.*

Phon zo, XXXIV, 18-20 :—香蒲, *Typha japonica*, Miq., and *T. angustifolia*, L.

SIEB., *Œcon.*, 7 :—*Typha angustifolia*, L. Japonice : *gama.* Sinice : 蒲. Usus pro fomite.

197.—菰 *ku.*　*P.*, XIX, 16.　*T.*, XL.

Comp. *Classics*, 350.

Pie lu :—*Ku.* Only the name. The root, the sprouts and the leaves are used in medicine.

HAN PAO-SHENG [10th cent.] :—The *ku ken* (root) grows in water. The leaves of the plant resemble those of the sugar-cane. The old root is contorted and thick. In the summer a fungus (菌 *kün*) is produced in it, which is edible and is called 菰菜 *ku ts'ai* (vegetable). In the third year a white stalk appears in the heart of the plant like *Nelumbium* root, resembling the forearm of a child. It is white and delicate and has black veins. It is likewise edible and is called | 首 *ku shou* (head of the *ku*).

CH'EN TS'ANG-K'I [8th cent.]:—The small *ku shou* when broken shows a black dust inside. It is called 烏鬱 *wu yü* and is eaten by the people.

SU SUNG [11th cent.]:—The *ku ken* (root) is common in rivers, lakes and swamps. Its leaves resemble those of the *p'u* (*Typha*) and the *wei* (common reed). They are cut for feeding horses, which it fattens. At the end of spring the root sends up white sprouts resembling bamboo-sprouts. They are called *ku ts'ai* [*v. supra*], also 茭白 *kiao pai*, and are used for food, both raw and cooked. They are of an agreeable sweet taste. In the heart of these sprouts is a central mass which resembles the forearm of a child. This is called | 手 *ku shou* (arm) not *ku shou* (首 head). This plant is mentioned in the *Rh ya* [88], and the commentator speaks of the fungus produced in it. Since that time the people in the south use the character 菰 *ku* as a synonym for 菌 *kün* (mushroom). The root of the *ku* is like that of the common reed. It is common in the marshes of the two Che [Che kiang, App. 10]. When the stem of this plant has become hard it is called | 蔣草 *ku tsiang ts'ao*. In autumn it bears fruit. The seeds are called 彫胡米 *tiao hu mi* and in times of scarcity are used as a substitute for corn.

K'OU TSUNG-SHI [12th cent.] says:—The *ku* is a kind of reed. The people on the northern border of China use it for feeding horses and for making mats. It flowers in the 8th month. The flowers resemble those of the common reed.

This plant, first mentioned in the *P.* [*l.c.*] as a water-plant, is there spoken of for a second time [XXIII, 15] among the cereals, under the name of 菰米 *ku mi* (grain). Comp. *Rh ya*, 88, *Classics*, 350.

T‘AO HUNG-KING :—The *ku mi*, also called *tiao hu* [*v. supra*] is employed for making cakes used as food.

CH‘EN TS‘ANG-K‘I [8th cent.]:—The *tiao hu* is the grain from the *hu tsiang* [*v. supra*] plant. It is mentioned in the Classics and is used as food.

SU SUNG [11th cent.]:—In ancient times the *ku mi* was much valued as food, but nowadays the people gather it only in times of scarcity.

K‘OU TSUNG-SHI [12th cent.]:—The seeds are green and about an inch long. The savages gather them and cook them mixed with millet as food.

LI SHI-CHEN :—The flowers of the *tiao hu* resemble those of the *wei* (common reed). The seeds are nearly an inch long. They are gathered after hoar-frost. They are as large as the *mao chen* [sprouts of *Imperata*. See 37]. Their outer skin is gray, but the flour within is very white, mucilaginous, and can be used for food. The young stalks of this plant are likewise eaten and are known under the name of 茭筍 *kiao sun*.

This plant is figured in the *Ch.* [XVIII, 13] under the names *ku* or *kiao pai*, and in the *Kiu huang* [LIII, 27] sub *kiao sun*. It is the *Hydropyrum latifolium*, Griseb. For further particulars see *Bot. sin.*, 350.

198.—水萍 *shui p‘ing.*   P., XIX, 18.   T., LXVII.

Comp. *Rh ya*, 113, *Classics*, 400.

Pen king:—*Shui* (water) *p‘ing*. The whole plant is used in medicine. Taste pungent. Nature cold. Non-poisonous.

Pie lu :—The *shui p'ing* grows in Lei tse [in Shan tung, App. 182] in ponds and swamps. It is gathered in the 3rd month and dried in the sun.

T'AO HUNG-KING :—This is the water-plant called 大萍 *ta* (large) *p'ing*, not that plant now called 浮萍子 *fou* (floating) *p'ing tsz'*. The *Lei kung yao tui* says that it has white flowers in the 5th month, but this does not agree with the *ta p'ing* which is now common in ditches and canals. The *ta p'ing* is the fruit of the water-plant which the king of Chu fell in with when crossing the Kiang (Yang tsz').[56]

CH'EN TS'ANG-K'I [8th cent.]:—There are two kinds of *shui p'ing*. The larger one is called 蘋 *p'in*. It has round leaves more than an inch in diameter. The small *p'ing tsz'* is that which is commonly met with in ditches and canals, and this latter is the *shui p'ing* mentioned in the *Pen king*.

SU SUNG [11th cent.] refers to *Rh ya*, 113, and notices that SU KUNG [7th cent.] distinguishes three kinds of *shui p'ing*,—the larger or *p'in*, an intermediate sort which is called 荇 *hang* [*Limnanthemum*. See *Classics*, 399] and a small kind, 浮萍 *fou* (floating) *p'ing*, which floats on the surface of the water. Now the *p'in* is seldom used in medicine, but the *fou p'ing* is commonly employed.

LI SHI-CHEN :—The *shui p'ing* used in the ancient pre- scriptions was the small *fou p'ing* not, as T'AO HUNG-KING asserts, the large *p'ing*. The *fou p'ing* is common in standing pools, where it appears at the end of spring. It is believed

---

[56] T'AO HUNG-KING alludes to a tradition related at length in the *Kia yu* (Family Sayings of Confucius): 楚昭王, the prince Chao of the state of Ch'u [B.C. 515-489], when once crossing the Kiang in a boat, met with a curious water-plant having a large fruit. It was sent to Confucius, who declared it to be the fruit of the *p'ing* plant, which appears only to princes destined to become leaders of the empire.

45

that it is produced by a metamorphosis of poplar flowers. The *fou p'ing* consists of numerous small leaves floating on the water. Its tender, hair-like roots proceed from the underside of the leaves. One kind has its leaves green on both sides, the leaves of another kind are green above and reddish purple, like blood, beneath. The latter is called 紫 | *tsz'* (purple) *p'ing*. It is much valued as a medicine.

Ch., XVIII, 5 :—*Shui p'ing* or *fou p'ing*. The figure represents *Lemna* (Duckweed).

LOUR., *Fl. cochin.*, 492 :—*Zala asiatica* [according to ROXBURGH, LOUREIRO'S plant is *Pistia stratioides*, L.]. Sinice : *fou peng*. Habitat fluctuans in fluminibus lenti cursus in Cochinchina et China.

TATAR., *Cat.*, 55 :—*Shui p'ing, Lemna gibba*. [The name *fou p'ing tsz'* has been confounded by TATARINOV [24] with *fu p'en tsz'*, *Rubus*. See 166.] The Peking Duckweed is *Lemna minor*, L., is very common there and known under the Chinese name *fou p'ing tsz'*.—P. SMITH, 131, *Lemna*.

Cust. Med., p. 358 (238):—*Fou p'ing* exported 1885 from Canton 3.89 piculs,—p. 306 (480), from Amoy 0.02 picul.[57]

Amœn. exot., 900 :—泙 *fe*, vulgo *ukingusa*, i.e. herba natans. Lenticula palustris vulgaris.

Phon zo, XXXIV, 1 :—水萍 *Spirodela* (*Lemna*) poly-rhiza, Schleid.—*Ibid.* :—青萍 (green duckweed) *Lemna minor*, L.—*Ibid.*, 3 :—紫萍 (purple duckweed) *Salvinia natans*, Hoffm.

---

[57] *Fou-p'ing* 浮萍. [*Customs Med.*, export from Canton.] A sample of this drug [in the Pharmaceutical Museum, London] from Hongkong is *Pistia stratiotes*, L.—A. HENRY.

The *p'in* or larger *shui p'ing* has a special article in the
P., XIX, 21.

Li ki [LEGGE], II, p. 432 (10):—Marriage ceremony:
the young lady offers a sacrifice to the ancestors, using fish
for the victim, and soups made of duckweed and pondweed,
苾之以蘋藻.

WU P'U [3rd cent.]:—The 蘋 *p'in* or *shui p'ing*, also
called 水廉 *shui lien*, floats upon the surface of the water.
The leaves are round and small. Each stalk bears one leaf.
The roots are at the bottom of the water, in the mud.
The plant produces a white flower in the 5th month. It is
gathered in the 3rd month and dried in the sun.

CH'EN TS'ANG-K'I [8th cent.]:—The leaves of the *p'in*
are round, and about one inch in diameter. Beneath the leaves
there is a speck-like foam. It is also called 萍荣 *fou ts'ai*.
It is dried in the sun and used as a medicine. The smaller
sort, the *siao p'ing*, grows in ditches and canals [*Lemna*,
v. supra].

CHANG YÜ-HI [11th cent.] quotes the *Rh ya* [114], and
says that the *p'in* is little used in medicine, the small sort
(*Lemna*) being preferred for medical use.

LI SHI-CHEN:—The *p'in* is the 四葉菜 *sz ye ts'ai*
(four-leaved vegetable) which floats on the surface of the
water. The root is at the bottom of the water. The stalks
(petioles) are more slender than those of the *shun* [*see
the next*] and the *hang* [*Limnanthemum, v. supra*]. The leaf
is as large as a finger-nail, green on the upper side, purple
underneath and finely veined, somewhat resembling the leaflets
of the *ma t'i küe ming* [*Cassia*. See 110]. Properly its
leaf consists of four leaves (leaflets) united to form a cross.
It is therefore also called 田字草 *t'ien tsz' ts'ao* [plant the
leaves of which resemble the character 田]. In summer

and autumn it produces small white flowers and is therefore also called 白 | *pai* (white) *p'in*.

The *Ch.* [XVIII, 6] figures sub *p'in* or *pai p'in*, *Marsilea quadrifolia*, L. The descriptions in the *P*. agree so far as the leaves are concerned. But *Marsilea* is a pseudo-fern and has no flowers.

Phon zo, XXXIV, 3 :—蘋, *Hydrocharis morsus rance*. Probably this may be the *p'in* with white flowers of the *P*. Comp. *Ch.*, XVIII, 2.

Ibid., 3 :—田字草, *Marsilea quadrifolia*.

199.—蓴 *shun*.　*P.*, XIX, 24.　*T.*, LXV.

Comp. *Classics*, 398.

Pie lu :—*Shun*. The whole plant used in medicine. Taste sweet. Nature cold. Non-poisonous.

HAN PAO-SHENG [10th cent.]:—The leaves of the *shun* resemble those of the *fu k'ui* [*Limnanthemum*. See *Classics*, 399]. They float on the water. The stem is edible. Flowers yellowish white. Seeds (or fruit) of a purple colour. The stem, which from the 3rd to the 8th month is as thick as a hair-pin, is yellowish red, and short or long according to the depth of the water. It is then called 絲 | *sz'* (floss silk) *shun* and is sweet and soft. In the 9th and 10th months it gradually becomes coarse and hard. In the 11th month the sprouts appear in the mud. They are coarse and short, and known under the name of 瑰 | *k'uai* (piece) *shun*. They are bitter and harsh. The people press out the juice and use it with other vegetables for soup.

LI SHI-CHEN :—The name is also written 蓴 *shun*, as in the *Ts'i min yao shu* [5th cent.]. The plant grows in South China in lakes and ponds. The people of Wu Yüe

[Kiang su, and An hui, App. 389] like it as food. The leaves resemble those of the *hang ts'ai* [*Limnanthemum, v. supra*] but are more round, and resemble in their outlines a horse's hoof. The stem is purple, resembles a tendon, is mucilaginous, tender, and can be boiled into soups. It blossoms in summer. Yellow flowers. The fruit is greenish purple, of the size of a small crab-apple. It contains small seeds. The young stems, before the leaves appear on them, are known under the name of 稚 | *chi* (young) *shun*, the plant with expanded leaves is 絲 | *sz'* *shun*. In autumn the old plant is called 葵 | *k'ui* (mallow) *shun*. It is also called 豬 | *chu* (pig) *shun*, for it is good for feeding pigs. It is the 茆 *mao* of the *Shi king*.

Ch., XVIII, 14 :—*Shun*. The figure represents a water-plant with peltate leaves. P. SMITH erroneously identifies it with *Scirpus*. See also *W.D.*, 783.

Amœn. exot., 828 :—蓴 *sjun*, vulgo *nonawa*. Sagitta aquatica minor latifolia. C. BAUH. radice eduli.

Although it appears from KÆMPFER's quotations of BAUHIN that he means *Sagittaria sagittifolia*, THUNBERG [*Fl. jap.*, 82], relying upon the Chinese and Japanese names quoted, identifies it with *Menyanthes nymphoides*, L. Subsequently he describes it as a new plant (*M. peltata*) which is *Villarsia peltata*, Roem. & Sch., and *Limnanthemum peltatum*, Griseb. But according to FRANCHET this is *Brasenia peltata*, Pursh. For further particulars see *Bot. sin.*, II, 398.

200.—海藻 *hai tsao*. *P.*, XIX, 26. *T.*, LXVI.

Comp. *Rh ya*, 197, 129.

Pen king:—*Hai tsao* (seaweed), 洛首 *lo shou*. The whole plant is officinal. Taste bitter and salt. Nature cold. **Non-poisonous.**

Pie lu:—The *hai tsao* grows in Tung hai [in Shan tung, App. 372. But Tung hai also means "Eastern Sea"], in ponds and marshes. It is gathered on the 7th day of the 7th month and dried in the sun.

T'AO HUNG-KING:—It grows on islands in the sea, is of a black colour and has the appearance of flowing hair. The leaves are large and resemble those of the *tsao* [*Potamogeton* and other water-plants. See *Classics*, 401].

CH'EN TS'ANG-K'I [8th cent.]:—There are two kinds of *tsao*. The 馬尾 | *ma wei* (horse's tail) *tsao* grows in shallow water. It looks like a short horse-tail, is fine leaved and black. Before use it must be steeped in water to remove the brackish taste. The other kind has large leaves and grows in the deep sea near the kingdom of Sin lo [S. Corea, App. 311]. The leaves are like those of the *shui tsao* [*Potamogeton* and other sweet water-plants] but larger. The sea people, having attached a rope to their waists, glide down to the bottom of the sea and so secure the seaweed. Owing to the appearance of a large fish, dangerous to man, it cannot be gathered after the 5th month. This plant is mentioned in the *Rh ya* [201].

SU SUNG [11th cent.]:—This seaweed now grows in the sea near Teng chou and Lai chou [both on the north coast of Shan tung].

LI SHI-CHEN:—The *hai tsao* is gathered on the sea-coast. The 海菜 *hai ts'ai* (sea vegetable) is prepared from it.

Ch., XVIII, 7:—*Hai tsao* or 顛髮菜 *t'ou fa ts'ai* (hair-of-the-head vegetable). The figure represents an *Alga* with verticillate leaves.

TATAR., *Cat.*, 6:—*Hai ts'ai. Sargassum*, etc. *Algæ.*— P. SMITH, 5, *Agar agar.*

The *hai tsao* procured from Tien tsin, and examined by Professor AGARDH in Sweden and Professor GOBI in

St. Petersburg, proved to be *Sargassum siliquastrum*, and the *hai ts'ai*, extensively used for food in China and brought from the coast of Manchuria, is *Laminaria saccharina*, L. [comp. 201].

Cust. Med., p. 202 (265):—*Hai tsao* exported from Ning po 107 piculs,—p. 374 (459), from Canton 0.18 picul.

Amœn. exot., 831 :—藻 *soo*, vulgo *momubah*. Herba marina sive Alga in genere.

Phon zo, XXXIV, 17 :—海藻 or 馬尾草 and [19] 大葉藻 (large-leaved), *Algæ*, not determined.

201.—昆布 *kun pu*.   P., XIX, 29.   T., LXXIV.

Comp. *Rh ya*, 201.

Pie lu:—The *kun pu* is produced in the Eastern Sea [Tung hai, App. 372].   Taste salt.   Nature cold.   Non-poisonous.

Wu P'u [3rd cent.]:—The *kun pu* is also called 綸布 *lun pu*.

T'ao Hung-king :—The *kun pu* is produced in Kao li [Corea, App. 116].   It is twisted into ropes like hemp. It is of a yellowish black colour, soft but tough and edible. The *Rh ya* calls it *lun*.

Ch'en Ts'ang-k'i [8th cent]:—The *kun pu* is produced in the Southern Sea.   The leaves are like a hand, large and of a purplish red colour.   The sort with fine (capillary) leaves is called *hai tsao* [see 200].

Li Sün [8th cent.]:—This plant undulates [in the sea]. That produced in [the sea of] Sin lo [S. Corea, App. 311] has fine (capillary) leaves of a yellowish black colour. The foreigners (Coreans) twist it into ropes, dry it in the shade and carry it by ship to China.

LI SHI-CHEN:—The *kun pu* produced in [the sea of] Teng chou and Lai chou [northern coast of Shan tung] has the appearance of twisted ropes. The sort which is brought from Min [Fu kien, App. 222] and Che [Che kiang] has large leaves and is used for food. All the different sorts of *hai ts'ai* (seaweed) resemble each other in quality and taste, and their medical virtues also are much alike.

Cust. Med., p. 202 (273):—*Kun pu* exported 1885 from Ning po 114 piculs,—p. 216 (97), from Wen chou 41 piculs,—p. 300 (402), from Amoy 7 piculs,—p. 334 (285), from Swatow 1.87 picul.

The *kun pu* is *Laminaria saccharina*, L. For further particulars see *Bot. sin.*, II, 201. Comp. also *supra*, 200.

202.—石斛 *shi hu*. P., XX, 1. T., CXXXIX.

Pen king:—Shi hu, 禁生 *kin sheng* and 林蘭 *lin lan* [*lan* of the forests. Regarding *lan* (orchid) see 62]. The stem is used in medicine. Taste sweet. Nature uniform. Non-poisonous.

Pie lu:—Other names: 石蓫 *shi chu*, 杜蘭 *tu lan*. The *shi hu* grows in Liu an [in An hui, App. 220] in mountain-valleys, along the edge of the water and on stones (rocks). The stem is gathered in the 7th and 8th months and dried in the shade.

T'AO HUNG-KING:—The drug now used comes from Shi hing [in Kuang tung, App. 289], where it grows on stones. It (the stem) is gold coloured, resembles the leg of a locust, and is met with also in Mid China. Another sort, inferior in quality, is produced in Süan ch'eng [in An hui, App. 315]. It grows on oak trees (*li*) and is called 木斛 *mu* (tree) *hu*. It has a long, hollow stem of a paler colour.

Su Kung [7th cent.]:—It is now produced in King and Siang [both in Hu pei, App. 145, 305], in Han chung [S. Shen si, App. 54] and in Kiang tso [An hui and Kiang su, App. 124]. There are two kinds. One resembles barley. It (the stem) consists of a series of joints. At the top is one leaf. This is called 麥 | mai (barley) hu. The other sort has a stem like the thigh-bone of a bird. The leaves are at the summit of the stem. It is called 雀髀 | tsio pi (bird's thigh) hu. There is another kind of hu, which resembles a bamboo. The leaves proceed from the joints. The shi hu is dried for use or it is steeped in wine.

Su Sung [11th cent.]:—The shi hu now grows also in King chou [in Hu pei, App. 146], in Kuang chou [in Ho nan, App. 163], in Shou chou and Lü chou [both in An hui, App. 290, 206], in Kiang chou [in Kiang si, App. 125], in Wen chou and T'ai chou [both in Che kiang, App. 385, 326]. But the best comes from Kuang nan [Kuang tung, App. 162]. It is a mountain plant. The stem looks like a small bamboo. Small leaves proceed from the joints of the stem. It blossoms in the 7th month and bears fruit in the 10th. The root is slender, long and of a yellow colour. That growing on rocks is the best.

Li Shi-chen :—The shi hu grows in bunches on rocks. Its root consists of numerous knots (bulbs). When dried it becomes white and delicate. The fresh stem and the leaves are green, but when dried they become yellow. It has red flowers. The rootlets which are produced upon the joints are broken off by the people and planted in coarse sand among stones. They suspend this plant in their houses, and when sprinkled with water it will not die for years. It is therefore called 千年潤 ts'ien nien jun (thousand-years moist). The shi hu, which grows on rocks, is a short plant and has a solid stem, but the mu hu, which grows upon trees, is long and has a hollow stem. They are very easily

46

distinguished. These are common plants in China. The
best sorts are produced in Shu (Sz ch'uan). As the *shi hu*
has a slender stem like a [Chinese] hair-pin it is also called
金釵石 | *kin ch'ai* (gold hair-pin) *shi hu.*

Although the descriptions of the *shi hu* in the *P.* are
vague and unsatisfactory, I agree with WILLIAMS [*Dict.*,
233] that this name and the other synonyms are applied to
orchidaceous plants.

Ch., XVI, 1, 2 :— *Shi hu.* Two figures. The first
seems to represent *Dendrobium moniliforme*, Swartz, and the
other is a larger Orchid, with a jointed stem.

LOUR., *Fl. cochin.*, 633 :— *Ceraja simplicissima* [*Dendrobium*, Benth. & Hook., *Gen pl.*, III, 498]. Caule simplicissimo,
parasitico, erecto. Habitat in sylvis Cochinchinæ et Chinæ
rupibus ac arboribus inhærens. Sinice : *xe (shi) hu.* In
medicina sinensi utitur.

TATAR., *Cat.*, 53 :— *Shi hu. Dendrobium Ceraja* et musci
ac lichenes varii.— GAUGER [34] describes and figures this
drug. He suggests that it may be the subterraneous stem
of a *Graminea.*— P. SMITH, 85.

HENRY, *Chin. pl.*, 424, 539 :— *Shi hu. Dendrobium
nobile*, Lindl. Exported from Sz ch'uan as a drug under
the name of 雅斗 *ya tou*, both the plants with still living
flowers and the young stems in a dried state.

Cust. Med., p. 80 (203, 200, 213) :— Han kow exported
1885 fresh 斛斗 121 piculs, 金斛斗 30 piculs, 雅斗
8 piculs,— p. 60 (44), I chang 雅斗 14 piculs,— p. 374
(453), 376 (487), Canton exported 金釵斛 30 piculs,
石斛 20 piculs,— p. 424 (145), *shi hu* exported from Pak
hoi 18 piculs,— from Kiung chou 4 piculs.

The 木斛 was exported : p. 202 (279), from Ning po
99 piculs,— p. 268 (141), from Takow 4.25 piculs,— p. 216
(96), from Wen chow 2 piculs,— p. 288 (203), from Amoy
0.04 picul.

So moku, XVIII, 17 :—石斛 and [18] 黃花石斛 [yellow-flowered] *Dendrobium moniliforme.*—See also *Phon zo*, XXXV, 2, 3.

So moku, XVIII, 20, and *Phon zo*, XXXV, 3 :—麥斛. A small orchid, not determined by FRANCHET. Pseudo-bulbs (?). One leaf on the summit as described in the *P.*

So moku, XVIII, 19 :—雀髀斛, *Dendrobium reptans*, Franchet. *Ibid.*, 32, and *Phon zo*, VIII, 21 :—釵子股, *Luisia teres*, Bl. [*Epidendron teres*, Thbg.].

KÆMPFER [*Amœn. exot.*, 864-867] describes and depicts the *Dendrobium moniliforme* under the Chinese name 風蘭, in Japanese *fu ran*. But in Chinese as well as in Japanese botanical works these names are applied to another Orchid.

So moku, XVIII, 25 :—風蘭, *Œcœoclades falcata*, Regel. [*Orchis falcata*, Thbg., *Fl. jap.*, 26 ; *Limodorum falcatum*, Thbg., *Icon. pl. jap.*, 6 ; *Aërides Thunbergii*, Miq., *Aërides japonicum*, Lindl. & Reich., *Bot. Mag.* (1869) tab. 5798]. The genus *Œcœoclades* is now included under *Saccolabium*, Benth. & Hook., *Gen. pl.*, III, 578.

The above Chinese name is not found in the *P.* But the *Ch.* [XVI, 37] gives a figure of the 風蘭 *feng lan* (wind or air Orchid). The drawing agrees well with that of the *So moku.* The Chinese text says that the *feng lan*, also called 弔蘭 *tiao* (suspended) *lan*, is a parasitic plant on rocks, found in Fu kien, Kuang tung and Kiang si. It resembles the *shi hu* and the *lan.* Stem and leaves drooping. The leaves are flat, two or more inches long. They roll up and do not open again. It flowers in the 5th month. The people place this plant in a bamboo basket and suspend it from the eaves of their houses, where it thrives and blossoms, subsisting only on the air.

This Chinese plant was first noticed by SEMEDO 250 years ago [see my *Early Europ. Res. Fl. Chin*, p. 7].

LOUR. [*Fl. cochin.*, 642] describes it under the name of *Ærides odorata* (air plant); sinice: *fum lan*. Planta parasitica . . . . folia linearia, crassa, magna, subincurva, reflexa . . . . racemis simplicibus longis . . . . Mirabilis hujus plantæ proprietas est, quod ex sylvis domum delata, et in ære libero suspensa, absque ullo pabulo vegetabili terreo, vel aqueo, in multos annos duret, crescat, floreat, et germinet.

Cust. Med., No. 1276 :—*Tiao lan*. *Dendrobium nobile*, Lindl.

203.—石韋 *shi wei*. *P.*, XX, 4. *T.*, CXXXIX.

Pen king:—*Shi wei* (rock thong). The leaves are officinal. Taste bitter. Nature uniform. Non-poisonous.

Pie lu:—Other name: 石皮 *shi p'i* (rock leather). The *shi wei* grows in Hua yin [in Shen si, App. 87], in mountain-valleys and on rocks. The leaves are gathered in the 2nd month and dried in the shade. The best kind is that which grows in places where neither the noise of water nor the human voice are heard.

T'AO HUNG-KING :—The plant creeps on rocks, and the leaves are like leather, whence the above names. It is a common plant. That from Kien p'ing [in Hu pei, App. 139] has large, long and thick leaves.

SU KUNG [7th cent.]:—It grows densely on the sides of rocks, in the shade, but does not creep [as T'AO HUNG-KING asserts]. The kind which grows on old brick walls is called 瓦韋 *wa* (brick) *wei*. It is useful in the treatment of urinary calculus.

SU SUNG [11th cent.]:—This plant is now found in Tsin chou and Kiang chou [both in Shan si, App. 358, 123],

Ch'u chou [in An hui, App. 25], Hai chou [in Kiang su, App. 48], Fu chou [in Fu kien, App. 46] and in Kiang ning [Nan king, App. 129], where it grows on rocks. The leaves are leathery, resemble willow-leaves in shape, are covered with hair on the under side and also show spots there. A peculiar kind, which grows in Fu chou, is called 石皮 *shi pi* (rock leather). It (the leaves) has hairs in the 3rd month. A decoction of it is used in the treatment of rheumatism.

Li Shi-chen :—The *shi wei* is also called 石䪥 *shi che* (leather) and 石蘭 *shi lan*. It grows plentifully on rocks and in crevices. Leaves nearly a foot long and one inch broad, soft and pliable like leather. They have yellow hairs on the under side, or golden stars (spore cases), whence one kind is called 金星草 *kin sing ts'ao* (golden star plant). Another sort has leaves like the apricot. It likewise grows on rocks.

Ch., XVI, 4 :—*Shi wei*. The figure represents a plant with long tongue-shaped leaves, probably *Polypodium (Niphobolus) Lingua*, Sw. *Ibid.*, XVI, 11, 12 :—*Kin sing ts'ao*. Two ferns represented, one with tongue-shaped leaves.

Lour. [*Fl. cochin.*, 825] applies the Chinese name *xe ui (shi wei)* to another fern with long lanceolate leaves, *Ophioglossum lusitanicum*, and the Chinese name *xi ui tan* to *O. scandens*.

Hanb., *Sc. pap.*, 266 :—*Shi wei*. Fronds of *Niphobolus Lingua.*—P. Smith, 155.

Cust. Med., p. 204 (283) :—*Shi wei* exported 1885 from Ning po 6.55 piculs,—p. 360 (274), from Canton 4.42 piculs,—p. 280 (109), from Amoy 0.20 picul.

Amœn. exot., 石韋 *secki ji* vulgo *iwanokawa, etotsba*, etc. Hemionitis petræa, folio oblongo majusculo simplici, ex

obtusa basi in longum mucronem, in forma venabuli excur-
rente.—According to Thunberg [*Fl. jap.*, 330] this is
Acrostichum Lingua [*Polypodium* or *Niphobolus* of other
authors].

Phon zo, XXXV, 7 :—石韋 and [8] 杏葉石韋
[apricot-leaved, comp. ' *supra*], *Polypodium Lingua*.—*Ibid.*,
11 :—瓦韋, fig. dextra, *Polypodium lineare*, Thbg. ; fig.
sinistra, *Vittaria lineata*, Sw.

204.—石長生 *shi ch'ang sheng*. P., XX, 6. T., CLXXV.

Pen king :—Shi ch'ang sheng [according to Li Shi-chen
the meaning of the name is "persistent plant growing on
rocks"], 丹草 *tan ts'ao* (cinnabar plant), properly 丹沙草
tan sha ts'ao. Stem and leaves used in medicine. Taste
saltish. Nature slightly cold. Poisonous.

*Pie lu :—*The *shi ch'ang sheng* grows in Hien yang [in
Shen si, App. 65] in mountain-valleys.

T'ao Hung-king :—It is not used now in medicine.
It is a fine, delicate plant about a foot high with purple
flowers, growing on the sides of rocks. Leaves resemble
those of the *küe* [*Pteris aquilina*. See *Classics*, 377], but
they are finer and black, like shining varnish.

Su Kung [7th cent.] :—This plant grows more than a
foot high. Its stem and leaves are gathered for medical
use in the 5th and 6th months. This is the plant 龄筋草
han kin ts'ao, now found in the druggists' shops, which
resembles the *ts'ing siang* [see 82] and has a slender but
strong purple stem. It is now used in sacrifices.[58]

P. Smith [142] may be right in identifying the *shi
ch'ang sheng* with the Maiden-hair plant (*Adiantum*), although

[58] 太常.

the vague descriptions in the *P.* do not seem to point to this plant. The *Ch.* [XVI, 5] figures, sub *shi ch'ang sheng*, a Fern but not *Adiantum*.

Phon zv, XXXV, 15:—石長生, *Adiantum monochlamys*, Eaton [*Adiantum æthiopicum*, Thbg., which has black petioles].

Referring to T'AO HUNG-KING's ststement that the *shi ch'ang sheng* has black leaves, I may observe that most of the species of *Adiantum* have black stipes. *See* BENTH., *Fl. hongk.*, 446, *Adiantum*:—Stipes usually slender, black and shining, etc.

205.—景天 *king t'ien*. *P.*, XX, 7. *T.*, CXXXVI.

Pen king:—*King t'ien* (brilliant heaven), 愼火 *shen huo* and 戒火 *kiao huo* [the meaning of both names is "guarding against fires"]. Flowers [and also leaves] used in medicine. Taste bitter. Nature uniform. Non-poisonous. [Subsequent writers say "slightly poisonous."]

Pie lu:—Other names: 救火 *kiu huo*, 據火 *kü huo* [both names have the same meaning as *shen huo, v. supra*], 火毋 *huo mu* (mother of fire). The *king t'ien* grows in T'ai shan [in Shan tung, App. 322] in river-valleys. It is gathered on the 4th day of the 4th month and on the 7th day of the 7th, and dried in the shade.

T'AO HUNG-KING:—The *king t'ien* is the most elegant of all plants. The people keep it in pots upon their houses, for it is reputed to be a protection against fires. Another name for it is 胖火 *pi huo* [same meaning as the former names]. Outside the city of Kuang chou (Canton) there stands a large tree, from three to four fathoms in girth, which is called 愼火樹 *shen huo shu*.

MA CHI [10th cent.] observes that in the accounts of Southern China no mention is made of this tree.

Su Sung [11th cent.]:—Now this plant is found in the southern as well as in the northern part of China. People cultivate it in their courtyards and also place it in pots upon the roofs of their houses. Its leaves resemble those of the *ma ch'i hien* (*Purslane*) but are larger and grow in several rows on the stem. The leaves are very fat and soft. The plant blossoms in summer. Small reddish purple flowers. After autumn it decays, but it has a perennial root. Stem, leaves and flowers are all used in medicine.

K'ou Tsung-shi [12th cent.]:—The plant is very easy to cultivate. A branch broken off and put into the ground, and watered, will soon thrive.

Li Shi-chen:—The *king t'ien* is much cultivated on artificial rocks in gardens. It is very fat. The stem is tinged with red and yellow, from one to two feet high. When broken it discharges a juice. Leaves pale green, shining, soft, spoon-shaped, thick, and not pointed. In summer it opens its small white flowers. The fruit (capsule) is similar to that of the *lien k'iao* [*Hypericum*. See 120] but smaller, and contains small black seeds of the size of millet. The leaves taste somewhat sweet and bitter. They can be eaten after scalding. The plant is also called 護火 *hu huo* [meaning as in the above names].

Ch., XI, 34:—*King t'ien* or 八寶兒 *pa pao rh* (the eight jewels,—a Buddhist term). Good drawing of *Sedum alboroseum*, Baker, a beautiful species, commonly cultivated in North China under the name of *pa pao rh*. It is nearly allied to the Japanese *S. erythrostictum*, Miq., and *S. spectabile*, Bor.

Tatar., *Cat.*, 58:—*King t'ien*. Filix?—P. Smith, 114, House-leeks, *Sedum*.

Amœn. exot., 912:—慎火 *sinqua*, vulgo *ikingusa* et *iwarenge*. Sedum majus vulgare in ollis hic culta. Datur et minus, singulari gaudens charactere.

So moku, VIII, 38 :—景天. Japonice: *benkeisoo. Sedum erythrostictum*, Miq.—Comp. also *Phon zo*, XXXVI, 1.

SIEB., *Icon. ined.*, IV :—Same Chinese name, *Sedum telephioides?*

SIEB., *Œcon.*, 337 :—*Sedum telephioides.* Japonice: *benkusoo* ; sinice : 景天草. Adhibetur in hæmorrhagiis.

Phon zo, XXXVI, 6, 7 :—慎火樹. The plant represented under this Chinese name looks like an *Euphorbia*.

206.—陟釐 *chi li*. *P.*, XXI, 1. *T.*, LXXIV.

Comp. *Rh ya*, 129.

Pie lu:—The *chi li* grows in Kiang nan [App. 124] in ponds and swamps.

In the *Lei kung yao tui* it is called 側梨 *tse li* and said to grow in rivers.—The *Shuo wen* calls it 水衣 *shui i* (water covering).

T‘AO HUNG-KING :—In the south the people use it for making paper. It is also used as a medicine.

SU KUNG [7th cent.] :—The *chi li* is the same as the 水苔 *shui t‘ai* (water moss or fucus), of which in the time of the Tsin dynasty [3rd cent.] a paper, called 側理紙 *tse li chi*, was made. It is a kind of coarse moss or fucus, of a dark green colour, growing in the water. The paper made of it is also called 苔紙 *t‘ai chi*. In the *Tung yang-fang* [4th cent.] it is stated that this moss grows on stones in water. It is green, has the appearance of hair and is also called 石髮 *shi fa* (stone hair).

K‘OU TSUNG-SHI [12th cent.] :—The *chi li* is now collected and dried for food by the people, under the name 苔脯 *t‘ai fu* (preserved fucus). The 菁苔 *ts‘ing t‘ai* (green moss) is similarly treated. Both are very nourishing. They are common in the market of Pien king [Kai feng fu, in Ho nan. See App. 248].

Su Sung [11th cent.]:—The dried *shi fa* is eaten salted or in soups. It forms an excellent article of food.

Li Shi-chen:—The *chi li* grows on stones in water, covering them densely, and has the appearance of human hair. That growing in stagnant water, where there are no stones, forms masses of intricated filaments, like floss silk, and is called 水綿 *shui mien* (water-floss).

Ch., XVIII, 10:—*Chi li* and [11] *shi fa. Algæ, Conferva?* Similar plants are figured in the *Phon zo* [XXXVII, 15] sub 陟釐 and [16] 水綿.

Amœn. exot., 833:—苔菜 *tai sei,* vulgo *aii nori.* Muscus marinus Corallinæ similis, multifidus, folio tenuissimo.—According to Martens [*Preuss. Exped. nach Ost Asien. Tange. China and Japan*] this is *Ceramium rubrum,* Huds.

207.—井中苔 *tsing chung t'ai.* P., XXI, 2. T., LXXIV.

Pie lu:—*Tsing chung t'ai* (moss growing in wells), also 苹藍 *p'ing lan.* Taste sweet. Nature very cold. Non-poisonous.

T'ao Hung-king:—It grows in disused wells. Used as a counter-poison.

Amœn. exot., 831:—苔 *tai,* vulgo *koki.* Muscus in genere.

208.—垣衣 *yüan i.* P., XXI, 4. T., CXL.

Pie lu:—*Yüan i* (covering of walls), 垣贏 *yüan ying* (abundance of the walls), 天韭 *t'ien kiu,* 昔邪 *si sie,* and 鼠韭 *shu kiu,* also 青苔衣 *ts'ing t'ai i* (green, mossy covering). It grows in shady places, on old walls and on houses (roofs). It is gathered on the 3rd day of the 3rd month and dried in the shade.

Su Kung [7th cent.]:—That which grows on the northern shady sides of old walls is called *ts'ing t'ai i*, that which grows on stones is called *si sie* or 烏韭 *wu kiu* [see 210], and that which grows on roofs is 屋遊 *wu yu* [*see the next*]. All these sorts resemble each other.

Ch., XVI, 53 :—Rude figure of the *yüan i*. Moss.

See also *Phon zo*, XXXVII, 20.

209.—屋遊 *wu yu*. P., XXI, 5. T., CLXXVIII.

Pie lu :—Wu yu (roof rambler). It grows in the shade on the tops of houses and is gathered in the 8th and 9th months.

T'ao Hung-king :—It grows on old tile-roofs. It is cut off for medical use.

The *Kia yu Pen ts'ao* [11th cent.] calls it 瓦苔 *wa* (tile) *t'ai*.

Li Shi-chen :—The plant is several inches long. It is also called 瓦松 *wa sung* (tile pine), 瓦蘚 *wa sien*, and 博邪 *po sie*.

Ch., XVI, 54 :—昨葉荷草 *tso ye ho ts'ao* or 瓦松 *wa sung*. Henry [*Chin. pl.*, 518] identifies this figure with *Cotyledon fimbriata*, Turcz., var. *C. ramosissima*, Max., which grows on old tile-roofs in Hu pei.—The same plant is common at Peking and is there called *wa sung*.

Tatar., *Cat.*, 62, 63 :—*Wa sung* or *wu yu*. *Umbilicus malacophyllus* (*Cotyledon malacophylla*). Order *Crassulaceæ*.

Comp. also *Phon zo*, XXXVII, 20 :—屋遊. Rude drawing.

210.—烏韭 *wu kiu*. P., XXI, 7. T., CLIX.

Pen king :—Wu kiu (black onion). Apparently the whole plant is used in medicine. Taste sweet. Nature cold. Non-poisonous. [A later writer says "poisonous."]

Pie lu:—The *wu kiu* grows on stones (rocks) in the mountains. The 鬼麗 *kui li* likewise grows on rocks. It is dried in the sun.

Su Kung [7th cent.]:—This is the 石苔 *shi t'ai*, also called 石髮 *shi fa* [rock hair. Comp. also 206]. It grows in shady rocky vales which never see the sun. It resembles the *küan po* [*see the next*].

Ta Ming [10th cent.]:—It is also called 石衣 *shi i* (covering of rocks), and is from four to five inches long. It is poisonous.

Ch., XVI, 51:—*Wu kiu.* Rude drawing.

Phon zo, XXXVIII, 3:—烏韭, a Fern. *Trichomanes japonicum*, Thbg.

211.—卷柏 *küan po*. *P.*, XXI, 8. *T.*, CXLVII.

Pen king:—*Küan po* (curled inward *Thuja*). The whole plant is officinal. Taste pungent. Nature uniform. Non-poisonous.

Pie lu:—Other names: 萬歲 *wan sui* (ten thousand years), 求股 *kiu ku* and 交時 *kiao shi*. The *yüan po* grows in Ch'ang shan [in Chi li, App. 8], among stones (rocks) in mountain-valleys. It is gathered in the 5th and 7th months and dried in the shade.

Wu P'u [3rd cent.] calls it 豹足 *pao tsu* (leopard's foot).

T'ao Hung-king:—This plant grows, densely crowded on rocks, in Mid China. Fine leaves (finely dissected) resembling those of the *po* [*Thuja*. See 300], and curved inward like the toes of a bird. The plant is of a greenish yellow colour.

Chang Yü-hi [11th cent.]:—It grows in Kien k'ang [Nan king, App. 137]. Fan tsz' ki jan says it is produced in San fu [in Shen si, App. 265].

Su Sung [11th cent.]:—It is now found in Kuan and Shen [Shan si and Shen si, App. 158, 284], and in I chou and Yen chou [both in Shan tung, App. 106, 404]. The plant has a perennial root of a purple colour and is covered with many hairs (radical fibres). The leaves are like *Thuja* leaves but smaller, curved inward like the toes of a chicken. The plant grows from three to five inches high, and has neither flowers nor seeds. It is common on stones.

Li Shi-chen says that this plant never dies.

Ch., XVI, 3 :—*Kŭan po.* The figure seems to represent *Selaginella involvens*, Spg. The description in the *P.* agrees. This curious plant, of the order *Lycopodiaceæ*, is very common in the Peking mountains, where it grows on stones and rocks. It has the fronds curled in and contracted when dry, in which condition it is of a yellowish brown colour, but it expands immediately and assumes a fresh, green colour when put into hot water. Its common name at Peking is 湯 湯 青 *t'ang t'ang ts'ing*, which means "it becomes green in hot water."

Tatar., *Cat.*, 60 :—*Kŭan po.* *Lycopodium hygrometricum.* —P. Smith [141] erroneously identifies *kŭan po*[59] with *Lycoperdon squalmatum*, which, he says, is a fungus with curved, compressed fronds [*sic* !].

Amœn. exot., 912 :—卷 柏 *kimpaku*, vulgo *iwagogi* et *iwasiba.* Muscus saxatilis Ericoides.—Thunberg [*Fl. jap.*, 386] comments :—Lichines fruticulosi.—*Iwahiba* is the Japanese name for *Selaginella involvens* [*Lycopodium circinale*, Thbg., *l.c.*, 341].—Matsumura, 177.

Phon zo, XXXVIII, 6 :—卷 柏, *Selaginella involvens.*

[59] In the Pharmaceutical Museum, London, there are samples of this drug from Hankow [sent by Porter Smith] and from Hongkong. These are *Selaginella involvens*, Spg.—A. Henry.

212.—玉柏 *yü po.* *P.,* XXI, 9. *T.,* CLXIII.

Pie lu :—Yü po (jade *Thuja*), also 玉遂 *yü sui.* It grows on stones and resembles a pine tree. It is only from five to six inches high, and has purple flowers. The leaves and the stem are used in medicine. Taste sour. Nature warm. Non-poisonous.

Li Shi-chen :—This is a small kind of *shi sung* [*v. infra*]. The people gather it and keep it in basins, where it lives for many years, whence the names 千年柏 *ts'ien nien po* (thousand years *Thuja*) or 萬年松 *wan nien sung* (ten thousand years pine).

Regarding the 石松 *shi sung* (stone pine) Ch'en Ts'ang-k'i [8th cent.] states :—It is a plant from one to two feet high, resembling a pine tree, which grows in the mountains of T'ien t'ai [in Che kiang, App. 340]. The mountain people use the root and the stem [as a medicine].— Li Shi-chen adds :—It is a large kind of *küan po.*

Ch., XVI, 42 :—*Wang nien po* and [43] *wan nien sung.* Rude drawings, but they seem to be intended for *Lycopodium.* —*Ibid.,* XVII, 3 :—*Shi sung.*

Cust. Med., p. 378 (497):—*Wan nien sung* exported 1885 from Canton 3.14 piculs,—p. 302 (435), from Amoy 0.03 picul.

Cl. Abel [*Journ. in the Int. of China,* 1816, 1817, p. 220] notices a *Lycopodium,* cultivated and spontaneous, in the Canton province, which might perhaps be best compared to a fir tree *en miniature.*—R. Fortune [*Wand.,* 84 ; *Tea Count.,* 8 ; *Res. am. Chin.,* 80] met on the hills of Hong kong a curious dwarf *Lycopodium* which takes the very form of a dwarf tree *en miniature.* He saw it also in Hong kong gardens. The Chinese, who prize it highly, call it

man neen chung [evidently he meant to write *wan nien sung*].
FORTUNE introduced it to England, where it was named
Lycopodium cæsium.

Phon zo, XXXVIII, 10:—玉柏, *Lycopodium japonicum,*
Thbg.—10:—千年松, *L. cernuum,* L.—11:—石松,
L. clavatum, L.

213.—馬勃 *ma pu.*   P., XXI, 10.   T., CLXV.

Pie lu:—*Ma pu.* It grows in gardens and neglected
places. Taste pungent. Nature uniform. Non-poisonous.

T‘AO HUNG-KING:—The popular name is 馬㞘勃 *ma
pi pu* [which has the indecent meaning "horse-fart"]. It
is of a purple colour, hollow and soft. It discharges a powder
(spores) like the *kou kan tan* [dog-liver ball,—probably the
name of a Fungus].

K‘OU TSUNG-SHI:—It grows on decayed wood in damp
places, and is gathered in autumn. It varies as to size, being
sometimes very large.

LI SHI-CHEN:—Other names: 灰菰 *hui ku* (ashes
fungus) and 牛尿菰 *niu sui ku* (ox-urine fungus).

Ch., XVI, 52:—*Ma pu.* Bad drawing. Fungus.

TATAR., *Cat.,* 37:—*Ma pu. Lycoperdon.*—P. SMITH,
140:—*Lycoperdon giganticum.*[60]

The name *Lycoperdon* is derived from *lykos* (a wolf) and
perdo (to break wind). In English this fungus is commonly
called Puff-ball.

DEBEAUX [*Florule de Shanghai,* 73] states that *Lycoperdon giganteum* [sinice: *ma po*] is frequent near Shang hai.

---

[60] PORTER SMITH'S specimen in the Pharmaceutical Museum, London,
was pronounced by M. C. COOKE to be a true species of *Polysaccum,* and
not *Lycoperdon.* See *Pharmac. Journal,* II, p. 160.—A. HENRY.

Cust. Med., p. 134 (210):—*Ma pu* exported 1885 from Chin kiang 2 piculs,—p. 386 (604), from Canton 0.30 picul, —p. 278 (71), from Amoy 0.18 picul.

214.—屈草 *k'ü ts'ao.* P., XXI, 13. T., CLXXXII.

Pen king:—*K'ü ts'ao.* Taste bitter. Nature slightly cold. Non-poisonous.

Pie lu:—It grows in Han chung [S. Shen si, App. 54], in swamps, and is gathered in the 5th month.

Obscure plant unknown to subsequent Chinese authors.

215.—別羈 *pie ki.* P., XXI, 13. T., CLXXXII.

Pen king:—*Pie ki.* Taste bitter. Nature slightly warm. Non-poisonous.

Pie lu:—The *pie ki* is also called 別枝 *pie chi.* It grows in Lan t'ien [in Shen si, App. 175] in river-valleys.

This is likewise an obscure plant.

I omit 78 names of vegetable drugs mentioned in the *Pie lu* with short notes on their medical virtues. [*P.*, XXI, 13-18.] They all refer to obscure plants unknown to subsequent authors.

216.—巨勝 *kü sheng.* P., XXII, 1. T., XXXVIII.

Pen king:—*Kü sheng.* Seeds officinal. Taste sweet. Nature uniform. Non-poisonous.

Pie lu:—胡麻 *hu ma* (hemp of the Western Barbarians) or 狗虱 *kou shi* (dog-louse), also called 巨勝 *kü sheng.* It is produced in Shang tung [S.E. Shan si, App. 275] in swamps. It is gathered in autumn. 菁蓂

tsing jang is the name for the leaves of the *kü sheng*. It grows in Chung yüan [Ho nan, App. 34].

T'AO HUNG-KING :—The *hu ma* is one of the most important cereals. The pure black kind (black seeds) is called 巨勝 *kü sheng*. The character *kü* means great [and *sheng* = superior]. The name *hu ma* [v. *supra*] refers to its being a native of Ta wan [ancient name for Fergana, 2nd cent. B.C. See *Bot. sin.*, I, 24]. The *hu ma* has a round stem and the *kü sheng* a square one.

CH'EN TS'UNG-CHUNG [an author of the Sung period] states that the *hu ma* was brought [in the 2nd cent. B.C.] by CHANG K'IEN from Ta wan to China, and that it is commonly called 油麻 *yu ma* (oil-hemp).

SU SUNG [11th cent.] proves that the *hu ma*, a foreign plant brought to China in the Han dynasty, has been erroneously identified by some early authors with the *kü sheng* of the *Pen king*, which is quite a different plant.

The *hu ma*, now more commonly called 脂麻 *chi ma* (fat or oil hemp), is the *Sesam* plant, extensively cultivated all over China for the oil of its seeds. Further particulars regarding this plant will be found in another part of the *Botanicon sinicum*.

I am not prepared to say what the *kü sheng* of the ancient Chinese Materia Medica was. In the *Cust. Med.* the 巨勝子 *kü sheng tsz'* appears only once [p. 158 (276)], where it is stated that 2.40 piculs have been imported from Han kow and other ports to Shang hai. It is identified there with *Sesamum indicum*, probably on the authority of P. SMITH [195], who relies on TATAR. [*Cat.*, 60]. But BRAUN [*Hank. Med.*, 12] states that the seeds exported from Han kow under the name of *kü sheng tsz'* bear no resemblance to *Sesam* seed. They are yellowish brown, oblong and have all the appearance of fennel seed. The druggist's shops

at Peking distinguish two sorts of *kü sheng tsz'*—the
black and the white. Both were examined about 10 years
ago by the late MAXIMOWICZ. The first kind—small,
triangular black seeds—seemed to belong to a *Nigella*, although no species of this genus of *Ranunculaceæ* has hitherto
been observed in China by our botanists. The white
(yellowish) seeds seemed to belong to *Ixeris* or *Mulgedium*
(*Compositæ*).

217.—大 麻 *ta ma.*   P., XXII, 11.   T., XXXIX.

Comp. *Rh ya*, 104, 140,—*Classics*, 388.

Pen king :—Ta ma (great hemp). The flowers when
they burst [when the pollen is scattered] are called 麻 賁
ma fen or 麻 勃 *ma pu*. The best time for gathering is the
7th day of the 7th month. The seeds are gathered in the
9th month. The seeds which have entered the soil are
injurious to man.[61] It grows in T'ai shan [in Shan tung,
App. 322]. The flowers, the fruit (seed) and the leaves are
officinal. The leaves and the fruit are said to be poisonous,
but not the flowers and the kernels of the seeds.

T'AO HUNG-KING :—The *ma fen* is the male hemp, which
does not bear seed. The people use the fibres for making
cloth and shoes.

SU KUNG [7th cent.] :—The *fen* is the seed of the *ma*,
not the flower [as the *Pen king* states]. The author refers
to the *Rh ya* and the *Classics*.

CH'EN TS'ANG-K'I [8th cent.] :—The hemp (*ma*) which
is sown early in spring is called spring hemp. The seeds
are small and poisonous. That which is sown late in spring
is called autumn hemp. Its seeds are used in medicine
and oil is expressed from them.

 " 八 土 者 損 人·

LI SHI-CHEN:—The *ta ma* (great hemp) is also called 火麻 *huo* (fire) hemp and 黃麻 *huang* (yellow) *ma*. The *Rh ya i* calls it 漢麻 *Han ma* (Chinese hemp). It is largely cultivated for the oil of its seeds as well as for its textile fibres. There is a male and a female hemp plant. Only the latter bears seed. The leaves are narrow and long, resembling those of the *i mu* [*Leonurus*. See 78]. From seven to nine leaves (leaflets) proceed from the top of a common stalk. It blossoms in the 6th month. Small yellow flowers in spikes (or racemes). The fruit is as large as that of the *hu sui* (*Coriander*).—LI SHI-CHEN observes that the early authors do not agree in their statements regarding the *ma fen*. The *Pen king* says that *ma fen* and *ma pu* both denote the flowers of the plant, and mentions besides this the seeds. But according to WU P'U [3rd cent.] *ma fen* and *ma pu* are not the same, *ma pu* being a name for the flowers, which are not poisonous, whilst the *fen*, called also 麻藍 *ma lan* and 青葛 *ts'ing ko*, is poisonous. WU P'U does not explain what he understands by *ma fen*, but evidently he means the seeds. SU KUNG, who wrote four centuries later, says *ma fen* is the seed. LI SHI-CHEN comes to the conclusion that *fen* is the covering of the seed. It is poisonous, whilst the 麻仁 *ma jen*, or the kernel within, is innoxious and can be eaten. This explains the statement in the *Chou li* that the *ma* seed was used for food and offered to the emperor. The leaves of the *ma* are poisonous.

The *ta ma* of the *Pen ts'ao* is the common hemp, *Cannabis sativa*. Further particulars regarding this important Chinese textile plant will be given in another part.

In the *Cust. Med.* the 火麻子 or 〡 〡 仁 is stated to be imported to many of the Treaty ports. It is said to come from Han kow, Shang hai, Tien tsin and Chefoo. *See also* P. SMITH, 111.

218.—小麥 *siao mai*. Wheat. *P.*, XXII, 17. *T.*, XXXII.
Comp. *Classics*, 339. P. SMITH, 230.

It is noticed in the *Pie lu*; only the name and medical
virtues. No explanation by T'AO HUNG-KING.

The following are used in medicine: 麥苗 *mai miao* (the
young plants), 麪 *mien* (flour), 麥麩 *mai fu* (the bran of
wheat), 浮麥 *fou mai* (floating wheat,—unknown to me)
and 麥奴 *mai nu*, the latter being described by K'OU
TSUNG-SHI as "black spots appearing on the ripe ears of
wheat" [produced probably by a small parasitical fungus].

The 浮麥 is mentioned in the *Cust. Med.* [p. 292 (283),
Amoy,—p. 366 (355), Canton].

219.—大麥 *ta mai*. Barley. *P.*, XXII, 23. *T.*, XXXII.
Comp. *Classics*, 340.

The *Pie lu* notices only the name and the medical virtues.

T'AO HUNG-KING:—The *ta mai* is also called 稞麥
k'o mai [see *W.D.*, 425] and 牟麥 *mou mai*. It resembles
the *kung mai* [*see the next*] but its husk is thinner.

The young plants and 麥蘖 *mai ye* (malt) are both used
in medicine. LI SHI-CHEN explains the character *ye* by
"barley steeped in water and made to germinate." The 大麥
奴 *ta mai nu* also is officinal. [Comp. 218]. P. SMITH, 33.

220.—穬麥 *kung mai*. *P.*, XXII, 25. *T.*, XXXII.

The *Pie lu* notices only the name and the medical virtues.

T'AO HUNG-KING:—The *kung mai* is used for feeding
horses. It is also, like wheat and barley, a nourishing and
strengthening food for man.

Further particulars regarding the *kung mai* will be found

in another part. It is the so-called "nacked Barley," the grain of which separates from the chaff scales after the manner of wheat.

221.—稻 *tao.* Rice. *P.,* XXII, 29. *T.,* XXV.
Comp. *Classics,* 337, 338.

The *Pie lu* notices only the name and the medical virtues.

T'AO HUNG-KING :—In Taoist prescriptions two kinds of rice are distinguished—the 稻米 *tao mi* and the 粳米 *keng mi.* The *tao mi* is very white. In Kiang tung [Kiang su, Che kiang etc., App. 124] they have no *tao mi*, and apply this name to the *keng mi* [*see the next*].

I have already referred to the fact [see *Bot. sin.,* II, 338] that in ancient times the character *tao*, now a common name for rice, was applied to the glutinous rice. *Keng* is the common rice.

The grain, the culm, the awns, and the flower of the rice plant are all used in medicine.

222.—粳 *keng* (the character is also written 秔). Common Rice, not glutinous. *P.,* XXII, 34.

Comp. *Classics,* 338.

The *Pie lu* notices only the name and the medical virtues.

T'AO HUNG-KING :—The 粳米 *keng mi* is the rice commonly eaten by the people. There are various sorts—the white, the red, the small, and the large.

223.—稷 *tsi.* The common Millet, *Panicum miliaceum,* L. *P.,* XXIII, 1. *T.,* XXIX.

Comp. *Classics,* 343.

The *Pie lu* notices only the name and the medical virtues.

224.—黍 *shu.*   A glutinous variety of *Panicum miliaceum.*
P., XXIII, 3.   *T.,* XXX.

   Comp. *Classics,* 341.

   The *Pie lu* notices only the name and medical virtues.

   The grain, the culm and the root are all used in medicine.

225.—粱 *liang.   Setaria italica.   P.,* XXIII, 7.   *T.,* XXIX.

   Comp. *Classics,* 344.

   The *Pie lu* notices only the name and the medical virtues.
The grain is used in medicine.

226.—粟 *su.   P.,* XXIII, 9.   *T.,* XXXI.

   Comp. *Classics,* 347.   *Setaria.*

   Pie lu :—Only name and medical virtues.   The grain is
used in medicine.

227.—秫 *shu.   P.,* XXIII, 12.   *T.,* XXXI.

   Comp. *Classics,* 348.   A glutinous *Setaria italica.*

   Pie lu :—Only names and medical virtues.   The grain
and the root are used in medicine.

228.—薏苡仁 *i i jen.   P.,* XXIII, 17.   *T.,* XL.

   Pen king :—I *i jen,* also 解蠡 *kiai li.*   The seed is
officinal.   Taste sweet.   Nature slightly cold.   Non-poisonous.
The root and the leaves are also used in medicine.

   Pie lu :—Other names : 芑實 *k'i shi,* 𧀸米 *kan mi.*   The
i mi jen grows in Chen ting [in Chi li, App. 11], in marshes,
in the plain and in fields.   The fruit is gathered in the 8th
month and the root at all times of the year.

T'AO HUNG-KING :—Chen ting hien belongs to the prefecture of Ch'ang shan [in Chi li, App. 8]. The plant is also common in Mid China, where it is much cultivated. The seeds produced in Kiao chi [*Cochinchina.* App. 133] are very large and are known there under the name 蘚珠 *kan chu* (bead). MA YÜAN [63] when he was in Kiao chi tasted these seeds and introduced the plant to China, where the fruit was then called *chen chu* (pearl). The kernels (仁 *jen*) are used in medicine.

MA CHI [10th cent.] :—Now the drug from Liang Han [Sz ch'uan, App. 188] is generally used. It is less efficacious than that from Chen ting. That of a greenish white colour is good. To obtain the kernels the fruit is steamed in a boiler, then dried and [the hard shell] broken off.

SU SUNG [11th cent.]:—The *i i* plant grows from 3 to 4 feet high. Leaves like those of the *shu* [*Panicum.* See 224]. Reddish white flowers in spikes. In the 5th or 6th month it produces fruit of a greenish white colour, resembling beads, but slightly oblong. It is therefore also called 薏珠子 *i chu tsz'* (*chu tsz'* = bead). Children perforate [the hard shell of] these globular fruits and string them together to play with. This fruit is gathered in the 9th and 10th months.

LEI HIAO [5th cent.]:—That with a larger fruit, the flour of which is not used, is called 粳糧 *keng kan*. It is tasteless.

The *i i jen* has a smaller fruit of a green colour. The taste [of the kernel] is sweet. When chewed it sticks to the teeth.

LI SHI-CHEN :—The *i i* is much cultivated. The plant has a perennial root. The young leaves resemble those of the

_____

[63] A renowned commander, † A.D. 49. See MAYERS, *Chin. Read. Man.*, 478.

pa mao.[63] In the 5th or 6th month it sends up the stem
which bears the flowers, and afterwards the fruit. There are
two kinds. That which sticks to the teeth [*v. supra*], and has
a pointed fruit with a thin shell is the *i i.* The kernel is white
like glutinous rice. Gruel can be made of it. It is
also ground into flour and used for food or for fermenting
liquors. The other sort is globular [the fruit] and has a
thick, hard shell. This is called 菩提子 *p'u t'i tsz'* [name
of the beads in Buddhist rosaries]. It furnishes but little
flour. This is the *keng kan* [mentioned by LEI HIAO]. It
is used for the beads of rosaries and is therefore also called
念珠 *nien chu* (prayers bead). Its root is white, as large as
the handle of a spoon, contorted, and of a sweet taste. The
Kiu huang Pen ts'ao terms this plant [the seeds] 回回米
Hui hui mi (rice of the Mohammedans), also 西番蜀秫
Si fan shu shu (*Si fan Sorgho. Si fan*=N.W. Sz ch'uan),
also 草珠兒 *ts'ao chu rh* (vegetable bead). The leaf is called
屋菼 *wu t'an.*

Ch., I, 5 :—*I i* or 草子兒 *ts'ao tsz' rh.* Good drawing
representing Job's-tears (*Coix Lachryma,* L.). This plant is
commonly cultivated at Peking under the name of *ts'ao tsz' rh.*
It has large, round, hard fruits. The hard covering of the
farinaceous seed is the ossified calyx. There are two varieties
at Peking with white or grayish covering of the fruit. I have
also seen in the druggist's shops a variety (or species) with
small, oblong pointed fruit.

In the *Kiu huang* [LII, 7, sub 川穀 *Ch'uan ku* (Sz
ch'uan corn)] a *Coix* with small fruit is represented. See the
same, *Ch.,* II, 9.

LOUR., *Fl. cochin.,* 673 :—*Coix Lachryma.* Sinice : *y y gin.*
TATAR., *Cat.,* 29 :—薏米仁 *i mi jen, Coix exaltata.*

---

[63] 芭茅 in Japan *Erianthus japonicus.*

[JACQUIN figured this species, *Elog. gram. rar. tab.*, 40, without describing it. Apparently it is only a variety of *C. Lachryma*.]—P. SMITH, 125.

In the Chinese Customs' Reports *i i jen* is generally erroneously identified with pearl-barley. The seed of *Coix* deprived of the shell has indeed some resemblance to pearl-barley.

Cust. Med., p. 198 (216):—*I mi jen* exported 1885 from Ning po 210 piculs,—p. 50 (68), from Che foo 98 piculs,—p. 162 (329), from Shang hai 46.37 piculs.—Exported also from Han kow. See *Hank. Med.*, 20.

Amœn. exot., 834 :—薏苡. Medicis et literatis *jokui* et *jokuinin*, vulgo *dsudsudama*, it. *fatsji koku*. Arundo granifera. Milium arundinaceum ; aliis *Lachrymo Jobi*.

Phon zo, XLII, 4-6 :—Above Chinese name. Representation of two varieties of *Coix Lachryma* with oblong and globular fruit.

MATSUMURA, 55:—薏苡, *Coix agrestis*, Lour., and 川穀, *Coix Lachryma*.

SIEB., *Œcon.*, 44 :—*Coix Lachryma*. Sinice: 川穀. Varietates :—

    (a). *susutama* }
    (b). *joosusutama* } pro orbiculis ad preces.
    (c). *toomuki*. Edulis ac medicæ usui.

229.—大豆 *ta tou*. Soy-bean, *Soja hispida*, Mœnch. *P.*, XXIV, 1. *T.*, XXXV.

Comp. *Rh ya*, 29, *Classics*, 355.

Pen king :—*Ta tou*. The seed of the 黑火豆 *hei* (black) *ta tou* is used in medicine. Taste sweet. Nature uniform. Non-poisonous. When eaten it causes the body to become heavy.

Pie lu:—The *ta tou* is produced in T'ai shan [in Shan tung, App. 322] in marshes. It is gathered in the 9th month.

Su Sung [11th cent.]:—The *ta tou* is now generally cultivated in two varieties—the white and the black. The latter is used in medicine.

Li Shi-chen :—There are many sorts of the *ta tou*—the black, white, yellow, gray, green and spotted [according to the colour of the seeds]. The black sort is used in medicine, is also a valuable food, and is employed in making 豉 *shi* (Soy. *V. infra*, 234, and *Bot. sin.*, II, 355). From the yellow sort oil is expressed and 醤 *tsiang* (sauces) and 腐 *fu* (beancurd. See *l.c.*) are prepared. These beans are also eaten roasted. Besides the seeds of the *ta tou*, and the oil expressed from them, other parts of the plant are likewise officinal, *viz.* the 大豆皮 *ta tou p'i* (the valves of the legume), the leaves, the carbonized straw and the flowers.

See P. Smith, 88, *Dolichos soja.*—

230.—大豆黃卷 *ta tou huang küan.*   P., XXIV, 7.

This drug is noticed in the *Pen king*.

As T'ao Hung-king and Li Shi-chen explain, this consists of the germs of the black Soy bean, produced by steeping the beans in water and causing them to germinate. These germs are used as food.

This is still an article of food at Peking, but produced from the yellow Soy bean and called 黃豆芽 *huang tou ya.*

231.—赤小豆 *ch'i siao tou.*   P., XIV, 9.   T., XXXVI.

Pen king:—*Ch'i siao tou* (red small bean). Seeds, germs and leaves are used in medicine.

This is a red variety of *Phaseolus Mungo.* See *Bot. sin.,* II, 356.

Cust Med., p. 366 (335):—*Ch'i siao tou* exported 1885 from Canton 7 piculs. Exported also from Han kow. See *Hank. Med.*, 16. Identified there with *Abrus precatorius*, L. At Peking also the name *ch'i siao tou* is applied to the seeds of *Abrus precatorius*. But in Chinese botanical works it is *Phaseolus*. Comp. P. SMITH, 1 ; TATAR., *Cat.*, 15.

232.—腐婢 *fu pi*. *P.*, XXIV, 13. *T.*, XXXVI.

Pen king :—*Fu pi*. Flowers. Taste pungent. Nature uniform. Non-poisonous.

Pie lu:—The *fu pi* is produced in Han chung (S. Shen si). It is the flower of the *siao tou* [see 231]. It is gathered in the 7th month and dried in the sun during 40 days.

T'AO HUNG-KING :—The medical virtues of the seeds and the flowers of the *ch'i siao tou* are not the same. Therefore the *Pen king* notices these drugs separately. Not used in medicine. T'AO HUNG-KING observes that there grows near the sea-shore a small tree resembling the *chi tsz'* [*Gardenia*. See 335]. Its stem and leaves are very crooked. It has a fetid smell. The people call it *fu pi*.

SU KUNG [7th cent.] means that the name *fu pi* refers to the flowers of the *ko* plant [*Pachyrhizus*. See 174].

SU SUNG [11th cent.]:—The name *fu pi* is applied to three different plants, *viz.* the small tree near the sea-shore [noticed by T'AO HUNG-KING], the *ko* flower and the flower of the *ch'i siao tou*.

Phon zo, XLIII, 13 :—腐婢. The figure represents a plant with yellow flowers.

SIEB., *Icon. ined.*, VI :—Same Chinese name applied to *Premna japonica*, Miq. Order *Verbenaceæ*. According to MAXIMOWICZ [*Diagn. Pl. asiat.*, VI, 510] this is *P. microphylla*, Turcz. Shrub with yellow flowers. See also *Ind. Fl. sin.*,

II, 256. In Hu pei this plant is called 臭 梁 子 *ch'ou*
(stinking) *liang tsz'*. See HENRY, *Chin., pl.*, 86.

233.—藊 豆 *pien tou*.   *P.*, XXIV, 21.   *T.*, XXXVII.

Pie lu :—Pien tou. Seeds, leaves and flowers official.
Taste of the seeds sweet. Nature slightly warm. Non-
poisonous.

T'AO HUNG-KING :—The *pien tou* is much cultivated. It
climbs on fences and walls. The pods are eaten steamed and
are very palatable.

SU SUNG [11th cent.]:—It is a climbing plant with large
leaves and small flowers. The latter are purple or white.
The pods are produced beneath the flowers. The seeds are
black or white. The white seeds are of a warm nature, and
the black, which are smaller, are cold. The white are used in
medicine. The black-seeded sort is also called 鵲 豆 *ts'io tou*
(magpie bean), for it (the seed) has a white road (rib) like that
seen on the wing of a magpie [evidently the *hilum* is meant].

LI SHI-CHEN :—The name *pien tou* is derived from 扁
pien (flat) and refers to the flat pods. It is a twining
plant with large, roundish, pointed leaves. The flowers
resemble a small butterfly with its wings and tail. There are
numerous varieties, according to the shape of the legume,
which is long or round, sometimes shaped like a dragon's or
tiger's claw, or like a pig's ear or a sickle. The young pods
are eaten as a vegetable. The ripe seeds are eaten boiled.
The seeds are, according to the varieties, black, white, red and
variegated. There is one sort with hard legumes, non-edible,
with coarse, round seeds of a white colour. These seeds are
used in medicine. Other popular names are : 蛾 眉 豆
ô mei tou (silkworm-moth's eyebrows bean,—referring to the
rib of the seed or *hilum*) and 沿 籬 豆 *yen li tou* (fence-
climber bean).

LOUR., *Fl. cochin.*, 534 :—*Dolichos purpureus.* Sinice : *tsu* (*tsz'* = purple) *pien teu.* Legumina tenera et recentia sapida sunt et salubria; and *D. albus.* Sinice: *pe* (white) *pien teu.* Tenerior et sapidior præcedente, nec forma valde differens.

TATAR., *Cat.*, 3 :—*Pien tou*, Semina *Lablab vulgaris.*— P. SMITH, 128.

At Peking *Lablab vulgaris* is much cultivated under the name of *pien tou*, especially the purple flowered, which has also a purple coloured legume ; also the white flowered, *pai* (white) *pien tou*, which has greenish white seeds that are used in medicine.

HENRY, *Chin. pl.*, 471 :—*Pien tou, Dolichos Lablab.*

Cust. Med., p. 162 (340):—*Pai pien tou* exported 1885 from Shang hai 306 piculs,—p. 200 (231), Ning po 72.82 piculs,—p. 130 (145), Chin kiang 2.08 piculs.—Exported also from Han kow. See *Hank., Med.*, 34.

Amœn. exot., 836 :—藕, vulgo *adsi mame*, it. *kaadsi mame.* Phaseolus arvensis, longis sarmentis repens, flore exili purpureo, siliquis brevibus latioribus caudatis ; semine ciceris rotundo, rubente.—This is evidently not *Lablab.* Not identified by THUNBERG.

Phon zo, XLIII, 25 :—藕豆, *Dolichos Lablab.*

So moku, XIII, 14 :—鵲豆, *Dolichos cultratus*, Thbg.

234.—大豆豉 *ta tou shi.   P., XXV, 2.*

Comp. *Bot. sin.*, II, 355 :—Soy. It is noticed in the *Pie lu.* LI SHI-CHEN says it is prepared from the black soy bean.

In the *Cust. Med.* the 豆豉 *tou shi* is mentioned as an article of import, p. 110 (183) Wu hu,—p. 164 (368) Shang hai,—p. 216 (89) Wen chou. It is said to come from Han kow and Ning po, and is identified there with salted black beans.—See also *Hank. Med.*, 45.

235.—蘖米 *ye mi.* *P.*, XXV, 24.

Mentioned in the *Pie lu.* According to the definitions
given by the ancient Chinese authors, this is grain which has
sprouted, rice-wheat, barley, millet, also beans, etc. Thus
ye mi may be translated by " Malt." Comp. also 219.

236.—飴餹 *i t'ang. P.*, XXV, 25.

Mentioned in the *Pie lu.* Taste sweet.

WILLIAMS [*Dict.*, 275] translates *i t'ang (t'ang* = sugar)
by "sugar-plums, sweetmeats." According to T'AO HUNG-KING
it is a preparation of sugar, called also 膠飴 *kiao* (gum glue)
i. One sort, which is tough and of a white colour, is called
餳餹 *sing t'ang.* Not used in medicine.

HAN PAO-SHENG [10th cent.]:—The 飴 *i* is soft sugar.
In North China it is called 餳 *sing* [Comp. *W.D.*, 809]. It is
made from glutinous and common rice, glutinous millet and
Sorgho. Hemp-seed, the peduncles of *Hovenia dulcis* and
some drugs, are sometimes added. That prepared of glutinous
rice is used in medicine.

LI SHI-CHEN :—For the preparation of the *i* or *sing*
malt of barley is used, or the sprouts of other grain.

At Peking comfits, bonbons, etc. made of the sugar
prepared from glutinous rice are sold in the streets. Comp.
also STAN. JULIEN et P. CHAMPION, *Industries de l'Empire
chinois,* p. 210.

237.—醬 *tsiang. P.*, XXV, 28.

Comp. *Bot. sin.*, II, 355, and *W.D.*, 968.

Mentioned in the *Pie lu.* According to LI SHI-CHEN
this is the name for various sauces made of wheat or barley
flour, or of the soy-bean and other beans with salt.

238.—醋 *ts'u*, Vinegar. *P.*, XXV, 30.

Comp. *Bot. sin.*, II, 349.

Mentioned in the *Pie lu*. According to SU KUNG the *ts'u* is made of various grains, also of sugar, grapes and other fruits.

239.—酒 *tsiu*. Wine. *P.*, XXV, 43. *Pie lu.*

Comp. *Bot., sin.*, II, 349.

240.—韭 *kiu*. *P.*, XXVI, 1. *T.*, LV.

Comp. *Classics*, 359.

Allium odorum, L. Mentioned in the *Pie lu*, but only the name and medical properties are noticed. Root, leaves, flowers and seeds are used in medicine.

Cust. Med., p. 74 (130):—韭 柔 子 *kiu*-seeds exported 1885 from Han kow 7.60 piculs.

Further particulars regarding this plant will be given in another part.

241.—葱 *ts'ung*. *P.*, XXVI, 7. *T.*, LVI.

Comp. *Classics*, 357 :—*Allium fistulosum*, L.

Pen king, Index :—葱 實 *ts'ung shi* (fruit). All parts of the plant are used in medicine. The lower, white part of the scape is called 葱 莖 白 *ts'ung heng pai.*

Further particulars in another part.

242.—薤 *hiai*. *P.*, XXVI, 15. *T.*, LV.

Comp. *Classics*, 360; *Rh ya*, 63, 3. *Allium.*

Pen king, Index :—薤 實 *hiai shi* (fruit). The scape, or rather the lower white part of it, is called | 白 *hiai pai.*

Pie lu:—The *hiai* grows in Lu shan [in Ho nan, App. 203] in swamps.

Cust. Med., p. 276 (38):—*Hiai pai* exported 1885 from Amoy 2.52 piculs.

243.—蒜 *suan.*   *P.,* XXVI, 18.   *T.,* LVII.

Comp. *Classics*, 358.   Garlic.

Pie lu:—The *suan* or 小蒜 *siao* (small) *suan* is gathered on the 5th day of the 5th month [apparently the bulb is meant]. Taste pungent. Nature warm. Slightly poisonous. The leaves are likewise used in medicine.

T'AO HUNG-KING :—The fresh leaves of the *siao suan* can be eaten mixed with boiled food. In the 5th month the leaves wither. The root is called 亂子 *luan tsz'.* It is much used as food and has a very strong smell.

HAN PAO-SHENG :—The small *suan* is frequently met with in a wild state. It is also called 亂, which character is to be pronounced *luan.*

[The above character is properly pronounced *wan,* and means a kind of reed. See *Rh ya,* 214. The old dictionaries [see *K.D.*] say that it also means garlic, and is then pronounced *luan.* It stands evidently for 卵 *luan* (testicle), an ancient name for garlic, referring to the bulbs, which resemble testicles.]

The *Ku kin chou* [4th cent.] states :—The 蒜 *suan* or 卵蒜 *luan suan* is commonly called 小蒜 *siao* (small) *suan.* The 大 | *ta* (great) *suan,* also called 胡 | *hu suan,* is a native of the western countries [*see the next*].

Further particulars will be given in another part.

244.—葫 *hu.*   *P.,* XXVI, 21.   *T.,* LVII.

Pie lu:—The *hu* is the *ta* (great) *suan.* It is taken out on the 5th day of the 5th month. That with a single seed (fruit)[61] is preferred for medical use. Taste pungent. Nature warm. It is poisonous. When constantly eaten it is injurious to the eyes.

T'AO HUNG-KING :—Now the people call the large *suan*, or Garlic, *hu*, and the common (Chinese) garlic *siao* (small) *suan*. The smell is the same in both kinds.

The *T'ang yün* Dictionary [7th cent.] says 'that the *hu* garlic was first brought from the Western countries by Chang K'ien [in the 2nd cent. B.C. See *Bot. sin.*, I, p. 24].

The *hu* is probably the Rocambole, *Allium scorodoprasum*. Further particulars in another part.

245.—蕸 *sung.* P., XXVI, 30. T., LIX.

Pie lu :—Sung. Leaves and seeds are used in medicine.

T'AO HUNG-KING :—The *sung* is a common vegetable, much used as food. An oil is expressed from the seeds.

LI SHI-CHEN says that *sung* is the vegetable that is commonly called 白菜 *pai* (white) *ts'ai* (vegetable).

Pai ts'ai is the Chinese cabbage, *Brassica chinensis*, L., extensively cultivated in the north of China for its leaves as well as for the oil expressed from the seeds.

Further particulars will be given in another part.

246.—芥 *kie.* P., XXVI, 31. T., LX.

Comp. *Classics*, II, 362. Mustard plant.

Pie lu :—Kie. The stem, the leaves and the seeds are used in medicine.

"獨子者入藥尤佳. Perhaps the bulb is meant. As we have seen, the character 子 is sometimes used for tuber.

50

T‘AO HUNG-KING :—The *kie* resembles the *sung* [see 245]
but the leaves are covered with hair and have a pungent taste.
They are eaten raw or pickled. The seeds are used for
preserving the *tung kua* [*Benincasa*. See 265].

Cust. Med., p. 50 (59):—*Kie tsz‘* (mustard seeds) exported 1885 from Chefoo 2,070 piculs,—p. 292 (263), from
Amoy 1.82 picul,—p. 162 (339), from Shang hai, *pai*
(white) *kie tsz‘* 1.20 picul.—Exported also from Han kow.
See *Hank. Med.*, 3.

More details regarding the Chinese mustard plants will
be found in another part.

247.—蕪菁 *wu tsing*.   *P.*, XXVI, 36.   *T.*, LXIX.

Comp. *Classics*, 361.   Rape.

Pie lu:—*Wu tsing*. The root, the leaves and the seeds
are officinal, also the flowers.

T‘AO HUNG-KING :—The *wu tsing* is akin to the *lu fu*
[radish. See *Rh ya*, 39]. The latter is nowadays also called
溫菘 *wen sung*. Its root is eaten but not the leaf. The root
of the *wu tsing* is smaller than that of the *wen sung*. The
leaves of the *wu tsing* resemble those of the *sung* [see 245]
and are good as food. It is cultivated in Si ch‘uan [in
Kan su, App. 296]. The seeds of the *wu tsing* are very
like those of the *wen sung*. They are not used now in
medicine, but are eaten. The root is much used as food,
steamed or pickled.

SU KUNG [7th cent.]:—The *wu tsing* is called 蔓菁
man tsing in the northern provinces. In its root, leaves
and seeds it resembles the *sung* (*Brassica chinensis*) rather
than the *lu fu* (Radish).

Further particulars in another part.

248.—生薑 *sheng kiang.* *P.,* XXVI, 45. *T.,* XLIII.

Comp. *Classics,* 381. Ginger.

Pie lu:—*Sheng kiang* (fresh ginger). Fresh ginger as well as the *kan kiang* [dried ginger. *See the next*] are produced in Kien wei [in Sz ch'uan, App. 140], in mountain-valleys, also in King chou [Hu pei, Hu nan, App. 146] and Yang chou [Kiang su, Che kiang, App. 400]. It (the rhizome) is taken up in the 9th month. Taste pungent. Nature slightly warm. Non-poisonous. The leaves also are used in medicine.

In my *Bot. sin.,* II, p. 195, Dr. FABER states that at Kew it has been found out that Chinese ginger is not *Zingiber* but *Alpinia.* But this is a mistake, for which neither Dr. FABER nor the botanists of Kew are responsible. *See* Mr. CH. FORD'S *Report of the Hong kong Botan., etc. Department for* 1890, 18 and 19.

Further particulars in another part.

249.—乾薑 *kan kiang.* *P.,* XXVI, 51.

Pen king:—*Kan kiang* (dried ginger).

T'AO HUNG-KING:—Dry ginger is prepared in many villages of the district of Chang an in the prefecture of Lin hai [in Che kiang, App. 3, 192]. The ginger of Shu Han [Sz ch'uan, App. 293] is famed since ancient times, that from King chou [Hu pei, App. 146] is also good, but it is not fit for preparing dry ginger. Dry ginger is made by macerating the root in water for many days, scraping off the skin, and then drying the root in the sun.

TATAR., *Cat.,* 53:—*Sheng kiang.* Radix *Zingiberis cruda.*—26:—*Kan kiang.* Rad. *Zingiberis.*—P. SMITH, 102,

Hank. Med., 36 :—*Sheng kiang* exported from Han kow. —*Cust. Med.*, p. 68 (32):—*Kan kiang* exported 1885 from Han kow, 853 piculs,—p. 58 (11), from I chang 3.15 piculs.

Further particulars in another part.

250.—水蘄 *shui k'in.* *P.*, XXVI, 58. *T.*, LXV.

Comp. *Rh ya*, 116, *Classics*, 370.

Pen king:—*Shui* (water) *kin* and 水英 *shui ying.* The stem is officinal. Taste sweet. Nature uniform. Non-poisonous.

Pie lu:—Other name : 芹菜 *k'in ts'ai.* The *shui k'in* grows in Nan hai [in Kuang tung, App. 228] in ponds and marshes.

T'AO HUNG-KING :—The name is more commonly written 水芹 *shui k'in.* In the 2nd and 3rd months, when the plant has put forth buds, it is pickled or eaten boiled.

SU KUNG [7th cent.]—The *shui k'in* is the same as the *k'in ts'ai* [*v. supra*]. There are two kinds. The 荻芹 *ti k'in* is white. Its root is used. The 赤 | *ch'i* (red) *k'in*, of which the stem and the leaves are eaten, pickled or in a fresh state, is red.

HAN PAO-SHENG [10th cent.]:—The 芹 *k'in* grows in water. Its leaves resemble those of the *kung k'iung* [*Angelica.* See 47]. It has white flowers but no fruit. The root is also white.

LI SHI-CHEN :—The character 蘄 [in the above name] is more correctly written 蘄 *k'in*, and this character was subsequently considered to be the same as 芹 *k'in.* The *Rh ya* [116] says that another name for the 芹 *k'in* is 楚葵 *Ch'u k'ui* [mallow of the country of Ch'u or Hu Kuang, App. 24]. The *Lü shi Ch'un ts'iu* [3rd cent. B.C.] speaks of the *k'in* in Yün meng [in Hu pei, App. 423]. Yün meng was in the country of Ch'u. In the same country lies also 蕲州 K'i chou [App. 121]. The *Rh ya i* [12th cent.] states that the 芹 *k'in* plant is very common there, and suggests that the name of that place may be derived from the *k'in* plant, for the character 蕲 in ancient times was pronounced 芹 *k'in*, as is expressly stated by Kuo P'o [see *Rh ya*, 5]. Li Shi-chen says there are two kinds of *k'in*— the 水芹 *shui k'in*, which grows in water, and the 旱 | *han k'in*, which grows in dry soil. The first is common on the margins of rivers and lakes and in marshes, the other is met with on the plain. There is a red and a white sort [the author apparently refers to the *shui k'in*]. The leaves spring from the joints of the stem and stand opposite, resembling those of the *kung k'iung* [*v. supra*]. The stem has ridges (is channelled) and is hollow. The plant is very fragrant. It blossoms in the 5th month. Small white flowers like those of the *she ch'uang* [*Cnidium*. See 49]. The people of Ch'u [Hu kuang] gather the plant in times of scarcity. It is very nourishing. It is mentioned in the *Shi king*.

The *shui k'in* is an umbelliferous plant, the *Œnanthe stolonifera*, DC. For further particulars see *Bot. sin.*, II, 370.

The 旱芹 *han* (dry soil) *k'in*, or simply 芹 *k'in* or | 菜 *k'in ts'ai*, is Celery, *Apium graveolens*. It is much cultivated at Peking. It is not clear whether by *han k'in* Li Shi-chen means celery. In the *P.* [XXVI, 59], after the *shui k'in*, the plant 菫 *kin* is treated of, and *han k'in* given as a synonym. But the character *kin* in the *Rh ya* and *Classics* seems rather to refer to a *Viola*. See *Bot. sin.*, II, 371.

The drawing sub *han k'in*, in the *Ch.* [III, 40] seems to represent Celery. PARKER [*Canton plants*, 18] has *han k'in ts'ai, Apium graveolens. See also* P. SMITH, 57.

Amœn. exot., 825 :—芹 *kin*, vulgo *seri*. Petroselinum folio Alsines, Morsus Gallinæ dictæ.—THUNBERG [*Fl. jap.*, 120] identifies this with *Apium petroselinum*, L. But *seri* is the Japanese name for *Œnanthe stolonifera*, and in SIEB. *Œcon.* [252] *inondo* is given as the Japanese name of *Anethum graveolens*. Rarius pro condimento in hortis cultum.

251.—蕏 *tsi.* *P.*, XXVII, 5. *T.*, LX.

Comp. *Rh ya*, 103, *Classics*, 367.

Pie lu:—*Tsi*. Leaves, flowers and fruit used in medicine. Taste sweet. Non-poisonous.

WU P'U [3rd cent.]:—The *tsi* grows in waste places. Its fruit, which is called 蒫實 *ts'o shi* [comp. *Rh ya*, 103], is gathered on the 3rd day of the 3rd month and dried in the shade.

T'AO HUNG-KING :—There are many sorts of *tsi*. Of the common sort, which the people now use for food, the leaves are pickled, and also boiled into soup. It is mentioned in the *Shi king*.

LI SHI-CHEN :—There are several sorts of *tsi*—the large, the small and others. The small *tsi* has the stem, the leaves and the flowers flat (thin, tender) and is very palatable. The smallest sort is called 沙 | *sha* (sand) *tsi*. The larger *tsi* has a less agreeable taste and its stem is hard. One sort. which is covered with hair, is called *si ming* [*see the next*] and is not good as food. All these kinds begin to grow after the winter solstice. In the 2nd or 3rd month the root sends up a stem from five to six inches high, and small white flowers appear. The fruit is a small pod (silicle) resembling the *p'ing* (*Lemna*, duckweed) and is three-horned (triangular). It contains small seeds like those of the *t'ing li* [see 114]. The fruit is called *ts'o* [*v. supra*] and is gathered in the 4th month. The stem of the plant is used for making staves for carrying lanterns[65] (?). The plant is said to drive away musquitoes and nocturnal moths, and is therefore called 護生草 *hu sheng ts'ao* (plant protecting living beings).

At Peking the name *tsi ts'ai* is applied to *Capsella bursa pastoris*, Mœnch. It is cultivated as a pot-herb and is also a common wild plant. Comp. also P. SMITH, 196.

A good drawing of it, sub *tsi*, is found in the *Ch.* [III, 46]. See also *Kiu huang*, LIX, 27. Under the same Chinese name it is figured in the *So moku* [XII, 2]. Japonice: *nadzuna.*

Amœn. exot., 897 :—薺 *sei*, vulgo *nadsuna.* Bursa pastoris major, folio sinuato. C. Bauh. P.

SIEB., *Œcon.*, 284 :—*Capsella Bursa pastoris.* Japonice: *natsna* ; sinice : 薺. Herba edulis.

252.—菥蓂 *si ming.*   *P.,* XXVII, 5.   *T.,* LX.
Comp. *Rh ya,* 18.

Pen king :—*Si ming,* 火蕺 *ta ts'i.* The leaves with the stem and the seeds are officinal. Taste pungent. Non-poisonous.

Pie lu :—Other name : 大薺 *ta* (large) *tsi.* The *si ming* grows in Hien yang [in Shen si, App. 65] in mountain-marshes and by road-sides. It is gathered in the 4th or 5th month and dried in the sun.

 " 其 莖 作 挑 燈 杖 ·

Wu P‘u [3rd cent.] gives the synonyms 析目 *si mu*, 榮目 *yung mu* and 馬駒 *ma ku*.—In the *Kuang ya* it is 馬辛 *ma sin*.

T‘ao Hung-king:—It is a common plant. It is also called *ta tsi tsz‘*. Little used in medicine.

Su Kung [7th cent.]:—The names *si ming* and *ta tsi* are from the *Rh ya*. Another name is 老薺 *lao* (old) *tsi*. Its taste is sweet, not pungent.

Li Shi-chen:—The *si ming* and the *tsi* [v. 251] are akin, the smaller being the *tsi* and the larger the *si ming*. The latter is covered with hair. The medical virtues of the seeds are the same in both. The *t‘ing li* [*Sisymbrium*. See 114] is likewise akin to the *si ming*, but the seeds of the latter are sweet and it has white flowers, whilst the *t‘ing li* has yellow flowers and bitter seeds. The *si ming* is sometimes called *t‘ien* (sweet) *t‘ing li*.

The *si ming* or *ta tsi* described in the *P.* is probably *Thlaspi arvense*, L. For further particulars see *Bot. sin.*, II, 18.

253.—繁縷 *fan lü*. *P.*, XXVII, 6. *T.*, CXLIV.

Comp. *Rh ya*, 81.

Pie lu:—Fan lü (entangled floss). It is gathered on the 5th day of the 5th month at mid-day. Taste sour. Nature uniform. Non-poisonous. Apparently the whole plant is used in medicine.

Su Kung [7th cent.]:—This is the 鷄腸 *ki ch‘ang* (chicken's bowels). The plant is common in damp places and on the margins of ditches and canals.

Han Pao-sheng [10th cent.]:—White flowers. The whole plant is officinal.

Su Sung [11th cent.]:—It is also called *ki ch'ang*. It is a common plant in the fields, near the water and in damp places. Its leaves resemble those of the *hang ts'ai* [*Limnanthemum. Bot. sin.*, II, 399] but are smaller. In summer and autumn it bears small white flowers. The stem is twining, and when broken it shows fibres like floss. It is hollow, whence the name *ki ch'ang* [*v. supra*]. The *Pie lu* considers that the *ki ch'ang* and the *fan lü* are not identical.

Li Shi-chen:—The *fan lü* is also called 鵝腸 *ô ch'ang* (goose's bowels), but it is not the same as the *ki ch'ang*. Another name for it is 滋草 *tsz' ts'ao* (plant drawn out in length). It is very common in damp places. Leaves as large as the end of a finger. Tender, twining stem, hollow in the centre. When broken it shows a filament like floss.[66]

It is a sweet, tender, palatable pot-herb. After the 3rd month it opens its flowers with small white petals. The fruit is also small, not larger than a grain of the *pai* (*Echinochloa*). It contains minute seeds resembling those of the *t'ing li* [*Sisymbrium.* See 114]. Wu Shui [an author of the Mongol period] says, that with yellow flowers is the *fan lü* and the white flowered is the *ki ch'ang*. These two are certainly distinct plants, although they resemble each other. Only the *ô ch'ang* (or *fan lü*) is of a sweet taste, has a hollow stem with a filament and white flowers, whilst the *ki ch'ang* is bitter and viscid, the stem has no filament within and is of a slightly purplish colour. The flowers are purple coloured.

Ch., IV, 7:—*Fan lü.* The figure seems to represent a *Stellaria*. See also *Kiu huang*, XLVIII, 7, *ô ch'ang.*— Henry, *Chin. pl.*, 524:—鵝兒腸 *ô rh ch'ang*, *Stellaria aquatica*, Fries.

斷之中空有一縷如絲·

Amœn. exot., 896 :—藥縷 [the second character is indecipherable, but evidently *lu* is meant] *fan ru*, vulgo *fa kobi*,
it. *fagu jera*. Morsus Gallinæ. Alsine vulg. I. Tabern.—
According to MAXIMOWICZ [*Decad.*, XIV, 42] this is *Stellaria
media*, Vill.

So moku, VIII, 66 :—Same Chinese name, *Stellaria
media*, and [65] *St. neglecta*, Weih.

Phon zo, XLVIII, 11:—鵁腸, *Malachium*.

254.—鷄 腸 草 *ki ch'ang ts'ao*.   P.,   XXVII, 7.
T., CLXVIII.

Pie lu :—Ki ch'ang ts'ao (chicken-bowels plant).   Apparently the whole plant is officinal.   Taste slightly pungent
and bitter.   Non-poisonous.

T'AO HUNG-KING :—It grows in gardens and court-yards.
Children knead the juice of this plant with spider's webs,
when this very sticky substance is good for catching cicadas.

LI SHI-CHEN :—The *ki ch'ang* grows in low damp places.
The leaves resemble those of the *ó ch'ang* [*Stellaria.* See
253] but are of a darker colour.   The stem is slender,
tinged with purple, not hollow, and does not show the peculiar
filament [as in the *ó ch'ang*].   It blossoms in the 4th month.
Small purple [or violet] flowers with a five-cleft corolla.
The fruit is likewise small and contains minute seeds.   The
plant is used as a pot-herb.   It is not to be confounded
with the *ó ch'ang*.   These two plants are already separated in
the *Pie lu*, but SU·KUNG says that they are identical.   The
ki ch'ang when chewed becomes viscous ; the juice is fit for
catching cicadas.   The *ó chang* does not possess this property.

Ch., IV, fol. 8 :—*Ki ch'ang*.   Rude drawing, only leaves
represented.

So moku, III, 25 :—鷄 腸, *Eritrichium pedunculare*,
A.DC.   Order *Boragineæ*.   The description in the *P.* seems
to agree.

255.—苜蓿 *mu su.* *P.,* XXVII, 8. *T.,* LXXIII.

Pie lu:—*Mu su.* The leaves and the root are used in medicine. Taste bitter and harsh. Non-poisonous.

T'AO HUNG-KING:—In Chang an [in Shen si, App. 6] the *mu su* is cultivated in gardens. It is much valued by the people in the north. In Kiang nan [Kiang su, An hui, etc., App. 124] it is not much eaten, because it is tasteless. There is a plant named *mu su,* growing in foreign countries, which is used in diseases of the eye, but that is a different plant.

MENG SHEN [7th cent.]:—Where the *mu su* grows the people use the root [as a medicine] and call this drug 土黃芪 *t'u huang k'i* [native *huang k'i. See above,* 2].

K'OU TSUNG-SHI [12th cent.]:—It abounds in Shen si [App. 284], where it is used for feeding cattle and horses. The young leaves are also eaten by man. The plant has a a perennial root, and when cut off it thrives again.

LI SHI-CHEN:—The ancient authors write the name also 牧宿 *mou su* and 木粟 *mu su.* The *Si king tsu ki* [written about our era] reports that the *mu su* was originally brought to China by CHANG K'IEN [in the 2nd cent. B.C. See *Bot. sin.,* I, p. 24] from Ta wan (Ferghana) and soon became a common wild plant. The people of Shen and Lung [Shen si and Kan su, App. 284, 216] cultivate it. It is cut thrice a year and grows again from the root. The leaves are used as food. One plant has twenty or more stems like the *hui t'iao* [*Chenopodium. See Bot. sin.,* II, 446]. Three leaves at the top of a common petiole (trifoliate leaves). The leaflets resemble those of the *küe ming* [*Cassia. See* 110], but are smaller, of the size of a finger-nail. It flowers from summer until autumn. Small yellow flowers and small, roundish, thin pods, curved, twisted,

prickly and black when ripe. The seeds resemble millet, are
edible and also fit for fermenting wine. This plant is also
called 懷風 *huai feng* and 光風 *kuang feng*. The people
of Mou ling [in Shen si, App. 225] call it 連枝草 *lien chi
ts‘ao*. In the *Kin kuang ming king* (a Buddhist book) it is
termed 塞鼻力迦 *sa-bi-li-ka*.

Ch., III, 56, and *Kiu huang*, LVIII, 34:—*Mu su.*
Rude drawings, but probably *Medicago sativa* is intended.
The description in the *P.* agrees in a general way. *M. sativa*,
the common *Lucerne* in Europe, has generally purple or violet
flowers, but sometimes they are yellow. At Peking *mu su*
is *M. sativa*, with violet flowers ; it is not cultivated there, but
is common in the neighbourhood. Father DAVID [*Journ.*, I,
64] saw it cultivated in Southern Chi li.

TATAR., *Cat.*, 40 :—*Mu su. Medicago sativa.*—P. SMITH,
145 :—*Mu su, M. radiata.* But this species is not known
from China.

Mu su is not Chinese but most probably a foreign name.
As to the Sanscrit name, *sa-bi-li-ka* [*v. supra*], I may
observe that BURNES mentions, among the grasses cultivated
for cattle in Kabul, the *Trifolium giganteum*, called *sibarga*,
and the *Medicago sativa*, called *rishka* [BALFOUR, *Cyclop. of
India*].

So moku, XIV, 14 :—苜蓿, *Medicago denticulata*, Willd.
Yellow flowers. Known also from China.

Phon zo, XLVIII, 16 :—Same Chinese name applied to
M. denticulata and *M. lupulina*.

256.—莧 *hien.* *P.*, XXVII, 9. *T.*, LXI.
　Comp. *Rh ya*, 107.
　Pen king :— 丨實 *hien shi* (fruit). Seeds, leaves and
root officinal. Taste of the seeds sweet. Nature cold. Non-
poisonous. The leaves are a nourishing vegetable.

Pie lu:—The *hien shi*, which is also called 莫 | *mu (mo)* *shi* and 細 | *si* (small) *shi*, grows in Huai yang [in Ho nan, App. 91] in marshes and fields. Its leaves resemble the *lan* leaves [*Polygonum.* See 123]. They are gathered in the 11th month.

Li Tang-chi [3rd cent.]:—*Hien shi* is the same as the vegetable | 菜 *hien ts'ai.*

T'ao Hung-king :—The *hien shi*, which according to the *Pie lu* is the same as the *si hien*, and the leaves of which resemble the *lan*, is the 白 | *pai* (white) *hien*. The *si hien* is the same as the 糠 | *k'ang hien*, and is the best sort for food. All the sorts of *hien* are valuable in the cold season. They ripen after hoar-frost, wherefore the *Pie lu* states that the *hien shi* is gathered in the 11th month. There is also the 赤 | *ch'i* (red) *hien*, with a purple stem, not fit for being used as food. Another kind is the 馬 | *ma* (horse) *hien*. It grows along the ground and has very small fruits (seeds). Its popular name is 馬 齒 | *ma ch'i* (horse's teeth) *hien*. But this is not akin to the *hien shi* [it is the *Portulaca oleracea*, L.].

Su Kung [7th cent.]:—The *ch'i* (red) *hien* is also called 蕢 *kui*. *Mu shi*, in the *Pie lu*, is a misnomer.

Han Pao-sheng [10th cent.]:—There are six sorts of *hien*, viz. the *ch'i* (red) *hien*, the *pai* (white) *hien*, the 人 | *jen* (man) *hien*, the 紫 | *tsz'* (purple) *hien*, the 五 色 | *wu se* (five colours) *hien* and the *ma* (horse) *hien*. Only the fruits (seeds) of the *jen hien* and the *pai hien* are used in medicine. The *ch'i* (red) *hien* is of a pungent taste and has a different effect.

Su Sung [11th cent.]:—The *jen hien* and the *pai hien* have great cooling properties. There are other sorts, such as the *k'ang hien*, the 胡 | *hu hien* and the *si hien*. They all have the same seeds. The largest sort is the *pai hien* and the smallest the *jen hien*. The seeds ripen after hoar-frost, and are small and black. The purple *hien* has a purple stem and leaves. The people of Wu [Kiang su, App. 389] use it for dyeing their nails. The red *hien* is also called 花 | *hua* (flowered or coloured) *hien*. Stem and leaves are of a dark red colour. The root and the stem are preserved for food and are of an agreeable pungent taste. The *wu se* (five colours) *hien* is now rarely used. The *si* (small) *hien* is also called 野 | *ye* (wild) *hien* or 豬 | *chu* (pig) *hien*. It is good for feeding pigs.

Lɪ Sʜɪ-cʜᴇɴ:—All sorts of *hien* are sown in the 3rd month. After the 6th month the plant cannot be eaten. The old plants attain the height of a man. The small flowers appear in spikes. The seeds are small, black and shining, just as the seeds of the *ts'ing siang tsz'* [*Celosia argentea*. See 82] and the *ki kuan tsz'* [*Celosia cristata*]. The seeds are gathered in the 9th month. The *si hien* is the wild *hien*. The northern people call it *k'ang hien*. It has a soft, weak stem and small leaves. The taste is more pleasant than that of the cultivated *hien*.

Ch., III, 9 :—*Hien, hien ts'ai, Amarantus Blitum*, L. Good drawing. See also *Kiu huang*, LVIII, 25.

At Peking *hien ts'ai* is *Amarantus Blitum*, a common weed, also cultivated as a pot-herb.

Ch., III, 11 :—*Jen hien*, and [9] verso, *ye hien*, species of *Amarantus*.

Lᴏᴜʀ., *Fl. cochin.*, 685 :—*Amarantus tricolor*, L. Sinice : *hum* (i.e. *hung*, red) *hien*. *Ibid.*:—*A. polygamus*, L. Sinice : *pe* (*pai*, white) *hien*. Habitat in Cochinchina et China tam cultus quam spontaneus. Ex omnibus Amaranti speciebus, quæ in India edi solent, hæc est salubrior et suavior ; proinde que præ aliis usitatior.—P. Sᴍɪᴛʜ, 59, sub *Chenopodium*.

Pᴀʀᴋᴇʀ., *Canton pl.*, 12, 107 :—假 | | *kia* (pseudo) *hien ts'ai, Amarantus spinosus*, also *Euxolus viridis*, Moq. Tend. This plant is common in N. China, and in its outer appearance bears a strong resemblance to *A. Blitum*.

FRANCHET refers the figures sub 莧 in the *So moku* [XX, 22], Japonice *shiou*, and likewise [XX, 19] sub 雁來紅, to *Amarantus melancholicus*, and *Phon zo* [XLVIII, 17, 18] sub 莧 to *Am. mangostanus*, L.—

雁來紅 *yen lai hung* in China is *Am. melancholicus*, also *Am. tricolor*. As to the identification of the Japanese drawing in the *So moku* [XX, 22:—莧], FRANCHET seems to be mistaken. I think it is *A. Blitum*, which is not found in FRANCHET's *Enum. Jap.*, but which has been reported from Japan [*Journ. Bot.*, 1877, 297].

SIEB., *Œcon.*, 124 :—*Amarantus oleraceus*, 莧 (a.)—*hiju* caule foliisque viridibus ; (b.)—*aka hiju*, caule foliisque purpurascentibus.

So moku [XX, 23]:—野莧, *Euxolus viridis*. [*Amarantus Blitum* in THUNB., *Fl. jap.*, 57]. Japonice : *no hiju*.

SIEB., *Œcon.*, 123. *Amarantus japonicus*, 野莧. Japonice: *no biju*.

Phon zo, XLVIII, 20 :—赤莧, *Amarantus melancholicus*, also *A. tricolor*.

257.—苦菜 *k'u ts'ai*.   P., XXVII, 14.   T., LIX.

Comp. *Rh ya*, 24, *Classics*, 365.

Pen king :—*K'u ts'ai* (bitter vegetable), 荼 *t'u*. Leaves, root and flowers used in medicine. Taste of the leaves bitter. Nature cold. Non-poisonous.

Pie lu :—Other name : 游冬 *yu tung*. The *k'u ts'ai* grows in I chou [Yün nan, Sz ch'uan, App. 102] in river-valleys, in the mountains and by waysides. It does not die in winter. Gathered on the 3rd day of the 3rd month and dried in the shade.

The ancient *Ts'ai yao lu* says :—The *k'u ts'ai* begins to grow in the 3rd month, in the 6th month it has yellow

flowers, and in the 8th month black seeds. Perennial root.
It does not die in winter.

Su Kung [7th cent.]:—The *k'u ts'ai* or *t'u* is mentioned
in the *Rh ya*. It is also called *yu tung* [*v. supra*]. Leaves
like those of the *k'u kü* (*Lactuca*) but smaller. The plant when
broken discharges a white juice. Yellow flowers resembling
the *kü* (*Chrysanthemum*).

Han Pao-sheng [10th cent.]:—It blossoms in spring
and the seeds are produced in summer, in autumn it blossoms
again but does not produce seeds. It does not wither in
winter.

K'ou Tsung-shi [12th cent.]:—It is mentioned in the
Yüe ling (*Li ki*). It is found in all parts of China. In the
north its leaves fall off in winter, but in the south they are
green in summer as well as in winter. The leaves resemble
those of the *k'u kü* [*v. infra*] but are narrower, of a paler
green colour, and contain a white, milky juice of a bitter
taste. The flowers are like those of the wild *Chrysanthemum*.
It blossoms from spring till autumn.

Li Shi-chen :—The *k'u ts'ai* is the same as the 苦藚
k'u mai. When cultivated in gardens it is called 苦苣 *k'u kü*.
There are two varieties—one with a red, the other with a
white stem—when the plant begins to grow. The stem is
hollow in the centre and soft. When broken it discharges a
white juice. The callous leaves resemble those of the radish.
They are of a green colour with a bluish tinge. The leaves
clasp the stem, the upper leaves being like the beak of a crane.
Each leaf has irregular lobes on the margin as if forcibly
lacerated. Flowers yellow, resembling those of a wild *Chrysanthemum* when beginning to expand. Seeds many together
like those of the *tung hao* (*Chrysanthemum Roxburghii*).
They are known under the name of 鶴虱 *kuan shi* (heron's
lice). After the plant has ceased blooming the seeds are

collected. These are provided with white soft hairs (pappus) and are carried away by the wind to distant places where they fall down and germinate. In the *Ji yung Pen ts'ao* [Mongol period] this plant is called 萹苣 *pien kü*. Another name is 天香菜 *t'ien hiang ts'ai*.

Ch., III, 15 :—苦菜 *k'u ts'ai, Lactuca versicolor,* Schlt. Bip. Good drawing.

Kiu huang, LVIII, 28 :—苦蕒 *k'u mai* or 老鸛菜 *lao kuan ts'ai, Lactuca,* perhaps *denticulata.*

Ch., III, 18 :—光葉 | | *kuang ye* (glabrous leaves) *k'u mai, Lactuca,* perhaps *denticulata.*

Ch., III, 21 :—野 | | *ye* (wild) *k'u mai, Lactuca denticulata,* var. *sonchifolia.*

Ch., III, 19 :—滇苦菜 *Tien* (Yün nan) *k'u ts'ai.* The figure seems to represent *Sonchus oleraceus,* L.

Ch., III, 20 :—苣蕒菜 *kü mai ts'ai, Sonchus,* and [22] 家苣蕒 *kia* (domestic) *kü mai,* resembles the figure on fol. 20.

Kiu huang, LVIII, 26 :—苦苣菜 *k'u kü ts'ai,* also 野苣 *ye kü* and 萹 | *pien kü.* Rude drawing, probably *Lactuca* intended.

There are at Peking four wild-growing species of *Lactuca* and their varieties, the leaves of which are eaten by the natives. Some of them are also cultivated.

1.—*Lactuca squarrosa,* Miq. 春不老 *ch'un pu lao.* It grows to the height of 6 feet. Leaves very irregularly shaped. Yellow flowers.

2.—*L. tatarica,* C. A. Mey. Blue flowers.

3.—*L. denticulata,* Max. Typical form. Common. Yellow flowers.

var. *sonchifolia.* Common. Sinice : *k'u dia rh* (popular name).

var. *ramosissima.* Common in the mountains.

4.—*L. versicolor,* Schl. Bip.   Small plant with yellow flowers. Wild and cultivated. Sinice: 山苦蕒 *shan k'u mai.*

The Chinese at Peking cultivate the *Sonchus arvensis,* L. under the names 苦蕒菜 *k'u mai ts'ai* or 苣 ⏐⏐ *kŭ mai ts'ai.* The leaves are eaten. It is also a common wild plant.

The genera *Lactuca* and *Sonchus* are closely allied. Both belong to the group *Cichoraceæ* of compound flowers. They resemble each other in the flower-heads, involucres, etc. The species of both contain a milky sap.

Lour., *Fl. cochin.,* 583 :—*Cichorium endivia,* L. Sinice: *khu tsai.* Habitat in locis borealibus imperii sinensis.

Tatar., *Cat.,* 33 :—*K'u mai ts'ai, Cichorium.*—P. Smith, 60, 62, *Chicory* and *Cichorium.*—Bunge [*Enum. pl. Chinæ bor.*] mentions *Cichorium intybus* as cultivated in Chinese gardens at Peking. I never met with this plant there.

Henry, *Chin. pl.,* 189 :—*K'u ts'ai* in Hu pei is *Lactuca squarrosa.*

Hank. Med., 22 :—*K'u ts'ai* exported from Han kow.

So moku, XV, 6 :—苦菜, *Sonchus oleraceus,* L.

Sieb., *Œcon.,* 219 :—*Sonchus oleraceus.* Japonice : *kesi asami.* Sinice : 苦菜.

Phon zo, XLIX, 2, 3 :—苦菜. 苦蕒菜. 苣, *Sonchus arvensis,* L.—*Ibid.,* 9 :—水苦蕒, *Lactuca denticulata.*

So moku, XV, 20 :—山苦蕒, *Lactuca Raddeana,* Maxim.

258.—落葵 *lo k'ui.*   *P.,* XXVII, 23.   *T.,* LXXXV. Comp. *Rh ya,* 148.

Pie lu :—*Lo k'ui.* Other names : 天葵 *t'ien k'ui* [67] and 繁露 *fan lu.* The leaves and the fruit are officinal. Taste of the leaves sour and mucilaginous. Non-poisonous.

---

[67] The name *t'ien k'ui* in the *P.* is likewise applied to an *Anemone,*—*t'u k'ui.* See *Bot. sin.,* II, 115.

T'AO HUNG-KING:—The *lo k'ui* is also called 承 露 *ch'eng lu*. It is largely cultivated. The leaves are prepared into a condiment with fish. They are cooling and mucilaginous. The fruit (berry) is of a purple colour. Ladies use it as a cosmetic. It is little used in medicine.

HAN PAO-SHENG [10th cent.]:—It is a twining plant. The leaves in shape resemble apricot-leaves, and are roundish and thick. The fruit resembles that of the *wu wei tsz'* [*Schizandra*. See 164], is at first green and becomes black when ripe.

MA CHI [10th cent.]:—The *lo k'ui* is also called 藤 葵 *t'eng k'ui* (twining mallow). Popular name 胡 藤 脂 *hu yen chi*.

LI SHI-CHEN:—The *lo k'ui* is planted (sown) in the 3rd month. The young leaves are eaten. It is a twiner. The leaves resemble apricot-leaves but are thick, succulent and mucilaginous. They are eaten as a vegetable together with meat. In the 8th or 9th month it opens its small purple flowers, which are arranged in spikes. Fruit (berry) as large as that of the *wu wei tsz'* [*v. supra*] and of a purplish black colour when ripe. The juice of these berries is red like the 胭 脂 *yen chi* (cosmetics, rouge). Ladies employ it for painting their faces and lips, it is also used for dyeing cloth. It is called *hu yen chi* [*v. supra*] and 染 絳 子 *jang kiang tsz'* (berry which dyes a red colour). But this colour is changeable. The plant is also called 丨 丨 菜 *yen chi ts'ai* and 御 菜 *yü ts'ai* (imperial vegetable).

Ch., IV, 6:—*Lo k'ui*. Good drawing of *Basella*. The description in the *P.* agrees. At Peking *Basella rubra*, L., Order *Chenopodiaceæ*, is cultivated under the name of 胭脂豆 *yen chi tou* (cosmetic pea). The berries are used as a cosmetic.

LOUR., *Fl. cochin.*, 229:—*Basella nigra* (=*B. rubra*). Sinice: *lo quei*. Usus esculentus.

The *Cust. Med.* [p. 146 (121) and 376 (493)] notices a
drug *t'ien k'ui tsz'* as exported in small quantities from
Shang hai and Canton, and in the *Hank. Med.* [44] it appears
also as an article of export from Han kow. It is there
arbitrarily identified with *Pyrola*. As has been stated above,
t'ien k'ui in the *P.* is given as a name for *Basella* and likewise
for an *Anemone*. Not having seen the drug *t'ien k'ui* of the
Cust. Med., I am unable to say to what plant it may belong.

So moku, II, 70 :—洛葵, *Basella rubra*, L.

SIEB., *Œcon.*, 119 :—*Basella rubra*. Japonice : *Tsuru
murasaki*. Sinice : 洛葵. *Baccæ tinctoriæ*.

259.—蕺 *ts'i*.   *P.*, XXVII, 24.   *T.*, LXIV.

Pie lu :— *Ts'i*. Leaves used in medicine. They are
slightly poisonous. Taste pungent. When eaten to excess
they cause shortness of breath.

SU KUNG [7th cent.]:—The *ts'i ts'ai* (vegetable) grows
in damp, shady places in the mountains. Sometimes it
creeps. Leaves like those of buckwheat, but fat (succulent).
The stem is of a purplish red colour. The people of Shan nan
[S. Shen si, App. 268] and Kiang tso [An hui and Kiang su.
App. 124] eat it raw as a salad. In Kuan chung [Shen si,
App. 158] they call it 菹菜 *tsū ts'ai* (pickled or salted
vegetable).—The *Kuang ya* [3rd cent.] gives 菹 *tsu* as a
synonym for *ts'i*.

HAN PAO-SHENG [10th cent.]:—Stem and leaves are of a
purplish red colour. When in flower the plant has a fetid
smell.

The *Cheng Ts'iao T'ung chi* [12th cent.], with respect
to the *ts'i*, refers to *Rh ya*, 144 [this is an error. Comp.
above, 106], and states that it is a twining plant with leaves
like the *kū tsiang* (Betel pepper leaf].

LI SHI-CHEN:—The *ts'i* has a frouzy[68] smell, and is therefore also called 魚腥草 *yü sing ts'ao* (plant having the odour of fish). The leaves resemble those of the *hang* [*Limnanthemum.* See *Classics,* 399], are three-horned (heart-shaped), red on one side and green on the other. The plant is good for feeding pigs.

Ch., IV, 9:—*Ts'i ts'ai* or *yü sing ts'ai.* The drawing represents *Houttuynia cordata,* Thbg. HENRY, *Chin. pl.,* 560:—*Yü sing ts'ai* in Hu pei is *Houttuynia cordata.*

So moku, II, 17:—蕺菜, *Houttuynia cordata,* Thbg. *Fl. jap.,* 234, tab. 26. Order of *Piperaceæ.* A common plant in China and Japan. It it the *H. fœtida* of LOUDON and the *Polypara cochinchinensis,* LOUR., *Fl. cochin.,* 78. Inter olera in acetariis edulis.

The plant is figured under the above Chinese name in SIEB. *Icon. ined.* [VII].

SIEB., *Œcon.,* 8:—*Houttuynia cordata.* Japonice: *dokudame.* Sinice: 蕺菜. Pro fomentatione in doloribus rheumaticis.

In the *Gardener's Chron.* [1882, II, p. 438] it is stated that the flowers of this plant exhale a "boiled snake-" like perfume.

260.—鹿藿 *lu huo.* *P.,* XXVII, 27. *T.,* XXXVII.
Comp. *Rh ya,* 96.

Pen king:—*Lu huo* (deer bean):—Apparently the leaves and the seeds are officinal. Taste bitter. Nature uniform. Non-poisonous.

Pie lu:—The *lu huo* grows in Wen shan [in Sz ch'uan, App. 388] in mountain-valleys.

T'AO HUNG-KING:—This drug is not used now in medicine. *Lu huo* is also another name for the *ko* plant [*Pachyrhizus.* See 174].

68 腥

Su Kung [7th cent.]:—This plant resembles the *wan tou*
(common pea) but it is longer and coarser. It is gathered
as food, and has the smell of beans. The mountain people
call it 鹿豆 *lu tou* (deer bean).

Han Pao-sheng [10th cent.]:—The *lu tou* can be eaten
raw. The leaves are gathered in the 5th and 6th months and
dried in the sun. Mentioned in the *Rh ya.*

Li Shi-chen:—The *lu tou* is the same as the 野綠豆 *ye*
(wild) *lü tou* [*Phaseolus Mungo. Bot. sin.*, II, 356]. It is
also called 虉豆 *lao tou.* It is frequently met with in wheat-
fields. In its leaves it resembles the *lü tou* [*v. supra*] but is
smaller and a twining plant. It is eaten both raw and
cooked. It blossoms in the third month. Flowers pale
purple. The fruit is a small pod with seeds as large as the
tsiao [*Zanthoxylon Bungeanum*] and black. They can be eaten
boiled, or they are ground into meal from which cakes are
made.

Ch., III, 45 :—*Lu huo.* The drawing represents a
leguminous plant with trifoliate leaves.

So moku, XIII, 33 :—鹿藿, *Rhynchosia volubilis*, Lour.,
Fl. cochin., 562, a leguminous plant.

261.—芋 *yü.   P.*, XXVII, 31.   *T.*, LIII.

Pie lu :—*Yü* or 土芝 *t'u chi.* The seeds and the leaves
with the stem are official. The seeds are said to be slightly
poisonous.

Yü or 芋頭 *yü t'ou* are names applied to several species
of *Colocasia* cultivated for their edible roots, viz. *Colocasia
antiquorum*, Schott. (*Arum Colocasia*, L.), *Colocasia esculenta*,
Schott. (*Arum esculentum*, L.), *Colocasia indica*, Kth. (*Arum
indicum*, L.) and others.

In the *Shuo wen* the character *yü* is written 芌 *yü.*—In
the *Kuang ya* [3rd cent.] it is said that the stalks of the

yü are called 蕨 *keng* [*W.D.*, 323.—*K.D.*].—The *yü* is frequently spoken of by authors of the Han period.

Further details regarding these plants will be given in another part.

262.—薯蕷 *shu yü.* *P.*, XXVII, 33. *T.*, LIV.

Comp. *Classics*, 379.

Pen king:—*Shu yü.* The root is officinal. Taste sweet. Nature warm. Non-poisonous.

Pie lu:—The *shu yü* grows in Sung kao [in Ho nan, App. 317], in mountain-valleys. The root is gathered in the 2nd and 8th months and dried in the sun.

By the *shu yü* of the ancient authors we have to understand various species of *Dioscorea*, as *D. japonica* and *D. Batatas* and others, cultivated for their edible roots and found also in a wild state in China. The popular name is 山藥 *shan yao.* Comp. HENRY, *Chin. pl.*, 412.

TATAR., *Cat.*, 51, 55.—P. SMITH, 86.

For medical use the roots of wild species of *Dioscorea* are employed. Several sorts of this drug are mentioned in the *Cust. Med.*,—p. 46, (27), *shan yao* exported 1885 from Chefoo 250.66 piculs,—p. 24 (24), 淮山藥 *huai shan yao* from Tien tsin 4,365 piculs,—p. 68 (25), the same from Han kow 1,858 piculs,—p. 26 (49), 山藥頭 *shan yao t'ou* from Tien tsin 843 piculs.

Further particulars regarding *Dioscorea* will be given in another part.

263.—百合 *po ho.* *P.*, XXVII, 39. *T.*, CXXIII.

Pen king:—*Po ho.* The root is officinal. Taste sweet. Nature uniform. Non-poisonous. Flowers and seeds also used in medicine.

Pie lu :—Other names : 摩羅 *mo lo*, 重箱 *chung siang*, 中逢花 *chung feng hua*, 彊罜 *kiang kü* and 蒜腦 *suan nao*. The *po ho* grows in King chou [Hu kuang, App. 146] in mountain-valleys. The root is gathered in the 2nd and 8th months and dried in the shade.

Wu P'u [3rd cent.]:—Other names : 重邁 *chung mai* and 中庭 *chung t'ing*.

T'ao Hung-king :—The *po ho* is also called 彊仇 *kiang ch'ou*. It is a common plant in Mid China. The root resembles the *hu suan* [foreign or great garlic. See 244]. It consists of a great number of flat pieces collected together. It is much used as food by the people, steamed or boiled. It is believed that the *po ho* is produced by the metamorphosis of a conglomeration of earth-worms.

Po ho [the Chinese name means "a hundred pieces together"] is the name applied in China and Japan to several lilies, the bulbs of which, formed by large, fleshy scales, are used as food. At Peking *Lilium tigrinum*, Ker., is called *po ho*.

Tatar., *Cat.*, 1.—*Po ho*. Bulbus *Lilii tigrini*.—Gauger [7] describes and figures these scales.—P. Smith, 134, sub *Lilium candidum*.—Henry, *Chin. pl.*, 357, 58 :—At I chang the cultivated *po ho* is *L. tigrinum*, the wild-growing *po ho* is *L. Brownii* and other species.

Cust. Med., p. 78 (169):—*Po ho* exported 1885 from Han kow 1,491 piculs,—p. 280 (93), from Amoy 4.66 piculs, —p. 348 (121), 386 (625), from Canton, fresh or dried bulbs, or bulbs ground into powder, about 80 piculs.

Further particulars in another part.

264.—苦䔞 *k'u hu*.   P., XXVIII, 6.   T., XLVII.
Comp. *Classics*, 384.

Pen king :—K‘u hu (bitter bottle gourd). The pulpa with the seeds are officinal. Taste bitter. Nature cold. Poisonous. Flowers and leaves likewise used in medicine.

In the *Kuo yü* [5th cent. B.C. See *Bot. sin.*, I, 408] it is called 苦匏 *k‘u* (bitter) *p‘ao*.

*Pie lu :—*The *k‘u hu* grows in the country of Tsin [Shan si, App. 353].

T‘AO HUNG-KING :—The *k‘u hu* is bitter like gall, not edible. It is not a distinct species of *hu*, but it happens that among the [sweet] *hu* there are some fruits which have a bitter taste.

The *hu* or *p‘ao* is the *Lagenaria* or Bottle gourd. Further details in another part.

265.—冬瓜 *tung kua.* *P.*, XXVIII. *T.*, XLIV.

Pen king :—Tung kua (winter gourd), 白瓜 *pai kua* (white gourd) and 水芝 *shui chi.* In the Index of the *Pen king* we have 白冬子 *pai tung tsz‘* (*tung kua* seeds). The pulpa of the fruit and the seeds are officinal. Taste sweet. Nature slightly cold. Non-poisonous. [Subsequently the rind of the fruit and the leaves were also used in medicine.]

*Pie lu :—*The *pai kua tsz‘*, or the kernels (仁) of the *tung kua*, are produced in Sung kao [in Ho nan, App. 317]. The plant grows in marshes. The seeds are gathered in the 8th month.

In the *Kuang ya* [3rd cent.] it is called 地芝 *ti chi.*

Tung kua is now the common name for *Benincasa cerifera*, Sav., the White Gourd of India, much cultivated all over China.

TATAR., *Cat.*, 22 :—*Tung kua tsz‘*. Semina *Benincasæ ceriferæ.*—P. SMITH, 36.

53

Cust. Med., p. 80 (191) :— *Tung kua jen* (seeds) exported 1885 from Han kow 12.45 piculs,—p. 200 (251), from Ning po 11 piculs,—p. 372 (437), from Canton 0.75 picul.

Ibid., p. 194 (151) :— *Tung kua p'i* (rind of the fruit) from Ning po 6.47 piculs,—p. 356 (216), from Canton 1.45 picul.

266.—芝 *chi.* P., XXVIII, 22. T., XLVIII.

Comp. *Rh ya,* 41, *Classics,* 380.

According to some ancient (especially Taoist) works the *chi* is a felicitous plant, the plant of immortality. The *Pen king* and the *Pie lu* apply the name *chi* to various mushrooms, of which six sorts are enumerated. They are distinguished according to their colour and called the *liu* (six) *chi* :—

1.—The 青 | *ts'ing* (green) *chi,* also called 龍 | *lung* (dragon) *chi* in the *Pie lu,* is said to grow on the T'ai shan mountain [in Shan tung, App. 322]. Taste sour. Non-poisonous.

2.—The 赤 | *ch'i* (red) *chi,* called 丹 *tan* (cinnabar) *chi* in the *Pen king,* grows on the Huo shan mountain [according to T'AO HUNG-KING the same as the Heng shan mountain in Hu nan, App. 100]. Taste bitter. Non-poisonous.

3.—The 黃 | *huang* (yellow) *chi,* which is called 金 | *kin* (gold) *chi* in the *Pen king,* grows on the Sung shan [in Ho nan, App. 317]. Taste sweet. Non-poisonous.

4.—The 白 | *pai* (white) *chi,* called 玉 | *yü chi* in the *Pen.king,* grows on the Hua shan [in Shen si, App. 86]. Taste pungent. Non-poisonous.

5.—The 黑 | *hei* (black) *chi*, called 玄 | *hüan* (dark) *chi* in the *Pen king*—it is also called 紫 | *su chi*—grows on the Ch'ang shan mountain [in Chi li, App. 8]. Taste saltish. Non-poisonous.

6.—The 紫 | *tsz'* (purple) *chi*, which is called 木 | *mu* (wood) *chi* in the *Pen king*, grows on the Kao hia shan [mountain, unknown to Tao Hung-king]. Taste sweet. Non-poisonous.

It is believed that the felicitous plant *chi*, or plant of immortality of the ancient Chinese, is the 靈芝 *ling chi* (divine *chi*), a branched *Agaricus* which is now sold in Chinese drug-shops [see *Bot. sin.*, II, 41]. In the *P.* [XXVIII, 35] *ling chi* is given as a synonym of 石耳 *shi rh* (mushroom which grows on stones). The name *ling chi* appears first in the *Ling yüan fang* [11th cent.]

In the *Cust. Med.* the drug *ling chi* is twice mentioned, [p. 206 (324)] as imported to Ning po from Han kow and [p. 308 (502)] as imported to Amoy. It is identified there with bird's excrement.

Further particulars regarding the *chi* in another part.

267.—木耳 *mu rh*. *P.*, XXVIII, 26. *T.*, LII.

Pen king:—*Mu rh* (woody ears,—mushrooms produced on trees). Taste sweet. Nature uniform. Slightly poisonous.

Pie lu:—There are five kinds of *mu rh*, which grow in Kien wei [in Sz ch'uan, App. 140]. They are gathered in the rainy season, in the 6th month, and dried in the sun.

T'ao Hung-king:—The *Pie lu* does not say on what trees these mushrooms grow. The 桑耳 *sang* (mulberry tree) *rh* grows on old mulberry trees and is of a green, yellow, red or white colour. The mushrooms called *mu rh* are gathered by the people and pickled for food, but not used in medicine.

SU KUNG [7th cent.]:—The five kinds of *mu rh* grow on five different trees, *viz.* the 桑 *sang* (mulberry tree), the 槐 *huai* (*Sophora japonica*), the 楮 *ch'u* (*Broussonetia papyrifera*), the 榆 *yü* (elm tree), and the 柳 *liu* (willow tree).—The mulberry mushroom is stated to be poisonous.

Comp. P. SMITH, 99, Fungus and, 152, Mushrooms.

The *mu rh* sold at Peking in the markets are *Exidia* (*Hirneola*) *auricula Judæ* or Jew's ear.

Further particulars in another part.

268.—翟 菌 *huan kün*. Mushroom. *P.*, XXVIII, 34. *T.*, LII.

Pen king:—*Huan kün* and 翟 蘆 *huan lu.* Taste saltish. Nature uniform. Slightly poisonous.

Pie lu:—The *huan kün* mushroom grows in Tung hai [in Shan tung, App. 372], in ponds and marshes, also in Chang wu (hien) belonging to Pu hai (kün) [in Chi li, App. 4, 262]. It is gathered in the 8th month and dried in the shade.

T'AO HUNG-KING:—It is a mushroom (*kün*) which is brought from the north. It has no shape. It is believed that the excrement of herons is metamorphosed into this mushroom, wherefore it is also called 鸛 菌 *kuan kün* (heron mushroom). Eaten in a soup of pork it has the property of expelling intestinal worms.

SU KUNG [7th cent.]:—The *huan kün* is produced in Pu hai [*v. supra*] in swamps among reeds and on brackish ground. T'AO HUNG-KING's statement that it originates from heron's excrement is incorrect, for this mushroom grows of itself. It is white, light, empty, of the same texture outside and inside, and quite different from other mushrooms. It is an excellent vermifuge.

HAN PAO-SHENG [10th cent.]:—It is produced in Ts'ang chou [in Chi li, App. 343]. It appears in autumn after the ordinary rains, but it is scarce in time of drought or of heavy rains. It ought to be dried in the sun.

See the Japanese drawing of the 蕈菌 in the *Phon zo* [LX, 23].

269.—地耳 *ti rh*.   *P.*, XXVIII, 35.   *T.*, LII.

Pie lu :—*Ti rh* (mushrooms which grow on the ground). Taste sweet.   Nature cold.   Non-poisonous.

LI SHI-CHEN says that the popular name of these mushrooms is 地踏菰 *ti ta ku.*

See P. SMITH, 99, Fungus and, 152, Mushrooms.

Comp. the Japanese drawing sub 地耳 in the *Phon zo* [LX, 24].

270.—李 *li*. The Plum. *Prunus domestica*, L. *P.*, XXIX, 1. *T.*, CCXX.

Pie lu :—The fruit, kernels of the stones, rind of the root, leaves, flowers, and the gum exuding from the tree (李膠 *li kiao*) are all officinal.

Comp. *Rh ya*, 269-271, *Classics*, 472.

P. SMITH, 174 :—Plum.

Cust. Med., p. 76 (156):—李仁 *li jen*, plum kernels, exported 1885 from Han kow 13.25 piculs,—p. 32 (149), from Tien tsin 5 piculs,—p. 160 (322), from Shang hai 1.36 picul.

Further particulars in another part.

271.—杏 *hing*, the Apricot.   *P.*, XXIX, 3.   *T.*, CCXIII.

Comp. *Classics*, 471.

Index of the *Pen king* 杏核仁 *hing ho jen* (kernels of apricot-stones).

Pie lu:—The *hing* grows in Tsin [Shan si, App. 353] on the plain and in mountain-valleys. Gathered in the 5th month.

The fruit, kernels of the stones [which are said to be slightly poisonous and of which an emulsion is made], flowers, leaves, twigs and root are all officinal.

P. SMITH [8] erroneously identifies *hing* with the almond, as do also the Customs reports.

Cust. Med., p. 50 (66):—*Hing jen* (apricot-kernels) exported 1885 from Chefoo 792 piculs.—Exported also from Han kow. See *Hank. Med.*, 16.

Further particulars in another part.

272.—梅 *mei*, *Prunus Mume*, S. & Z. *P.*, XXIX, 11.— *T.*, CCV.

Comp. *Rh ya*, 227, *Classics*, 473. The character was originally written 某 *mei.*

Pen king:—梅實 *mei shi* (fruit).

Pie lu:—The *mei* fruit is produced in Han chung [S. Shen si, App. 54] in mountain-valleys. The fruit is gathered in the 5th month and dried by fire.—The sour fruit, pellicle of the stone, kernel, flowers, leaves and root are all officinal.

P. SMITH, 8, 174, sub Almond and Plum.

Cust. Med., p. 372 (439):— *Wu* (black) *mei* exported 1885 from Canton 1,530 piculs,—p. 368 (361), medicated (salted) *mei* from Canton 0.10 picul,—p. 330 (231), *wu mei* from Swatow 362 piculs,—p. 216 (92), from Wen chow 4.40 piculs,—p. 202 (254), from Ning po 1.75 picul.

Further particulars in another part.

273.—桃 t'ao. The Peach. P., XXIX, 16. T., CCXV.

Comp. Rh ya, 266-268, 170, Classics, 470.

Pen king, Index:—桃核仁 t'ao ho jen (peach-stone kernels).

Pie lu:—The t'ao grows in T'ai shan [in Shan tung, App. 322].

The fruit and the kernels of the stones are both officinal. The latter are said to be bitter and sweet and non-poisonous. The 桃毛 t'ao mao (the down which covers the fruit) is said to be slightly poisonous.

桃蟲 t'ao hiao, mentioned in the Pen king, is the peach-fruit which has remained on the tree during the whole winter. It is gathered in the 1st month. The Pie lu calls it 桃奴 t'ao nu, also 桃景 t'ao king and 神桃 shen t'ao. It is bitter. Somewhat poisonous.

The flowers, leaves, bark of the trunk, bark of the root, 桃膠 t'ao kiao, or gum exuding from the tree, are also used in medicine.

MENG SHEN [7th cent.] mentions the 桃符 t'ao fu, slips of peach-wood used as a charm. Comp. Bot. sin., II, 456, peach-wand used in ancient times to expel evil spirits. 桃橛 t'ao küe (poles of peach-wood) are used for the same purpose.

P. SMITH, 168, 169 :—Peach.

Cust. Med., p. 52 (82):—桃仁 t'ao jen (peach-kernels) exported 1885 from Chefoo 91.43 piculs,—p. 62 (60), from I chang 33.65 piculs,—p. 130 (153), from Chin kiang 13.47 piculs. The same exported also from Han kow. See Hank. Med., 43.—

Cust. Med., p. 360 (280):—Peach-leaves exported from Canton 0.21 picul,—p. 356 (212), bark of the peach tree from Canton 0.40 picul.

Further particulars in another part.

274.—栗 *li*. The Chestnut. *P.*, XXIX, 28. *T.*, CCXXII.
Comp. *Classics*, 494.

Pie lu:—The *li* grows in Shan yin [in Che kiang,
App. 271]. It is gathered in the 9th month.

The 栗莌 *li fu*, or thin inner skin of the nut, the 栗殼
li k'o, or involucre of the fruit, the flowers, bark of the tree,
and root are all officinal.

P. SMITH, 60 :— Chestnut.—*Hank. Med.*, 24, *li k'o*,
exported from Han kow.

275.—棗 *tsao*. The Jujube. *Zizyphus vulgaris*, Lam.
P., XXIX, 31. *T.*, CCXXIII.
Comp. *Rh ya*, 272-282, 331, *Classics*, 484.

Pen king:—棗 *tsao*. Index of the *Pen king* 大棗
ta (large) *tsao*.

Pie lu:—The *tsao* grows in Ho tung [in Shan si, App.
80]. The *ta tsao* (or large Jujube) is also called 乾 | *kan*
(dried) *tsao*, 美 | *mei tsao* and 良 | *liang tsao* (excellent
jujube). It is gathered in the 8th month and dried in the
sun. [Probably the large jujubes of the Shan tung province
are meant.]

The raw, fresh fruit, dried fruit, kernels of the stones,
especially those which are three years' old, leaves, centre of
the wood, bark and root are all used in medicine.

P. SMITH, 125 :—Jujube.

Cust. Med., p. 34 (177) :—*Tsao jen* (jujube kernels)
exported 1885 from Tien tsin 826 piculs,—p. 78 (185), from
Han kow 200 piculs,—p. 132 (157), from Chin kiang 48.56
piculs,—p. 52 (84), from Chefoo 9 piculs.

Ibid., p. 194 (147) :—*Tsao p'i* (date-peel) exported from
Ning po 15.60 piculs,--p. 104 (89), from Wu hu 203 piculs.

Further particulars in another part.

276.—梨 *li.* The Pear. *P.*, XXX, 1. *T.*, CCXXXI.

Comp. *Rh ya*, 301, 331, *Classics*, 481.

Pie lu :—Li. Fruit, flowers and bark of the tree used in medicine.

P. SMITH, 169 :—Pear.

Cust. Med., p. 294 (312):—**梨乾** *li kan* (translated by "dried pears") exported 1885 from Amoy 1.05 picul,—p. 368 (385), from Canton 0.49 picul. [I do not think that pears are produced in Amoy or Canton.]—*Ibid.*, p. 354 (198) :—**梨皮** *li p'i* (translated by "peel of Chinese pears") exported from Canton 2.60 piculs.

Further particulars in another part.

277.—木瓜 *mu kua.* Quince. *Cydonia sinensis*, Thouin. *P.*, XXX, 6. *T.*, CCLXXX.

Comp. *Rh ya*, 231, *Classics*, 478.

Pie lu :—Mu kua. Only the name.

The fruit, seeds, twigs with the leaves, bark and root are all officinal.

P. SMITH, 181 :—*Pyrus Cydonia.*

Cust. Med., p. 78 (164) :—*Mu kua* exported 1885 from Han kow 536 piculs. *Ibid.*, 178 (552):—*Mu kua tsiu* (wine) from Shang hai 3.25 piculs.

Further particulars in another part.

278.—柰 *nai. P.*, XXX, 15. *T.*, CCXXI.

Pie lu:—Nai. Only the name. The fruit is used in medicine.

The *Shuo wen* [1st cent.] says the *nai* is a fruit. The name is still in use and is applied to a Chinese fruit not yet

54

identified. WILLIAMS [*Dict.*, 613] states that it is a kind of bullace, a large yellow plum.

Cust. Med., p. 370 (399, 400):—柰仁 *nai jen* (kernels) imported to Canton from Ning po, Shang hai and Han kow.

279.—柿 *shi*. *Persimmon*. *Diospyros Schitze*, Bge. *D. chinensis*, Bl. *P.*, XXX, 17. *T.*, CCXXXIV.

Comp. *Classics*, 491.

Pie lu:—Only the name. The fruit, peduncle, bark and root are all officinal.

P. SMITH, 86 :—*Diospyros Kaki.*

Cust. Med., p. 388 (652):—*Shi t'i* (peduncles) exported from Canton 2.88 piculs,—p. 310 (537), from Amoy 1.58 picul,—p. 296 (338) and 310 (536), the dried fruit and cakes (*shi shuang*) made of it 3.70 piculs, from Amoy.

Further particulars in another part.

280.—安石榴 *an shi liu*. The Pomegranate. *P.*, XXX, 22. *T.*, CCLXXXII.

Pie lu:—*An shi liu.* Only the name. The pulp, rind of the fruit, root and flowers are all officinal.

The pomegranate (*Punica granatum*, L.) is not indigenous in China. It was introduced to China by the famous general CHANG K'IEN about B.C. 120. [See *Bot. sin.*, I, p. 24.]

P. SMITH, 176 :—Pomegranate.

Cust. Med., p. 354 (207):—*Shi liu p'i* (pomegranate-peel) exported 1885 from Canton 7.22 piculs,—p. 150 (171), from Shang hai 3.40 piculs,—p. 284 (166), from Amoy 1.74 picul.

Further particulars in another part.

281.—橘 *kü*. The Orange. *P.*, XXX, 25. *T.*, CCXXVII.

Comp. *Classics*, 486.

Pie lu:—The 橘 *kü* and the 柚 *yu* (*Citrus decumana*) grow in Kiang nan [Kiang si, App. 124] and Shan nan [S. Shen si, App. 268]. They are gathered in the 10th month.

The pulp of the fruit, peel,—known under the name of 陳 皮 *ch'en p'i* [the peel of the unripe fruit is 青 皮 *ts'ing p'i*]—seeds and leaves are all officinal.

TATAR., *Cat.*, 12:—青 皮 *ts'ing p'i*. Fructus *Citri microcarpæ.*—HANB. *Sc. pap.*, 239.

P. SMITH, 164:—Orange.

Cust. Med., p. 284 (159):—橘 皮 *kü p'i* (orange-peel) exported 1885 from Amoy 1.52 picul.—*Ibid.*, 192 (123):—橘 紅 *kü hung* (orange-peel) exported from Ning po 27.19 piculs. The same exported from Han kow. See *Hank. Med.*, 11.

Ibid., p. 214 (71):—青 皮 *ts'ing p'i* exported from Wen chow 18.40 piculs,—p. 284 (158), from Amoy 10.22 piculs,—p. 192 (121, 122), from Ning po 36 piculs,—p. 366 (344), from Canton 8.43 piculs.

Ibid., p. 352 (176):—陳 皮 *ch'en p'i* exported from Canton 1,987 piculs,—p. 72 (81), from Han kow 895 piculs,—p. 322 (108), from Swatow 721 piculs,—p. 226 (107), from Foo chow 720 piculs.

Ibid., p. 192 (125):—橘 白 *kü pai* (inner white skin of oranges) exported from Ning po 3.38 piculs.

Ibid., p. 76 (132):—橘 絡 *kü lo* (orange fibres around the flesh) exported from Han kow 91.14 piculs,—p. 62 (50), from I chang 4.83 piculs,

Ibid., p. 198 (197):—橘核 *kü ho* (orange-seeds) exported from Ning po 11.7 piculs,—p. 216 (72), from Wen chow 2 piculs.—Exported also from Han kow. See *Hank. Med.*, 11.

Further particulars in another part.

282.—枇杷 *p'i p'a,* *Eriobotrya japonica,* Lindl. *P.,* XXX, 38. *T.,* CCLXXVIII.

Pie lu:—*P'i p'a.* Only the name.

The fruit, leaves, flower and bark are all officinal.

P. SMITH, 93 :—*Eriobotrya japonica.*

Cust. Med., p. 360 (265):—*P'i p'a* leaves exported 1885 from Canton 49.39 piculs,—p. 152 (202), from Shang hai 20.12 piculs,—p. 288 (208), from Amoy 2.40 piculs.— Exported also from Han kow. See *Hank. Med.*, 33.

Further particulars in another part.

283.—櫻 桃 *ying t'ao.* *Prunus pseudocerasus,* Lindl. Chinese Cherry. *P.,* XXX, 41. *T.,* CCLXXIX.

Comp. *Rh ya,* 266, *Classics,* 477.

Pie lu:—*Ying t'ao.* Only the name and medical virtues.

The fruit, leaves, twigs and flowers are all officinal.

P. SMITH, 58 :—*Cerasus pseudo-cerasus.*

Further particulars in another part.

284.—山櫻桃 *shan ying t'ao.* Mountain Cherry. *Prunus tomentosa,* Thbg. *P.,* XXX, 43. *T.,* CCLXXIX.

Comp. *Classics,* 477.

Pie lu:—*Shan ying t'ao,* 朱桃 *chu t'ao,* 英豆 *ying tou.* This cherry is of the size of a wheat-grain. It is covered with hair. Gathered in the 4th month and dried in the shade.

WU P'U calls it 麥櫻 *mai ying* (wheat cherry).

Further particulars in another part.

285.—龍眼 *lung yen*. *Nephelium longan*, Camb. *P.*, XXXI, 4. *T.*, CCLXXVIII.

Pie lu:—*Lung yen*, also 益智 *I chi*.[69] It grows in Nan hai [Kuang tung, App. 228] in the mountains. The fruit of the larger kind resembles the *pin lang* [Betel-nut. See 287].

Wu P'u calls it 龍目 *lung mu* (dragon's eye).

The fruit and the seed are used in medicine.—P. SMITH, 155. Nowadays the leaves and the flowers of the Lungan are officinal and exported in small quantities from Canton. See *Cust. Med.*, p. 358 (256), 364 (311).

286.—榧實 *fei shi*.   *P.*, XXXI, 11.   *T.*, CCXXXV.

In the Index of the *Pen king* there is the name 彼子 *pi tsz'* or 柀子 *pi tsz'*, which the ancient Chinese authors believe to be identical with the *fei shi* (fruit).

Pie lu:—The *fei shi* grows in Yung ch'ang [W. Yün nan, App. 426]. The *pi tsz'* grows in Yung ch'ang, in mountain-valleys. Taste sweet. Poisonous.

Besides this the *Pie lu* notices the 桃華 *pai hua* (flower). Taste bitter.   CH'EN TS'ANG-K'I [8th cent.] says that it means the flowers of the *fei shi*.

榧 *fei* is now the *Torreya nucifera*, S. & Z., order *Taxaceæ*, the fruit (nut) of which is edible. But 柀 *pi*, according to the *Rh ya* [228] is another name for the *shan* or *Cunninghamia sinensis*.

* The *Kuang ya* gives likewise *i chi* as a synonym for *lung yen*. The authors of the Sung period, however, apply the name 益智子 *i chi tsz'* to the bitter-seeded Cardamom. See *P.*, XIV*a*, 40.—HANB., *So. pap.*, p. 252.

Cust. Med., p. 158 (289):—*Fei tsz'* exported 1885 from Shang hai 28.62 piculs,— p. 366 (353), from Canton 1.30 picul.

Further particulars in another part.

287.—檳榔 *pin lang.* The Betel-nut, *Areca Catechu,* L. *P.,* XXXI, 14. *T.,* CCLXXXV.

Pie lu:—The *pin lang* grows in Nan hai [Canton, App. 228]. The fruit is used in medicine. Taste bitter, pungent and harsh. Nature warm. Non-poisonous.

Cust. Med., p. 404 (152):—*Pin lang* exported 1885 from Kiung chow 14,068 piculs,— p. 370 (409), from Canton 3,882 piculs,—p. 354 (204), betel-nut husk from Canton 1,312 piculs,— p. 400 (81), from Kiung chow 1,154 piculs.

Further particulars in another part.

288.—秦 椒 *Ts'in tsiao. P.,* XXXII, 1. *T.,* CCL.

Comp. *Rh ya,* 259, *Classics,* 497.

Pen king :—*Ts'in tsiao.*

Pie lu:—The *Ts'in tsiao* grows in the mountains of 秦 Ts'in [Shen si] and especially in the 秦 嶺 Ts'in ling range [S. Shen si, App. 358], also in Lang ye [in Shan tung, App. 178]. The fruit is gathered in the 8th and 9th months. It is called 椒 紅 *tsiao hung* (red carpels of the *tsiao*) and is of a pungent taste. The nature of the unripe fruit is warm, while that of the ripe fruit is cold. It is poisonous.

T'AO HUNG-KING :—The drug *tsiao* which is brought from Western China resembles the *tsiao* [of Mid and Eastern China] in taste and smell, but is larger and of a yellowish black colour. It is not to be confounded with the *kiu tsz'* [see 290].

Su Kung [7th cent.]:—The *Ts'in tsiao* in its leaves, trunk and fruit resembles the *Shu tsiao* [see 289], but it is smaller and less pungent in taste. It grows plentifully in the Ts'in ling mountains in the district of Lan t'ien [in Shen si, App. 175].

Su Sung [11th cent.]:—It is common in Ts'in chou [in Kan su, App. 358], in Feng chou, Kin chou, and Shang chou [all in Shen si, App. 39, 143, 278] and in Ming chou and Yüe chou [both in Che kiang, App. 224, 418]. It flowers at the beginning and produces fruits towards the end of autumn. The fruit is gathered in the 9th or 10th month. In the *Rh ya* it is called *ta* (great) *tsiao*. Also mentioned in the *Shi king*. The *Ts'in tsiao* has large fruits.

K'ou Tsung-shi [12th cent.]:—The *Ts'in tsiao* is produced in the country of Ts'in, whence the name. The various *tsiao* trees resemble each other in their trunks, but the *Ts'in tsiao* has larger leaves and the fruit is also larger as in the *Shu tsiao*, and the wrinkles are different. The *Shu tsiao* is likewise found in the country of Ts'in.

Li Shi-chen :—The *Ts'in tsiao* is the same as the 花椒 *hua tsiao*. Its native country is the land of Ts'in, but it is now common all over China and is easily cultivated. Its leaves grow opposite and are pointed. It is covered with spines. It blossoms in the 4th month. Small flowers. The fruit is produced in the 5th month. It is at first green, becomes red when ripe and is larger than the fruit of the *Shu tsiao*, but the eyes (目, or seeds) are smaller, shining and black. The *Fan tsz' ki jan* says :—The *Shu tsiao* is produced in 武都 Wu tu [in Kan su, App. 395]. That of a red colour is the best. The *Ts'in tsiao* grows in Lung si and T'ien shui [both in Kan su, App. 216, 339]. That with small seeds is the best. Su Sung's statement regarding its flowering in autumn is incorrect.

Ch., XXXIII, 40 :—*Ts'in tsiao* or *hua tsiao*. The figure represents a *Zanthoxylum*, probably *Z. Bungeanum*, Pl., for *hua tsiao* is the common name applied to this tree at Pe king. It seems, however, that in other parts of China *Z. piperitum*, DC., and other species bear the same Chinese name. The reddish brown carpels covered with prominent tubercles, which include the black, shining seeds, are used in medicine and for pickling vegetables.

TATAR., *Cat.*, 9 :— *Hua tsiao, Xanthoxylum.*—HANB., *Sc. pap.*, 228 :—Fruits of the *hua tsiao* described and figured.—P. SMITH, 234.

HENRY, *Chin. pl.*, 42 :—*Hua tsiao, Zanthoxylum Bungei*. Occurs [in Hu pei] in both wild and cultivated states.

According to the *Hank. Med.* [18] the *hua tsiao* fruit is exported from Han kow. In the *Cust. Med.* it figures only as an article of import, to Wu hu, Chin kiang, Ning po, Wen chow, Fu chow, Amoy, Canton, etc.

As to the Chinese names applied to *Zanthoxylum* in Japan, see *Bot. sin.*, II, 497.

289.—蜀 椒 *Shu tsiao*.   P., XXXII, 2.   T., CCL.

Pen king:—*Shu* (Sz ch'uan) *tsiao*. The fruit is officinal. The 椒 紅 *tsiao hung* (the red carpels) are of a pungent taste. Nature warm. Poisonous. The 椒 目 *tsiao mu* (eyes or seeds) are bitter. Nature cold. Non-poisonous. The leaves and the root are also used in medicine. The latter is said to be slightly poisonous.

Pie lu :—Other name : 巴 椒 *Pa* (Sz ch'uan) *tsiao*. The *Shu tsiao* grows in Wu tu [in Kan su, App. 395] in mountain-valleys, also in Pa (E. Sz ch'uan). The fruit is gathered in the 8th month and dried in the shade.

T'AO HUNG-KING :—It is cultivated in Shu (Sz ch'uan) and in Pei kün [in Hu pei, App. 243]. The rind and the flesh (of the carpels) is thick. The inner side is white. Strong smell and taste. It is also found in Kiang yang [in Sz ch'uan, App. 130], in Tsin k'ang [in Kuang tung, App. 355] and Kien p'ing [in Hu pei, App. 139], but this drug is small, red, pungent, not fragrant and less potent than that from Pa.

The *P'ao chi lun* [5th cent.] calls it 南椒 *nan* (southern) *tsiao.*

SU KUNG [7th cent.]:—Now that produced in Si ch'eng, depending on Kin chou [in Shen si, App. 295, 143] is the best.

The *Ji hia Pen ts'ao* [10th cent.] calls it 漢椒 *Han tsiao.*

SU SUNG [11th cent.]:—In Kui chou and Hia chou [both in Hu pei, App. 169, 64], in Shu ch'uan (Sz ch'uan, App. 292] and in Shen and Lo [both in Ho nan, App. 283, 201] it is much cultivated in gardens. It is a tree, from 4 to 5 feet high, which resembles the *chu yü* [*Boymia*, see 291] but is smaller and provided with spines. The leaves are hard (coriaceous) and shining. A beverage is made by boiling them. In the 4th month it produces fruits. It does not flower. The fruits appear in the axils of the twigs and leaves, resemble small peas, are globular and have a purplish red skin. They are gathered in the 8th month and dried by fire. This tree is also found in Kiang and Huai [Kiang su and An hui, App. 124, 89] and in North China. It is similar to that growing in Shu, but the drug yielded by it is different, it has a thick rind, white on the inner side, and has an ardent taste.

LI SHI-CHEN :—Other names : 川椒 *Ch'uan* (Sz ch'uan) *tsiao* and 點 l *tien tsiao.* The *Shu tsiao* (the fruit) has a thick, fleshy, wrinkled (warty) rind (carpels) which contains a shining, black seed resembling the pupil of the eye, wherefore the seeds are also called 椒目 *tsiao mu* (eye).

Ch., XXXIII, 40 :—*Shu tsiao.* The figure represents a *Zanthoxylum* with winged petioles.

LOUR., *Fl. cochin.*, 38 :—*Piper pinnatum* [obscure plant, probably a *Zanthoxylum*]. Sinice : *xu* (*shu*) *tsiao.*

TATAR., *Cat.*, 16 :— *Ch'uan tsiao.* Fructus *Fagaræ piperitæ* (*Zanthoxylum piperitum*).

In the *Cust. Med.* the *Ch'uan tsiao* is mentioned as a drug imported to Tam sui, Ta kow, Swatow, Kiung chow, and said to come from Sz ch'uan.

The 蜀 椒 figured in the *Phon zo* [LXX, 3] looks like *Zanthoxylum piperitum*, DC.

Another *Zanthoxylon* is noticed in the *P.* [XXXII, 8] under the name 崖 椒 *yai tsiao.* The *T'u king Pen ts'ao* [11th cent.], in which it is first mentioned, describes it as growing in Shi chou [in Hu pei, App. 288]. The leaves are larger that those of the *Shu tsiao.* The people there collect the bark at all times of the year and use it as a medicine.

LI SHI-CHEN :—Its popular name is 野 ‖ *ye* (wild) *tsiao.* It is not very aromatic. The seeds are gray, not black nor shining. The savages add it when roasting chicken and duck.

Ch., XXXIII, 41 :—*Yai tsiao.* The figure represents a *Zanthoxylum.*

Phon zo, LXX, 4 :—崖 椒, *Zanthoxylum shinifolium*, S. & Z.—Hoffm. & Schlt., 635.

290.—蔓 椒 *man tsiao.* *P.*, XXXII, 8. *T.*, CCL.

Pen king :—Man (climbing) *tsiao.* The fruit, root and stem are officinal. Taste bitter. Nature warm. Non-poisonous.

Pie lu :—Other names : 豬 | *chu* (pig) *tsiao,* 豕 | *shi* (pig) *tsiao,* 羝 | *chi* (sow) *tsiao* and 狗 | *kou* (dog) *tsiao.* The *man tsiao* grows in Yün chung [in the Ordos, App. 422] in the mountains. The root and the leaves are used for fermenting wine.

T‘AO HUNG-KING :—It is a common mountain-plant. Its popular name is 樛 子 *kiu tsz‘.* It resembles the *tsiao tang* (*Zanthoxylum*) but is smaller and not aromatic. Other name : 豨 | *hi* (pig) *tsiao.* Used as a diaphoretic.

In the *T‘u king* [11th cent.] it is called 金 | *kin* (gold) *tsiao.*

LI SHI-CHEN :—The *man tsiao* is a climbing plant. It exhales an offensive odour, wherefore it is called " pig *tsiao."* It has weak branches. The fruits (seeds) and the leaves resemble those of the *tsiao.* Used as food by the mountain people. The term *kiu tsz‘,* given to it by T‘AO HUNG-KING, is a general name for the fruit of all sorts of *tsiao.*

Ch., XXXIII, 53 :—*Man tsiao.* It does not seem that the drawing represents a *Zanthoxylum.* But the 蔓 椒 in the *Phon zo* [LXX, 4, 5] is a *Zanthoxylum.*

291.—吳茱萸 *Wu chu yü* [comp. 339]. *P.,* XXXII, 13. *T.,* CCXLIX.

Comp. *Rh ya,* 329, *Classics,* 498.

Pen king :—*Wu chu yü* (*chu yü* of the kingdom of Wu). The fruit, the leaves and the root with the white rind are used in medicine. Taste of the fruit pungent. Nature warm. Slightly poisonous. The leaves and the root are non-poisonous.

Pie lu :—The *Wu chu yü* grows in Shang ku [in Chi li, App. 272] and in Yüan kü [in Shan tung, App. 415]. It is gathered on the 9th day of the 9th month and dried in the shade. The old (not fresh) drug is the best,

Su Sung [11th cent.]:—It is a common tree, especially in Kiang [Kiang su, An hui, App. 124], Che [Che kiang, App. 10] and Shu Han [Sz ch'uan, App. 293]. It grows 10 feet and more high, has a green bark, leaves like those of the ch'un (Cedrela), but broader and thicker and of a purplish colour. It blossoms in the 3rd month. Small, purplish red flowers. In the 8th month it produces fruit which resembles the tsiao fruit (Zanthoxylum). When young it is of a yellow colour and when ripe is dark purple coloured. That with small seeds which remain green for a long time is the 吳 ｜ ｜ Wu chu yü, while that with larger seeds which are yellowish black for a long time is called 食 ｜ ｜ shi (edible) chu yü [this is Zanthoxylum ailanthoides. See Bot. sin., II, 498].

Li Shi-chen :—The chu yü has weak, succulent twigs. The leaves are long and wrinkled. Fruit in clusters, different from the tsiao fruit. There are two sorts—one with large and the other with small seeds. The latter is used in medicine. Some ancient writers says that the chu yü tree is planted near wells to make the leaves fall into the well. Those who drink this water will never be afflicted with contagious diseases. The fruit is also suspended in the houses to expel evil spirits. The white poplar and the chu yü tree when planted east of the house bring prosperity and prevent evil.

Ch., XXXIII, 37 :—Wu chu yü. Rude drawing.

Tatar., Cat., 63 :—Wu chu yü, Fructus Zanthoxyli.— P. Smith, 234 :—Xanthoxylum piperitum.

The Wu chu yü is the Boymia (Evodia) rutæcarpa. See Henry, Chin. pl., 96, 212. Further particulars in Bot. sin., II, 498.

Cust. Med., p. 80 (194):—Wu chu yü exported 1885 from Han kow 261 piculs,—p. 424 (135), from Pakhoi 4.40 piculs.

292.—瓜 蒂 *kua ti.* *P.,* XXXIII, 1. *T.,* XLV.

Kua is a general term for the fruits of cucurbitaceous plants. In the Classics *kua* generally means gourds, but sometimes also melons. See *Bot. sin.,* II, 382. *Ti* is the footstalk of a flower or fruit.

The drug *kua ti* is noticed in the *Pen king.* It is stated to have a bitter taste and to be poisonous. Subsequent writers refute its poisonous properties. SU SUNG [11th cent.] explains that by *kua ti* the footstalks of the 甜 瓜 *t'ien* (sweet) *kua* or melon are meant.

Pie lu:—The *kua ti* is produced in Sung kao [in Ho nan, App. 317] in marshes. It is gathered on the 7th day of the 7th month and dried in the shade.

In the *T'ang pen ts'ao* [7th cent.] the melon is first distinguished by a distinct name—甘 瓜 *kan* (sweet) *kua.* The name 甜 瓜 *t'ien* (sweet) *kua,* now in general use, appears first in the *Kia yu Pen ts'ao* [11th cent.]. WANG CHENG [13th cent.] says there are two classes of *kua.* One is distinguished by its sweet fruits and termed 果 瓜 *kuo kua.* It comprises the *t'ien kua* (melon) and the 西 瓜 *si kua* (water-melon). The other class, called 菜 瓜 *ts'ai kua,* produces fruits which are used as vegetables, cucumbers, pumpkins and gourds.

The 瓜 瓤 *kua jang,* or pulp of the melon, and the 瓜子仁 *kua tsz' jen,* or kernels of melon-seeds, are likewise used in medicine.

Cust. Med., p. 48 (42) and 230 (146):—*Kua ti* noticed as imported to Chefoo and Fu chow from Canton.

The *Hank. Med.* [22] mentions 瓜子 *kua tsz'* (melon-seeds) as exported from Han kow. The seeds are slightly fired and eaten for pastime, chiefly in Chinese tea-houses.

I may observe that the *kua tsz‘* or melon-seeds of the
Customs reports are not melon-seeds, but the larger seeds of
the water-melon [melon-seeds are small]. They are largely
exported from New chwang. Probably the seeds of some
gourds and squashes go under the same name.

Further particulars in another part.

293.—葡萄 *p‘u t‘ao*. The Vine. *Vitis vinifera*. Grapes.
P., XXXIII, 7. *T.,* CXIII. *Pen king*.

Pie lu:—The *p‘u t‘ao* grows in the mountains of Lung
si [in Kan su, App. 216], Wu yüan [North of the Ordos.
App. 397] and Tun huang [in Kan su, App. 371].

The Han History states that the Chinese General CHANG
K‘IEN [see *Bot. sin.,* I, p. 24] first introduced the vine to
China from Western Asia, about B.C. 120. The name was
originally written 蒲桃 *p‘u t‘ao*, probably the rendering in
Chinese of a foreign name. If this statement be true it is
inconsistent with the notice of the grape in the earliest
Chinese Materia Medica. LI SHI-CHEN therefore supposes
that the vine has always been indigenous in Lung si, but
was not cultivated in China proper.

I may observe that several species of wild vine with
edible berries grow abundantly in the mountains of North
China,—*Vitis ficifolia*, Bge., *Vitis bryoniæfolia*, Bge. Comp.
Classics, 492. Further particulars in another part.

294.—甘蔗 *kan che*. The Sugar-cane. *P.,* XXXIII, 11.
T., CXIII.

Pie lu:—*Kan* (sweet) *che*. Only the name.

The earliest Chinese account of the sugar-cane is found
in TUNG FANG-SO's *Shen i king* [B.C. 2nd cent.]. It is
described there under the name of 䓊蔗 *kan che* as a reed

growing in Nan fang (Southern China) and containing a sweet juice. The medical quality ascribed to it is that of preventing the developement of intestinal worms.—Sz MA SIANG JU [† B.C. 126] in one of his poems alludes to the property of the juice of the *kan che* of dissipating intoxication occasioned by wine.

In the *Nan tu fu*, a poem written by CHANG HENG [A.D. 78-139] the sugar-cane is termed 諸蔗 *chu che*, and under the same name it is mentioned in the *Shuo wen*.

Further particulars in another part.

295.—蓮藕 *lien ou*. *Nelumbium speciosum*. The Lotus. *P.*, XXXIII, 16. *T.*, XCIII.

Comp. *Rh ya*, 90-104, 191, *Classics*, 395.

Pen king :—Lien ou (藕 *ou* is the name for the root), 蓮實 *lien shi* (fruit) and | 莖 *lien heng* (stalks).

*Pie lu :—*The 藕實莖 *ou shi heng* grows in Ju nan [in Ho nan, App. 110] in ponds and marshes. It is gathered in the 8th month.

TATAR., *Cat.*, 34 :—蓮花 *lien hua*, *Nelumbium*, | 鬚 *lien su*, Stamina *Nelumbii speciosi*, | 房 *lien fang*, Carpophorum *Nelumbii speciosi.—Ibid.*, 6 :—荷葉 *ho ye*, Folia *Nelumbii speciosi.—Ibid.*, 42 :—藕節 *ou tsie*. Articuli radicis *Nelumbii*, and [41] 藕粉 *ou fen*. Amylum radicis *Nelumbii*.

HANB., *Sc. pap.*, 240 :—蓮子 *lien tsz'*. Nuts of *Nelumbium speciosum*.

P. SMITH, 139 :—Lotus.

Cust. Med., p. 30 (112):—*Lien su* (Lotus-stamens) exported 1885 from Tien tsin 69 piculs,—p. 74 (116), from Han kow 38 piculs,—p. 128 (105), from Chin kiang 14 piculs.

Ibid., p. 354 (200) :— | 蓬 *lien p'eng* (fruit receptacle) from Canton 3 piculs,—p. 308 (501), | 房 *lien fang* (fruit receptacle) from Amoy 0.22 picul.—*Ibid.*, p. 384 (589), | 子 心 *lien tsz' sin* (germ of Lotus-seed) from Canton 1 picul.

Ibid., p. 358 (251):—*Lien ye* (Lotus-leaves) from Canton 0.85 picul,—p. 288 (199), from Amoy 0.46 picul.

Ibid., p. 278 (83):—*Ou tsie* (rhizomes of Lotus) exported from Amoy 20.48 piculs,—p. 346 (101, 102), from Canton 12 piculs.—*Ibid.*, p. 386 (619), *ou fen* (flour prepared from the rhizomes) from Canton 0.83 picul.

Further particulars in another part.

296.—菱 實 *ki shi.* The fruit of the Water-caltrop, *Trapa.* *P.*, XXXIII, 26. *T.*, XCIX.

Comp. *Rh ya*, 124, *Classics*, 397.

Pie lu :—Ki shi, also 薐 *ling.* Fruit and flowers used in medicine. The *Feng su t'ung* [2nd cent.] calls it 水 栗 *shui li* (water-chestnut).

Further particulars in another part.

297.—芡 實 *k'ien shi.* Fruit of *Euryale ferox*, Salisb. *P.*, XXXIII, 27. *T.*, XCIX.

Comp. *Classics*, 396.

Pen king :—K'ien shi, 雞 頭 *ki t'ou* (cock's head) and 鴈 喙 *yen hui* (goose's beak). Besides the fruit (seeds) the stem and the root are used in medicine.

Pie lu :—The *ki t'ou shi* grows in Lei ch'i [see App. 181] in ponds and marshes. It is gathered in the 8th month.

TATAR., *Cat.*, 57 :—*Ki t'ou. Euryale ferox.*—P. SMITH, 95.

Cust. Med., p. 106 (127):—*K'ien shi* exported 1885 from Wu hu 2,737 piculs,—p. 128 (113), from Chin kiang 1,172 piculs,—p. 90 (59), from Kiu kiang 7.36 piculs,—p. 30 (121), from Tien tsin 4 piculs.

Further particulars in another part.

298.—烏芋 *wu yü*. Tubers of *Scirpus tuberosus*, Roxb. *P.*, XXXIII, 29. *T.*, CXIV.

Comp. *Rh ya*, 59.

Pie lu:—The *wu yü* [black *Colocasia*. See 261], also called 藉姑 *tsie ku*, has leaves like the *yü* (*Colocasia*). The root is gathered on the 3rd day of the 3rd month and dried in the sun.

Li Shi-chen observes that the *Pie lu* is mistaken in identifying the *wu yü* with the *tsie ku*, for the latter is the *ts'z' ku* or *Sagittaria* [*see the next*], whilst the *wu yü*, called also 地栗 *ti li* (ground chestnut) and 荸臍 *pi ts'i*, is a quite different plant the leaves of which do not resemble those of the *yü*. The tubers of the *wu yü* are eaten and also used in medicine.

Comp. P. Smith, 92:—*Eleocharis tuberosa.*—Further particulars in another part.

299.—藉姑 *tsie ku* in the *Pie lu* [see 298]. *P.*, XXXIII, 31, sub 慈姑 *ts'z' ku*. *T.*, CXIV.

These are the tubers of *Sagittaria*. See P. Smith, 189.

Further particulars in another part.

300.—栢 *po*. *Thuja orientalis*, L. *P.*, XXXIV, 1. *T.*, CCIII.

Comp. *Rh ya*, 225, *Classics*, 505.

56

Pen king :—栢實 *po shi* (fruit).   The fruit, leaves and white rind of the root are used in medicine.

Pie lu :—The *po shi* is produced in T'ai shan [in Shan tung, App. 322] in mountain-valleys.   That which the *Pie lu* says regarding the *po* leaves is unintelligible.[70]

TATAR., *Cat.*, 3, 5.—P. SMITH, 216.

Cust. Med., p. 78 (170):—栢子仁 *po tsz' jen* (kernels of *Thuja*) exported 1885 from Han kow 173.84 piculs,— p. 34 (164), from Tien tsin 16.84 piculs.—*Ibid.*, p. 360 :— *Po* leaves exported from Canton 17 piculs,—*Ibid.*, p. 388 (629):—栢香碎 *po hiang sui* (said to be the powdered root of *Thuja*) exported from Canton 5.28 piculs.

Further particulars in another part.

301.—松 *sung.   Pinus sinensis*, Lamb.   *P.*, XXXIV, 3. *T.*, CXCVII.

Comp. *Rh ya*, 324, *Classics*, 504.

Pen king :—松膏 *sung kao* and 松肪 *sung fang*, Resin [more commonly called 松香 *sung hiang* (pine fragrance)].

Pie lu :—The 松脂 *sung chi* (resin) is produced in T'ai shan [in Shan tung, App. 322] in mountain-valleys. It is gathered in the 6th month.

Besides the resin, the leaves and the white bark of the root, bark of the trunk, excrescences, cones, seeds and flowers are used in medicine.

TATAR., *Cat.*, 50.—P. SMITH, 97.   Fir.

Cust. Med., p. 336 (310):—*Sung hiang* (resin) exported 1885 from Swatow 40.67 piculs,—p. 310 (543), from Amoy 0.80 picul.

"柏葉尤良四時各依方面采陰乾.

Ibid., p. 390 (657):—松節 *sung tsie* (knots) exported from Canton 4.30 piculs,—p. 288 (216), from Amoy 0.35 picul.

Ibid., p. 280 (113):—Pine root exported from Amoy 0.36 picul.

Ibid., p. 196 (190):—Pine flowers exported from Ning po 26.58 piculs,—from Amoy 0.15 picul.

Further particulars in another part.

302.—杉 *shan* (*sha*). *Cunninghamia sinensis.* *P.*, XXXIV, 12. *T.*, CCLXI.

Comp. *Rh ya,* 228.

Pie lu:—*Shan.* Only the name.

The wood, bark, seeds and leaves are all used in medicine.

Further particulars in another part.

303.—桂 *kui* and 牡桂 *mou kui.* Chinese Cinnamon. *Cassia* bark. *P.*, XXXIV, 13. *T.*, CCXLI.

Comp. *Rh ya,* 247, *Classics.*

Pen king:—*Mou kui* (male cinnamon). Taste pungent. Nature warm. Non-poisonous.

Pie lu:—The 桂 *kui* grows in Kui yang [S.E. Hu nan, App. 167] and the 牡桂 *mou kui* in Nan hai [in Kuang tung, App. 228] in the mountains. The bark is gathered in the 2nd and 8th months and dried in the shade. The taste of the *kui* is sweet and pungent. Nature very hot. Slightly poisonous.

We read in the *Nan fang ts'ao mu chuang* [3rd cent. See *Bot. sin.*, I, p. 38]:—The *kui* is found in Ho p'u [in Kuang tung. See App. 70], where it grows on the summits of high mountains. It is an evergreen. There are forests formed exclusively of *kui* trees. In Kiao chi [Cochinchina, App. 133] the *kui* is cultivated in gardens. There are three sorts of *kui*. That which has leaves resembling those of the *po*[71] and a red bark is called 丹桂 *tan* (cinnabar red) *kui*. That with leaves like the *shi* (*Diospyros shitze*) is the 箘桂 *kün kui*. The third kind, with leaves like the *p'i p'a* [*Eriobotrya*. See 282], is the 牡桂 *mou kui*. The *San fu huang tu* [Han period] reports that one of the imperial palaces had pillars of fragrant *kui* wood.

T'AO HUNG-KING [referring to the statement of the *Pie lu*] says :—Nan hai is now called Kuang chou [App. 160]. The *Shen nung Pen king* mentions only the *mou kui* and the *kün kui* [see 304]. The sort commonly used is the *mou kui*. The drug is flat, large and very thin. The outer coat is yellow. It has but little resin and flesh. It smells like the *mu lan* [*Magnolia*. See 305] and tastes like the *kui*. The author is not sure whether it is the bark of an old *kui* tree or that of a quite different tree.—The *kün kui* bark is round (cylindrical) and has the appearance of a bamboo-cane. That which is threefold (rolled up) is the best.[72] But this drug is not seen in the markets. The people commonly use the bark of young branches, which roll up into tubes.—There is a third sort, which is only halfway rolled up.[73] [I understand : not rolled up as a tube, but each side curled inward, forming a channel.] This is called simply 桂 *kui* and is much used in medicine. The *kui* produced in Kuang chou is of a superior quality. That of Kiao chou [S.W. Kuang tung, App. 132] and Kui chou [in Kuang si, App. 164] is very small but has much resin and flesh, and is also valued. The *kui* from Siang chou [in Hu nan, App. 307] and Kui yang hien in the prefecture

[71] 柏 *Thuja*. This character is most probably a mistake.

[72] 三重者良.   [73] 半卷.

of Shi hing [in Kuang tung, App. 167, 289] is called
the 小桂 *siao* (small) *kui*. It is inferior to the drug from
Kuang chou. The *King* [*Sien king,* see further on] says :—
The leaves of the *kui* are like those of the *po* (*Thuja*), shining
and dark. The bark is yellow and the heart is red. At
the time of Emperor Wʋ TI of the Tsʻi dynasty [A.D. 483-
494] *kui* trees from Siang chou were sent to the capital
[present Nan king] and planted in the imperial garden Fang
lin yüan. In the eastern mountains (Tung shan) there grows
a kind of *kui* the bark of which has a strong smell. It
has peculiar persistent leaves. Perhaps it is the *mou kui.*
The people commonly call it 丹桂 *tan kui* [*v. supra*], for
it has a red bark. In North China the *kui* bark is an
important ingredient of food. In the *Li ki* the aromatic *kui*
is mentioned together with ginger.

SU KUNG [7th cent.]:—There are two sorts of *kui.*
TʻAO HUNG-KING quotes the *King,* which says that the leaves
of the *kui* resemble those of the *po.* SU KUNG does not understand from what source this erroneous statement is derived.
That which the *Pie lu* says regarding the *kui* is likewise
incorrect. The 桂 *kui,* simply so called, is the same as the
牡 | *mou kui,* or male *kui,* the *tsʻin* or *mu* (wood) *kui* of
the *Rh ya* [247]. Its leaves are more than a foot long.
In its flowers and fruit it does not differ from the *kün kui.*
The bark of the large as well as of the small branches go
all under the name of *mou kui,* but there is a difference.
The bark of the larger branches is of a coarse ligneous
texture, and it has but little flesh. Its taste is poor. This drug
is also called 木 | *mu* (wood) *kui* or *ta* (large) *kui.* It
is much inferior in quality to the bark taken from the small,
young branches, which has much flesh and is half-way rolled
up. It has small wrinkles on the inner surface. Its taste
is pungent and pleasant. This latter (*i.e.* the bark from
the small branches) is also called 肉 | *jou* (flesh) *kui,* or

桂枝 *kui chi* (branch) and 桂心 *kui sin* (heart). Now
the drug produced in Yung chou and Kui chou [both in
Kuang si, App. 430, 164] and in Kiao chou [in Kuang
tung, App. 132] is much valued.—The other sort, the 箇桂
kūn kui, has leaves resembling those of the *shi* (*Diospyros
shitze*). The leaves have three roads (nerves), are glabrous
on both sides and shining. The bark of the larger as well
as of the small branches forms tubes. Only the old, hard
bark does not roll up, and appears in a flat form. It has
but little aroma and is not used in medicine. But the
bark of the slender branches is thin and rolls up. That
rolled up two or three fold is much valued. It is also known
under the name 筒桂 *t'ung* (tube) *kui*. This is the *siao*
(small) *kui* of T'ao Hung-king. It is now produced in Shao
chou [in Kuang tung, App. 279].

HAN PAO-SHENG [10th cent.]:—There are three sorts
of *kui*. The 箇 | *kūn kui* has leaves resembling those of
the *shi* [*v. supra*] but they are pointed, narrow and shining.
The flowers are white and the stamens yellow. It blossoms in
the 4th and produces fruit in the 5th month. The bark of
this tree is greenish yellow and thin. It rolls up into tubes,
whence the name *t'ung kui*. The thick and hard bark, which
has but little aroma, is called 版 | *pan* (board) *kui* [for it
does not roll up]. It is not used in medicine.—The 牡 |
mou kui has leaves like the *p'i p'a* [*v. supra*]. They are
narrow and twice or thrice as long as those of the *kūn kui*.
The bark taken from the young branches curls inward on
both sides. That of a purplish colour, with wrinkles on
the flesh and of a delicate structure is called 桂枝 *kui chi*,
also 肉 | *jou kui*. After the outer coat of the bark has
been scraped off, the drug is called | 心 *kui sin*. The thick
bark is called 木 | *mu* (wood) *kui*. The leaves are likewise
valuable.—T'ao Hung-king says:—The drug which is half-
way rolled up, and which contains much resin, is from the

桂 *kui* tree, and [relying upon the *Sien king*,[74] states that] it has leaves like those of the *po* tree. This is the third sort. Su Kung is wrong in stating that there are only two sorts.

Ch'en Ts'ang-k'i [8th cent.]:—The *kün kui, mou kui* and *kui sin* are drugs different in appearance but all derived from the same tree. Kui lin [anciently name of a province, modern Kuang si, of which Kui lin fu is now the capital. See App. 165] and Kui ling [mountain chain in Kuang si, App. 166] derive their names from the *kui* trees growing there. Now the *kui* tree grows abundantly in all the prefectures south of that mountain chain and down to the borders of the sea. It grows especially plentifully in Liu chou and Siang chou [both in Kuang si, App. 199, 308].

Su Sung [11th cent.] notices the various forms of *kui* bark which in his time were brought to market from Ling piao [Kuang si, App. 197]. The *kün kui* described by the earlier authors then was called 筒 | *t'ung* (tube) *kui*. The *mou kui*, a thin yellow bark with but little resin and flesh, was called 官 | *kuan kui*. That with the sides rolled half-way up was 版 | *pan kui*. Su Sung then states that these drugs are also produced in Kuan chou, Pin chou, I chou [all in Kuang si, App. 159, 252, 105], Shao chou and K'in chou [both in Kuang tung, App. 279, 144]. A *Cassia* bark which answers the ancient description of the *kün kui* is met with in Pin chou; another, which seems to be that anciently described as *mou kui*, grows in I chou and Shao chou. The people there call this bark 木蘭皮 *mu lan p'i* and the flesh *kui sin* [*v. supra*]. A third kind, growing in K'in chou [*v. supra*], seems to agree with the description of the tree which the ancient authors call simply *kui*. The *kui* tree grows from 30 to 40 feet high and is found in the depths of the mountains.

[74] 仙經. Evidently a Taoist book.

In Man tung [App. 217] the people cultivate it in gardens.
The bark produced north of the mountain range has but little
aroma and pungency, and cannot be used in medicine. It
blossoms in the 3rd or 4th month. The flowers resemble
those of the *chu yü* [*Boymia.* See 291]. The fruit is
produced in the 9th month. The leaves are very fragrant.
An excellent beverage is made of them. The bark is collected
in the 2nd and 8th months and the flowers in the 9th month
[evidently a mistake]. Now the flowers and the fruits are
much used for decorative purposes at festivities.

FAN CH'ENG-TA in his description of the southern provinces [end of the 12th cent.] says:—The *kui* is a remarkable
tree of Southern China. It furnishes an important medicine.
The name of [the ancient province] Kui lin is derived from
the *kui* tree, but it does not grow there now. The *kui* is
produced in Pin chou and I chou [*v. supra*].[75]

LI SHI-CHEN admits two kinds of *kui*—the *mou kui* and
the *kün kui*. The *mou kui* has long hard leaves, covered with
hair and serrated on the margin. The flowers are white.
The bark contains much resin. The *kün kui* has leaves like the
shi (*Diospyros*) but narrower, shining, with three nerves, and
not serrated. The flowers are yellow or white. The bark is
thin and rolled up. The druggists also distinguish these two
sorts—the *kün kui*, which rolls up entirely, and the *mou kui*,
which is rolled up partly or not at all.

Besides the bark, the leaves and flowers of the *kui* are
officinal.

Ch., XXXIII, 7:—桂 *kui*. Rude drawing. The drawing
fol. 8, 蒙自桂樹 *kui* tree from Meng tsz' hien [in Yün nan],

<hr>

[75] The *Kuang si T'ung chi*, sub Kui lin fu, quotes the 百粤風土記,
a modern work, it seems, in which it is stated that there are, in Kui lin fu.
kui trees of enormous size. Ten men are required to encompass one of
them. They perfume the air with delightful fragrance. I suppose the
kui hua, or *Olea fragrans*, is meant. Comp. MARTINI's notice of Kui lin in
my *Early Europ. Res. Fl. China* [p. 11].

is better. It represents a *Cinnamomum* with large leaves having three nerves.

Lour., *Fl. cochin.*, 305 :—*Laurus Cinnamomum*, L. Sinice : *kuei xu (kui shu)*. Habitat agrestes in altis montibus Cochinchinæ, ad Occidentem, versus Laosios : ubi quæcunque invenitur arbor truncatur et excorticatur. Rami crassissimi dant vile *Cinnamomum*, quod plerumque abjicitur, quia longi itineris expensæ pretium superant. Tenuior cortex, a supremis ramis avulsus, *Zeylanico* crassitie æqualis, odore et sapore acerrimus non magni æstimatur, eoque utuntur indigenæ ad condiendos cibos. At mediorum ramorum cortex, lineam fere crassus, optimum et pretiosum præbet *Cinnamomum*, quæ utuntur in Medicina, et multo altiori pretio venditur, quam *Zeylanicum*. Hujus oleum per destillationem abundantius extrahitur : color est rubro-fuscus, acrifudo minus acuta, sapor multo dulcior *Cinnamomo Zeylanico.*—*Laurus Cinnamomum*, L., is the *Cinnamomum Zeylanicum*, Br. But Loureiro's plant is a new species—*Cinnamomum Loureiri*, Nees, in *DC. Prodr.*, XV, i, 16.

J. Reeves, in his Account of Chinese Materia Medica, 1828, writes :—Vast quantities both of Cassia seeds and *Cassia lignea* are annually brought from Kwang si [whose principal city derives its name from the forests of Cassia around it] to Canton and thence shipped off at about 24 dollars per picul to England, while the Chinese themselves use a much thicker bark, unfit for the European market.

Williams [*Chinese Commercial Guide*, p. 113] notices the 桂皮 *kwei p'i*, the decorticated bark of the *Cinnamomum Cassia*,—the ｜｜油 *kwei p'i yu*, Cassia oil, obtained from the leaves and twigs of the Cassia tree by distillation,—the 桂枝 *kwei chi*, Cassia twigs, the extreme and tender ends of the branches, such as are used in distilling the oil,—the 桂子 *kwei tsz'*, Cassia buds, the fleshy ovaries of the seeds, and obtained from the same tree as the bark.

TATAR., *Cat.*, 28 :— | 皮 *kui p'i*, | 支 *kui chi*, | 心 *kui sin*.   Variæ species *Cinnamomi*.

P. SMITH, 52, 53, 64 :—Cassia bark, Cassia buds, Cassia leaves, Cassia twigs.   The thickest barks are called 肉桂 *jou* (fleshy) *kui*.   WILLIAMS [*Commercial Guide*, 85] places this among the articles of import, and states that this drug is produced in Cochinchina and in Sin chou, in the province of Kwang si.—

The *jou kui* in the Chinese druggists' shops is a thick bark and resembles the 官 | *kuan kui*, which I received from a Corean druggist's shop.

According to the *Cust. Med.* all the Chinese Cassia bark in various forms is exported from Canton, viz.:—

P. 354 (196, 197):—*Kui p'i*, Cassia bark, 6,432 piculs,— p. 384 (584), *kui p'i yu*, oil of Cassia bark, 0.31 picul.

P. 358 (250):—*Kui chi*, Cassia twigs, 5,007 piculs,— p. 354 (195), *kui chi p'i*, bark of Cassia twigs, 1,098 piculs.

P. 368 (380):—*Kui tsz'*, Cassia buds [*v. supra*], 2,803 piculs.

P. 354 (192):—*Jou kui*, fleshy Cassia bark, 212 piculs. Shipped to other Chinese ports.   Said to be produced in Kuang si and Annam.—p. 382 (569), *jou kui yu*, Cinnamon oil, 0.03 picul.

According to Mr. KOPSCH's translations from the *Kuang si T'ung chi*, regarding Cassia bark [*China Rev.*, IX (1881), 318], Cassia is produced in the 紫荊山 Tsz' king hills in the prefecture of Sin chou fu in Eastern Kuang si.   The Cassia from the 青花山 Ts'ing hua hills in Annam is also highly esteemed.—I may observe that the same prefecture of Sin chou fu was noticed, nearly 250 years ago, by Father MARTINI, as producing the best Cassia bark.   See my *Early Europ. Res. Bot. Chin.* [p. 13].

Mr. M. Moss, in his Narrative and Commercial Report of an Exploration of the West River to Nan ning fu, 1881, states that Cassia is only grown in Lo ting chou [W. Kuang tung] and in the districts surrounding the town of Tai wu, the produce being respectively known in the trade by the name of Lo ting and Tai wu Cassia. [Tai wu is probably the 大烏 of Chinese maps, S.E. of Sin chou fu in Kuang si].

In 1882 Mr. CH. FORD, Superintendent of the Botan., etc. Department, Hong kong, visited the Chinese districts where Cassia bark is produced, and published a very valuable article on the results of his investigations. Before FORD decided the question on the botanical origin of the Chinese Cinnamon or *Cassia lignea*, it was customary to refer the Chinese drug to *Cinnamomum Cassia*, first described by Blume in 1825, from a plant cultivated in Java [not to be confounded with *C. Zeylanicum*, or *Laurus Cinnamomum*, L., from Ceylon, known in Europe from an early period]. FORD proved by personal observation that the plant yielding the Chinese Cassia bark was indeed the plant described by Blume. He found that there are three chief districts in which it is produced, *viz.* Tai wu [*v. supra*], Lo ting and Luk po [not found on the map, but according to the geographical position given by FORD it lies N.E. of Lo ting]. FORD describes the peeling of the bark by the Chinese. By distillation the leaves afford Cassia oil. What is called Cassia buds are the immature fruits.

SIEB., *Œcon.*, 139 :—*Cinnamomum Cassia* (this is *C. dulce*, Nees). Japonice: *nikkei*; sinice: 桂. E China introducta, hinc ac inde colitur in usum medicum.—*Ibid.*, 140 :—*Cinnamomum Maruba.* Japonice: *maruba nikkei.* E China introductum, in usum medicum cultum. [This seems to be *C. sericeum*, SIEB. *See* FRANCH. & SAV., *Enum.*, I, 411.]

Kwa wi, 97 :—*Nikkei.* Sinice: 榎 or 桂 [the first

character is from the *Rh ya*, 247]. *Cinnamomum Loureiri,* Nees. Introduced from China.

Phon zo, LXXX, 2:—牡桂, *Cinnamomum pedunculatum,* Nees.

HOFFM. & SCHLT., 144:—*Cinnamomum Loureiri,* 桂 or 肉桂.

Phon zo, LXXX, 3:—肉桂, *C. Loureiri.*

SIEB., *Icon. ined.,* VI:—交趾桂 (Cassia from Cochinchina), *Cassia daphnoides,* S. & Z. =*C. sericeum,* SIEB.—See also the drawing under the same Chinese name in *Phon zo* [LXXX, 4, 5].

304.—箘桂 *kün kui. P.,* XXXIV, 21. *T.,* CCXLI.

Pen king:—Kün kui. The taste of this bark is said to be pungent. Nature warm. Non-poisonous.

Pie lu:—The *kün kui* grows in Kiao chi [Cochinchina, App. 133] and in Kui lin [in Kuang si, App. 165] in mountain-valleys and on steep rocks. The drug (bark) is hollow [76] and round (a tube), like a bamboo. It is gathered in the beginning of autumn.

The Chinese authors repeat what has already been said [sub 303] regarding this kind of Cassia bark, which appears in the form of quills.

SU KUNG [7th cent.] says that 箘 *kün* is the name of a bamboo [comp. *Bot. sin.,* II, 564], and that the *kün kui* derives its name from the tubes into which the bark is rolled up.

LI SHI-CHEN adds that one kind of *kün kui* is cultivated under the name of 巖桂 *yen* (rock) *kui* or 木樨 *mu si.* This is the *Olea fragrans.* For further particulars see *Bot. sin.,* II, 552.

[76] 無骨 no bone.

There are two more kinds of *kui* noticed in the *Pen ts'ao kang mu.*

天竺桂 *T'ien chu kui.* P., XXXIV, 22. T., CCXLI.

It is first mentioned by Li Sün [8th cent.] as growing in Nan hai [in Kuang tung, App. 228] in mountain-valleys. Its bark is used like that of the common *kui.* It is thin and not very pungent or ardent.

K'ou Tsung-shi [12th cent.]:—This bark resembles that of the *mou kui* [see 303] but is thinner.

Li Shi-chen:—This is' the same as the 山桂 *shan* (mountain) *kui* which is found in Kuang tung, Fu kien and Che kiang. It grows plentifully in T'ien chu in the prefecture of T'ai chou, whence the name. [77] It is a large tree which flowers abundantly. The fruit is of the size of a Lotus-nut. The Indian [T'ien chu] Buddhist priests believe that it is the *yüe kui* [see further on].

P. Smith [63] identifies *T'ien chu kui* with *Cinnamomum Tamala,* a kind of Cinnamon, he says, which is spoken of in the *Pen ts'ao* as of Indian origin, etc.—T'ien chu, indeed, means India, but in the above-quoted passage this name seems to refer to a place, perhaps a monastery, in Che kiang.

Sieb., *Œcon.,* 138 :—天竺桂, *Cinnamomum japonicum* (i.e. *C. pedunculatum,* Nees). Japonice : *kurotsusu.* E fructibus exprimitur oleum. In Sieb. *Icon. ined.* [VI] the same Chinese name is applied to *Litsæa glauca,* S., and *L. foliosu,* S. (*Laurineæ*).—See also *Phon zo* [LXXX, 11], the figure under the same Chinese name.

月桂 *yüe kui.* P., XXXIV, 22.

Ch'en Ts'ang-k'i [8th cent.] relates that all over Kiang tung, in the 4th or 5th month, the fruits of the *yüe kui* are found on the roads. They are as large as the *li tou* (fox bean)

[77] 台州天竺最多故名

and of an aromatic, pungent taste. There is an old tradition
that they fall down from the moon (*yūe* = moon). These
fruits are used in medicine.

LI SHI-CHEN says that the legends regarding the
cinnamon tree growing in the moon date from the T'ang and
Sung dynasties. It is reported in the T'ang History that in
A.D. 688, at T'ai chou in Che kiang, these *yūe kui* berries
fell down during 10 days. The same phenomenon took place
in the Sung period, in the reign of T'IEN SHENG (1023-1032),
when at the monastery of Ling yin, at Hang chou in Che
kiang, a rain of *yūe kui tsz'* fell down during 15 moonlight
nights.

Comp. also regarding the legend of the Cassia in the
moon, MAYERS' *Chin. Read. Man.*, 300.

P. SMITH [53] identifies the 月桂子 *yūe kui tsz'*
erroneously with Cassia buds.

SIEB., *Icon. ined.*, VI :—*Litsœa glauca*, SIEB. Sinice :
月桂.

305.—木蘭 *mu lan*. P., XXXIV, 23. T., CCXCIII.

Comp. *Bot. sin.*, II, 551.

Pen king :—Mu lan (tree *lan*) and 林 | *lin* (forest) *lan*.
The bark of the tree is officinal. Taste bitter. Nature cold.
Non- poisonous.

Pie lu :—Other name : 杜 | *tu lan*. The *mu lan* grows
in Ling ling [in Hu nan, App. 196] in mountain-valleys,
also in T'ai shan [in Shan tung, App. 322]. The bark
resembles that of the *kui* (cinnamon) and is fragrant. It is
gathered in the 12th month and dried in the shade.

T'AO HUNG-KING :—It is found in all the districts of
Ling ling. The tree resembles the *nan* tree [*Persea nan mu.*
See 310]. The bark is very thin, of a pungent taste and
aromatic. There is one sort in I chou [in Sz ch'uan,

App. 102] which has a thick bark and resembles the *hou p'o* [*Magnolia*. See 316]. It is superior in aroma and taste. The people in the East (Eastern China) use the bark of the *shan kui* (mountain cinnamon), which is akin to the *mu lan*. The Taoists use the *mu lan* as a perfume.

The *Yu yang tsa tsu* [9th cent.] calls it 木 蓮 花 *mu lien hua* (tree Lotus-flower) for its flowers resemble the Lotus. The leaves are like the leaves of the *sin i* [*Magnolia yü lan*. See 306].

HAN PAO-SHENG [10th cent.]:—The tree is 30 feet and more high. It resembles the *kĭn kui* [*Cinnamomum*. See 304]. The leaves have three nerves and are not so aromatic and pungent as those of the *kui* (cinnamon). The bark is like the *pan kui* [a thick sort of Cassia bark. See 303]. It shows perpendicular and horizontal lines. It is gathered in the 3rd and 4th months and dried in the shade.

SU SUNG [11th cent.]:—It is found in all the prefectures of Hu[nan], Ling[nan] and Shu ch'uan [Sz ch'uan, App. 83, 197, 292]. It is not at all like the *kui*, but there is in Shao chou [in Kuang tung, App. 279] a sort of the *kui* the bark of which is called *mu lan* by the people and the inner flesh *kui sin* [comp. above, sub 303].

LI SHI-CHEN :—The branches and leaves of this tree are scanty. Its flowers are white inside and purple outside. There are varieties which flower in all seasons. The tree grows in the depths of the mountains. It attains vast dimensions, and then is fit for building boats. LO T'IEN [9th cent.] says :—The *mu lien* grows in the mountains of Pa [Sz ch'uan, App. 235] and Hia [in Hu pei, App. 64]. The people there call it 黃 心 *huang sin* (yellow heart). It grows from 50 to 60 feet high. Persistent leaves. The trunk resembles a poplar and has white veins. Leaves like those of the *kui* (cinnamon), but thicker, larger and without ridges.

Beautiful, glossy, fragrant flowers resembling the Lotus-flower. There is however a difference in the stamens and the receptacle. The flowers appear in the 4th month and last for 24 days, but are not followed by fruits. This is the true *mu lan*. The flowers are red, yellow and white. The wood is fine-grained and has a yellow heart. It is much prized by wood-carvers. The tree from Shao chou, mentioned by Su Sung, is the *mou kui* [Cinnamon. See 303], not a *mu lan*. The *mu lan* tree does not die when its bark has been stripped off. Lo T'ien [*v. supra*] says that it blossoms in winter. The fruit is like a small *shi* (persimmon), sweet and pleasant. This latter statement is doubtful. The name *mu lan* is derived from the fragrance of its flowers. For *lan* is a fragrant flower (an orchid).

Ch., XXXIII, 14 :—*Mu lan.* The drawing represents a *Magnolia.*

Amœn. exot., 845 :—木蘭 *mokkwuren.* Frutex tulipifer ramis raris incondite divaricatis ; foliis plerumque nudus cito deciduis......flore lilionarcissi rubente. Figured in BANKS' *Icones Kœmpf. sel. tab.*, 43. This is *Magnolia obovata*, Thbg. Cultivated only in Japan. Said to be introduced from China. It has been recorded from Sz ch'uan and Kiang su. See *Ind. Fl. sin.*, I, 23.

The *Phon zo*, LXXX [13, 14] represents, under the above Chinese name, *Magnolia conspicua*, Sal.

HOFFM. & SCHLT., 96 :—木蓮花, *Buergeria (Magnolia) obovata*, S. & Z.

MATSUMURA, 118 :—木蘭, *Magnolia obovata*, and 黃心樹, *Magnolia compressa*, Maxim.

306.—辛夷 *sin i.*　*P.*, XXXIV, 25.　*T.*, CCXCIII.

Pen king :—*Sin i*, 辛雉 *sin chi*, 侯桃 *hou t'ao* and 房木 *fang mu.* The flower-buds are used in medicine.

Pie lu :—The *sin i* grows in Han chung, Wei hing [both in Shen si, App. 54, 384] and Liang chou [App. 187]. The tree resembles the *tu chung* [see 317] and is more than 10 feet high. The fruit [the unopened flower-buds are meant, *v. infra*] is like the *tung t'ao* (winter peach) but smaller. It is gathered in the 9th month. The heart and the outer hairs (the down) are removed [before use], for they are injurious to the lungs and excite coughing.

T'ao Hung-king :—It is now produced in Tan yang [in An hui, App. 328] and in Mid China. It resembles a small peach, is of a pungent taste and aromatic.

Su Kung [7th cent.]:—T'ao Hung-king does not mean the fruit of the *sin i* tree, but its unopened flower-buds, which are gathered in the first and second months. The *Pie lu* is mistaken in stating that the fruit is gathered in the 9th month.

Ch'en Ts'ang-k'i [8th cent.]:—The unopened flower of the *sin i* forms a globe, not unlike a small peach. It is covered with hair. It is also called 猴桃 *hou t'ao* (monkey peach). When the flower first opens it resembles a hair-pencil (the pencil-like tuft of stamens), whence the name 木筆 *mu pi* (tree pencil). As the flowers appear very early in the year the people in the south call them 迎春 *yin ch'un* (welcoming the spring).

Han Pao-sheng [10th cent.]:—It is a tree of great dimensions. Its leaves resemble those of the *shi* (*Diospyros*), but are narrower and longer. It blossoms in the first and second months. The unopened flower resembles a small downy peach. The flowers are white, tinged with purple. The tree does not produce seeds, but at the end of summer it flowers again. There is one kind which has leaves and flowers like the common *sin i*, but its flowers appear in the 3rd month and fall off in the 4th. This tree produces red

seeds like those of the *siang sz' tsz'* (*Abrus precatorius*).
Both kinds are common in the mountains.

CHANG YÜ-HI [11th cent.] gives a similar description
of the tree and states that it is much cultivated in gardens.
The leaves appear after the flowers have fallen off. It very
seldom produces fruit.

K'OU TSUNG-SHI [12th cent.]:—The *sin i* is a common
tree, much cultivated. The unopened flowers are used in
medicine. The opened flowers have no medical virtues.

LI SHI-CHEN gives a good description of the flower-buds
(苞 *pao*) and then says:—The opened flowers resemble the
Lotus-flower but are smaller and very fragrant. Their
fragrance resembles that of the *lan* (an orchid). The white-
flowered *sin i* is commonly called 玉 蘭 *yü* (jade) *lan*.
There is also a double-flowered variety.

Ch., XXXIII, 16 :—*Sin i* or *yü lan*. The figure seems
to represent *Magnolia conspicua*, Salisb. (*M. Yülan*, Desf.).
Good drawing, This tree is cultivated at Peking under the
name of *yü lan*.

TATAR., *Cat.*, 47 :—*Sin i*, gemmæ *Magnoliæ*.—P. SMITH,
142.

HENRY, *Chin. pl.*, 297 :—木 筆 *mu pi* in Hu pei =
Magnolia Yülan.

Cust. Med., p. 74 (115):—*Sin i* flowers exported 1885
from Han kow 145 piculs,—p. 90 (55), from Kiu kiang
25 piculs.

Amœn. exot., 845:—辛 夷 *sini* et *confusi*, vulgo *kobus*.
Arbor sylvestris tulpifera . . . . . folio Mespili; floribus
primo vere ex lanuginoso folliculo ante folia nascentibus
singulis . . . . Figured in BANKS' *Icon. Kœmpf. sel. tab.*
[42]. This is *Magnolia Kobus*, DC.

Phon zo, LXXX, 16, 17 :—辛 夷, *Magnolia Kobus*.

Kwa wi, 96 :—玉 蘭, *Magnolia Kobus*.

307.—沈香 *ch'en hiang*. P., XXXIV, 26. T., CCCXVI.

The name means "fragrant [wood] that sinks [in water]." This is the *Lign aloes*, prized for its fragrance. According to LOUREIRO [*Fl. cochin.*, 327] it is the produce of *Aloëxylon agallochum*, a tree of Cochinchina. It is mentioned in the *Pie lu*.

TATAR., *Cat.*, 14.—P. SMITH, 133.—HANB., *Sc. pap.*, 263.

Further particulars in another part.

308.—鶏舌香 *ki she hiang* (fowl's-tongue spice). This drug is mentioned in the *Pie lu*. Taste pungent. Nature warm. Non-poisonous.

In the *P.* [XXXIV, 30] *ki she hiang* is given as a synonym for 丁香 *ting hiang* (cloves, which have been long known to the Chinese). But according to the *Nan fang ts'ao mu chuang* [3rd cent.] the name *ki she hiang* was also applied to the fragrant fruits of the tree 蜜香 *mi hiang*, which seems to be a kind of Aloewood.

Further particulars in another part.

309.—檀香 *t'an hiang*. Sandalwood. Mentioned in the *Pie lu*. P., XXXIV, 35. T., CCCXVI.

The ancient Chinese authors state that this tree does not grow in China. Its fragrant wood is brought from the countries of the Southern Sea, where it is called 旃檀 *chan tan* or 白 | | *pai* (white) *chan tan*. The Sandalwood is yielded by *Santalum album*, L., a tree of Malabar. Its Sanscrit name is *chandana*.

Further particulars in another]part.

310.—楠 *nan.* Mentioned in the *Pie lu.* *P.*, XXXIV, 37. *T.*, CCLIX. Comp. *Bot. sin.*, II, 512 :—*Persea nanmu,* Oliv. Wood and bark used in medicine. Further particulars in another part.

311.—釣樟 *tiao chang.* Mentioned in the *Pie lu.* *P.*, XXXIV, 39. *T.*, CCLIX. Comp. *Bot. sin.*, II, 513 :—A tree of the order *Lauraceœ.* The bark of the root, leaves and flowers are used in medicine.

312.—薰陸香 *hün lu hiang* or 乳香 *ju hiang.* *P.*, XXXIV, 45. *T.*, CCCXVI.

The *Pie lu* notices the drug under the above names and speaks of its use in medicine.

Su Kung [7th cent.]:—The drug *hün lu hiang* resembles the *pai kiao hiang.*[78] That produced in T'ien chu (India) is of a white colour. The drug from Tan yü[79] is of a greenish colour and not very aromatic.

Li Sün [8th cent.]:—It is also called 馬尾香 *ma wei hiang.* The *Kuang chi* [6th cent.] states that the *hün lu hiang* exudes from a tree with a scaly bark. The 乳頭香 *ju t'ou hiang* (nipple fragrance) is brought from the Southern Sea. It is the resin of a kind of fir tree which grows in Po sz' (Persia). It is red like a cherry. That which is pellucid is the best.

Ch'en Ts'ang-k'i [8th cent.] says :—The *ju hiang* is a kind of *hün lu.*

[78] 白膠香, the white fragrant gum of the 楓 *feng* tree, *Liquidambar formosana,* Hce.

[79] 單于. Unknown to me.

CHANG YÜ-HI :—The *Nan fang i wu chi* [4th or 5th cent.] states that the *hün lu* is produced in the kingdom Ta ts'in (the Roman Empire in Europe and Asia) by a large tree which is found near the seashore, and in its leaves and branches resembles an old fir tree. It grows plentifully in a sandy soil. In summer the resin flows out from the trunk into the sand. It resembles the gum of the peach tree. It is collected by the natives and sold to the traders. The natives eat the inferior sorts.[80]

K'OU TS'UNG-SHI [12th cent.]:—The *hün lu* is also called *ju hiang*, for it runs down [from the trunk] in drops which have the form of a nipple. It is fragrant. There is in Nan Hin tu (Southern India), in the country of O-ch'a li, a tree called *hün lu*, which in its leaves resembles the *t'ang li* (a pear tree).[81] This tree yields the 西 香 *si hiang* (western perfume). The *ju hiang*, which is brought from the southern countries, is of a higher quality.

CH'EN CH'ENG [11th cent.]:—That which is called the "western perfume" comes from T'ien chu (India). The southern drug is produced in Po sz' (Persia) and other countries. The first is of a yellowish white colour and the second is purplish red. The resin taken fresh from the tree is more highly valued than that [collected from the ground and] mixed with sand. *Hün lu* is a general name. The name *ju hiang* (nipple perfume) is applied to the nipple-shaped drug. The resin from firs and from the *feng* tree (*Liquidambar*) sometimes shows pieces of the same shape.

LI SHI-CHEN :—The *ju hiang* is frequently adulterated with the *feng hiang* or *Liquidambar* resin. But the drugs can be distinguished by burning them. There are various sorts of *ju hiang* distinguished in commerce by peculiar

---

[80] This account is partly borrowed from the *Nan fang ts'ao mu chuang*.

[81] This is taken from HÜAN TSANG's travels. *See* BEAL'S *Buddh. Rec. of the Western World*, II, 265.

names. In Buddhist books the *ju hiang* is called 天澤香 *t'ien tse hiang* (heavenly, shining fragrance), also 多伽羅香 *to-ka-lo* perfume, 杜嚕香 *tu-lu* perfume and 摩勒香 *mo-le* perfume.

The drug *ju hiang* has been correctly identified by A. CLEYER in his *Specimen Medicinæ sinicæ* (1682), 210 :— *Ju hiam* est *Thus*, acrodulce, ulceribus medetur, creat carnem, sistit dolores, eximitur illi oleum.

WILLIAMS, *Chin. Commercial Guide*, 93 :—*Olibanum, ju hiang* (*i.e.* milk perfume). Article of import.

TATAR. [*Cat.*, 65] identifies *ju hiang* with *Resina San-darac*, but the drug *ju hiang* which I procured from a Chinese drug-shop, and which has been examined by Professor FLÜCKIGER, was *Olibanum* or Frankincense. The ancient Chinese descriptions agree.—*See also* P. SMITH, 161.

According to FLÜCKIGER and HANBURY [*Pharmacographia*, 120] *Olibanum* is obtained from the stem of several species of *Boswellia* growing in Eastern Africa and Southern Arabia. Other species from India are used in the country as incense. Comp. also my *Knowl. Anc. Chin. of the Arabs*, p. 19.— MARCO POLO (II, 442), speaking of Dufar on the Arabian coast, says that the white Incense (Frankincense) grows there. It resembles a small fir tree. Comp. on the subject YULE's note, *l.c.* 446.

Regarding the Sanscrit names for *Olibanum* as given in the *P.*, I may observe that Dr. EITEL [in his *Handb. of Chin. Buddhism*] identifies the Chinese *to-ka-lo* with the Sanscrit *tagara*, meaning "perfume". *Olibanum* in Sanscrit is *kunduru* [comp. above *tu lu* perfume] and *luban* in Hindustani, by which name it is also known to the Arabs. It is *lebonah* in Hebrew, signifying "milk." Modern travellers who have seen the frankincense trees state that the fresh juice is milky and hardens when exposed to the air.

313.—蘇合香 *su ho hiang*. *P.*, XXXIV, 53. *T.*, CCCXVI.

Pie lu:—The *su ho hiang* (perfume) is produced in Chung t'ai [see App. 32] in river-valleys.

T'AO HUNG-KING :—People say that the *su ho hiang* is the excrement of lions, but the foreigners assert that this is not true. Now this drug is brought to China from the western countries. It is not used as a medicine, but rather as a perfume.

The *Kuang chi* [6th cent.] says :—This fragrant substance is produced in the country of Su ho, whence the name. In Sanscrit books it is 篤魯瑟劍 *tu lu se kien*.

SU KUNG [7th cent.]:—Now this drug is brought from Si yü (Western Asia) and K'un lun [Pu lu Condore. *See* App. 171]. It is of a purplish red colour, resembles the true *tsz' t'an*,[82] is hard, very fragrant, and heavy like a stone. When of a good quality the ashes left after burning it are of a white colour. [The author seems to speak of a wood.]

SU SUNG [11th cent.]:—There is now in Kuang chou a *su ho hiang* which is a kind of *su mu*,[83] not fragrant at all. But the *su ho hiang* which is used in medicine is a substance of the consistence of an ointment, very fragrant and hot. Regarding this T'AO HUNG-KING states that it is the excrement of lions. According to the History of the Liang dynasty [A.D. 502-557] the *su ho hiang* comes from Chung T'ien chu (Mid India), and is not an original product but is prepared by mixing together and boiling several fragrant things. Others say that it is a natural drug collected by the people of Ta Ts'in, who by boiling the sap prepare a fragrant substance like an ointment

---

[82] 紫檀. Now the name of a precious wood—*Dalbergia*?

[83] 蘇木 Sapan-wood.

and sell it to the traders from all countries.[84] When this
drug reaches China it is not very fragrant.

LI SHI-CHEN :—The *Huan yü chi* [10th cent.] mentions
the 蘇合油 *su ho yu* (oil) as produced in An nan (Annam)
and San fo ts'i (Eastern Sumatra), by a tree which exudes
this substance. It is used in medicine. The best sort is
a thick liquid without sediment. The *Hiang pu* [a treatise on
perfumes, 11th cent.] says that the *su ho yu* is produced in
the country of the Ta shi (Arabs) and is a kind of *tu nou
hiang*.[85] The *Meng k'i pi t'an* [11th cent.] says :—The *su ho
hiang* is of a red colour like a hard wood, whilst the *su ho yu*
is a viscid resin which is commonly used.

As to the etymology of the term *su ho yu*, *su* is a
fragrant Labiate plant—the *Perilla ocimoides* [see 67], *ho* =
to unite, to mix, *yu* = oil.

WILLIAMS [*Chin. Commercial Guide*, 101] states :—Rose-
maloes, *su ho yu*, is a thick, scented, gummous oil of the
consistence of tar, obtained by pressure from beans, and
called *gurmala* in Bombay ; it is brought from Persia and
Upper India to Bombay, and when good has a pearly
appearance. It is used in making plasters among the Chinese,
and frequently also as a purge.—WILLIAMS is correct in
identifying the *su ho yu* with rose-maloes, but he is mistaken
as to the origin of the drug *gurmala*, which, according to
DYMOCK [*Veget. Mat. Med. of W. India*, 209] is *Cassia
fistula*.

Rassamala is the Javanese and Malay name for *Liquidambar altingiana*, Bl., a lofty and most valuable tree of Java,
with a fragrant wood which yields from incisions in the

---

[84] This passage is taken from the History of the Later Han [article on
Ta Ts'in (Roman Empire)] and refers to the end of the first century, when
the name *su ho* first appears. *See* Dr. F. HIRTH, *China and the Roman
Orient*, p. 42.

[85] 篤耨香. Noticed *P.*, XXXIV, 55, as a *Styrax*, like resin of Cambodja.

bark a honey-like, sweet-scented resin, which hardens by exposure to the air. The term rose-maloes is probably derived from the above name, but the drug so called is, as the late D. HANBURY has conclusively shown [*Sc. pap.*, 143], imported into Bombay from Aden, the Persian Gulf and the Red Sea, being probably brought thither from Alexandria. He has also established by comparison its identity with the substance known as Liquid Storax, obtained from *Liquidambar orientalis*, L., in Asia minor.—The *su ho yu* procured from a Chinese drug-shop at Peking was of the consistence of tar, of a light gray colour and scentless. It was sent to Professor FLÜCKIGER and proved to be Liquid Storax. GARCIA AB ORTA [writing at Goa, 1534-63] says that *Roça malha* is the Chinese name for Liquid Storax. Very probably the Javanese name *rossamala* (rose-maloes) was originally, and is still, applied to the Storax obtained from *Liquidambar altingiana* and other trees of S.E. Asia. *See* FLÜCK. & HANB., *Pharmacographia*, 247. MARTINI [*Atlas sinensis*, p. 25, written 200 years ago] states :—Regnum Annam, oleum illud seu liquor suavissimus quam Lusitani *rosamaliam* vocant, hic stillet ex arboribus. This may be the *Amyris ambrosiaca* in LOUREIRO's *Fl. cochin.* [283]:—Cochinchinese balsamum quod ex arbore agresti modice stillat, colore cinereo et fragrantia eximia non multum differt a styrace liquida, quæ Liquidambar a Linnæo vocatur.

I can make nothing of the name given in the *P.* as a Sanscrit name of the *su ho yu.*

Comp. also P. SMITH, 187 :—Rose-maloes.

There is another kind of Storax, mentioned in ancient Chinese records on foreign countries as a product of Southern and Western Asia, termed 安息香 *An si hiang.* *P.*, XXXIV, 52. *T.*, CCCXVI.

The name *An si* was applied in the 2nd cent. to the

kingdom of the Parthians in W. Asia. *An si* properly means "tranquility," and, as this perfume is reputed for expelling evil spirits, LI SHI-CHEN tries to explain the name in this way. He adds that the Sanscrit name of the drug is 拙貝羅香 *cho (ku) pei lo hiang*. It seems that in the accounts of foreign countries, as found in the Histories of the Chinese Dynasties, the *An si hiang* is not mentioned before the 7th century. It is frequently noticed as a product of Western Asia and Arabia, as well as of Siam, Sumatra and Cochinchina, in the Ming period [15th cent.]. See my article on the Arabs, pp. 19, 20.

SU KUNG [7th cent.]:—The *An si hiang* comes from the country of the western barbarians. It resembles the *sung chi* (common resin), is of a yellowish black colour and appears in lumps. The fresh resin is soft.

LI SÜN [8th cent.]:—It is produced in the countries of the Southern Sea and in Po sz' [Persia]. It is the resin of a tree, resembles the gum of the peach tree and is collected at the end of autumn.

CHANG YÜ-HI [11th cent.]:—The *Yu yang tsa tsu* [8th cent.] states:—The tree which yields the *An si hiang* grows in Po sz' [Persia]. It is also called 辟邪樹 *p'i sie shu* (tree which drives away evil). It grows from twenty to thirty feet high. Its bark is yellowish black. The leaves spread out into four corners and do not fall off in the cold. It blossoms in the 2nd month. Yellow flowers. The heart of the flower is green. It does not produce fruit. When the bark of the tree is scraped off the resin appears, like sugar. It is called *An si hiang*. In the 6th or 7th month, when it has become hard, it is fit for being burned as incense. It has the property of expelling all sorts of evil things.

LI SHI-CHEN:—This drug is found in An nan, San fo ts'i (Eastern Sumatra), and other foreign countries, and is much used in China as a perfume.   There is a preparation called 安息油 *An si yu*, a mixture of various fragrant substances.

WAN KI [16th cent.]:—By burning the true *An si hiang* incense rats can be allured[86] (?).

WILLIAMS, *Chin. Commercial Guide,* 93 :—安息香 *ngan si hiang, i.e.* the quieting perfume, *Benzoin* or *Benjamin,* the concrete juice of the *Styrax benzoin,* which is cultivated in Borneo and Sumatra.   It is almost tasteless, but when rubbed or heated gives off an extremely agreeable odour.

The drug *An si hiang* which I procured from S. China was *Benzoin,* but that which is sold at Peking under the above name in the perfume-shops is a composition of various perfumes.

314.—詹糖香 *chan t'ang hiang.*   P., XXXIV, 55. *T.,* CCCXVI.

This drug is mentioned in the *Pie lu,* but only the name.

T'AO HUNG-KING :—It is produced in Tsin an [in Fu kien, App. 354] and Ch'en chou [App. 12].   It is difficult to procure this drug in a pure state, for it generally contains particles of the bark and the dirt of the wood grub. The soft drug is the best.   It is used as a perfume but not employed in medicine.

SU KUNG [7th cent.]:—The *chan t'ang* tree resembles an orange tree.   The twigs and the leaves when burnt emit a fragrant smell.   The drug resembles sand sugar and is black.[87]   It is produced in Kiao and Kuang [in Kuang tung, App. 132, 160] and farther south.   It grows also in Tsin an [*v. supra*] and is commonly used.

" 燒之能集鼠者爲眞.
" 詹糖樹似橘煎枝葉爲香似砂糖而黑.

LI SHI - CHEN :—The name *chan t'ang hiang* means
" viscid sugar perfume." Its flowers are also fragrant.
The smell resembles that of the *mo li hua* (*Jasminum
grandiflorum*).

The *Ch.* [XXXIII, 56] figures, sub *chan t'ang hiang*, a
tree with oblong, pointed leaves and berries [red]. It is
said to grow in Hu nan and to be a kind of camphor tree
with fragrant leaves.

315.—蘗木 *po mu*.　P., XXXVa, 1.　T., CCCV.

Pen king :—*Po mu*. The root is called 檀桓 *t'an huan*.
Taste bitter. Nature cold. Non-poisonous.—Subsequent
writers say that the bark of the tree is officinal.

Pie lu :—Other name : 黃蘗 *huang* (yellow) *po*. The
po mu grows in Han chung [S. Shen si, App. 54], in
mountain-valleys, also in Yung ch'ang [W. Yün nan, App.
426].

T'AO HUNG-KING :—The drug produced in Shao ling
[in Hu nan, App. 280], which is light, thin and of a dark
colour, is the best. That from Tung shan [*see* App. 375] is
thick and of a pale colour. The *t'an huan* [said to be the root
of the *po mu*] is, according to the Taoists, a mushroom. There
is one kind of *po mu*, a small tree resembling the pomegranate,
with a bitter yellow bark. It is called 子蘗 *tsz' po*. The
bark is useful in curing a sore mouth. Another sort, likewise
a small tree, is covered with spines. Its bark is also yellow
and used for the same purpose as the *tsz' po* bark.

SU KUNG [7th cent.] :—The *tsz' po* is also called 山石榴
shan shi liu (mountain pomegranate). It resembles the *nü
cheng* [*Ligustrum*. See 342]. Its bark is white, not yellow.
This is the 小蘗 *siao* (small) *po* [*see further on*]. The
spiny *tsz' po* [mentioned by T'AO HUNG-KING] is called 刺蘗
ts'z' (spiny) *po*. It is different from the *siao po*.

CH'EN TS'ANG-KI [8th cent.]:—The 檀桓 *t'an huan* is the root of a hundred years' old *po* tree. It resembles the *t'ien men tung* [*see* 176], is 3 or 4 feet long and has on one side small lateral roots (tubers?) called 丨丨芝 *t'an huan chi* (fungus).

CHANG YÜ-SI [11th cent.]:—In the illustrated Herbal of Shu (Sz ch'uan) it is stated :—The 黄蘗 *huang* (yellow) *po* tree is 30 and more feet high. Leaves resembling those of the *Wu chu yü* [*Boymia* or *Evodia*. See 291] and the *tsz' ch'un* (*Cedrela?*). They do not fall off in winter. The outer bark is white, the inner dark yellow. Its root produces nodular masses resembling the *fu ling* which grows beneath fir trees [*Pachyma*. See 350]. This tree now grows in Fang chou [in Hu pei, App. 35], Shang chou [in Shen si, App. 278] and Ho chou [in Sz ch'uan, App. 69*b*] in mountain-valleys. The bark is tight, two-tenths or three-tenths of an inch thick, and of a bright yellow colour. It is gathered in the 2nd and 5th months and dried in the sun.

SU SUNG [11th cent.]:—It is a common tree. The best drug comes from Shu (Sz ch'uan). The flesh [of the bark] is of a dark [yellow] colour.

LI SHI-CHEN :—The name *huang po* is commonly but erroneously written 黄柏 *huang po* (yellow *Thuja*). The lateral tubers, which according to CH'EN TS'ANG-K'I grow on the root (*t'an huan*), are a kind of fungus.

Ch., XXXIII, 20 :—*Po mu* or *huang po*. Rude drawing of a tree with pinnate leaves. It grows in Hu nan and is used for dyeing.

LOUR., *Fl. cochin.*, 525 :—*Pterocarpus flavus*. Sinice : *hoam pe mo.* Arbor magna, in sylvis Sinensibus, cortice glabro, intus succoso, flavo, amaro. Folia impari-pinnata. Flos flavus papilionaceus. Florem non vidi, nisi pictum. Legumen breve, compressum, seminibus 2-3. Virtus corticis :

resolvens, vulneraria. Decocto corticis tinguntur serica colore flavo permanente, nec injucundo. LOUREIRO refers to RUMPH. *Amb. tab.*, 117. *Malaparius.*

TATAR., *Cat.*, 9 :—黄栢 *huang po.* Cortex *Pterocarpi flavi* [TATARINOV evidently relies upon LOÙREIRO].—P. SMITH, 180.—HANB., *Sc. pap.*, 266.

The drug *huang po* of the Peking drug-shops is a yellow bark and very bitter.

Cust. Med., p. 10 (75):—*Huang po* bark exported 1885 from New chwang 322.81 piculs,—p. 192 (131), from Ning po 125.87 piculs.

Dr. A. HENRY writes me :—The 黄柏 *huang po* of which the bark is used here (at I chang) as a drug is *Phellodendron amurense*, Rupr. The northern *huang po* is probably the same.

Phellodendron amurense has a yellow inner bark. *See* RUPRECHT's original description of this tree [first discovered by R. MAACK, on the Amur river] in *Mél. biolog. Acad.* (1857), p. 526.

The 蘗木 figured in the *Phon zo* [LXXXII, 2, 3] and the *santo* or *kivada*, sinice: 黄蘗, in the *Kwa wi* [100] have not been identified by FRANCHET. According to the late Dr. GEERTS [see my paper, *Bot. quest. conn. Export Trade China*, p. 4] the latter Chinese name in Japan is applied to *Evodia glauca*, Miq. The bark of this tree is used there for tinctorial purposes. This identification has been confirmed at London, where authentic herbarium specimens of the tree with the bark has been received from Japan. See *Pharmac. Journ.* (1888), p. 785.[88]

[88] *Evodia glauca* is not the source of the yellow bark in China or Japan : and the observations of GEERTS, of the Japanese, and of the *Pharmac. Journal* are wrong. The mistake arose from the fact that *Evodia glauca* and *Phellodendron amurense* are trees with very similar foliage [the fruit is very different]. Mr. E. M. HOLMES, of the Pharmeceutical Museum, London, was convinced, by comparing specimens of the commercial bark with herbarium specimens of *Evodia* and *Phellodendron* at Kew, that the bark of China and Japan is the product of one tree only, namely *Phellodendron amurense*, Rupr.—A. HENRY.

Phon zo, LXXXII, 4, 5 :—檀桓. A yellow root figured.

SIEB., *Œcon.*, 260 :—*Zanthoxylon kibada*. Sinice : 蘗木. Lignum ad luteum tingendum maxime æstimatum. Cortex habetur febrifugus.—According to ITO KEISKE [FRANCH. & SAV., *Enum. Jap.*, II, 693] *kiwada* is the Japanese name for *Phellodendron amurense*.

Thus, it seems, the Chinese names *po mu* and *huang po*, in China as well as in Japan, are applied to both the *Evodia glauca* and the *Phellodendron amurense*. Both belong to the same order *Rutaceæ*.

The 小蘗 *siao* (small) *po* [*v. supra*] has a special article in the *P.* [XXXVa, 6].

T'AO HUNG-KING calls it 子蘗 *tsz' po* and says it is a small tree resembling the *shi liu* (pomegranate). Bitter yellow bark. Another kind, which is provided with spines, has likewise a yellow bark.

SU KUNG [7th cent.]:—The *siao po* grows between rocks in the mountains. That which grows in Siang yang [in Hu pei, App. 306], east of the Hien shan[89] mountain, yields the best drug. It is also called 山石榴 *shan shi liu* (mountain pomegranate), for its branches and leaves are not unlike those of the pomegranate, but the flowers are different. The fruit is small, black and globular like that of the *niu li tsz'* [*Rhamnus*. See 341] or the *nü cheng tsz'* [*Ligustrum*. See 342]. But the bark of the tree is white, not yellow as T'AO HUNG-KING asserts. It is now kept in store in the Court of Sacrifices.[90] The spiny *tsz' po*, noticed by the same author, is the 刺蘗 *ts'z' po* (spiny *po*). It has small leaves and differs from the *siao po*.

[89] 睍山 to the S.E. of Siang yang fu.

[90] 太常

CH'EN TS'ANG-K'I [8th cent.]:—All the various sorts of the *po* tree have a yellow bark, and likewise the *siao po* which resembles the pomegranate. It has red fruits like those of the *kou k'i* [*Lycium*. See 345], pointed at both ends. The people cut off the branches and use them for dyeing yellow. SU KUNG is wrong in stating that the berries of the *siao po* are black and round. He had probably another plant in view.

LI SHI-CHEN:—The *siao po* is a small mountain-tree. Its outer bark is white and the inner bark yellow like that of the *po*, but thinner. The name *shan shi liu* is also applied to the *kin ying tsz'* (a Rose) and the *tu yüan hua* [*Rhododendron*. See 155].

The *siao po* is not figured in the *Ch.*—The name 刺蘗 is given there [XXXVII, 43] as a synonym for 黃蘆木 *huang lu mu*. The figure, a spiny tree, is possibly intended for a *Berberis*.

SIEB., *Icon. ined.*, I:—*Berberis chinensis*, Desf. Sinice: 小蘗

Phon zo, LXXXII, 5, 6:—Same Chinese name. Figure not identified by FRANCHET.

HOFFM. & SCHLT., 85:—Same Chinese name. *Berberis Thunbergii*, DC.

I may observe that the inner bark of the European *Berberis vulgaris*, and also the root, afford a bright yellow dye.

316.—厚朴 *hou p'o*. P., XXXVa, 7. T., CCCVIII.

Pen king:—Hou (thick) *p'o*. The bark of a tree. Taste bitter. Nature warm. Non-poisonous.

Pie lu:—The *hou p'o* is also called 厚皮 *hou p'i* (thick bark) and 赤皮 *ch'i pi* (red bark). The tree is called 榛 *chen*,[91]

---

[91] This is properly the name for the hazel-nut. See *Classics*, 496.

the fruit is 逐 折 *chu che* [*see further on*]. The *hou p'o* is produced in Kiao chi (Cochinchina) and Yüan kü [in Shan tung, App. 415]. The bark is gathered in the 3rd, 9th and 10th months.

In the *Kuang ya* it is called 重 皮 *chung p'i* (heavy bark).

T'AO HUNG-KING:—Now the drug produced in Kien p'ing and I tu [both in Hu pei, App. 139, 104] is considered the best. It (the bark) is very thick and the flesh is of a purple colour. That with a thin, white outer coat is not valued. It is much used in common prescriptions. The Taoists do not employ it.

SU SUNG [11th cent.]:—It is common in the mountains of Lo yang [in Ho nan, App. 201], in Shen si [App. 284], Kiang and Huai [Kiang su and An hui, App. 124, 89], Hu nan [App. 83] and Shu ch'uan [Sz ch'uan, App. 292]. That from Tsz' chou and Lung chou [both in Sz ch'uan, App. 366, 210] is also good. The tree is from 30 to 40 feet high, and from 1 to 2 feet in diameter. Leaves like those of the *hu* (*Quercus obovata*, large leaves) and persistent. Flowers red, fruit green. The bark is very scaly and shrivelled, thick, of a purple colour and succulent. That which is thin and white is not fit for use.

K'OU TSUNG-SHI [12th cent.]:—It grows also in I yang hien [in Ho nan, App. 108] and in Shang chou [in Shen si, App. 278], but the drug there is thin, of a pale colour and far inferior to that from Tsz' chou [*v. supra*], which is thick, of a purple colour and oily (resinous).

LI SHI-CHEN:—The *p'o* tree has the outer bark white and the flesh purple-coloured. The leaves resemble those of the *hu* [*v. supra*]. It blossoms in the 5th or 6th month. Small flowers. The fruit resembles that of the *tung ts'ing* [*Ilex.* See 342], is green at first, and red when ripe. It

60

has kernels. The fruit is gathered in the 7th or 8th month and is of a sweet pleasant taste.

Ch., **XXXIII**, 30 : — *Hou p'o*. The figure represents only leaves. A *Magnolia* may be intended.—See also **XXXVIII**, 4, *t'u* (native) *hou p'o*.

TATAR., *Cat.*, 8 :—*Hou p'o*. Cortex ?

HANB., *Sc. pap.*, 266 :—*How puh*, a rough, thick bark of a bitterish pungent aromatic taste. *Magnolia hypoleuca*, S. & Z.—For this identification HANBURY relies upon HOFFM. & SCHLT., 355. He does not mention the colour of the bark. The drug *hou p'o* obtained from a Peking druggist's shop was of a reddish brown colour.—P. SMITH, 142.

Father A. DAVID [1869] states [*Nouv. Arch. Mus. d'Hist. nat.*, IX, Bull. 28] that the Chinese in Sz ch'uan cultivate *Magnolias* not for the flowers but for the bark of the trees, which is much prized by the Chinese as a medicine. They call it *ho po*. In his *Journ. trois. royage Emp. Chin.* [II, 360] the same author notices that in 1873 he saw in Kiang si a splendid plantation of a *Magnolia* with very large leaves, the same as he had previously met with in Sz ch'uan. It much resembled the American *M. macrophylla*. The bark is sold at very high prices as a medicine.

PARKER [*Chin. Rev.*, XI, 22] mentions the drug *hou p'o* in Sz ch'uan.

HENRY, *Chin. pl.*, 120 :—*Hou p'o*, *Magnolia sp. nova*,[92] the bark of which is a famous Chinese drug, largely exported from Sz ch'uan. Two varieties, one with red and the other with white flowers, are cultivated in the mountains of the Pa tung district. The leaves are very large. This seems to be the tree figured in *Ch.* [XXXVIII, 4] with the name

---

[92] The *Hou-p'o* tree of Hupeh turned out at Kew to be *Magnolia hypoleuca*, S. & Z., and not a new species, as was at first supposed.— A. HENRY.

t'u hou p'o. The description in *Ch.* [XXXIII, 30], *hou p'o,* seems to point to another tree.

Cust. Med., p. 72 (90):—*Hou p'o* (bark) exported 1885 from Han kow 117.52 piculs,—p. 60 (36), from I chang 3.15 piculs, and p. 62 (45), of *hou p'o* flowers 8.30 piculs.

Kwa wi, 86:—厚朴 Japonice: *tan pa cou, fonoki. Magnolia hypoleuca,* S. & Z.—Same identification in SIEB., *Icon. ined.,* I, and *Phon zo,* LXXXII, 7, 8.—*M. hypoleuca* has not been recorded from China.

SIEB., *Œcon.,* 272:—*Magnolia hoo noki* [i.e. *M. hypoleuca*]. Sinice: 浮爛羅勤. E ligno conficiuntur gladiorum vaginæ, carbones ad lævigandum liguum æsque adhibentur.

The above name is pronounced in Chinese *fou lan lo le.* In the *P.* [XXXVa, 11] there is a short note by CH'EN TS'ANG-K'I [8th cent.] regarding this tree. The author states that it grows in the kingdom of K'ang.[93] Its bark resembles the *hou p'o* bark. Taste sour. Non-poisonous. It is used a medicine.

The *Pie lu* first says that the fruit of the *hou p'o* tree is called 逐折 *chu che* [v. supra] but afterwards the same work gives the following short and obscure account of the same drug:—The *chu che* cures ulcers, strengthens the breath and clears the eyes. It is also called 百合 *po ho* [properly a name for lilies] and 厚實 *hou shi.* It grows between trees and has a yellow stem. In the 7th month it bears a black fruit like the Soy-bean.

T'AO HUNG-KING says:—The name *chu che* is also applied to the fruit of the *tu chung* [see 317].

---

[93] 康國, an ancient name for Samarkand.

317.—杜仲 *tu chung*. *P.*, XXXVa, 11. *T.*, CCCVII.

Pen king:—*Tu chung*, 思仙 *sz' sien*. The bark of the tree is officinal. Taste pungent. Nature uniform. Nonpoisonous.

Pie lu:—Other name : 思仲 *sz' chung*. The *tu chung* grows in Shang yü [*v. infra*] in mountain-valleys and in Shang tang [in Shan si, App. 275] and Han chung [S. Shen si, App. 54]. The bark is gathered in the 2nd, 5th, 6th and 9th months.

WU P'U [3rd cent.] calls it 木綿 *mu mien* (properly a name of the Cotton tree).

T'AO HUNG-KING :—The Shang yü mentioned in the *Pie lu* is not the district of this name in Hui ki [Che kiang, App. 98] but the Yü situated near Kuo[94] in Yü chou [Ho nan, App. 413]. The drug which is now used comes from Kien p'ing and I tu [both in Hu pei, App. 139, 104]. It resembles the bark of the *hou p'o* [*Magnolia*. See 316]. When broken it shows white filaments like floss silk.

HAN PAO-SHENG [10th cent.]:—It grows in mountain recesses. It is a tree about 30 feet high. Leaves like those of the *sin i* [*Magnolia*. See 306].

SU SUNG [11th cent.]:—It is now produced in Shang chou [in Shen si, App. 278], in Ch'eng chou [in Kan su, App. 18] and in Hia chou [in Hu pei, App. 64], on high mountains. In its leaves it is akin to the *che* (*Cudrania triloba*). Its bark when broken shows numerous fibres like white floss silk. In Kiang nan they call it 檰 *mien*. The young leaves are eaten and are known under the name of │芽 *mien ya*. The flowers and the fruit are of a bitter

---

[94] 虞 Yü and 虢 Kuo, near the Yellow River, mentioned in the *Ch'un ts'iu*.

and harsh taste. They are used in medicine. The wood is fit for making pattens.

Ch., **XXXIII**, 18 :—*Tu chung.* Rude drawing.

TATAR., *Cat.*, 21 :—*Tu chung.* Cortex tenuis arboris.— I have seen the drug *tu chung.* It is a bark. As P. SMITH [94, sub *Evonymus*] correctly describes, on breaking it, and drawing the fractured edges asunder, a delicate, silvery, silky fibre is seen, which may be drawn out to the length of almost an inch without breaking.

HENRY [*Chin. pl.*, 477] states that *tu chung* in Hu pei is a new species of *Ulmus.* But subsequently he informed me that he had sent fruits of the *tu chung* to Kew, where the botanists considered it to be an *euphorbiaceous* plant.[95]

Cust. Med., p. 72 (101):—*Tu chung* exported 1885 from Han kow 1,707 piculs,—p. 60 (39), from I chang 7 piculs.

Phon zo, **LXXXII**, 14,15 :—杜仲. Japonice : *totchiou. Evonymus Sieboldianus*, Bl.—**HOFFM.** & **SCHLT.**, 238, same Chinese name, *Evonymus japonicus*, Thbg.—**SIEB.**, *Icon. ined.*, III, same Chinese name, *Evonymus totsju* (i.e. *Sieboldianus*).

SIEB., *Œcon.*, 269 :—*Evonymus japonicus*, Thbg. 杜仲. Japonice : *masaki.* Pro sepibus vivis.

The plant which yields the drug *tu chung* has been described and figured in **HOOKER'S** *Icones. Plant.* [tab. 1950] under the name of *Eucommia ulmoides* [a new genus]. The most singular feature about the plant is the extraordinary abundance of an elastic gum in the bark, the leaves, the petioles and the pericarp; any of these snapped across, and the parts drawn asunder, exhibit the silvery sheen of innumerable threads of this gum.

318.—漆 *ts'i.* Varnish yielded by the Chinese Lacquer

[95] Specimens sent from Hu peh of the true *Tu chung* tree, *i.e.* the one affording the peculiar bark used as a drug, were at first thought to be a new species of *Ulmus*, but OLIVER now thinks differently and has described and figured the tree as *Eucommia ulmoides*, Oliv. HOOKER, *Ic. Plantarum*, table 1950.—A. HENRY.

tree, *Rhus vernicifera*, L.　*P.*, XXXV*a*, 17.　*T.*, CCLVII.
Comp. *Classics*, 517.

Pen king :—Ts'i. The Index of the *Pen king* has 乾漆
kan ts'i (dried varnish). It is stated to be non-poisonous.
But subsequent writers say it is poisonous. The leaves and
fruit of the tree are also used in medicine.

*Pie lu :—*The *kan ts'i* grows in Han chung [S. Shen si,
App. 54] in mountain-valleys. It is collected after midsummer and dried.

Cust. Med., p. 380 (523):—漆乾 *ts'i kan* exported 1885
from Canton 0.72 picul.

Further particulars in another part.

319.—梓 *tsz'.*　*P.*, XXXV*a*, 20.　*T.*, CCXL.

Comp. *Rh ya*, 293, *Classics*, 508.

Pen king :—T'sz'. The white rind of the tree, 梓白皮
tsz' pai p'i, is officinal. Taste bitter. Nature cold. Non-
poisonous.

*Pie lu :—*The *tsz' pai p'i* (white rind of the *tsz'*) is
produced in Ho nei [S.E. Shan si, App. 77] in mountain-
valleys.

T'AO HUNG-KING :—This drug is the rind of the *tsz'* tree.
There are three sorts. That employed for medical use ought
to be firm, not rotten.

The *Ts'i ming yao shu* [5th cent.], Chap. V, under the
head of 楸梓, states :—That of a white colour (white bark
or timber) and with horns (long slender capsules) is called
梓 *tsz'*,—that which resembles the *ts'iu*, and likewise has
horns is the 角楸 *kio* (horn) *ts'iu*, also called 子楸 *tsz' ts'iu*
(*i.e.* which bears fruits),—that of a yellow colour, and which
does not bear fruit, is the 柳楸 *liu* (willow) *ts'iu*. As it has
a yellow wood it is also called 荆黄楸 *King huang ts'iu*,
yellow *ts'iu* from King chou [Hu pei].

Ta Ming [10th cent.]:—There are many sorts of *tsz'*. Only the bark of the 楸梓 *ts'iu tsz'* is used in medicine.

Su Sung [11th cent]:—This tree is common in Mid China. Its timber is much used for building palaces, temples, and pavilions in gardens. It is also much cultivated. It resembles the *t'ung* [*Paulownia*. See 320], but the leaves are smaller. Purple flowers. The *Rh ya* [293] identifies the *tsz'* with the 椅 *i*. Kuo P'o says it is the same as the 楸 *ts'iu* (*Catalpa*). It is also mentioned in the *Shi king*. One kind is called 鼠] *shu tsz'* or 櫻 *yü* [*v. infra*, 341, and *Rh ya*, 260]. This is also a kind of *ts'iu*.

Wang ki [16th cent.]:—The ancient dictionary *Shuo wen* says:—The 椅 *i* and the 梓 *tsz'* are the same. It also says that the 梓 is the same as the 楸 *ts'iu*, and further on that the 楸 is identical with the 檟 *kia*. We may therefore assume that 椅 *i*, 梓 *tsz'*, 楸 *ts'iu* and 檟 *kia* are four names designating the same tree. The *tsz'* tree has long slender horns (capsules) like quills. Late in winter, when the leaves have fallen off, the horns are still seen hanging on the tree. The fruit is called 豫章 *yü chang*.[96] [This latter statement is from the *Ku kin chu*, 4th cent.].

Li Shi-chen:—The *tsz'* tree is common in China. There are three sorts. That with a white wood is called 梓 *tsz'*, that with a red wood is 楸 *ts'iu*, that with a beautifully grained wood is 椅 *i*. There is a small sort of *ts'iu* which is called 榎 *kia* [see *Rh ya*, 292].

The names *tsz'* and *ts'iu* are now applied to *Catalpa*, of which two Chinese species are known from China, viz. *C. Bungei*, C. A. Mey., and *C. Kæmpferi*, S. & Z. For further particulars see *Bot. sin.*, II, 508.

---

[96] This is properly an old name for the Camphor tree. See *Classics*, 518.

After the *tsz'* the *P.* [XXXV*a*, 21] notices the 楸
ts'iu tree. LI SHI-CHEN says that there are two kinds—the
common *ts'iu*, which yields an excellent timber, and for which
he refers to the *tsz'*, and the 刺 楸 *ts'z'* (thorny) *ts'iu*, which
is described and figured in the *Kiu huang Pen ts'ao* [LIV,
31]. A large tree. Its bark is greenish white with yellowish
white spots. The trunk and the branches are covered with
large thorns. The leaves resemble those of the common
ts'iu but are thinner and of a sweet taste. In their young
state they can be eaten when boiled.—The drawing of the
Kiu huang is reproduced in the *Ch.* [XXXIV, 16].

Kwa wi, 89 :—*Acanthopanax ricinifolium*, Seem. Japonice : *favodara* ; sinice : 刺 楸. The Japanese drawing
agrees with the drawing in the *Ch.*

HENRY, *Chin. pl.*, 79 :—*Ts'z' ts'iu*. *Acanthopanax
ricinifolium*, a large, very thorny tree with leaves resembling
somewhat the *Catalpa*, whence the Chinese name.

320.—桐 *t'ung*. *P.*, XXXV*a*, 23. *T.*, CCXXXVII.

Comp. *Rh ya*, 309, *Classics*, 515.

Pen king :—*T'ung*. In the Index of the *Pen king* :—
桐葉 *T'ung ye* (leaves). The leaves, bark and the flowers are
used in medicine. Taste of the leaves bitter. Nature cold.
Non-poisonous.

Pie lu :—*T'ung ye*. Produced in T'ung po [in Ho nan,
App. 379] in mountain-valleys.

The *t'ung* is the *Paulownia imperialis*, S. & Z. For
further particulars see *Bot. sin.*, II, 515.

Cust. Med., p. 150 (180) :—*T'ung p'i* (bark) exported
1885 from Shang hai 0.33 picul.

321.—棟 *lien.*   *P.,* XXXV*a*, 28.   *T.,* CCLXI.

Pen king :—Lien. In the Index of the *Pen king :—*
| 實 *lien shi* (fruit). The fruit, root, bark, flowers and
leaves are officinal. The taste of the fruit, root and bark
is bitter. Nature cold. Slightly poisonous.

*Pie lu :—*The *lien shi* grows in King shan [in Hu pei,
App. 149] in mountain-valleys.

The character *lien* is explained in the *Shuo wen* by "name
of a wood."

T'AO HUNG-KING :—The *lien* is a common tree. On the
5th day of the 5th month the people gather the leaves and
wear them in their girdles. These leaves are said to avert
evil.

SU KUNG [7th cent.]—There are two sorts—the female
and the male *lien ;* the male does not bear fruit. The root
is of a red colour, poisonous, and when taken internally it
provokes intensive vomiting and even causes death. The
female, fruit-bearing *lien* has a white root which has only
slightly poisonous properties. Only the female root is used
in medicine.

SU SUNG [11th cent.]:—The best *lien shi* (fruit) comes
from Shu ch'uan [Sz ch'uan, App. 292]. The tree is more
than 10 feet high. The leaves (leaflets) are close together
like those of the *huai* (*Sophora*) but longer. It flowers
in the 3rd and 4th months. The flowers are reddish purple
and very fragrant. The fruit is a small ball, green at first and
yellow when ripe. It is gathered in the 12th month.
There is no fixed time for taking out the root.

The *Rh ya i* [12th cent.] gives a similar account
of the *lien,* and adds :—The fruit is like a little bell. It is
yellow when ripe. Its popular name is 苦棟子 *k'u* (bitter)
lien tsz' and also 金鈴子 *kin ling tsz'* (golden bell). It is

61

mentioned in *Huai nan tsz'*,—*lien*, the tree of the 7th month. The people of Ch'u [Hu kuang, App. 24] are accustomed to hold a festival on the 5th day of the 5th month in commemoration of K'ü Yüan's suicide [B.C. 314]. Bamboo-sprouts and rice-cakes enveloped in the leaves of the *lien*, with silken thread of five colours tied around, are cast into the river to propitiate the water-spirits. The women put *lien* leaves in their hair, etc.[97] According to the *Feng su t'ung* [2nd cent.] the phœnix and the unicorn eat the *lien*, but the dragon abhors it.

Li Shi-chen :—The *lien* tree grows very rapidly. In three or four years it is fit for beams. The fruit resembles a round jujube. The best is produced in Sz ch'uan.

Ch., XXXIII, 45 :—*Lien*. Good drawing representing a *Melia*. The description in the *P.* agrees.

Lour., *Fl. cochin.*, 329 :—*Melia azedarach*, L. Sinice : *xun (shun) lien*.

Bridgm., *Chrest.*, 441 (44) :—*Melia*, Pride of India, at Canton 森木 *shen mu* or 苦楝 *k'u lien*. Same identification in Parker's *Canton plants* [169].

Tatar., *Cat.*, 34 :—楝 樹 *lien shu*. Arbor?—*Ibid.*, 15 :—川 楝 子 *Ch'uan lien tsz'* (*lien* from Sz ch'uan) and [59]:—金林子 *kin lin tsz'* [the second character is a mistake for 鈴 *ling*]. Both drugs are identified with Fructus *Mespili japonicæ.*—Gauger, 54 :—*Ch'uan lien tsz'*. The fruit figured and described. Gauger means it is a *Diospyros*.

The drugs which I obtained under the names of *Ch'uan lien tsz'* and *kin ling tsz'* from a Chinese drug-shop at Peking were, undoubtedly, the dried fruits of a *Melia*, yellowish brown ; five-celled stony endocarp. But the *Ch'uan lien* [in Thibetan *barura*] sold in the Thibetan

---

[97] This is taken from the *Ts'i hiai ki* or Record of Marvels, 5th cent.

drug-shop at Peking was *Terminalia belerica* [determined at Kew].

HANB. [*Sc. pap.*, 244] figures and describes the Chinese drug *Ch'uan lien tsz'* from Shang hai. He is not sure whether it is *Melia*, for the stony endocarp was from six to eight celled.

P. SMITH, 145, 146, sub *Melia* and *Medlar*.

According to PARKER [*China Rev.*, X, 169] *k'u lien tsz'* in Sz ch'uan is *Melia azedarach*, L.

HENRY, *Chin. pl.*, 240 :—*Lien shu. Melia azedarach.*— *Ibid.*, 241 :—*K'u lien tsz'.* At I chang this name is applied to *Melia azedarach* and also sometimes to *Picrasma quassioides*, Benn.—The *k'u lien tsz'* or *Ch'uan lien tsz'*, a drug largely exported from Sz ch'uan, is the fruit of a species of *Melia* not yet identified.[98] [Comp. *Cust. Med.*, No. 251:— *Ch'uan lien tsz'. Melia Toosendan*, S. & Z.]

Melia azedarach is a common tree in Mid and South China. See *Ind. Fl. sin.*, I, 113. It is known from India, where the root, bark and fruit of the tree are used medicinally, that the fruit has poisonous properties. It has very fragrant flowers. It is a highly valued timber at Canton.

Cust. Med., p. 76 (134) :—*Ch'uan lien tsz'* (fruit) exported 1885 from Han kow 527 piculs.—*Ibid.*, 368 (375):—*K'u lien tsz'* exported from Canton 3.11 piculs.—*Ibid.*, p. 344 (67):—*K'u lien ken* (root) exported from Canton 0.17 picul,—p. 278 (59), from Amoy 0.03 picul.—*Ibid.*, p. 354 (193):—*K'u lien p'i* (bark) exported from Canton 0.27 picul.

Amœn. exot., 788 :—楝 *den*, it. *ootz*, vulgo *sendam*, aliis *kindeis.* Azadarach Avicennæ.

Kwa wi, 122 :—Same Chinese name. *Melia azedarach.*

---

[98] Now determined as *Melia Toosendan*, S. & Z.—A. HENRY.

Phon zo, LXXXIII, 7, 8 :—棟, *Melia japonica*, G. Don. This species is reduced to *M. azedarach* in the *Ind. Fl. sin.*

SIEB., *Œcon.*, 274:—*Melia azedarach. Sendan.* 棟. E fructibus exprimitur oleum (THBG.), id quod ignoro, sed fructus in variolis, ac corticem in vermibus esse remedium mihi relatum.

322.—槐 *huai.* P., XXXVa, 31. T., CCLV.

Comp. *Classics*, 546.

Pen king:—Huai. In the Index of the *Pen·king :—* | 實 *huai shi* (fruit). Taste of the fruit bitter. Nature cold. Non-poisonous. The flowers, leaves, bark of the trunk, root, and the gum exuded by the tree are all used in medicine.

Pie lu:—The *huai shi* is produced in Ho nan [App. 76] in marshes. It is fit for making sacrificial candles.[99]

TATAR., *Cat.*, 10 :—槐 花 *huai hua.* Flores *Sophoræ japonicæ,* ii, | 角 *huai kio.* Fructus *Sophoræ japonicæ.*

HANB., *Sc. pap.*, 237 :— | 實 *huai shi.* Legumes of *Sophora japonica.*—P. SMITH, 201.

Cust. Med., p. 154 (227):— | 米 *huai mi* (unopened flower-buds) exported 1885 from Shang hai 599 piculs,—*ibid.* (226) :— | 花 *huai hua* (flowers) exported 87 piculs,—p. 196 (178), from Ning po 10.80 piculs.—*Ibid.*, p. 368 (365) :— | 角 *huai kio* (pods) exported from Canton 8.52 piculs,—p. 276 (42), from Amoy 0.47 picul.—*Ibid.*, p. 32 (141):— | 子 *huai tsz'* (seeds) exported from Tien tsin 899 piculs.

Further particulars in another part.

[99] 神 燭.

323.—秦 皮 *ts'in p'i.*   *P.*, XXXV*b*, 1.   *T.*, CCCVI.

Pen king:—*Ts'in p'i* (bark), 樗 皮 *tsin p'i* and 樿 木 *sin mu.* The bark of the tree is officinal.   Taste bitter. Nature slightly cold.   Non-poisonous.

The *Shuo wen* [see *K.D.*] explains the character 樿 by 青皮木 (a tree with a green bark).   HUAI NAN TSZ' says it is of a green colour and is used as a medicine for the eyes.   KAO YU, of the Han dynasty, in commenting upon HUAI NAN TSZ', identifies the *tsin* with the 苦櫪 *k'u* (bitter) *li.*   The *li* is said by some ancient authors [see *K.D.*, and *W.D.*, 537] to be the same as 檪 *li* (an oak).—Regarding the 樿 the *T'ang yün* dictionary says that it resembles the *huai* (*Sophora*).

Pie lu:—Other name : 石檀 *shi t'an.*   The *ts'in p'i* is produced in Lü kiang [in An hui, App. 207] in river-valleys and in Yüan kü [in Shan tung, App. 415] by river-sides. The bark is gathered in the 2nd and 8th months and dried in the shade.

T'AO HUNG-KING :—The popular name of the drug is 樊 槻 皮 *fan kui p'i.*   When steeped in water it yields a bluish indelible ink.

SU KUNG [7th cent.]:—The leaves of this tree are like those of the *t'an* (*Cæsalpinia*) but smaller, wherefore it is also called *shi t'an* [*v. supra*].   As the bark is of a bitter taste the tree is also called 苦 樹 *k'u shu* (bitter tree). The bark shows white spots and is not coarsely veined.[100]   By steeping the bark in water an indelible blue ink is prepared.

SU SUNG [11th cent.]:— This tree grows in all prefectures of Shen si [App. 284] and in Ho yang [in Ho nan, App. 81].   It is a tall tree resembling the *t'an* (*Cæsalpinia*).

[100] 皮 有 白 點 而 不 粗 錯.

Trunk and branches are all of a green colour. Leaves like the head of a spoon, large and not shining. It bears neither flowers nor seed. The root resembles that of the *huai* (*Sophora*). The popular name of this tree is 白楊木 *pai* (white) *sin mu*.

LI SHI-CHEN explains that the name 秦皮 *ts'in p'i* refers to the country of Ts'in [Shen si], where the tree grows.

Ch., XXXIII, 31 :—*Ts'in p'i.* The figure represents a tree, but the drawing is not characteristic.

TATAR., *Cat.*, 13 :— *Ts'in p'i.* Cortex. This drug, obtained from a Peking druggist's shop, was sent to Professor FLÜCKIGER for examination. An infusion of the bark mixed with iron produced indeed an ink, but it was not blue.

The name *k'u li* is applied in the Peking mountains to *Fraxinus Bungeana*, DC., upon which the wax insect lives.

Cust. Med., p. 284 (156):—*Ts'in p'i* exported 1885 from Amoy 0.22 picul,—p. 148 (151), imported to Shang hai 145 piculs,—p. 352 (180), to Canton 118 piculs. Said to be shipped from Han kow, Ning po and Tien tsin.—In the *Hank. Med.* the drug *ts'in p'i* is not mentioned.

HOFFM. & SCHLT., 250:—秦皮樹, *Fraxinus longicuspis*, S. & Z. According to GEERTS the same Chinese name is also applied to *Fr. Sieboldii*, Bl.

Phon zo, LXXXIII, 17, 18:—Same Chinese name. Tree not identified by FRANCHET.

MATSUMURA, 82 :—樺, *Fraxinus puhinervis*, Bl.

324.—合歡 *ho huan.* P., XXXVb, 3. T., CCXCVI.

Pen king :—*Ho huan.* This tree grows in Yü chou [Ho nan, App. 413] in mountain-valleys. It resembles the *kou ku* tree [which is an *Ilex*. See 342]. The bark is used in medicine. Taste sweet. Nature uniform. Non-poisonous. The leaves and the flowers likewise are officinal.

Pie lu :—It grows in I chou [in Yün nan, App. 102].

TATAR., *Cat.*, 6 :—*Ho huan. Acacia Nemu.*—P. SMITH, 2.

Cust. Med., p. 154 (223) :—*He huan* (flowers) exported 1885 from Shang hai 0.18 picul.

Further particulars regarding this well-known tree will be given in another part.

325.—皂荚 *tsao kia. P.*, XXXV*b*, 4. *T.*, CCCIV.

Pen king :—*Tsao kia* (black pod). The pod is officinal. Taste pungent and salt. Nature warm. Slightly poisonous. The seeds, thorns, bark of the trunk and of the root, and leaves are likewise used in medicine.

Pie lu :—The *tsao kia* grows in Yung chou [Shen si, App. 424] and in the district Tsou hien in Lu [Shan tung, App. 365, 202]. The pod looks like the tusk of a boar. It is gathered in the 9th and 10th months and dried in the shade.

The *Kuang chi* [5th cent] calls it 雞栖子 *ki si tsz'* (cock's perch).

T'AO HUNG-KING :—It is a common tree. The pod which is two feet long is the best. It is frequently worm-eaten, and then is injurious to man.

SU KUNG [7th cent.] :—There are three kinds of *tsao kia*. One is called 猪牙皂荚 *chu ya* (boar's tusk) *tsao kia*. It is an inferior sort. The pod is crooked, thin, uncomely and not succulent. When used for washing it does not remove the dirt. That which is two feet long [*v. supra*] is coarse and dry. The best is that which is only from six to seven inches long. It is round, thick, jointed (*i.e.* contracted between the seeds), has a thin skin and much flesh and is of a strong taste.

Su Sung [11th cent.]:—Now the best is produced in Huai chou and Meng chou [both in Ho nan, App. 93, 220]. The tree is tall. The *Pen king* recommends the *chu ya tsao kia*, T'AO HUNG-KING the pods which are two feet long, and SU KUNG those which are only six inches long. All these sorts are used in medicine; the boar-tusk pods are useful in tooth-ache. The young sprouts (leaves) are eaten as a vegetable.

LI SHI-CHEN:—The *tsao* is a tall tree. The leaves resemble those of the *huai* (*Sophora*). They are thin, long and pointed. Many thorns in the axils of the branches. It blossoms in summer. Small yellow flowers. There are three kinds, distinguished according to the pods. One kind has small pods resembling the tusk of a boar, another has long, thick and fleshy pods containing much fat and viscid matter. The third sort is long, thin, dry and meagre. It does not contain any viscid matter. The fat and fleshy sort is the best. As the tree is beset with thorns it is difficult to ascend. The people therefore at the proper time surround it with bamboo baskets. Then during one night all [it is not clear whether the thorns or the ripe pods] will drop. A strange thing! When sometimes a tree does not produce fruit, the people bore a hole in the trunk, fill it with three or five pounds of cast iron, and cover it with mud. Then it will produce fruit. Other names for the tree: 烏犀 *wu si* (black rhinoceros) and 縣刀 *hüan tao* (suspended sword). The thorns are known under the name 天丁 *t'ien ting* (clavus cœlestis).

Ch., XXXIII, 33:—*Tsao kia.* Good drawing. Leaves and long pods. *Gleditschia sinensis*, Lam.—See also *Kiu huang*, LVI, 3.

This beautiful tree is called 皂角 *tsao kio* (black horn) at Peking. The trunk [of old trees] where the branches begin is surrounded with a formidable crown of enormous branched thorns. Small, greenish yellow scented flowers, and large flat, fleshy, black pods about one foot long. These pods are used as soap.

LOUR., *Fl. cochin.*, 801 :—*Mimosa fera* [very probably he means *Gleditschia*]. Sinice : *tsao kie.*

TATAR., *Cat.*, 57 :—皂角 *tsao kio.* Legumen *Gleditschiæ chinensis.*—*Ibid.*, 56 :—皂刺 *tsao ts'z'.* Spinæ *Gleditschiæ.*— *Ibid.*, 29 :—牙皂. Legumen *Gleditschiæ.*

HANB., *Sc. pap.*, 248 :—牙皂 *ya ts'ao* (tusk pod). Legumes of *Prosopis.*[101] Legumes of the same Chinese name which I obtained from the Thibetan drug-shop at Peking, consisting of small, curved pods about three inches long and one-third inch broad, and which were examined at Kew, proved to belong to a *Prosopis.*

P. SMITH, 105, 179 :—*Gleditschia* and *Prosopis.*

Comp. also HENRY, *Chin. pl.*, 499, 500.

The tree, with leaves, flowers and pods, is figured in HOOKER's *Icones. Plant.*, tab. 1412.

Cust. Med., p. 296 :—*Ta* (great) *tsao* exported 1885 from Amoy 0.55 picul,—p. 294 (291), *siao* (small) *tsao* from Amoy 0.30 picul.—*Ibid.*, p. 372 (430) :—*Tsao* seeds exported from Canton 19 piculs.—*Ibid.*, 126 (96) :—*Tsao* thorns exported from Chin kiang 57.23 piculs,—p. 356 (226), from Canton 11.62 piculs,—p. 288 (220), from Amoy 0.55 picul.

Ibid., p. 62 (61) :—*Ya tsao* exported from I chang 28.79 piculs,—p. 80 (196), from Han kow 42 piculs,— p. 132 (166), from Chin kiang 7.80 piculs,—p. 166 (389), from Shang hai 0.70 picul.

---

[101] Specimens of *ya-tsao* from Szechuan sent to Kew, with flowers, leaves and pods, have been described as *Gleditschia officinalis*, Hemsl. *Decades Kewenses*, I, Kew Bulletin, No. 64, p. 82.—A. HENRY.

Amœn. exot., 841 :—皂莢 *sokio*, vulgo *kawara fudsi.*
Arbor vasta, foliis impariter pinnatis ; siliquis (quas non
vidi) longis, multisque, quod dicunt, interstitiis intus dis-
tinctis. Ex Sina adducta arbor, rara hic est, fructu imperfecto
vel nullo. An arbor Cassiæ fistulæ ?

Kwa wi, 88 :—Same Chinese name. *Gleditschia japonica*,
Miq.

Phon zo, LXXXIII, 23, 24 :—Same Chinese name.
Gl. heterophylla, Bge., and *Gl. japonica.—Ibid.*, 25 :—猪牙
皂莢. Only pods figured.

After the *tsao kia* the *P.* describes [XXXV*b*, 13] the
肥皂莢 *fei* (fat) *tsao kia.* No ancient author quoted.
LI SHI-CHEN states :—The *fei tsao kia* grows on high mountains. It is a tall tree. Leaves like those of the *t'an*
(*Cæsalpinia*) and the common *tsao kia.* It blossoms in the
5th or 6th month. White flowers. The pods are from three
to four inches long and resemble those of the *yün shi*
[*Cæsalpinia.* See 140] but are thick, fleshy and fat. Each
pod contains several black seeds as large as the end of a
finger, not exactly globular. They are black like varnish,
very hard, with a white kernel within like a chestnut, which
can be eaten when roasted. The tree is also cultivated. The
pods are gathered in the 10th month, boiled and roasted,
then pounded to powder and mixed with wheaten flour and
perfumes. This composition, formed into balls, is used [instead
of soap] for washing the body and the face, to cleanse them
of dirt. It is richer in fat than the pods of the *tsao kia.*
It is said that the water from the *fei tsao kia* kills gold-
fish and drives away ants.

The large black, hard seeds of this tree are brought to
Peking from Mid China and used by women in washing
the head and hair. They are called *fei tsao.* The same

have been described and figured in HANB. *Sc. pap.*, 238. HANBURY means that they belong to a *Dialium* (*Leguminosæ*). It seems to me that this is the same tree as that of which BAILLON in 1875 described [*Journ. Soc. cent. d'hortic.*, p. 164-168] the pods received from Shang hai, and upon which he established the new species *Gymnocladus chinensis*. These pods are stated there to be used as soap by the Chinese.

Leaves and ripe pods of the *fei tsao* tree were procured for me from Wu hu by Mr. T. L. BULLOCK, in 1881, and sent to the Botan. Garden, St. Petersburg. The late MAXIMOWICZ considered them to belong to a *Cæsalpinia*. But in the *Ind. Fl. sin.* [I, 203] the Chinese soap tree, *fei tsao*, is said to be *Gymnocladus chinensis*. Mr. HEMSLEY determined it probably from complete specimens in flower and in fruit. Comp. also HENRY, *Chin. pl.*, 500.

P. SMITH, 1 :—*Acacia concinna. Fei tsao kia.*

Cust. Med., p. 198 (200):—*Fei tsao* exported 1885 from Ning po 56.20 piculs,—p. 366 (352), from Canton 0.50 picul.—Exported also from Han kow. See *Hank. Med.*, 13.

326.—欒 華 *luan hua.* P., XXXV, 15. T., CCCVIII.

See *Classics*, 550.

Pen king :—*Luan hua.* The flowers are officinal. Taste bitter. Nature cold. Non-poisonous.

Pie lu :—The *luan hua* grows in Han chung [S. Shen si, App. 54] in river-valleys. The flowers are gathered in the 5th month.

SU KUNG [7th cent.]:—The leaves of this tree resemble those of the *mu kin* [*Hibiscus syriacus*. See *Rh ya*, 6] but are thinner and smaller. Yellow flowers resembling the *huai*

flowers (*Sophora*) but larger. The seeds are enclosed in a
bladder like the *suan tsiang* [*Physalis*. See 106]. They are
black, round like peas, and hard. The people use them much
as beads. The flowers are gathered in the 5th or 6th month.
The people in the south use them for dyeing a bright yellow
colour. They are also employed for curing sore eyes.

Su Sung [11th cent.]:—The tree is cultivated in gardens
in the south as well as in Pien [in Ho nan, App. 248].

K'ou Tsung-shi [12th cent.]:—This tree is found in the
mountains of Ch'ang an [in Shen si, App. 6]. The seeds are
called 木 欒 子 *mu luan tsz'*. They are brought to the
capital, where they are used as beads. They are not employed
in medicine.

This is the *Kœlreuteria paniculata*, Laxm. For further
particulars see *Bot. sin.*, II, 550.

327.—欅 *ku*. *P.*, XXXV*b*, 20. *T.*, CCLXIII.

Comp. *Rh ya*, 238.

Pie lu:—*Ku*. The bark is used in medicine. Taste
bitter. Nature very cold. Non-poisonous.

T'ao Hung-king:—The *ku* is a common tree in the
mountains. Its bark resembles the bark of the *t'an*
(*Cæsalpinia*) and the *huai* (*Sophora*). Leaves like those of
the *li* and the *hu* (oaks). It is well known.

Su Kung [7th cent.]:—It grows by the sides of rivulets.
The leaves resemble those of the *ch'u* [*Ailantus*. See *Rh ya*,
224] but are narrower and longer. It is a large tree, several
fathoms in girth. The bark is very coarse and thick and
does not resemble the *t'an* bark.

K'ou Tsung-shi [12th cent.]:—The *ku* tree is now more
commonly called 欅柳 *ku liu* (willow), for its leaves resemble

the willow. Large specimens are from 50 to 60 feet high and from 2 to 3 fathoms in girth. It is frequently met with in Hu nan and Hu pei, but is not used for timber and is not fit for making utensils. The young bark is employed for making buckets and hoops for sieves.

Li Shi-chen:—The wood of the *yü* is reddish purple. It is highly valued for making boxes and tables. Cheng. Ts'iao [12th cent.] says:—The *kü* is a kind of *yü* (elm). Its branches are more hardy. Its fruits are like elm-fruits, which look like small coins. The villagers gather the leaves and prepare a sweet tea therefrom.

Ch., XXXIII, 63:—*Kü*. According to Henry [*Chin. pl.*, 247] this is the *Pterocarya stenoptera*, Cas. (Order *Juglandeæ*), a common tree in Hu pei. But it seems that in other parts of China the name *kü* is applied to an elm-like tree, as also in Japan where it is *Ulmus keaki*.

For further particulars see *Bot. sin.*, II, 238.

328.—柳 *liu.* *Salix babylonica*, L. *P.*, XXXVb, 21. *T.*, CCLXIII.

Comp. *Classics*, 524.

Pen king:—*Liu.* In the Index of the *Pen king*:—柳華 *liu hua* (flowers). The *Pen king* explains the latter term by 柳絮 *liu sü* (willow-wool, *i.e.* the cottony down of the seeds). Taste bitter. Nature cold. Non-poisonous. The leaves, branches, white bark of the root, and the gum exuding from the tree are all used in medicine.

Pie lu:—The *liu hua* is produced in Lang ye [in Shan tung, App. 178] in marshes.

P. Smith, 231:—Willows.

Cust. Med., p. 288 (202):—*Liu*, twigs and leaves, exported 1885 from Amoy 4 piculs,—p. 358 (252), from Canton 0.53 picul,—p. 344 (80), willow-root exported from Canton 0.01 picul.

Further particulars in another part.

329.—榆 *yü. Ulmus.* Elm tree. P., **XXXV**b, 30. T., CCLXIX.

Comp. *Rh ya*, 304, *Classics*, 528.

Pen king:—榆 *yü*, 零榆 *ling yü.* The white bark of the tree is officinal. Taste sweet. Nature uniform. Mucilaginous and nourishing. Non-poisonous. The leaves, flowers and fruit [*see the next*] are all used in medicine.

Pie lu:—The 榆皮 *yü p'i* (bark) is produced in Ying ch'uan [S.E. Ho nan, App. 408] in mountain-valleys. The white, inner bark is gathered in the 2nd month and dried in the sun. The fruits are gathered in the 8th month. Both drugs (the bark and the fruit) should be kept dry, otherwise they become poisonous.

P. SMITH, 92 :—Elm-bark.

Cust. Med., p. 212 (49):—*Hiang yü p'i* (fragrant elm-bark) exported 1885 from Wen chow 502 piculs.

Further particulars in another part.

330.—蕪荑 *wu i.* P., **XXXV**b, 33. T., CCLXIX.

Comp. *Rh ya*, 263.

Pen king:—無姑 *wu ku*, in the Index of the *Pen king* 蕪荑 *wu i.* It is a fruit (fruit of an elm). Taste pungent. Nature uniform. Non-poisonous.

Pie lu:—The *wu i* grows in Tsin shan [App. 356] in river-valleys. The fruit is gathered in the 3rd month and dried in the shade.

T'AO HUNG-KING:—This drug is now brought from Kao li [Corea, App. 116]. It resembles the fruit of the common elm,[102] and has a fetid odour. The people there prepare it in sauces for food. It is a vermifuge and is also used to drive away moths.

SU KUNG [7th cent.]:—Now the best drug comes from Yen chou and T'ung chou [both in Shen si, App. 403, 378].

CH'EN TS'ANG-K'I [8th cent.]:—The *wu i* has a strong, fetid smell. It is the fruit of the mountain-elm.

MA CHI [10th cent.]:—It is common in Ho tung and Ho si [Shan si, App. 80, 79].

SU SUNG [11th cent.]:—It is produced in Mid China. The best comes from T'ai yüan [in Shan si, App. 325]. It is a kind of small elm, the fruit of which ripens earlier than that of the common elm, and has a fetid odour. Mentioned in the *Rh ya*. The fruit is dried for use. The people pound it also and cook it as food. It is also preserved with salt. The salt destroys the disagreeable odour. It is not used as a medicine.

LI SHI-CHEN:—There are two kinds of *wu i*, the large and the small. The latter is the same as the 榆荚 *yü kia*, or fruit of the common elm, which the people prepare into sauces for food. The larger *wu i* is that used in medicine.

The large *wu i* is probably *Ulmus macrocarpa*, Hance.

[102] 榆荚

Hank. Med., 7:—臭 無 荑 *ch'ou* (stinking) *wu i*.
Exported from Han kow. Mr. BRAUN says:—A small,
lentil-shaped seed of a very disagreeable and strong odour.
The flesh of the berry generally adheres to the seed.

The *Cust. Med.*, p. 489 (1457) calls the *wu i* a medicine-
cake.—*Ibid.*, p. 74 (123), *wu i* exported 1885 from Han kow
5.12 piculs.

331.—巴豆 *Pa tou*. *P.*, XXXV*b*, 43. *T.*, CCCVI.

Pen king:—巴豆 *Pa tou*, 巴菽 *Pa shu*. The seed is
officinal. Taste acrid. Nature warm. Poisonous. This is
one of the five principal poisons mentioned by SHEN NUNG.
See above, sub 132.

Pie lu:—The *Pa tou* grows in Pa [E. Sz ch'uan,
App. 235], whence the name, in river-valleys. Gathered in
the 8th month. Before use the heart and the skin are
removed. Very poisonous.

T'AO HUNG-KING:—The *Pa tou* is a violent purgative.
It should be taken in a fresh state.

SU SUNG [11th cent.]:—It is now found in Kia chou,
Mei chou and Jung chou [all in Sz ch'uan, App. 122, 219,
112]. It is a tree from 10 to 20 feet high. Leaves like
those of the *ying t'ao* (cherry) tree but thicker and larger.
They are green at the beginning but gradually change to
yellowish red. They wither in the 12th month and shoot
again in the 2nd month. In the 4th month the old leaves
fall off and are replaced by fresh leaves. Flowers of a
yellowish colour and in racemes. In the 5th or 6th month
the fruit (a capsule) is produced, which is green at the
beginning but yellow when ripe. It resembles the capsule of
the *pai tou k'ou* [*Amomum Cardamomum. See* 58]. Each
capsule consists of two divisions (cells), and in each division is

one seed, sometimes three seeds. The seed has an outer coat which is removed before use. The drug from Jung chou has on the outer coat from one to three perpendicular lines (small furrows) like thread. The people there call it 金線巴豆 *kin sien* (gold thread) *pa tou*. This is considered the best sort, but it is rare.

Li Shi-chen :—The *Pa tou* is produced in Pa [E. Sz ch'uan], and the seed resembles the soy-bean, whence the above names. Lei Hiao [5th cent.] says :—[There are three sorts.] One [the seed], which is contracted, small and of a yellow colour, is the 巴 *pa*, that which is three-edged and black is the 豆 *tou*, and that which is small and pointed on both ends is called 剛子 *kang tsz'*. The *pa* and the *tou* can be used in medicine, but not the *kang tsz'*, which kills man. Li Shi-chen observes that this account is not clear. He thinks that Lei Hiao's contracted and small sort is the female—the three-edged, and that with pointed ends is the female *pa tou*. The male is violent and the female has a slow effect. The capsule of the *pa tou* is like the integument of the *ta feng tsz'* [103] but thinner. The kernel within is like the *hai sung tsz'*.[104] The capsule cannot be compared to the *pai tou k'ou*.—One name for the *pa tou* is 老陽子 *lao yang tsz'*.

Besides the seeds and the integument, the oil of the seeds and the root are used in medicine.

Ch., XXXIII, 54 :—*Pa tou.* A poor drawing, from which nothing can be determined.

Lour., *Fl. cochin.,* 714 :—*Croton tiglium,* L. Sinice : *pa teu.* Habitat incultum in Cochinchina et China.—Loureiro gives the same Chinese name to *Croton congestum.*

Tatar., *Cat.,* 1 :—*Pa tou.* Semina *Crotonis Tiglii.*— Hanb., *Sc. pap.,* 230.—P. Smith, 79, 159.

---

[103] 大風子, *Chaulmoogra. See* Hanb., *Sc. pap.,* 244.

[104] 海松子, the large seeds of *Pinus coraënsis.*

According to PARKER [*China Rev.*, IX, 329] the *pa tou* or Croton-seeds are used in Sz ch'uan for catching fish.

Cust. Med., p. 78 (167):—*Pa tou* exported 1885 from Han kow 2,039 piculs,—p. 62 (57), from I chang 31.70 piculs,—p. 364 (328), from Canton 10.35 piculs,—p. 292 (261), from Amoy 0.36 picul.

Phon zo, LXXXVI, 6, 7 :—巴豆. *Croton*.

EBN BAITHAR [transl. by SONTHEIMER, I, 427]:—*Dend, Croton tiglium*. The Chinese drug mentioned.—F. WATSON, *Native & Scient. Names of East. œcon. plants*, p. 51 :— *Croton Tiglium*, in Arabic *ba too* [probably derived from the Chinese *pa tou*].—SCHLIMMER, *Term. med. pharm. franc. persane* :—One of the Persian names for *Croton Tiglium* is *bidendjireh khatai* (*Ricinus* from China).

332.—桑 *sang*. The Mulberry tree. *P.*, XXXVI, 1. *T.*, CCXLVI.

Comp. *Rh ya*, 303, *Classics*, 499.

Pen king:—桑 *sang*. The Index of the *Pen king* has 桑根白皮 *sang ken pai p'i* (the white bark of the root of the mulberry tree). It is of a sweet taste. Nature cold. Non-poisonous. The leaves also are officinal and said to be slightly poisonous. Subsequent writers say, "non-poisonous." The ashes of the wood are used in medicine. The fruit also is officinal.

The *Pie lu* asserts that the root of the mulberry tree in its upper part, at the surface of the soil, has poisonous properties.

TATAR., *Cat.*, 43 :—*Sang p'i*. Radix *Mori*.—P. SMITH, 151 :—Mulberry bark.

Cust. Med., p. 354 (205):—*Sang pai p'i*, bark of the root, exported 1885 from Canton 157.40 piculs,—p. 72 (98), from Han kow 105 piculs,—p. 284 (168), from Amoy 25.15 piculs,—p. 262 (68), from Ta kow 6.56 piculs.

Ibid., p. 360 (271, 284) :—Exported from Canton, mulberry leaves 22 piculs, twigs 14 piculs.

Ibid., p. 200 (235):—*Sang chen*, mulberry fruit, exported from Ning po 5.15 piculs,—p. 370 (412), from Canton 0.36 picul.

Further particulars in another part.

333.—楮 *ch'u*, *Broussonetia papyrifera*, Vent. The Paper mulberry. *P.*, XXXVI, 10. *T.*, CCLXII.

Comp. *Classics*, 503.

Pie lu:—*Ch'u*, 楮 實 *ch'u shi* (fruit), also 穀 | *ku shi*. It is produced on the Shao shi mountain [App. 281]. The fruit is gathered in the 8th and 9th months and dried in the sun during 40 days. Taste sweet. Nature cold. Non-poisonous.

For other Chinese accounts regarding this tree see another part.

HANB. *Sc. pap.*, 231 :—楮 實 子 *ch'u shi tsz'*. The small seed-like nuts or achenes of *Broussonetia papyrifera*, Vent. (*Moreæ*).

P. SMITH, 167 :—Paper mulberry.

Cust. Med., p. 198 (198):—*Ch'u shi tsz'* exported 1885 from Ning po 0.47 picul.—Exported also from Han kow. See *Hank. Med.*, 8.

334.—枳 *chi.　P.,* XXXVI, 13.　*T.,* CCLIV.

Classics, 488.

Pen king :—枳實 *chi shi* (fruit).　Taste bitter.　Nature cold.　Non-poisonous.—The bark of the root and the young leaves are likewise used in medicine.

Pie lu :—The *chi shi* grows in Ho nei [in Shan si and Ho nan, App. 77] in marshes.　It is gathered in the 9th and 10th months and dried in the shade.

In the *K'ai pao Pen ts'ao* [10th cent.] it is called 枳殼 *chi k'io* (*k'io*=peel).　For Chinese descriptions of this shrub and its fruit, which is an *Aurantiacea,* see another part.

LOUR., *Fl. cochin.,* 571 :—*Citrus fusca.*　Sinice : *chi keu.*

TATAR., *Cat.,* 16 :—*Chi k'io.*　Fructus *Citri decumanæ* [an erroneous identification].

HANB. *Sc. pap.,* 238 :—*Chi k'io.*　The drug described.— P. SMITH. 66 :—*Citrus fusca.*—The *chi k'io* is probably *Ægle sepiaria,* DC.

P. SMITH says that *chi shi* is the unripe fruit, and *chi k'io* the ripe fruit.　It has a very thick peel.

Cust. Med., p. 232 (175):—*Chi shi* exported 1885 from Foo chow 94.37 piculs,—p. 74 (129), from Han kow 80 piculs,—p. 62 (49), from I chang 50.17 piculs,—p. 366 (337), from Canton 3.76 piculs.

Ibid., p. 72 (85):—*Chi k'io* exported from Han kow 4,309 piculs,—p. 60 (35), from I chang 510.62 piculs.

335.—巵子 *chi tsz'* [the first character is now generally written 梔].　*P.,* XXXVI, 21.　*T..* CCCII.

Pen king :—*Chi tsz',* 木丹 *mu tan* (wood red).　The fruit is officinal.　Taste bitter.　Nature cold.　Non-poisonous.

Sz ma Siang ju [2nd cent. B.C.] in one of his poems calls it 鮮支 *sien* (bright) *chi*.

Pie lu :—Other name : 越桃 *Yüe t'ao.* The *chi tsz'* grows in Nan yang [in Ho nan, App. 231] in river-valleys. The fruit is gathered in the 9th month and dried in the sun.

T'ao Hung-king :—It is a common plant. There are two or three kinds, which differ slightly one from another. The best drug is the seven-edged [he refers to the edges of the fruit]. It is gathered when hoar-frost first appears. It is more generally used as a dye than for medical purposes.

Su Sung [11th cent.] :—It is common in all prefectures of South China and in Western Shu [Sz ch'uan, App. 292]. It is a tree from 7 to 8 feet high. Leaves resembling those of the *li* (plum tree) but, thick and hard, also similar to the *ch'u p'u tsz'.*[105] It blossoms in the 2nd and 3rd months. White flowers with six-cleft corolla and very fragrant. Some believe that this is the 薝蔔[106] *tan p'u* flower produced in Western countries. The fruit, which appears in summer and autumn, resembles the *ho tsz'* (*Terminalia chebula, Myrobalan*). It is of a yellow colour when ripe. The kernels within are dark red. It is much cultivated by the people in the South. After this the author repeats the quotation from the *Shi ki* [182].

For medical purposes the 山] | *shan* (mountain or wild) *chi tsz'* is employed. To physicians it is also known under the name of 越桃 *Yüe tao* [peach of the kingdom of Yüe or Che kiang. See App. 418]. It (the fruit or capsule) is globular and has a thin skin. It contains small kernels. The best sort is that with a seven-edged or nine-edged fruit. The

---

[105] 樗蒲子. Unknown to me. Mentioned by Ma Yung [first cent.]. See *K.D.*, sub *p'u*. Palladius [*Chin. Russ. Dict.*] says :—*Ch'u p'u*, a kind of game.

[106] The first character is evidently a mistake for 薝 *chan*. In Chinese Buddhist works the above name is intended for *Michelia Champaka.*

large and oblong sort is called 伏 尸 | | *fu shi chi tsz'* in the
P'ao chi lun [5th cent.]. It is not efficacious as a medicine.

LI SHI-CHEN :—The leaves of the *chi tsz'* resemble a
hare's ear. They are thick, of a dark green colour, and
wither in autumn. The flowers are as large as a wine-cup.
White petals and yellow stamens. The fruit (berry) has a thin
skin. The seeds are small and have a beard. It is collected
after hoar-frost. In Shu (Sz ch'uan) there is a red-flowered
sort. The fruit of the *chi tsz'* is employed for dyeing an
orange colour.

Ch., XXXIII, 43 :—*Chi tsz'*. The drawing seems to
be intended for *Gardenia florida*, L.

LOUR., *Fl. cochin.*, 183 :—*Gardenia florida*, L. Sinice :
chy tsu. Baccarum recentium pulpa succosa et ruberrima
tinguntur eleganter serica.

TATAR., *Cat.*, 17 :—*Chi tsz'*. Fructus *Gardeniæ floridæ*.
—P. SMITH, 101.—HANB., *Sc. pap.*, 241-242 :—Fruit of the
chi tsz' figured and described. The berries are six-ribbed.
HANBURY examined also the *shan chi tsz'* with smaller fruits,
and the 黃 | | *huang* (yellow) *chi tsz'*, which latter fruit
seemed to be identical with the common *chi tsz'*, but had been
referred to *G. radicans* by Prof. MARTIUS. The name *huang
chi tsz'* is not found in the *P.*, but it is noticed in the
Chung shu shu [8th cent. *Bot. sin.*, I, p. 79]. In BRIDGM.
Chin. Chrest. [p. 453 (31)] and in PARKER'S *Canton plants*
[117] *huang chi* is given as a name for *Gardenia florida*.

PARKER, *Sz ch'uan plants*, 27, and HENRY, *Chin. pl.*,
64:—*Chi tsz' hua* in Sz ch'uan and Hu pei is *Gardenia florida*.

Cust. Med., p. 366 (338):—*Chi tsz'* exported 1885 from
Canton 256 piculs,—p. 232 (176), from Foo chow 13 piculs,
—p. 420 (96), from Pak hoi 1.81 picul.

Cust. Med., p. 74 (127):—建 | | *kien chi tsz'* (*kien*
probably means Fu kien province) exported from Han kow
6.86 piculs.

Cust. Med., p. 328 (188):—*Huang chi tsz'* exported from Swatow 301.88 piculs,—p. 216 (75), from Wen chow 100.64 piculs,—p. 368 (366), from Canton, where it is also called 水 | | *shui* (water) *chi tsz'*, 15 piculs,—p. 198 (212), from Ning po 7.87 piculs.

Cust. Med., p. 200 (239):—*Shan chi tsz'* exported from Ning po 45.51 piculs,—p. 92 (73), from Kiu kiang 6.95 piculs.—Exported also from Han kow. See *Hank. Med.*, 35.

The 紅 | | *hung* (red) *chi tsz'* is noticed in the *Hank. Med.* [p. 20].

Cust. Med., 372 (418):—山紅 | | *shan hung chi tsz'* exported from Canton 8 piculs.

Cust. Med., 372 (417):—山黑 | | *shan hei chi tsz'* (black mountain *Gardenia*) exported from Canton 524 piculs.

Amœn. exot., 808:—梔 *ssi*, vulgo *kutsjinas*, Mespilus, vulgari similis, folio majori, flore suaviter fragrante niveo, tubulato, in sena labia, longa, angusta, ad rosæ magnitudinem expanso; fructu turbinato, hexagono, senis striis protuberantibus & in alabastra desinentibus conspicuo; pulpa intus crocea, infectoribus expetita, saporis ingrati, innumeris referta seminibus Sesamino similimis.

Phon zo, LXXXVII, 10-12:—梔子, *Gardenia florida*. Same Chinese name, *G. maruba*, S., and *G. radicans*, Thbg.

Kwa wi, 121:—Same Chinese name, *Gardenia florida*.

HOFFM. & SCHLT., 254:—*Gardenia florida.* Same Chinese name, also 黃梔子. *Ibid.*, 255:—*G. radicans*, 水梔花.

SIEB., *Icon. ined.*, V:—*Gardenia radicans*, 山梔.

336:—酸棗 *suan tsao*. *Zizyphus vulgaris*, var. *spinosa*. *P.*, XXXVI, 24. *T.*, CCXXIII.

Comp. *Rh ya*, 275, *Classics*, 484, 485.

Pen king:—*Suan tsao.* The fruit is officinal. Taste sour. Nature uniform. Non-poisonous.

Pie lu:—The *suan tsao* grows in Ho tung [in Shan si, App. 80] in marshes. The fruit is gathered in the 8th month and dried in the shade during 40 days. It is useful in heat of the heart, in sleeplessness and in other complaints.

T‘AO HUNG-KING:—It grows in the eastern mountains, where it is called 山棗 *shan tsao* (mountain-jujube). The fruit is like the *tsao* from Wu ch‘ang [in Hu pei, App. 392] but it is very sour. The people of Eastern China eat the fruit in order to keep them awake,—not to cause sleep, as the *Pie lu* states.—But LI SHI-CHEN agrees with the *Pie lu,* stating that the kernels of the stones act as a soporific.

LOUR., *Fl. cochin.,* 196 :—*Rhamnus soporifer.* Sinice : *soan tsao.* Habitat in provinciis borealibus Sinarum. Virtus hypnotica, paregorica. Uti solent nucleis excorticatis, et diu coctis. Somnum leniter conciliat, dolores mitigat.

TATAR., *Cat.,* 50 :—酸棗仁 *suan tsao jen.* Nuclei Jujubæ. P. SMITH, 44, sub Buckthorn.

Further particulars in another part.

337.—白棘 *pai ki. P.,* XXXVI, 26. *T.,* CCLXXI.

Comp. *Classics,* 485.

Pen king:—*Pai* (white) *ki.* The spines of the tree are officinal. Taste pungent. Nature cold. Non-poisonous. Besides the thorns, the twigs, flowers, fruit and leaves are used in medicine.

Pie lu:—Other names : 棘刺 *ki ts‘z‘* (thorn), 棘鍼 *ki chen* (needle) and 菥蓂 *si ming* [107] ; the flowers are termed

---

[107] Properly a name applied to *Thlaspi.* See 252.

戟原 *ki yüan.* The *pai ki* grows in Yung chou [in Shen si, App. 424] in river-valleys. The 丨丨花 *ki ts'z' hua* (flower) grows by roadsides. It [whether the flower or the thorns it is not clear] is gathered 120 days after the winter solstice. The fruit is gathered in the 4th month.

LI TANG-CHI [3rd cent.]:—*Pai ki* is the name for the needles (thorns) of the *suan tsao* (jujube) tree [see 336]. Now the people substitute for this drug the *t'ien men tung* [*Asparagus lucidus.* See 176].

SU KUNG [7th cent.]:—There are two kinds of *ki*—the red and the white. The *pai* (white) *ki* has a stem as white as flour, but in its fruits and leaves it resembles the *ch'i* (red) *ki.* The thorns of the white kind are valued as a medicine, but it is scarce. There are also two kinds distinguished according to the shape of the thorns. One has straight thorns which have strengthening properties, the other has hooked (recurved) thorns which are useful in the cure of abscesses [causes them to discharge]. The *ki ts'z' hua* (flowers) are from the same plant, not, as the *Pie lu* intimates, a distinct plant. In the south the *t'ien men tung* is substituted for the *ki* needles, and called therefore *tien ki.*[106]

HAN PAO-SHENG [10th cent.]:—There are two sorts of *ki*—the red and the white. The Dictionary *Ts'i yün* [6th cent.] says:—The 棘 *ki* is a 小棗 *siao tsao* (small jujube). It is a common wild shrub, from two to three feet high, which grows thickly about and which in its flowers, leaves, stem and fruit resembles the *tsao* (jujube).

The white *ki* of the ancient Chinese authors is perhaps a *Paliurus*, belonging to the same order of *Rhamnaceæ* as *Zizyphus. Paliurus Aubletia,* Roem & Schult., of South China, has straight spines. The young branches and leaves are

[106] 顓 棘 See 176.

more or less tomentose. *See* BENTHAM, *Fl. hongk.*, 66.
The red *ki* may be a *Zizyphus*. The *Z. vulgaris*, var. *spinosa*,
a very common shrub in North China, has a reddish brown
bark.

338.—薤核 *jui ho*. *P.*, XXXVI, 28. *T.*, CCLI.

Comp. *Rh ya*, 300.

Pen king:—*Jui ho*. The kernel of the fruit is officinal.
Taste sweet. Nature warm. Non-poisonous.

Pie lu:—The 薤核仁 *jui ho jen* (kernel of the fruit-
stone) grows in Han ku [in Ho nan, App. 55] in
river-valleys, also in Pa si [in Sz ch'uan, App. 236].

T'AO HUNG-KING:—It grows in P'eng ch'eng [in Kiang
su, App. 247]. It (apparently the fruit-stone) is as large as
a black bean, globular, flattened, veined, and resembles a
walnut. The people use it (the kernel) together with the
shell; they break the shell and then weigh the kernel.

HAN PAO-SHENG [10th cent.]:—It is produced in Yung
chou [in Shen si, App. 424]. It is a tree with small
leaves resembling those of the *kou k'i* [*Lycium*. See 345],
but narrow and long. White flowers. The fruit is produced
on the stem, is of a purplish red colour and of the size of
the *wu wei tsz'* [*Schizandra*. See 164]. The stem is covered
with small spines. The fruit ripens in the 5th or 6th month
and is then gathered and dried in the sun.

SU SUNG [11th cent.]:—It now grows in Ho tung and
P'ing chou [both in Shan si, App. 80, 253]. It is a small
tree, from four to five feet high. The stem has spines.

LI SHI-CHEN refers it to *Rh ya*, 300, 棫 *yü* or 白桵
pai jui, and observes that the character *yu* in later times
was applied to the *tso*[109] tree (oak). The flowers and fruits

[109] 柞.

are drooping [from the stem], whence the name, for 蕤 means drooping leaves or flowers.

The drawings sub *jui ho* in the *Kiu huang* [LV, 1] and the *Ch.* [XXXVII, 35, 37] seem to represent *Berberis*. See also *Ch.*, XXXIII, 29, same Chinese name, representation of a quite different plant.

TATAR., *Cat.*, 50 :—蕤 仁 *jui jen*, Fructus ?—Under this name, in the Peking drug-shops, a small fruit-stone is sold, identical with TATARINOV's drug in the Bot. Museum of the Academy. Probably a *Prunus*.

Cust. Med., p. 479 (1206) :—*Jui jen* (also *sui jen*), seeds of an unknown shrub. In the *Hank. Med.* [21] it is identified with Bamboo-seeds, apparently on the authority of P. SMITH [32], who states that *jui jen* consists of the stones of a bamboo or *Polygonum* fruit [*sic* !].

Cust. Med., p. 78 (179) :—Same drug [identified there with Bamboo rhizome (*sic* !)] exported 1885 from Han kow 58.12 piculs,—p. 62 (59)], from I chang 13.75 piculs.

Phon zo, LXXXVII, 14 :—蕤核 given as a synonym of 白棘. The figure represents a seed or kernel.

339.—山茱萸 *shan chu yü* [comp. 291].   *P.*, XXXVI, 29. *T.*, CCXLIX.

Comp. *Classics*, 498.

Pen king :—*Shan* (mountain) *chu yü* and 蜀酸棗 *Shu suan tsao* (sour jujube from Sz ch'uan). The fruit is officinal. Taste sour. Nature uniform. Non-poisonous.

Pie lu :—Other name : 魆實 *k'i shi*. The *shan chu yü* grows in Han chung [S. Shen si, App. 54], also in Lang ye, Yüan kü and Ch'eng hien, a district in Tung hai [all in Shan tung, App. 178, 415, 20, 372]. The fruit is gathered in the 9th and 10th months and dried in the shade.

Wu P'u [3rd cent.] calls it 雞足 *ki tsu* (cock's foot) and 鼠矢 *shu shi* (rat's excrement).

T'ao Hung-king :—It grows in all mountains of Mid China. It is a large tree. The fresh ripe fruit is red and resembles the *hu t'ui tsz'*.[110] It is edible. When dried the skin becomes very thin, and the berry is used together with the stone.

Su Kung [7th cent.]:—Its leaves are like those of the *mei* [*Prunus Mume.* See 272]. The tree has spines. It blossoms in the 2nd month. The flowers resemble apricot-flowers. The fruit is produced in the 4th month. It is red, resembles the *suan tsao* [small jujube. See 336] and is gathered in the 5th month.

Su Sung [11th cent.]:—It is now produced in Hai chou [in Kiang su, App. 48] and in Yen chou [in Shan tung, App. 404]. It is a tree about 10 feet high with leaves resembling elm-leaves. White flowers. The *P'ao chi lun* [5th cent.] says that there is one kind [of the *shan chu yü*] which strongly resembles the *tsio rh su*,[111] but it has an eight-edged stone and is not used in medicine.

K'ou Tsung-shi [12th cent.]:—The *shan chu yü* differs far from the *Wu chu yü* [*Boymia.* See 291]. The medical properties in each are also very different. Why the name *chu yü* has been applied to each of them is not manifest.

Li Shi-chen :—It is also called 肉棗 *jou tsao* (fleshy Jujube).

Tatar., *Cat.,* 52 :—*Shan chu yü.* Drupæ *Corni.*— P. Smith, 74 :—*Cornus officinalis.*

In the *Cust. Med.* it is called 萸黃肉 *chu yü jou* (flesh), p. 202 (258), exported from Ning po 1,841.11 piculs,—p. 166

---

[110] 胡頹子 *Elæagnus* in Japan.

[111] 雀見蘇 same as the *hu t'ui tsz'* (*Elæagnus*) v. *supra.*

(392), from Shang hai 14.74 piculs. It is also exported from Han kow. See *Hank. Med.*, 52.

This seems to be the fruit of *Cornus officinalis*, S. & Z. For further particulars see *Bot. sin.*, II, 498.

340.—郁李 *yü li*. *P.*, XXXVI, 33. *T.*, CCLXXX.

Comp. *Rh ya*, 306, *Classics*, 474.

Pen king:—Yü li (elegant plum) and 爵李 *tsio li* (bird plum). The fruit and the kernel of the fruit are used in medicine. Taste sour. Nature uniform. Non-poisonous. The root also is officinal.

Pie lu:—Other names: 薁李 *yü li* and 車下李 *ch'e hia li* (plum under the cart). The *yü li* grows in Kao shan [in Kiang su, App. 118] in river-valleys and in the mountains. The root is gathered in the 5th and 6th months.

T'AO HUNG-KING:—It is common in the mountains. The ripe fruit is of a red colour and edible.

HAN PAO-SHENG [10th cent.]:—It is a tree from five to six feet high. In its leaves and flowers it resembles the great *li* (plum tree), but the fruit is small, like a cherry, of a sweetish sour taste, slightly harsh, and aromatic.

CHANG YÜ-HI [11th cent.] refers it to *Rh ya*, 306.

SU SUNG [11th cent.]:—The people of Pien and Lo [both in Ho nan, App. 248, 201], cultivate the *yü li* in gardens. It has long branches, produces an abundance of small flowers and has a luxuriant foliage. It is not used in medicine.

K'OU TSUNG-SHI [12th cent.]:—The *yü li* resembles the 御李子 *yü li tsz'* (imperial plum). The fruit is red, edible, but somewhat harsh. It can be prepared as sweetmeats. It abounds in Shen si.

LI SHI-CHEN:—Its flowers are of a pale red colour. The fruit is like a small plum.

TATAR., *Cat.*:—郁李仁 *yü li jen.* Nuclei *Cerasi.*—P. SMITH, 58, sub *Cerasus communis.*

Cust. Med., p. 16 (141):—*Yü li jen* exported 1885 from New chwang 68.87 piculs.—Exported also from Han kow. *Hank. Med.*, 24.

This is the *Prunus japonica*, S. & Z. For further particulars see *Bot. sin.*, II, 474.

341.—鼠李 *shu li.* P., XXXVI, 35. T., CCXX.

Pen king:—*Shu li* (rat plum). The fruit is officinal. Taste bitter. Nature cold. Non-poisonous. The bark is also used in medicine.

Pie lu:—Other names: 鼠梓 *shu tsz'*[112] and 牛李 *niu li* (ox plum). The *shu li* grows in the fields. It (the bark) is gathered at all times of the year.

SU KUNG [7th cent.]:—It is also called 皂李 *tsao li* (black plum) and 趙李 *Chao li.* Bark and fruit slightly poisonous.

SU SUNG [11th cent.]:—This is the 烏巢子 *wu* (black) *ch'ao tsz'* and also 山李子 *shan li tsz'.* It is common in Shu ch'uan [Sz ch'uan, App. 292]. Its branches and leaves resemble those of the *li* (common plum). The fruit is like the *wu wei tsz'* [*Schizandra.* See 164] and is of a beautiful black colour. It contains a purple juice. The fruit is gathered when ripe and dried in the sun. The bark is gathered at all times of the year.

K'OU TSUNG-SHI [12th cent.]:—This is the 牛李 *niu li*, a tree from seven to eight feet high. Leaves like those of

---

[112] Properly this name is applied to another tree. See *Rh ya*, 260.

the common plum tree but narrower and not smooth. The fruits are produced along the branches. They are at first green and become purplish black when ripe. In autumn, after the leaves have fallen off, the fruits are all still on the branches. It is common in Kuan and Shen [Shen si, App. 158, 284], also in Hu nan and in the northern part of Kiang nan.

LI SHI-CHEN :—It grows by roadsides. The fruits are produced on the branches, forming a kind of spike. The juice of the young fruits is used for dyeing a green colour. Other names : 楮李 ch'u li, 牛皂子 niu tsao tsz', 烏樐 wu ts'o and 椑 pei.

Ch., XXXIII, 52 :—Shu li. Rude drawing. Tree with berry-like fruits. Probably Rhamnus.

In the Peking mountains the name 牛李子 niu li tsz' is applied to Rhamnus arguta, Maxim. It has black berries containing a purplish black juice. Branches used for dyeing a green colour. The same Chinese name is applied to Rhamnus virgata, Benth.—Comp. also HENRY, Chin. pl., 484.

Phon zo, LXXXVIII, 3, 4 :—鼠李, Rhamnus japonica, Max.

342.—女貞 nü cheng.[113]   P., XXXVI, 37.   T., CCCVII.

Shan hai king :—The 楨木 cheng mu (tree) grows on the T'ai shan mountain [in Shan tung, App. 322]. KUO P'O comments :—This is the nü cheng, the leaves of which do not fall off in winter.

The 女貞 nü cheng is mentioned by Sz' MA SIANG JU [2nd cent. B.C.], in his Shang lin yuan fu, as growing in the Imperial Garden at Ch'ang an.

---

[113] The fruits deposited under this name in the Pharmaceutical Museum, London, from Hongkong, are those of Ligustrum lucidum, Ait.—A. HENRY.

The ancient Dictionary *Shuo wen* explains the character 楨 *cheng* by 剛木 solid tree.

Pen king:—*Nü cheng.* In the Index of the *Pen king* 女貞實 *nü cheng shi* (fruit). Taste bitter. Nature uniform. Non-poisonous. The leaves are also used in medicine.

Pie lu:—The *nü cheng shi* is produced in Wu ling [in Hu nan, App. 394] in river-valleys. It is gathered at the beginning of winter.

T'ao Hung-king :—It is a common, handsome evergreen tree with luxuriant foliage. The bark is green, the flesh (inner bark) is white. It is like the *Ts'in p'i* [see 323]. It is a handsome evergreen tree. It (the fruit) is recommended in the prescriptions to promote longevity, but commonly it is not used in medicine.

Su Kung [7th cent.]:—The leaves of the *nü cheng* resemble those of the 冬青 *tung ts'ing* tree and also the 枸骨 *kou ku* [*Ilex*, see further on]. The fruit ripens in the 9th month and is black like the *niu li* [*Rhamnus.* See 341]. T'ao Hung-king is wrong in comparing it to the *Ts'in p'i.* The latter has small leaves which wither in winter, the *nü cheng* has large evergreen leaves.

Su Sung [11th cent.]:—The *nü cheng* is a common tree. It is mentioned in the *Shan hai king* [*v. supra*]. Evergreen leaves like those of the *kou ku* and the *tung ts'ing* [*v. supra*]. It blossoms in the 5th month. Small, greenish white flowers. The fruit is produced in the 9th month. When ripe it is black like that of the *niu li tsz'.* Some say that the *nü cheng* and the *tung ts'ing* are identical. But the *tung ts'ing* is distinguished by its wood being as white as ivory. Its fruit is also used in medicine. In Ling nan [South China, App. 197] there is one kind of *nü cheng* which produces red flowers in great profusion. But this is quite different and is not used in medicine.

Li Shi-chen:—The *nü cheng* with its evergreen leaves is an emblem of chastity, whence the name [*nü* = girl, *cheng* = chastity]. The *nü cheng*, the *tung ts'ing* and the *kou ku* [these names are frequently confounded] are three distinct trees. The 女貞 *nü cheng* is the tree which is now commonly called 蠟樹 *la shu* (wax tree). The people in the East call the *nü cheng* with luxuriant foliage also 冬青 *tung ts'ing*. But the name *tung ts'ing* is properly applied to another tree. They resemble each other in the facility with which they are raised from seeds and in their having thick, pliable, long leaves, dark green on the upper side and paler underneath ; but the leaves of the *nü cheng* are oblong, from four to five inches long, and its fruit is black, whilst the *tung ts'ing* has roundish leaves and red berries. It (the *tung ts'ing*) produces a profusion of flowers, and in autumn the whole tree is covered with berries of which thrushes are very fond. Its wood is white. Nowadays the name *nü cheng* is little known. The people more commonly call this tree *la shu* (wax tree), for in summer the wax insect which produces the *pai la*, or white wax, lives upon the branches of it.

The 冬青 *tung ts'ing* is treated of in a special article in the *P.* [XXXVI, 39] and *T.* [CCCVIII]. The name means "green in winter, evergreen." The fruit, leaves and bark are used in medicine.

Ch'en Ts'ang-k'i [8th cent.]:—The people of Kiang tung [Kiang su, An hui, App. 124] write the name 凍青 *tung ts'ing* (green in the cold season). The wood of the *tung ts'ing* is white, veined and fit for making ivory-like tablets.[114] The leaves can be used for dyeing a dark red colour. One author says :—The *tung ts'ing* grows in the Wu t'ai shan mountains [in Shan si] and resembles the *ch'un* (*Cedrela*). It has red berries like the *yü li* [*Prunus japonica*,

[114] 作象齒笏, tablets held before the breast by officers at audiences.

65

See 340], of a sourish taste, but smaller. This is another
sort.

Li Shi-chen :—The *tung ts'ing* is akin to the *nü cheng*.
It is a mountain-tree. The leaves are roundish and the berries
red. The *nü cheng* has oblong leaves and black berries.
The *Kiu huang Pen ts'ao* [LIV, ii] says :—The *tung ts'ing*
is a tree about 10 feet high and resembles the *kou ku tsz'*
[*v. infra*]. It has luxuriant foliage. The leaves resemble
those of the *lu tsz'* [115] tree but are smaller, also those of the
ch'un (*Cedrela*) but they are rounded, not pointed. It
blossoms in the 5th month. Small white flowers. The berry
is of the size of a pea and of a red colour. The young shoots
are used for food.

枸骨 *kou ku.*   P., XXXVI, 40.   T., CCLI.

Ch'en Ts'ang-k'i [8th cent.]:—The *kou ku* tree
resembles the *tu chung* [see 317]. Its wood is white like the
bones of a dog,[116] whence the name. It is the *kou* of the
Shi king [See *Classics*, 490]. Certain musquitoes are produced
in the leaves of this tree.

Su Sung [11th cent.]:—It grows abundantly in Kiang
and Che [An hui and Che kiang, App. 124, 10]. In South
China its wood is highly valued by turners for making boxes.

Li Shi-chen :—The *kou ku* tree resembles the *nü cheng*.
Its wood is very white. The leaves are several inches long,
of a beautiful green colour, thick, hard (leathery) and evergreen. Each leaf has five horns terminating in spines. It
blossoms in the 5th month. Small white flowers. The
fruit resembles that of the *nü cheng*, also that of the *pa k'ia*
[*Smilax*. See 179]. When ripe it is of a dark red colour.
It has a thin skin and is of a sweet taste. The kernel (seed)

---

[115] 欅子, *Mespilus.*   [116] 狗骨.

consists of four parts [comp. *infra*]. The people gather the bark of the tree and by boiling it prepare a bird-lime. The bark and the leaves are used in medicine. Another name for the tree is 貓兒刺 *mao rh ts'z'* (cat-thorn).

Ch., XXXIII, 25 :—女貞 *nü cheng* or 蠟樹 *la shu* (wax-insect tree). Rude drawing. Probably a *Ligustrum* is intended.—*Ibid.*, XXXV, 51 :—冬青 *tung ts'ing*. Rude drawing. Perhaps an *Ilex* is meant. The *tung ts'ing* described in the *P.* is without doubt an *Ilex*, probably *Ilex cornuta*, Lindl. See also *Kiu huang*, LIV, ii.

TATAR., *Cat.*, 41 :—女貞 *nü cheng*. *Rhus succedanea*. This is an erroneous identification based upon an article on Chinese insect-wax, by STAN. JULIEN [1840], in which it is stated that, according to A. BROGNIART, the *nü cheng* is *Rhus succedanea*. This mistake is perpetuated in P. SMITH, 185.

That which I received from a Peking drug-shop under the name of *nü cheng* consisted of the dried berries of a *Ligustrum*. In the Thibetan drug-shop at Peking the same berries were sold under the name of 冬青 *tung ts'ing*. Under the same name *Ligustrum Ibota* is cultivated at Peking.—TATAR., *Cat.*, 22 :—*Tung ts'ing*, Fructus *Ligustri vulgaris*.—P. SMITH, 134, 229 :—*Tung ts'ing*, *Ligustrum lucidum*.—HANB., *Sc. pap.*, 67.—Father HEUDE, S.J., informed me that in Mid China *tung ts'ing* is a common name for *Ligustrum lucidum*, Ait.—At New chwang *tung ts'ing* is the name commonly given to the Mistletoe [*Notes & Quer. on Chin. & Jap.*, 1869, p. 175].—HENRY, *Chin. pl.*, 483 :—凍青樹 *tung ts'ing shu* in Hu pei is *Xylosma racemosum*, Miq. (order *Bixineæ*).

Cust. Med., p. 370 (404):—*Nü cheng tsz'* exported 1885 from Canton 53.51 piculs,—p. 296 (321), from Amoy 0.75 picul. Exported also from Han kow. See *Hank. Med.*, 30.

Ch., XXXVI, 49 :—蠟樹 *la shu* (wax tree). Drawing
not characteristic, but from the description of the tree it
would seem that a *Fraxinus* is meant. Fruit shaped like flat
horns resembling the fruit of *Ailantus*. The tree is said to be
cultivated in Kui chow for feeding the wax insect. But the
小 | | *siao la shu* (small wax tree) [*Ch.*, XXXVII, 18]
seems to be a *Ligustrum*. The description states :—There are
two kinds. One is the 水 | | *shui la shu* (water wax
tree). This is also called *nü cheng*. The other, the 魚 | |
yü (fish) *la shu*, is smaller. It is also called 水冬青 *shui
tung ts'ing*. The wax insect feeds on both of these trees.

HENRY, *Chin. pl.*, 205-208 :—At I chang the names
la shu and *shui la shu* are applied to *Ligustrum lucidum*, and
shan la shu and *siao la shu* to *Ligustrum chinense*, Lour.—
Peh-la shu is *Fraxinus*.

HOFFM. & SCHLT., 325 :—女貞, *Ligustrum japonicum*,
Thbg.,—[327, 328]:—Same Chinese name, *Lig. obtusifolium*,
S. & Z., and *Lig. ovalifolium*, Hassk,—[170]:—Same Chinese
name, *Cornus alba.*—*Ibid.*, 326 :—水蠟樹, *Lig. Ibota*, S. & Z.

Phon zo, LXXXVIII, 5, 6 :—女貞. Only leaves
represented.

SIEB., *Œcon.*, 202 :—*Ligustrum Ibota*. Vivit in hoc
frutice insectum ceram proferens. Quoque sub hoc cœlo
hujus cera usitata.

Amœn. exot., 907 :—冬青 *too sei*, vulgo *mots noki*.
Arbor mediocris, incondita, ramis tortuosis, foliis integris,
asperis, ovatis. Ex cortice tuso Japones viscum conficiunt.—
According to MAXIMOWICZ this is *Ilex integra*.

Phon zo, LXXXVIII, 7, 8 :—冬青 Japonice : *to sei*.
FRANCHET identifies this drawing with *Olea fragrans*. But
he seems to be mistaken.[117]

---

[117] *Olea fragrans* is represented in the *Phon zo* [LXXX, 8] under the
Chinese name 木犀.

GEERTS, *Japan woods:*—冬青, *Ilex integra*, also *Ilex Oldhami.*

Ch., XXXV, 50 :—枸骨 *kou ku.* Good drawing. *Ilex cornuta*, with the characteristic spiny leaves. It agrees well with the description in the *P.*,—white wood, horned spiny leaves and red berries. According to the description in the *P.*, the kernel consists of four parts, which are the four bony seeds of the berry joined together. In Europe *Ilex aquifolia* is known to have a very white wood. Bird-lime is made of the bark.—P. SMITH, 114, sub Holly.

Amœn. exot., 781 :—枸骨 *ojo*, vulgo *tsuge.* Buxus arborescens, folio ovato, majusculo, extremitate cuspidata, ora raris aculeis serrata ; flosculis in foliorum sinu plurimis calyculatis tetrapetalis albis, ad seminis Coriandri ambitum patentibus, petalis rotúndis, baccis atropurpureis, rotundis, pisi magnitudinis, succo purpureo sylvestri turgidis, seminibus intus in orbem compactis duobus, tribus, vel quatuor, seminis Carvi magnitudinis & figuræ.—THUNBERG [*Fl. jap.*, 77] identifies this with *Buxus virens.*—MAXIMOWICZ [*de Ilice*, p. 45] means that KÆMPFER probably describes *Ilex subpuberula*, Miq.

Phon zo, LXXXVIII, 10, 12 :—枸骨, *Olea aquifolium*, S. & Z.

The same identification in SIEB., *Icon. ined.*, V.

343.—衛矛 *wei mou.* *P.*, XXXVI 40. *T.*, CCCVI.

Pen king:—*Wei mou* (arrow wing). Apparently the branches are officinal. Taste bitter. Nature cold. Non-poisonous.

Pie lu:—Other name: 鬼箭 *kui tsien* (devil's arrow). The *wei mou* grows on the Huo shan mountain [in Hu nan, App. 100]. It is gathered in the 8th month and dried in the shade.

The *Kuang ya* [3rd cent.] calls it 神箭 *shen tsien* (divine arrow).

T'AO HUNG-KING :—It is common in the mountains. The bark, which has wings, is stripped off for use. It is seldom used in medicine.

SU SUNG [11th cent.]:—It is found in all the prefectures of Kiang and Huai [Kiang su and An hui, App. 124, 89]. The stem is from four to five feet and more high. It is provided with three wings like the wings of an arrow. The leaves resemble those of the *shan ch'a* (mountain tea). The branches and the stem are gathered in the 8th, 11th and 12th months. The wood of the tree is called 狗骨 *kou ku* (dog's bone).

K'OU TS'UNG-SHI [12th cent.]:—It is common in the mountains but not met with in the plain. The leaves are scanty. The stem is of a yellowish gray colour like that of the *po* tree [*Evodia.* See 315]. The bark has ridges on three sides like the edges of a knife. The people use it freely for fumigating to expel evil spirits. It is seldom employed as a medicine.

LI SHI-CHEN :—The *kui tsien* grows in the mountains among rocks. It has a small trunk. Along the young branches run three wings. The leaves resemble those of the *ye ch'a* (wild tea). They stand opposite and are of a sour, harsh taste. It blossoms in the 3rd or 4th month. Small yellowish green flowers. The fruit is as large as that of the *tung ts'ing* [*Ilex.* See 342]. The mountain people use this tree only for fuel.

Ch., XXXIII, 42 :—*Wei mou.* The figure represents a tree with leaves and winged branches. HENRY [*Chin. pl.,* 321] may be right in identifying it with *Evonymus alatus,* Thbg.

Phon zo, LXXXVIII, 11, 12:—衞矛, *Evonymus alatus*, Thbg. Same identification in SIEB., *Icon. ined.*, III.

Evonymus alatus is a shrub with quadrangular winged branches, frequent in the Peking mountains, where it is known by the name 四稜樹 *sz' leng shu* (four-edged tree), also 茶葉 | *ch'a ye shu*. An infusion of the flowers is employed as a substitute for tea.

344.—五加 *wu kia*. P., XXXVI, 44. T., CCCVII.

Pen king:—Wu kia, | | 皮 *wu kia p'i* and 犳漆 *ch'ai ts'i* (wolf varnish). The bark of the root is officinal. Taste pungent. Nature warm. Non-poisonous.

Pie lu:—犳節 *ch'ai tsie*. The five-leaved *wu kia p'i* is the best. It is produced in Han chung [South Shen si, App. 54] and in Yüan kü [in Shan tung, App. 415]. The stem is gathered in the 5th and 7th months and the root in the 10th and dried in the shade.

T'AO HUNG-KING:—It is common in Mid China. In Eastern China there is the four-leaved sort. It is also good.

LEI HIAO [5th cent.]:—The *wu kia* tree is properly the 白楸樹 *pai* (white) *ts'iu shu*.[118]

SU SUNG [11th cent.]:—It is common in all prefectures of Kiang and Huai [Kiang su and An hui, App. 124, 89] and Hu nan. It is a climbing plant with a red stem, from three to five feet high, and with black spines. The leaves are quinate. This is the best sort. Frequently there are only four or three leaflets on a common petiole. These are inferior sorts. At the base of every leaf is a spine. It blossoms in the 3rd or 4th month. White flowers. The fruits are at first green and become black in the 6th month.

[118] Comp., sub 319, 刺樹楸 *ts'z' ts'iu shu*, *Acanthopanax ricinifolium*.

The root resembles that of the *King* [*Vitex*. See 349].
The bark is yellowish black, the flesh (inner bark) is white,
and the bone (the centre of the stem) is hard. There are
several sorts. The drug from Pien king [in Ho nan, App.
248] and Pei ti [App. 245] is large and flat, resembles the
Ts'in p'i [see 323] and the *huang po* [*Evodia*. See 315],
is of a white colour, odourless and tasteless. It is noted as
a cure for rheumatic complaints.—In Wu [Kiang su, App.
389] the people strip off the bark of the root of the wild
ch'un tree (*Cedrela*) and call it *wu kia*. It is soft and
tasteless. One sort, which grows in Kiang and Huai, is
called 追風使 *chui feng shi*. Steeped in wine it cures
rheumatism. In K'i chou [in Hu pei, App. 121] the *wu kia*
is called 木骨 *mu ku* (tree bone).

LI SHI-CHEN:—The name 五加 *wu kia* means "five
[leaves] united." The name is also written 五佳 *wu kia*.
It is sometimes also called 五花 *wu hua* (five flowers).
The people in Shu (Sz ch'uan) term it 白刺 *pai ts'z'* (white
spine) and also 文章草 *wen chang ts'ao*. LI SHI-CHEN
quotes a memoir in praise of the wine prepared from the
wen chang. In the *Sien king* (a Taoist work) it is called
金鹽 *kin yen*. The *wu kia* in spring shoots forth young
twigs from the old branches. The mountain people eat them
as a vegetable. It is like the *kou k'i* [*Lycium*. See 345].
The *wu kia* which grows in the north, in a sandy soil, is
a tree, whilst that produced in South China, in a hard soil, is
an herbaceous plant. In the T'ang period the drug *wu kia*
was presented as tribute from Hia chou [in Hu pei, App. 64].

Ch., XXXIII, 24:—*Wu kia p'i*. Representation of a
spiny climbing plant with digitate leaves and fruits in
umbels.

LOUR., *Fl. cochin.*, 233:—*Aralia palmata*. [In *DC.
Prodr.* (IV, 264) LOUREIRO's plant is referred to *Hedera*

scandens.] Sinice: *u kia pi.* Caule scandente, aculeato, foliis 5 lobatis. Usus corticis in scabie et in hydrope.

TATAR., *Cat.*, 64 :— *Wu kia p'i.* Cortex *Araliæ palmatæ.* —GAUGER [53]:—The same drug described and figured.— P. SMITH, 20.

HENRY, *Chin. pl.*, 529 :— *Wu kia p'i, Eleutherococcus Henryi*, Oliv., and *E. leucorrhizus*, Oliv. Shrubs which grow on the cliffs at Pa tung. The root-bark is used as a drug, the former being distinguished as the red kind and the latter as the white kind. The drug is exported from Sz ch'uan. HENRY says:—The figure of *wu kia p'i* given in the *Ch.* may be intended for *Eleutherococcus Henryi* or may be *Acanthopanax spinosum*, Miq., which occurs at I chang and is called by the same native name.

Cust. Med., p. 28 (92):— *Wu kia p'i* exported 1885 from Tien tsin 123.86 piculs,—p. 10 (83), from New chwang 91.17 piculs,—p. 48 (47), from Chefoo 55 piculs,—p. 72 (82), from Han kow 27.53 piculs,—p. 194 (153), from Ning po 13 piculs.

Amœn. exot., 777 :—五 加 *kooki*, vulgo *kuko*, aliis *numi gussuri.* Ligustrum spinosum, etc. According to THUNBERG [*Fl. jap.*, 94] this is *Lycium barbarum.* But the Chinese name given by KÆMPFER is wrong. The plant to which it belongs is the 樗[119] *kio*, vulgo *dara.* Frutex sylvestris arborescens spinis horridus, etc.

Kwa wi, 89 :—五 加, *Acanthopanax spinosum*, Miq. (*Aralia pentaphylla*, Thbg.).—According to HOFFM. & SCHLT. [403] *Panax divaricatum* (*Acanthopanax divaricatum*, S. & Z.) in Japan is known by the same Chinese name.

Phon zo, LXXXIX, 2, 3 :—五 加, *Acanthopanax spinosum.* Japonice: *koka. Ibid.*, 1, 2 :—五 加, *Panax sessiliflorum*, Rupr. & Max.

---

[119] Character erroneously applied to this plant. Comp. *Bot. sin.*, II, 518.

SIEB., *Œcon.*, 243:—*Aralia pentaphylla*, 五架. Japonice: *wu kogi*.   Folia tenera edunt.

345.—枸杞 *kou k'i* and 地骨皮 *ti ku p'i*.   *P.*, XXXVI, 47.   *T.*, CCLXXXIII.

Comp. *Rh ya*, 257, *Classics*, 526.

Pen king:—*Kou k'i* and *ti ku p'i* (earth bone skin). *Ti ku* is the name of the root.   Taste bitter.   Nature cold. Non-poisonous.   The leaves and the fruit are also used in medicine.   Another name is 地節 *ti tsie*.

Pie lu:—Other names: 枸忌 *kou ki*, 却老 *k'io lao*, 羊乳 *yang ju* (goat's nipple) and 仙人杖 *sien jen chang* (staff of the immortals).   The *kou k'i* grows in Ch'ang shan [in Chi li, App. 8] in the plain, in marshes, and in the mountains.   The root is of a very cold nature.   The fruit is slightly cold.   Non-poisonous.   The root is taken up in winter, the leaves are gathered in spring, and the stem and the fruit in autumn.

SU SUNG [11th cent.]:—The *kou k'i* is a common plant. In its leaves it resembles the *shi liu* (pomegranate), but they are softer and thinner and can be eaten.   They are known by the name of 甜菜 *t'ien ts'ai* (sweet vegetable).   The stem grows from three to five feet high, in a bushy manner. It blossoms in the 6th or 7th month.   Small, reddish purple flowers.   The fruit is oblong like the stone of a jujube. The root is called *ti ku* [*v. supra*].   It is mentioned in the *Shi king*.   There are two sorts.   One has an oblong fruit and the branches are without spines.   This is the true 枸杞 *kou k'i*.   The other has a globular fruit and the plant is provided with spines.   This is the 枸蕀 *kou ki* (spine).   The former is the larger kind, and this yields the drug for medical use.   The other (the spiny) is not used in medicine.   The name *sien jen chang* (staff of the immortals) is sometimes applied to the *kou k'i*. But the same name is also given to two other plants. One is a vegetable resembling the *k'u kü* (*Lactuca*) and the other a kind of black bamboo.

K'OU TSUNG-SHI [12th cent.]:—There is no foundation for the statement that the *kou k'i* and the *kou ki* are distinct plants. The only difference is that one is the old and the other the young plant. The latter is abundantly provided with spines, whilst the old plant is unarmed or has only few spines. The *suan tsao* [thorny jujube. See 336] shows the same peculiarity with respect to the *Ki* [see 337].

LI SHI-CHEN :—In ancient times the best *kou k'i* and *ti ku* were produced in Ch'ang shan [in Chi li, App. 8], and the plant is still found there. But subsequently the drug from Shen si became famous and that from Kan chou [Kan chou fu in Kan su] was considered the best sort. The *kou k'i* which now grows in Lan chou and Ling chou [both in Kan su] and west of Kiu yüan [north of the Ordos, App. 157] is a large tree with thick leaves and a coarse root. But the *kou k'i* of Ho si [west of the Yellow River, App. 79] and Kan chou [in Kan su] is distinguished by a globular fruit, like a cherry, which shrinks up when dried in the sun. It has but few kernels. The dried berry is red and of a sweet, agreeable taste like raisins. It is used for making sweetmeats. This is quite different from the first-mentioned *kou k'i*.

The name *kou k'i*, in China as well as in Japan, is applied to *Lycium chinense*, L., and probably also to other species. For further particulars see *Bot. sin.*, II, 526.

TATAR., *Cat.*, 26 :—*Kou k'i ts:'.* Baccæ *Lycii* chinensis. *Ibid.*, 21 :—*Ti ku p'i.* Cortex radicis *Lycii.*—P. SMITH, 37 :—*Kou k'i*, erroneously identified with *Berberis Lycium.*

Cust. Med., p. 76 (150):—*Kou k'i tsz'* (fruit) exported
1885 from Han kow 1,262 piculs,—p. 30 (120), from Tien
tsin 376 piculs,—p. 130 (131), from Chin kiang 29.83
piculs,—p. 152 (195), from Shang hai 8.80 piculs,—p. 62
(55), from I chang 7.20 piculs.

Ibid., p. 72 (93):—*Ti ku p'i* exported from Han kow
435.77 piculs,—p. 28 (87), from Tien tsin 82.40 piculs,—
p. 126 (85), from Chin kiang 22.90 piculs,—p. 284 (170),
from Amoy 13.43 piculs,—p. 418 (65), from Pakhoi 7.69
piculs.

The sort of *kou k'i* with globular red edible berries,
described by LI SHI-CHEN as produced in Kan su, is, I suspect,
Nitraria Schoberi, L., the fruit of which, according to
PRZEWALSKI, forms an important article of food to the
natives in Kan su and N.E. Thibet. The fruit of *Lycium* is
not edible.

The *Phon zo* [LXXXIX, 3, 4] figures, sub 枸杞 or
仙人杖, *Lycium chinense*, forma inermis, and [4, 5], sub
枸檵 or 地骨皮, the same, forma spinosa.

346.—溲疏 *shou shu*. *P.*, XXXVI, 54. *T.*, CCCVIII.

Pen king : — *Shou shu*. The bark is officinal. Taste
pungent. Nature cold. Non-poisonous.

Pie lu :—Other name : 巨骨 *kü ku*. The *shou shu* grows
in Hiung rh [in Ho nan, App. 69] in river-valleys, fields
and burial wastes. It is gathered in the 4th month.

LI TANG-CHI [3rd cent.]:—The *shou shu* is also called
楊櫨 *yang lu*, 牡荆 *mou king*[120] and 空疏 *k'ung* (hollow)
shu. The bark is white, and it [the stem?] is hollow. It has

---

[120] Properly a name for *Vitex*. See 348.

joints. The fruit resembles the *kou k'i tsz'* [fruit of *Lycium*. See 345]. It ripens in winter and is then of a red colour and a sweet and bitter taste. It is sometimes confounded with the *kou k'i tsz'*. This is not the *yang lu* which the people use for forming hedges.

Su Kung [7th cent.]:—The *shou shu* resembles the *k'ung shu* [*v. supra*]. It is a tree above 10 feet high with a white bark. Its fruit ripens in the 8th or 9th month, is of a red colour and resembles the *kou k'i*. The berries grow in pairs. Taste bitter. They do not resemble the fruit of the *k'ung shu*. The *k'ung shu* is the same as the *yang lu*. Its fruit is a pod.

Ma Chi [10th cent.] :—The *shou shu* resembles the *kou k'i*, but the *shou shu* has spines whilst the *kou k'i* is unarmed.

Li Shi-chen says that the above statements are not clear. The *shou shu* tree seems to be unknown to him.

Ch., XXXIII :—*Shou shu*. The figure represents a tree or shrub with oblong berries.

Amœn. exot., 855 :—楊櫨. *Korei utsugi*. Sambucina ramorum facie frutex Coræensis etc. Figured in Banks' *Icon. Kœmpf. sel.* [45]. This is *Diervilla grandiflora*, Sieb. & Zucc. *Flora. jap.*, I, 71, tab. 31. Siebold states that Kæmpfer is mistaken in writing *Korei utsugi*. It should read *joro utsugi*, *joro* being the Japanese pronunciation of the above Chinese characters.

Amœn. exot., 854 :—高麗 *joro*, vulgo *utsugi*. Sambuci facie frutex etc. According to Siebold, *l.c.*, instead of *joro* vulgo *utsugi* we have to read *Korei utsugi*. The above Chinese characters mean Corea. This is *Deutzia scabra*, Thbg. See Sieb. & Zucc., *Fl. jap.*, I, 20, where *su so* is given as the Chinese name, *i.e.* 溲疏. But in Siebold's *Icon. ined.* [IV] this Chinese name is applied to *Deutzia gracilis* and *D. crenata* and likewise to *Diervilla hortensis*, whilst 楼骨木[121] is given as the Chinese name for *Deutzia scabra*.—Sieb., *Œcon.*, 336 :—*Deutzia scabra*. Japonice : *utsugi* ; sinice : 溲疏. Folia ad lævigandum lignum.

In the *Phon zo* [LXXXIX] the Chinese name 溲疏 is applied [5, 6] to *Deutzia gracilis*, [8] to *Staphylea Bumalda* and [9] to *Philadelphus coronarius*, L.

The 楊櫨 [Japanese pronunciation *joro*], in Sieb. & Zucc, *Fl. jap.*, I, 74, is *Diervilla versicolor* (*Weigela japonica*, Thbg.). In the *Phon zo* [LXXXIX, 9, 10] the same Chinese name denotes *Diervilla floribunda*, S. & Z.

347.—石南 *shi nan*.   *P.*, XXXVI, 55.   *T.*, CCCVI.

Pen king :—*Shi nan*. The leaves are officinal. Taste acrid and bitter. Poisonous. The fruit also is used in medicine.

Pie lu :—The *shi nan* grows in Hua yin [in Shen si, App. 87] in mountain-valleys. The leaves are gathered in the 3rd and 4th months, the fruit in the 8th month, and dried in the shade.

T'ao Hung-king :—It is common in Eastern China. Its leaves resemble the *p'i p'a* [*Eriobotrya*. See 282]. They are rarely used in medicine.

Su Kung [7th cent.]:—Its leaves are like the *kien ts'ao*.[122] They do not wither in winter. The fine-leaved sort from Kuan chung [Shen si, App. 158] is the best. In South China the *shi nan* has long, large leaves like those of the *p'i p'a* [*v. supra*]. They are odourless and tasteless, and are not used in medicine.

---

[121] Compare above [121], *Sambucus*.

[122] 繭草, Cocoon plant. Unknown to me.

HAN PAO-SHENG [10th cent.]:—It is common in Chung nan and Sie ku [both in Shen si, App. 28, 309] in rocky places. Dealers in drugs sometimes confound it with the *shi wei* [*Polypodium lingua*. See 203].

SU SUNG [11th cent.]:—It is now found on rocks in South China as well as in the North, and is sometimes a large tree. That which grows in Kiang and Hu [Kiang si and Hu kuang, App. 124, 83] has leaves like the *p'i p'a* with small prickles. They do not fall off in winter. It blossoms in spring. White flowers in clusters. In autumn it bears small red fruits. The sort which is produced in Kuan and Lung [in Shen si and Kan su, App. 158, 216] has leaves like the *mang ts'ao* [see 158], of a greenish yellow colour with purple spots underneath. When rain is abundant they grow from 2 to 3 inches long. The slender root is a horizontal creeper and is of a purple colour. The tree has neither flowers nor fruit, but its foliage is luxuriant. In the North as well as in the South it is planted freely in court-yards. It is a handsome tree and affords ample shade. For medical use the small-leaved sort from Kuan chung [Shen si, App. 158] is employed.

The *Wei wang hua mu chi* says:—In South China the *shi nan* tree grows wild. It blossoms in the 2nd month. The fruit is like the *yen fu tsz'* [*Akebia?* See 184]. It ripens in the 8th month. The people gather it, take out the kernels, boil them together with fish, and so make a soup. It is not used now [in medicine].

K'OU TSUNG-SHI [12th cent.]:—The leaves of the *shi nan* are like the *p'i p'a* leaves, but smaller, glabrous, not downy underneath nor wrinkled. It blossoms in the first or second month. In winter a spathe can be seen consisting of two leaves. When the spathe bursts, 15 or more larger or smaller flowers appear like those of the *ch'un* (*Cedrela*). The

flowers are with six leaves (petals) of a red colour, in bunches.
There are numerous stamens which conceal the flowers. After
the tree has shed its flowers, the old leaves fall off and new
leaves appear. The *shi nan* is rarely seen in the northern
provinces, but it is common in Hu nan and Hu pei, in
Kiang si and in the two Che [Che kiang and Kiang su.
App. 10], where it is much employed by the people [as a
drug].

LI SHI-CHEN :—The *shi nan* grows on the sunny side of
rocks, whence the name (*shi* = rock, *nan* = south). In Kui
yang chou [in Hu nan, App. 167] it is called 鳳藥 *feng yao*,
and [the leaves] is used as a substitute for tea. Steeped in
wine it is useful in curing head-ache.

It is impossible to decide from the above descriptions
what tree is meant. Probably several plants are known by
the name *shi nan* in different parts of China.

Ch., XXXIII, 50 :—*Shi nan*. The figure represents a
plant with berries.

Amœn. exot., 877 :—石南 *sekki nan*, vulgo *saku nange*.
Frutex perennis orgyjam altus, etc.—This is *Rhododendron
Metternichü*, S. & Z. [*Fl. jap.*, I, 23, tab. 9].—Same identifi-
cation in the *Phon zo* [LXXXIX, 13, 14],—*Kwa wi*, 103.

HENRY [*Chin. pl.*, 368] says that in Hu pei *Rhododendron
Fortunei*, Ldl., is called 野枇杷 *ye p'i p'a*. Comp. above
the statement of the ancient Chinese authors that the leaves
of the *shi nan* resemble the *p'i p'a* leaves (*Eriobotrya*).

348.—牡荊 *mou king*. P., XXXVI, 56. T., CCLXX.
Comp. *Classics*, 521.

The *Pen king* calls it 小荊 *siao* (small) *king*. The fruit
is officinal. Taste bitter. Non-poisonous.—The leaves, root,
and the sap of the tree also are used in medicine.

Pie lu:—The *mou* (male) *king shi* (fruit) is produced in Ho kien [in Chi li, App. 75], Nan yang [in Ho nan, App. 231] and Yüan kü [in Shan tung, App. 415], also in P'ing shou [in Shan tung, App. 256] and Tu hiang [in Chi li, App. 369] in the high mountains, also in the fields. The fruit is gathered in the 8th and 9th months and dried in the shade.

T'AO HUNG-KING:—The name *siao* (small) *king* is improperly applied to the *mou king*, for it is a tree and its fruit is larger than that of the *man king* [*see the next*]. The *man king* is the sort of *king* of which staves are made. Its fruit is small, in appearance like that of hemp, and is of a greenish yellow colour. But the fruit of the *mou king*, which is found in North China, is as large as a pea, globular and of a black colour. It is much used in prescriptions for promoting longevity. The leaves also are used in medicine. The twigs and the leaves of the *mou king* all stand opposite.

SU KUNG [7th cent.]:—The *mou* (male) *king* is improperly so called, for it bears fruit. The name therefore probably refers to its being a tree, whilst the *man king* is a creeper. The latter has a large fruit, whilst the *mou king* has a small fruit and is therefore termed *siao king*. The *mou king* is fit for making sticks and lances. The fruit is small and of a yellow colour. It has a strong arborescent stem. In the History of the Han dynasty [chapter on Sacrifices] it is stated that the *mou king* is used for flag-staves for the funeral banners, not the *man king* [as T'AO HUNG-KING intimates]. There are two sorts of *mou king*—the green and the red. The former is the best. The *mou king* is frequently confounded with the *man king*, but they are quite different.

SU SUNG [11th cent.]:—The *mou king* is now found in Mei chou and Shu chou [both in Sz ch'uan, App. 219, 292] and in Pien king [in Ho nan, App. 248]. It is commonly called 黃荆 *huang king*. The wood of the stem and the branches is hard. It is upright. The leaves resemble those of the *pi ma* (*Ricinus communis*) but are more dissected and thinner. Flowers red and in panicles. Small yellow fruit of the size of the seed of hemp, whence the name *siao* (small) *king*.

Li Shi-chen:—The *mou king* is a common plant, especially in the mountains where it is used for fuel. If not cut for many years it becomes a tree of considerable size. The heart of the wood is square. The leaves are opposite, and each petiole bears five leaflets (digitate leaves), sometimes even seven. The leaflets are like elm-leaves, long and pointed, with the margin serrated and toothed. In the 5th month panicles of reddish purple flowers are produced in the axils. The fruit is as large as that of the *hu sui* (*Coriander*). It has a white inner skin.[123] Su Sung is wrong in asserting that the leaves of the *mou king* resemble those of the *pi ma*. There are two sorts—the green and the red. The green is called 荆 *king*, the red is 楛 *hu* [comp. *Classics*, 543]. The young flexible shoots of both kinds are employed in basket making. In ancient times poor women used the *king* for hair-pins.

The *mou king* is a *Vitex*. Further particulars sub. 349.

349.—蔓荆 *man king*. P., XXXVI, 60. T'., CCLXX.

Pen king:—*Man* (creeping) *king*. The fruit is officinal. Taste bitter. Nature slightly cold. Non-poisonous.

Pie lu:—Only the name.

Su Kung [7th cent.]:—It is a creeping plant, whence the name. The *man king* grows along the edge of the water. The stem is about 10 feet long. In spring small new leaves shoot forth from the old branches. In the 5th month the

[123] 有白膜皮裹之.

leaves resemble apricot-leaves. It blossoms in the 6th month. Flowers reddish white with yellow filaments. In the 9th month it bears fruit. This is as large as the seed of the *wu* (*Sterculia platanifolia*), has black spots and is light. The leaves fall off in winter. It is frequently confounded with the *mou king*.

Su Sung [11th cent.] :—It is common in Pien king [in Ho nan, App. 248], in Ts'in chou [in Kan su, App. 358], Lung chou [in Shen si, App. 215], and in Ming chou and Yüe chou [both in Che kiang, App. 224, 418]. The stem is from 4 to 5 feet high. The leaves proceed from the joints and are opposite. It looks like a small *lien* tree [*Melia*. See 321]. It blossoms in summer. Flowers of a pale red colour. Filaments yellowish white. Below the flower is the green receptacle of which the fruit is formed. The ancient authors named it *man* (creeping) *king*, but it does not creep.

Li Shi-chen says its branches are slender and weak, whence the name (creeping *king*).

In the *Ch.* [XXXIII, 27, sub 蔓荊 *man king* or 荊條 *king t'iao* (twigs)] is a good drawing of *Vitex incisa*, Lam. The same is figured in the *Kiu huang* [LV, 4, sub 荊子 *king tsz'*]. The description in the *P.* agrees. At Peking *king t'iao* is the common name for *Vitex incisa*.

Lour., *Fl. cochin.*, 474 :—*Vitex negundo*, L. Sinice: *muen kim* (*man king*).—*Ibid.*, 475 :—*V. spicata*, Lour. Sinice: *u chu kim* [probably 五指荊 *wu chi king* (five fingers *king*), which according to Parker is the Canton name for *V. negundo*].

Tatar., *Cat.*, 59 :—荊條 *king t'iao*, *Vitex incisa*.—*Ibid.*, 38, 蔓荊子 *man king tsz'*, semina *Viticis incisæ*.—P. Smith, 227.

According to PARKER [*China Rev.*, X, 377], in Sz ch'uan *Vitex negundo*, a common road shrub, is called 黃荊 *huang king*.—Same identifiation in HENRY's *Chin. pl.* [132].

The *Ind. Fl. sin.* [II, 257] enumerates six species of *Vitex* for China.

Cust. Med., p. 216 (81):—*Man king tsz'* (fruits) exported 1885 from Wen chow 3.10 piculs,—p. 294 (317), from Amoy 0.11 picul. Exported also from Han kow. See *Hank. Med.*, 27.

Phon zo, LXXXIV, 15, 16 :—牡荊, *Vitex cannabifolia*, S. & Z. [according to Japanese botanists introduced into Japan from China].

HOFFM. & SCHLT., 622 :—*Vitex cannabifolia*, 牡荊 or 黃荊. Same identification in the *Kwa wi* [111].

Phon zo, LXXXIX, 17, 18 :—蔓荊子, *Vitex trifolia*, L.—Same identification in the *Kwa wi* [88]. Introduced into Japan.

SIEB., *Icon. ined.*, V1 :—Same Chinese name, *Vitex obovata*, Thbg. (same as *V. trifolia*).

350.—茯苓 *fu ling*. P., XXXVII, i. T., CXCVII.

Pen king :—*Fu ling*, 伏菟 *fu t'u*. Taste sweet. Nature uniform. Non-poisonous.

Pie lu :—The *fu ling* which clings to the root [of the fir tree] is called 伏神 *fu shen*. The *fu ling* and the *fu shen* grow in T'ai shan [in Shan tung, App. 322], in mountain-valleys under large fir trees. It is dug up in the 2nd and 8th months and dried in the shade.

In the ancient Historical Records *Shi ki* [in the chapter on Divination] the name of this drug is written 伏靈 *fu ling*. It is there said to be produced by the divine spirit of the fir tree.

The *Sien king* says that the *fu ling* is as large as a man's fist. When worn in the girdle it will discomfit evil spirits.

T‘AO HUNG-KING :—The drug brought from Yü chou [in Kuang si, App. 412] is of a large size, like a vessel with a capacity of three or four *sheng*. The outer skin is black and has small wrinkles. The inner substance is hard and of a white colour. The best drug is that which has the appearance of a bird or a beast or a tortoise, etc. The red *fu ling* is less valued. It does not decay and is not eaten by insects. Even after remaining underneath the ground for thirty years it will not change its colour and texture.

SU KUNG [7th cent.] :—Now the *fu ling* which is produced in T‘ai shan is compact and finely veined. It is not much used. The best comes from Hua shan [in Shen si, App. 86]. This is very coarse in texture and massive. It is also found in the Southern mountains of Yung chou [Shen si, App. 424], but this also is inferior to the Hua shan drug.

In the *Ki shi chu* [T‘ang period] this drug is called 不死麫 *pu sz‘ mien* (undying flour).

HAN PAO-SHENG [10th cent.]:—It is found in places wherever large fir trees grow. It abounds in Hua shan, where it is found under dried-up fir trees. It appears in lumps. The drug which has the appearance of a tortoise or a bird is especially valued.

CHANG YÜ-SI [11th cent.]:—The *Fan tsz‘ ki jan* [*Bot. sin.*, I, p. 145 (104)] says :—The *fu ling* is produced on the Sung shan [in Ho nan, App. 317] and in San fu [in Shen si, App. 265]. HUAI NAN TSZ‘ says that the *fu ling* is found under fir trees a thousand years old. The *t‘u sz‘* (*Cuscuta*. See 163] grows above. The *Tien shu* says that after the resin of the fir tree has entered the ground and remained there a thousand years it is changed into *fu ling*. When you see the fir tree

turn red, the *fu ling* is beneath it.   The *Kuang chi* states that
the *fu shen* [*v. supra*] is a product of the resin of fir trees and
is superior to the *fu ling*.   It is brought from Pu yang hien
[in Chi li, App. 263].

SU SUNG [11th cent.]:—It is found on the mountains
T'ai shan [in Shan tung, App. 322], Hua shan [in Shen si]
and Sung shan [*v. supra*].   It clings to the roots of large
fir trees.   It produces neither leaves, flowers nor fruit, and
forms underground nodular masses as large as a man's fist,
which sometimes weigh several pounds.   There are two
sorts—the red and the white.   Some say that the *fu ling* is
the metamorphosed resin of the fir trees, and others that
it grows from the spurious vapors of the fir tree [?].[124]   The
largest lumps which do not adhere to the root are called
fu ling.   Those which clasp the root, and which are light and
of a loose texture are called 伏神 *fu shen*.   They are produced
by the spurious vapors of the tree and are of a superior quality.
In the chapter on Divination in the *Shi ki* [*v. supra*] it is
stated that the *fu ling* grows beneath the *t'u sz'* [*Cuscuta*,
comp. 163].   In appearance it resembles a bird.   The place
where the *fu ling* lies underground, sometimes from 4 to 7 feet
deep, is discovered by burning the *t'u sz'*.

LI SHI-CHEN :—The *fu ling* is also called 松腴 *sung yü*
(fat).   By the *t'u sz'* mentioned by the ancient authors in
connexion with the *fu ling* we are not to understand,
LI SHI-CHEN says, the plant of this name (*Cuscuta*) but a
kind of subtile vapor hanging above the spot where the
fu ling lies underground.   The mountain people know it.
The best sort is that in large lumps and as hard as a stone.
The light sort of a loose texture is not much valued.

The heart of the *fu shen* is called 神木 *shen mu* (divine
wood).   It is likewise used in medicine, as also the bark of
the *fu ling*.

[124] 假松氣而生.

Ch., XXXIII, 6 :—*Fu ling*. The drawing represents large nodular masses.

Father MARTINI, about 240 years ago, mentions the Chinese drug *fu lin* produced in Sz ch'uan [see my *Early Eur. Res. Bot. Chin.*, pp. 19, 20].

A CLEYER, *Specimen Medicinæ sinicæ* (1682), 189 :—*Pe fo lim* (white *fu ling*), est radix insipida subdulcis temperata etc. . . . Est idem quod Lusitanice dicitur *Pao de China* (China wood), nisi quod album et multo melius sit rubeo illo, et etiam carius multo.—CLEYER'S red *fo ling* is *Radix Smilacis*. Comp. *supra*, 179.

DU HALDE, *la Chine*, I, 30, III, 647.—GROSIER, *la Chine*, III, 324, 328.

LOUR., *Fl. cochin.*, 710 :—Ad radices Pinorum sylvestrium magnæ longævitatis in provincia boreali Chinensi *su chuyen* [Sz ch'uan] gigni solent quædam tubera, subrotunda, magna, scabra, fusca, intus albissima, quæ ab Europæis vocantur *Radix sinensis alba*, ab ipsis vero Sinensibus *Pe fu lin*. Horum tuberum decocto feliciter utuntur in praxi medica, præcipue in morbis pulmonum et vesicæ. Radix *Sinensis rubra* provenit ex diversa planta, quæ a Linnæo dicitur *Smilax Chinæ*.

TATAR., *Cat.*, 23, 2 :—*Fu ling* or *pai* (white) *fu ling*, *Pachyma pinetorum*. Fungus maximus.

GAUGER, 18 :—*Fu ling*, described and figured. But GAUGER is mistaken in supposing that it is the root of a *Dioscorea* or *Tamus*.

WILLIAMS, *Chin. Commerc. Guide*, 114, sub China root.

In 1859 the Rev. M. J. BERKELEY published in the *Journ. Proc. Linn. Soc. Bot.* [III, 102] an interesting article on some Tuberiform Vegetable Productions from China, in which the *pe fu ling* is determined as *Pachyma Cocos*, Fries, a Fungus-like substance.

HANBURY, *Sc. pap.*, 267 :—Good description of the Chinese drug *fu ling* :—Large, ponderous tuberiform bodies consisting internally of a compact mass of considerable hardness, varying in colour from cinnamon-brown to pure white. They are an altered state of the root of the tree, probably occasioned by the presence of a Fungus. It is the *Pachyma Cocos*, Fries, occurring in N. America, Japan and China. In America it is called "Indian Bread."—See also HANB., *Sc. pap.*, 200, where this drug is figured.

P. SMITH, 165 :—*Pachyma Cocos.*—HENRY, *Chin. pl.*, 478.

Cust. Med., p. 66 (18):—*Fu ling* exported 1885 from Han kow, 13,149.45 piculs.—*Ibid.*, p. 354 (184), *fu ling p'i* (bark) from Canton 27.49 piculs,—p. 212 (48), from Wen chow 1.70 picul.—*Ibid.*, p. 220 (25), *fu shen* imported to Foo chow 0.05 picul, from Hong kong. Said to be produced in Kuang tung.

Amœn. exot., 832 :—*Sjooro* (no Chinese characters). Tubera esculenta, terrestria, sub abietibus crescentia.— THUNBERG [*Fl. jap.*, 349] identifies this with *Lycoperdon Tuber*, L. Fungus magnitudine pruni majoris.

See in *Phon zo* [XCIII] the drawings, 2r, sub 白伏苓 (white),—2v, 赤伏苓 (red),—3r, 伏神.

351.—琥珀 *hu p'o.* Amber. *P.*, XXXVII, 7.

Pie lu:—The *hu p'o* is produced in Yung ch'ang [in Yün nan, App. 426].

T'ao Hung-king :—The ancients say that the *hu p'o* is the resin of the fir tree, which, being embedded in the soil during a thousand years, turns into amber. When burned it emits an odour like that of resin. It sometimes incloses insects. An imitation of the *hu p'o* is produced by boiling hen's-eggs with fish-roe. The genuine *hu p'o*, when rubbed between the hands till it becomes hot, will attract straw. Now all the *hu p'o* in China is brought from foreign countries.

Tatar., *Cat.*, 9 :—*Hu p'o*. Succinum.—Williams, *Chin. Comm. Guide*, 79 :—Amber, article of import.—P. Smith, 12.

352.—豬苓 *chu ling.*   P., XXXVII, 10.   T., CLXVII.

Pen king :—*Chu ling* (pig's tubers), 豭豬屎 *kia chu shi* (boar's excrement). Taste sweet. Nature uniform. Non-poisonous.

Pie lu :—The *chu ling* grows on the mountain Heng shan [in Hu nan, App. 61], also in Tsi yin and Yüan kü [both in Shan tung, App. 347, 415]. It is gathered in the 2nd and 8th months and dried in the shade.

T'ao Hung-king :—This drug appears in black lumps resembling pig's excrement. Chuang tsz' [4th cent. B.C.] mentions the 豕橐 *shi t'o*, and Sz' ma Piao [3rd cent.], in commenting upon Chuang tsz', says that it is the 苓 *ling*, the root of which resembles pig's excrement. That which is called *chu ling* consists of the 楓樹苓, tubers produced on the *feng* tree [*Liquidambar Formosana*. See *Bot. sin.*, II, 261]. The best sort has a black skin and white flesh. The skin is removed before using the drug.

Su Sung [11th cent.]:—It is now found in Shu chou [in Sz ch'uan, App. 292] and in Si chou [App. 304]. It

68

grows underground, but not always under the root of the *feng* tree. It is also called 地 烏 桃 *ti wu t'ao* (black ground peach).

LI SHI-CHEN :—The *chu ling* are excrescences produced by the superfluous vapors of trees, in the same way as the *fu ling* is produced by the fir tree. The *feng* tree produces the *chu ling* in the greatest abundance.

Ch., XXXIII, 55 :—*Chu ling*. The figure represents a plant with pinnate leaves, not tubers.

TATAR., *Cat.*, 17 :—*Chu ling*. Not identified.

HANB., *Sc. pap.*, 204, 269 :—*Chu ling*. Production similar to the *Pachyma Cocos* [see 350] but smaller.—See also BERKELEY, *l.c.* [*supra*, sub 350].

Cust. Med., p. 66 (10):—*Chu ling* exported 1885 from Han kow 1,337 piculs,—p. 22 (13), from Tien tsin 379.91 piculs,—p. 58 (3), from I chang 123.93 piculs.

Phon zo, XCIII, 6r, 豬 苓.

353.—雷 丸 *lei huan* (*wan*). *P.*, XXXVII, 12.   *T.*, CXXXVI.

Pen king : —*Lei huan* (thunder - ball).   Taste bitter. Nature cold.   Slightly poisonous.

Pie lu :—Other names : 雷 實 *lei shi* (thunder-fruit), 雷 矢 *lei shi* (thunder-dirt).   The *lei huan* is produced in Shi ch'eng [in An hui, App. 285] in mountain-valleys, also in Han chung [S. Shen si, App. 54].   Produced underground. The root is gathered in the 8th month.

T'AO HUNG-KING :—It is produced in Kien p'ing and I tu [both in Hu pei, App. 139, 104], and appears as small balls joined together.

Su Kung [7th cent.]:—The *lei huan* is the *ling*[125] produced by the bamboo. The balls are not joined together. It is produced in Fang chou [in Hu pei, App. 35] and Kin chou [in Shen si, App. 143].

Li Shi-chen:—The *lei huan* varies in size. It is like a chestnut and sometimes like the *chu ling* [see 352]. It is round, has a black skin, white flesh and is very hard and compact. The *lei huan*, like the 雷斧 *lei fu* (thunder-axe) and the 雷楔 *lei sie* (thunder-pile), are productions of the thunder-clap, and metamorphoses of the subtle vapors of plants. It is produced in the ground and is without leaves. It has the power of destroying worms and driving out evil spirits. That produced on the bamboo is called 竹苓 *chu ling*.

Tatar., *Cat.*, 34:—*Lei huan, Mylitta lapidescens.* Fungus asporus?—Gauger, 26:—The *lei huan* figured and decribed.—

Hanb., *Sc. pap.*, 205, 269:—The *lei huan* figured and described:—Small, round nodules of a dark brownish grey colour and very hard.

P. Smith, 154.

Cust. Med., p. 76 (155):—*Lei huan* exported 1885 from Han kow 68.43 piculs,—p. 64 (72), from I chang 4.53 piculs.
Phon zo, XCIII, 6v, 雷丸 and, 7, 竹苓.

354.—桑上寄生 *sang shang ki sheng.* *P.*, XXXVII, 13. *T.*, CLXXX.
Comp. *Rh ya*, 262, *Classics*, 449.
Pen king:—*Sang shang ki sheng* (lodging on the mulberry tree. Parasite), 寓木 *yü mu* (lodging on trees), 寄屑 *ki sie*, 宛童 *wan t'ung.* The stem, leaves and fruit are officinal. Taste of the stem and the leaves bitter; non-poisonous. The fruit is sweet and non-poisonous.

Pie lu:—The *sang shang ki sheng* grows on mulberry trees in Hung nung [in Ho nan, App 99] in river-valleys. The stem and the leaves are gathered on the 3rd day of the 3rd month and dried in the shade.

T‘AO HUNG-KING :—This parasitic plant grows upon fir trees, the poplar and the *feng* tree (*Liquidambar Formosana*). It is the same kind on all these trees, only the roots differ according to the tree upon which the plant lives. These roots are embedded in the substance of the joints of the branches of the tree. Leaves roundish, greenish red, thick, glossy, easily broken and evergreen. They spring from the joints of the plant. It blossoms in the 4th month. White flowers. The fruit, which is produced in the 5th month, is of a red colour and of the size of a pea. It is common. The best drug comes from P‘eng ch‘eng [in Kiang su, App. 247]. It is commonly called 續斷 *su tuan*. But this name in the *Pen king* is applied to a quite different plant [see 84, *Dipsacus*].

SU KUNG [7th cent.]:—This plant grows upon the *feng*, the *hu* (oak) and upon elms, willows and other trees. The leaves are like small willow-leaves, but thick and easily broken. The stem is coarse and short. The fruit is yellow and resembles a small jujube. There is one kind of this parasitic plant, growing in Kuo chou [in Ho nan, App. 173] on mulberry trees, the fruit of which contains a very viscid juice. The kernel is of the size of a small pea. The fruit ripens in the 9th month and is then of a yellow colour. It does not ripen in the 5th month, is not red and is not of the size of a small pea, as T‘AO HUNG-KING asserts. The people of Kiang nan employ the stem and call it *su tuan* [*v. supra*] which is properly the name of another plant.

HAN PAO-SHENG [10th cent.] : — This parasitic plant grows on various trees. People say that it is propagated by birds which eat the fruits and drop their excrement upon trees. The leaves resemble orange-leaves but are thick and soft. The best is that growing upon the mulberry tree.

TA MING [10th cent.]:—The people gather the plant which grows upon the *ku* tree [see 327] instead of that living upon the mulberry tree, which is very scarce. They resemble each other but are not identical. That growing upon the *feng* tree is an inferior sort which equals that produced on the *ku* tree. It is of a yellow colour and is gathered in the 6th or 7th month.

K'OU TSUNG-SHI [12th cent.]:—The *sang ki sheng* is said [by previous authors] to be a common plant. But nowadays it is difficult to obtain, in the north as well as in the south, for the plant is gradually becoming extinct.

CHU CHEN-HENG [14th cent.]:—The *sang ki sheng* is an important medicine.

LI SHI-CHEN :—This parasitic plant is from 2 to 3 feet long. Its leaves are round, slightly pointed, thick, soft, green and glossy on the upper side, and of a pale purplish colour and downy underneath. People say that this plant is common in Chuan Shu [Sz ch'uan, App. 26], where the mulberry tree abounds and where this plant can be taken direct from the tree and employed in a fresh state. It grows plentifully also on other trees, but then its medical virtues are not the same and it is sometimes injurious to life. The *Cheng Tsiao T'ung chi* [12th cent.] says that there are two kinds. One of them, the larger sort, has leaves like the *shi liu* (pomegranate). This is the 蔦 *niao* [of the *Shi king*]. The smaller kind has leaves like the *ma huang* [*Ephedra*. See 97]. This is the 女蘿 *nü lo* [of the *Shi king*]. The fruits are the same in both kinds. [Comp. *Classics,* 449, 450.]

Ch., XXXIII, 35 :—*Sang shang ki sheng.* Rude draw-ing. Probably a *Loranthus* is intended.—*Ibid.*, XXXVI, 24 :—栗 寄 生 *li ki sheng.* Rude drawing of a *Viscum* or *Loranthus,* said to grow on chestnut trees (*li*) in Yün nan.

TATAR., *Cat.*, 44 :—*Sang ki sheng. Viscum?*—The drug of this Chinese name which I obtained from a Peking drug-shop—yellow stems without leaves—and which was examined at Kew, proved to be the common *Viscum album,* L.

P. SMITH, 150 :—Mistletoe, *li hu.* *Li* and *hu* are names for oaks. Evidently the characters *ki sheng* (parasites) are omitted in the above name. *Ibid.*, 93 :—*Epiphytes. Ibid.*, 232 :—Willow-*Epiphyte, liu ki sheng.*

HENRY [*Chin. pl.*, 35, 392]:—*Sang ki sheng* in Hu pei, *Loranthus Jadoriki,* Sieb., and other species. These parasites, when they occur on the mulberry, are highly valued as drugs.

Cust. Med., p. 360 :—*Sang ki sheng* exported 1885 from Canton 41.13 piculs. Same drug exported from Hankow. See *Hank. Med.*, 35.

Ibid., p. 286 (184):—*Ki sheng* exported from Amoy 14.23 piculs.—*Ibid.*, 360 (282), *tsa ki sheng* (*Viscum* growing on various trees) exported from Canton 2.05 piculs.

The *Hank. Med.* [25, 43] mentions the *liu ki sheng* [growing on willows] and the *t'ao ki sheng* [on peach trees] as exported from Han kow. Both are noticed in the *P.* [XXXVII 16, 17].

The propagation of the mistletoe by birds eating the fruit, as noticed by the Chinese authors, is also mentioned by THEOPHRASTUS [*de causis plant*, 2, 17].

Phon zo, XCIII, 8, 9 :—桑寄生, *Viscum album.*

355.—松蘿 *sung lo.* *P.*, XXXVII, 15. *T.*, CL.

Comp. *Classics*, 450.

Pen king :—*Sung lo* (parasite on fir trees). Taste bitter. Non-poisonous.

Pie lu :—Other name : 女蘿 *nü lo.* The *sung lo* grows on the Hiung rh mountain [in Ho nan, App. 69] on *sung* (fir) trees. It is gathered in the 5th month and dried in the shade.

T'AO HUNG-KING :—The plant is common in the eastern mountains, where it grows on various trees, but the genuine drug is that produced upon fir trees. In the *Shi king* the *niao*, together with the *nü lo*, is mentioned as growing on the fir tree. The *niao* is the true *ki sheng* growing upon mulberry trees. The other parasite growing on the fir tree is different from that found on the mulberry tree [see 354] and is not used in medicine.

LI SHI-CHEN :—The *sung lo* is also termed 松上寄生 *sung shang ki sheng.* The *nü lo* has been variously identified by the ancient authors. MAO, in commenting upon the *Shi king*, says it is the *t'u sz'* [*Cuscuta*. See 163]. WU P'U [3rd cent.] says the *t'u sz'* is the same as the *sung lo* [*v. supra*]. T'AO HUNG-KING suggests that the *niao* of the *Shi king* is the plant growing upon the mulberry tree and *sung lo* the sort which grows upon fir trees. The *P'i ya* [11th cent.] states :—The *niao* is a parasite (*Viscum*) upon fir trees and *Thuja*, whilst the *nü lo* is a twining plant which climbs upon the fir tree. Others say :—The plant is called *nü lo* when it climbs on trees and *t'u sz'* when it twines about herbaceous plants. The *Cheng Tsiao T'ung chi* [12th cent.] says :—There are two kinds of *ki sheng*,—the large is called *niao* and the small *nü lo*.

The *sung lo* or *sung shang ki sheng* of the Chinese authors seems to be a species of *Viscum* or *Loranthus*.

Phon zo, XCIII, 11, 12 :—松蘿檜 [the third character means creeper], *Loranthus Kœmpferi*, Maxim. (*Viscum Kœmpferi*, DC.). See FRANCHET & SAV., *Enum. pl. Jap.*, I, 403, II, 482. It has been found on *Larix*, *Pinus Massoniana*, *Abies firma*.—

Amœn. exot., 785 :—寄生 *ksei*, vulgo *jodoroki*. *Viscum* baccis rubentibus etc. Crescit in Larice. Rusticorum vulgus id appellabat *gomi maatz* i.e. *Viscum lariceum*.

356.—占斯 *chan sz'*. P., XXXVII, 17. T., CCCXI.

Pie lu :—Chan sz', 炭皮 *t'an p'i* (charcoal skin). It grows in T'ai shan [in Shan tung, App. 322] in mountain-valleys. Gathered at any time of the year.

T'AO HUNG-KING :—LI TANG-CHI [3rd cent.] says it is a *ki sheng* (parasite) which grows upon the *chang* (Camphor tree). The people now erroneously call the skin (fleshy husk) of the walnut *chan sz'*. According to T'UNG KÜN this drug is produced in Shang lo [in Shen si, App. 274] and is a bark resembling the *hou p'o* bark [*Magnolia*. See 316].

LI SHI-CHEN :—Other ancient names : 茛無極 *liang wu ki*, 木占斯 *mu chan sz'*.

357.—竹 *chu*, the Bamboo. P., XXXVII, 18. T., CLXXXIX.

Comp. *Classics*, 563, 564.

*Pen king :—*竹 *chu*, 竹葉 *chu ye* (leaves), 竹實 *chu shi* (fruit). The *Pie lu* says the *chu shi* is produced in I chou [App. 102]. It mentions also the 竹筍 *chu sun* (bamboo-sprouts) as a medicine.

The *Pen king* and the *Pie lu* notice several peculiar kinds of bamboo as officinal.

Of the 箽竹 *kin chu*, the leaves, sap (瀝) and root are used; of the 淡竹 *tan chu*, the leaves and the root; of the 苦竹 *k'u* (bitter) *chu*, the leaves and the sap.

In the *P.* there appears also a drug derived from the bamboo and termed 竹茹 *chu ju*. It is mentioned in the *Pie lu* and in other ancient works.

TATAR. [*Cat.*, 17] gives the name but does not identify it. P. SMITH [31] says that bamboo-roots are meant. I have seen TATARINOV's drug. It seemed to be bamboo-shavings, probably the scraped tender epidermis of the skin, which in the dictionaries is called 筎 *ju*.

Cust. Med., p. 194 (156):—Bamboo-leaves exported 1885 from Ning po 259 piculs,—p. 152 (188), from Shang-hai 181 piculs.

Ibid., p. 380 (539):—*Chu ju* exported from Canton 17.91 piculs,—p. 308 (507), from Amoy 2.42 piculs.—Bamboo-roots are exported from Han kow. See *Hank. Med.*, 8.

358.—濰木 *huai mu.* *P.*, XXXVII, 26.

Pen king:—*Huai mu*, 百歲城中木 *po sui ch'eng chung mu.*

Pie lu:—The *huai mu* grows in Tsin yang [in Shan si, App. 357], in marshes. The 城裏赤柱 *ch'eng li ch'i chu* in P'ing yang [in Shan si, App. 257].

LI SHI-CHEN:—WU P'U [3rd cent.] says:—The *huai mu* grows in P'ing yang, in the country of Tsin [Shan si,

App. 353], and in Ho tung [in Shán si, App. 80], in marshes.
This is the same as the *ch'eng li ch'i chu* of the *Pie lu* and
the *po sui ch'eng chung mu* of the *Pen king*. It is a tree
growing within an old city, as the above names indicate.
The people of Tsin used it as a drug.

APPENDIX.

CHINESE GEOGRAPHICAL NAMES MENTIONED IN THE *PEN TS'AO KANG MU.*

As is well known, the modern political division of China proper is into 18 省 *sheng* or provinces with the subdivisions 府 *fu* (prefectures) 181, 直隸州 *chi li chou* (independent departments) 67, 州 *chou* (departments, dependent on a *fu*) 143, and 縣 *hien* (districts, the lowest division of a province, dependent on a prefecture or an independent department) 1,279.

The meaning of the character 州 *chou* has varied greatly in course of time. Originally the *nine* provinces into which ancient China was divided by Emperor YAO [B.C. 2360] were termed *chou.* His successor SHUN [B.C. 2255] divided the Empire into twelve *chou.* YÜ, the first Emperor of the 夏 Hia dynasty which reigned in China B.C. 2205–1766, re-established the division into *nine* provinces, and these nine *chou* continued during the 商 Shang (or 殷 Yin) [1766–1122] and 周 Chou [1122–249 B.C.].

The China of the 周 Chou dynasty lay between the 33rd and 38th parallels and occupied only about two-thirds of the present China proper, reaching to the south nearly half-way from the Yellow River to the Yang tsz'. It consisted of the royal state held by the kings (王 *wang*) themselves,

which was situated along the Wei and Ho rivers [in Shen si
and Ho nan], and a number of larger and smaller feudal
states, surrounding the royal dominions. WU-WANG, the
founder of the Chou dynasty, resided at 豐 Feng and 鎬
Hao [both in Shen si, near present Si an fu]. He built
also another residence in the east on the river 洛 Lo, which
was called 洛 邑 Lo i (afterwards 洛 陽 Lo yang, near
present Ho nan fu); but it was not until many centuries
later, since PING WANG [770–717], that the royal residence
was fixed at Lo yang.

The Chou dynasty was overthrown in the middle of the
3rd cent. B.C. by the princes of the powerful state of
Ts'in. One of them, Cheng, who ruled B.C. 249–210,
reduced all the petty states to his sway and in 221 took
the title SHI HUANG-TI (Emperor). His dynasty, which
lasted only 20 years, is called 秦 Ts'in. SHI HUANG-TI
succeeded in establishing his authority over the greater
part of China proper, with the exception of the south-western
regions (Kui chou and Yün nan). He fixed his residence at
咸 陽 Hien yang (now Hien yang hien, N.W. of Si an fu,
Shen si) and divided the empire, including the vast exten-
sions he had annexed towards the south, into 40 郡 kün or
provinces.

The next dynasty was the 漢 Han, which reigned in
China more than four centuries. The Chinese historians
distinguish the Earlier and the Later Han.

The 前 漢 Ts'ien Han or Earlier Han, called also the
Western Han [B.C. 202–A.D. 25]. In the reign of WU
TI [140-86] China, which then had about the same limits
as what is now called China proper, was divided into 13
州 chou or 部 pu. The subdivisions were 103 郡 kün or
prefectures on which depended 1,314 縣 hien (districts) and
邑 i (towns). Besides these, there were scattered over the

empire 241 侯國 *hou kuo* (small feudal states). The capital of China was at 長安 Ch'ang an, in Shen si, near present Si an fu. A *hien* comprised in the prefecture of Si an fu still bears the name Ch'ang an.

The 後漢 Hou Han or Later Han, called also Eastern Han from its capital 洛陽 Lo yang near present Ho nan fu. This dynasty reigned A.D. 25–220. The political division of the empire was not essentially changed.

三國 *San kuo*, the Three Kingdoms or Dynasties established in China after the downfall of the house of Han :—

1.—蜀 Shu or 蜀漢 Shu Han [A.D. 221-264]. This was regarded as the legitimate dynasty from its affinity with the Han. It ruled over Sz ch'uan (then called Shu), Kui chou and Yün nan. The capital was at 成都 Ch'eng tu (now Ch'eng tu fu in Sz ch'uan). The kingdom comprised 22 *kün* (prefectures) and *kuo* (feudal states).

2.—吳 Wu [A.D. 229-264]. This kingdom occupied the eastern part of Mid China,—Kiang su, Southern An hui, Kiang si, Fu kien, Hu kuang. Political division : 5 *chou* or provinces, 43 *kün* and *kuo*. Capital at 建鄴 Kien ye (Nan king).

3.—魏 Wei [A.D. 221-264]. This dynasty ruled over modern Ho nan, Shen si, Shan si, Shan tung and Chi li. 13 *chou* or provinces, with 91 *kün* and *kuo*. Capital at 洛陽 Lo yang.

The 晉 Tsin dynasty [A.D. 265-420] reigned again, till 317 at least, over the whole of China. Political division : 19 *chou* or provinces, 173 *kün* and *kuo* and 1,109 *hien*. The capital was at first at Lo yang, but owing to the invasions of Northern China by the Hiung nu and other Tartar and Tungus tribes, it was moved in 317 to 建康 Kien k'ang (present Nan king). 16 small foreign kingdoms or dynasties were then established in the northern provinces. They were all subsequently destroyed by the Pei Wei.

The 北魏 Pei (Northern) Wei dynasty [386-558] was of Tungus origin. It swayed Southern Mongolia and the northern part of China, and, encroaching upon the dominions of the Southern Sung and the other Southern Dynasties [v. infra], finally occupied also Mid China as far as the Yang tsz'. Political division of the empire : 113 州 chou (corresponding to the present fu or prefectures), 519 郡 kün (departments) and 1,352 hien (districts). The capital was at first at 代 Tai (now Ta t'ung fu in Shan si). In 495 it was moved to Lo yang. In 532 the Pei Wei dynasty split into two branches—the Tung or Eastern Wei [532–550], capital at 鄴 Ye (present Chang te fu in N. Ho nan) and the Si or Western Wei [532-558] with the capital at Ch'ang an. These two Wei dynasties were finally overthrown by two other short-lived dynasties :—

The 北齊 Pei (Northern) Ts'i dynasty [550-577] replaced the Eastern Wei ; the 北周 Pei (Northern) Chou dynasty [558-581] replaced the Western Wei and in 577 overthrew the Pei Ts'i. The Pei Chou was itself destroyed by the Sui [v. infra].

The Pei Wei, Pei Ts'i and Pei Chou are known in Chinese history under the general name of 北朝 Pei ch'ao (Northern Dynasties).

In the Southern half of China the Tsin dynasty was replaced by the 宋 Sung dynasty, more generally termed 南宋 Nan (Southern) Sung dynasty, to distinguish it from the Sung dynasty which reigned in China from the 10th to the 13th century. The dominions of this Nan Sung dynasty at first comprised South and Mid China and present Shan tung. Political division : 22 chou or provinces with 277 kün and 1,357 hien. The capital was at 建康 Kien k'ang (Nan king). Gradually the Northern Wei encroached upon the empire of the Nan Sung and drove them back beyond the Huai river. The Nan Sung dynasty was overthrown by the—

南齊 Nan (Southern) Ts'i dynasty [479-502], and this in turn was replaced by the—

梁 Liang dynasty [502-557]. The Pei Wei in the north continued their encroachments upon the southern empire, and finally the Yang tsz' formed the boundary between the northern and the southern empire. The dominions of the Liang were divided into 23 *chou* or provinces with 350 *kün* or prefectures and 1,203 *hien* or districts. The capital was at Kien k'ang (Nan king). The Liang dynasty was overthrown by the—

陳 Chen dynasty [557-589], which swayed the southern half of China and in turn was destroyed by the Sui [see *infra*].

The Nan Sung, Nan Ts'i, Liang and Chen dynasties are known in Chinese history as the 南朝 Nan ch'ao or Southern Dynasties.

The 隋 Sui dynasty [581-619]. All China, after it had been divided more than three centuries and a half, was again re-united and brought under the sway of this dynasty. The old division of China into 9 *chou* or provinces was again adopted with the old names of the Yü kung. The subdivisions were 190 *kün* and 1,255 *hien*. The capital was first at Ch'ang an ; in 605 the court moved to Lo yang.

The 唐 T'ang dynasty [618-907]. The second emperor of this celebrated dynasty [in 627] established a new political division of China. The empire was divided into 10 道 *tao* (circuits, or provinces which comprised 360 州 *chou* or prefectures). The larger *chou* were termed 府 *fu*. There were 1,557 縣 *hien* or districts. During the ruling of the T'ang the capital was at Ch'ang an.

In the reign of MING TI [713-756] another division of China proper took place. Some of the larger provinces were divided and there were then 15 *tao*. The term *chou* for prefecture was replaced by the older term *kün*. The 15 provinces comprised 328 *kün*, of which 49 were termed 都督府 *tu tu fu* (seats of a governor-general) and 12 大都督府 *ta* (great) *tu tu fu*. There were 1,573 *hien* or districts.

After the downfall of the T'ang dynasty five short-lived dynasties followed in succession. None of them ruled over the whole of China, for ten smaller independent kingdoms had risen in different parts of China, and the Ki tan or Liao penetrated China from the north. This period is known in Chinese history under the name of 五代 Wu tai, the Five Dynasties [907-960], *viz.* :—

1.—The 後梁 Hou (later) Liang [907-923]. The capital was at 汴 Pien (now K'ai feng fu in Ho nan).
2.—The 後唐 Hou T'ang [923-936]. Capital at Lo yang.
3.—The 後晉 Hou Tsin [936-946]. Capital at Pien.
4.—The 後漢 Hou Han [946-951].
5.—The 後周 Hou Chou [951-960].

The 宋 Sung dynasty [960-1280]. This dynasty succeeded in destroying all the small kingdoms and reuniting all China into one empire. In 997 China was divided into 15 路 *lu* (circuits or provinces). In the beginning of the 12th century there were 22 *lu* with 10 *tu tu fu*, 40 *fu* (larger prefectures), 245 *chou* (smaller prefectures), 1,221 *hien* (districts) and 69 軍 *kün* (military districts). The Sung had their capital at 開封府 K'ai feng fu, also called 大梁 Ta Liang (now K'ai feng fu in Ho nan). But as in 1126 the Sung were forced to abandon the northern part of their empire, as far south as the Han and Huai rivers, to the Kin, the capital was moved to 杭州 Hang chou, also called 臨安府 Lin an fu (now Hang chou fu in Che kiang). In 1276 the Mongols took Lin an fu, and in 1280 the whole empire of the Sung was in the hands of the Mongols.

Since the first half of the 10th century Northern China was detached from native rule and subject to foreign dynasties,—first to the 契丹 Ki tan, a people of Tungus origin who conquered Mongolia and in 936 wrenched from the emperor KAO TSU of the Hou Tsin dynasty the northern part of the present provinces of Chi li and Shan si. The dynasty of the Ki tan, known in Chinese history as the 遼 Liao, subsisted from 916 to 1125, when it was displaced by another Tartar dynasty, the Churche or 金 Kin [1125-1234], who, having subdued the whole of Mongolia, succeeded also in conquering from the Sung all the provinces of North China as far south as the Han and Huai rivers [1127]. The capital of the Kin was in 中都 Chung tu (now Peking). The Kin dynasty was destroyed by the Mongols in 1234.

The 元 Yüan or Mongol dynasty in China, after the destruction of the Kin and the Sung, ruled over China proper from 1260 (or 1280 when the south had also been subdued) to 1368. China was then divided into 12 provinces or governorships (行中書省 hing chung shu sheng). The capital was at 大都 Ta tu (now Peking). There were 187 路 lu or prefectures. In some of them were one or several 府 fu, altogether 42. There were 381 chou or departments and 1,132 hien.

The 明 Ming dynasty [1368-1644]. The political division of China proper in this period was similar to that of nowadays, but there were only 12 provinces. The first two, with the two capitals Peking and Nan king, were termed 北直隸 Pei chi li and 南直隸 Nan chi li (northern and southern

70

independent administration), the other provinces 布政使司
pu cheng shi sz‘ or governorships. Pei chi li is the present
Chi li,—Nan chi li, also called 江南 Kiang nan, comprised
the present provinces Kiang su and An hui. The present
Kan su was included in Shen si. The present Hu pei and
Hu nan formed the province 湖廣 Hu kuang. The other
names of the provinces were as nowadays. *140 fu, 193 chou*
and *1,138 hien.*

The following identifications of ancient geographical
names occurring in the *Pen ts‘ao kang mu* are for the greater
part based upon the geographical sections of the Chinese
dynastic histories. The Chinese geographical dictionary
歷代地理志 *Li tai ti li chi* [see *Bot. sin.,* I, p. 69] was
compiled from the same sources, but it refers only to the
prefectures, departments and districts. The names of the
ancient provinces of China, so frequently noticed in the *Pen
ts‘ao kang mu*, are not included. BIOT in his *Dictionnaire
des noms anciens et modernes des villes, etc. dans l'empire
chinois* (1842), translated from the 廣輿記 *Kuang yü ki*
[see *Bot. sin.,* I, p. 69], also does not give the names of the
provinces, nor are they given in PLAYFAIR'S *Cities and Towns
of China* (1879).

1.—安陸 An lu. *Pie lu.*
    Early Han : *hien.*   Now : Hu pei, Te an fu.

2.—安東 An tung. CH‘EN TS‘ANG-K‘I.
    Sung : *chou.* Now : Kiang su, Huai an fu, An tung
hien.

3.—章安 Chang an. T‘AO HUNG-KING.
    Later Han : *hien.* Tsin dynasty : *hien,* depending on
Lin hai kün [see *infra,* 192]. Now : Che kiang, T‘ai chou fu.

4.—章武 Chang'wu. *Pie lu.*

Later Han : *hien*, depending on Pu hai kün [see *infra*, 262]. Now : Chi li, T'ien tsin fu, Ts'ang chou.

5.—彰德 Chang te. SU SUNG.

Wu tai : *fu.* Now : Ho nan, Chang te fu.

6.—長安 Ch'ang an. T'AO HUNG-KING and K'OU TSUNG-SHI.

Ancient capital of China during the Han and T'ang dynasties. Now : Shen si, Si an fu.

7.—昌陽 Ch'ang yang. *Pie lu.*

Early Han : *hien.* T'ang : *hien.* Now : Shan tung, Teng chou fu, Lai yang hien.

8.—常山 Ch'ang shan. *Pie lu.*

Han and Tsin : *kün.* Now : Chi li, Cheng ting fu. There is also a mountain of this name. See *Medicinal plants,* 141.

9.—朝鮮 Chao sien. *Pie lu.*

Ancient name for Corea since the Han dynasty and still in use.

10.—浙 Che. SU SUNG and K'OU TSUNG-SHI.

We meet also with the terms 二浙 Rh Che or 兩浙 Liang Che, both meaning "the two Che," two provinces (*lu*) of China in the Sung period, *viz.* 浙西路 Che si lu (Western Che) and 浙東路 Che tung lu (Eastern Che), corresponding to present Che kiang and a part of Kiang su.

11.—眞定 Chen ting. *Pie lu.*

Early Han : *kuo.* Now : Chi li, Cheng ting fu.

12.—岑州 Ch'en chou. T'AO HUNG-KING [*Med. plants,* 314]. Not ascertained.

13.—陳留 Ch'en liu. *Pie lu.*

Early Han : *kün.* Now : Ho nan, K'ai feng fu, Ch'en liu hien.

14.—陳倉 Ch'en ts'ang. *Pie lu.*

Early Han : *hien.* Now : Shen si, Feng siang fu, Pao ki hien.

15.—郯 州 Cheng chou. Su Kung and Su Sung.

T'ang : *chou*, Sung : *chou*. Now : Ho nan, K'ai feng fu, Cheng chou.

16.—郯 山 Cheng shan. *Pie lu.*

According to T'ao Hung-king [*Med. plants*, 12] same as Nan cheng. See 226.

17.—儕 Ch'eng. T'ao Hung-king.

Ancient name for Ho nan. See *W.D.*, 31.

18.—成 州 Ch'eng chou. Su Kung and Su Sung.

T'ang, Sung : *chou*. Now : Kan su, Kie chou, Ch'eng hien.

19.—成 德 軍 Ch'eng te kün. Su Sung.

Not ascertained.

20.—承 縣 Ch'eng hien. *Pie lu.*

Early Han : *hien*, in Tung hai kün. See 372.

Now : Shan tung, Yen chou fu, I hien.

21.—池 州 Ch'i chou. Su Sung.

T'ang and Sung : *chou*. Now : An hui, Ch'i chou fu.

22.—朱 厓 Chu yai. *Pie lu* [*Med. plants*, 118].

Later Han : *hien*. Now : Island of Hai nan, K'iung chou fu.

23.—處 州 Ch'u chou. Su Sung.

T'ang and Sung : *chou*. Now : Che kiang, Ch'u chou fu.

24.—楚 Ch'u (Ts'u) and 楚 地 Ch'u ti (country). *Pie lu.* Wu P'u.

A large feudal state in the Chou dynasty occupying present Hu nan and Hu pei. Same as 荆 King. See 145.

25.—滁 州 Ch'u chou. Su Sung.

T'ang and Sung : *chou*. Now : An hui, Ch'u chou.

26.—川 Ch'uan. Su Sung and Li Shi-chen.

Su Sung writes also 川 蜀 Ch'uan Shu or 蜀 川 Shu Ch'uan and 川 西 Ch'uan si. All these names denote the western part of present Sz ch'uan. Comp. also 蜀 Shu [292] and 巴 Pa [235], which denotes the eastern part of present

Sz ch'uan and is also termed 川東 Ch'uan tung. Li Shi-chen [*P.*, XXXII, 2, article 蜀椒] states:—川 Ch'uan is a general name for Pa and Shu. This country is also called 四川 Sz ch'uan (the four rivers), for four large rivers run through it, *viz.* the 岷水 Min shui, 沱水 T'o shui, 黑水 Hei (black) shui and 白水 Pai (white) shui. These are the four principal affluents of the Yang tsz' from the north. The Hei shui (otherwise called Kia ling) flows into the great river at Ch'ung k'ing fu [or according to modern maps it is an affluent of the Kia ling], the T'o shui at Lu chou and the Min shui at Sü chou fu.—The Pai shui seems to be the Ya lung kiang.

27.—莊浪 Chuang lang. Li Shi-chen.

Now: Kan su, Liang chou fu, Chuang lang t'ing.

28.—終南 Chung nan. Han Pao-sheng.

A famous mountain in Southern Shen si, mentioned in the *Shi king.* Also Sung: *hien.* Now: Shen si, Si an fu, Chou chi hien.

29.—鍾山 Chung shan. Wu P'u.

Name of a mountain in Kiang su, Kiang ning fu.

Also: Sui, T'ang: *hien.* Now: Ho nan, Ju ning fu, Sin yang chou.

30.—中牟. Chung mou. *Pie lu* and T'ao Hung-king.

Early Han: *hien.* Now: Ho nan, K'ai feng fu, Chung mou hien.

31.—中山 Chung shan. *Pie lu.*

This place is mentioned in the *Tso ch'uan.*—Early Han: *kuo.* Now: Chi li, Ting chou.—T'ao Hung-king [*Med. plants*, 138] says it was in Tai [321], which was in N. Shan si.

32.—中臺 Chung t'ai. *Pie lu* [*Med. plants*, 313]. Not ascertained.

33.—中岳 Chung yo. *Pie lu.*

One of the five celebrated mountains, same as Sung shan [*infra*, 317].

34.—中原 Chung yüan. *Pie lu.*

A name for Ho nan. See *W.D.*, 1133.

35.—房州 Fang chou. Su Kung, Chang Yü-si and Su Sung.

T'ang and Sung: *chou.* Now: Hu pei, Yün yang fu, Fang hien.

36.—房陵 Fang ling. *Pie lu.*

Early Han: *hien.* Now: Hu pei, Yün yang fu, Fang hien.

37.—飛鳥 Fei wu. *Pie lu.*

According to the *Li tai*, etc. the name appears first in the Sui dynasty: *hien.* Now: Sz ch'uan, T'ung ch'uan fu, Chung kiang hien.

38.—汾州 Fen chou. Ch'en Ts'ang-k'i and Su Sung.

T'ang and Sung: *chou.* Now: Shan si, Feng chou fu.

39.—鳳州 Feng chou. Su Sung.

T'ang: *chou.* Now: Shen si, Han chung fu, Feng hien.

40.—馮翊 Feng i. *Pie lu.*

One of the three prefectures of Ch'ang an, to the N.E. of the capital. See *infra*, 265, San fu.

41.—奉高 Feng kao. *Pie lu.*

Early Han: *hien.* Now: Shan tung, T'ai an fu.

42.—涪都 Fou tu. Su Sung.

Not ascertained. Perhaps 涪州 Fou chou, which name exists since the Wu tai period. Now: Sz ch'uan, Ch'ung king fu, Fou chou.

43.—鄜州 Fu chou. Su Kung and Su Sung.

T'ang and Sung: *chou.* Now: Shen si, Fu chou.

44.—扶風 Fu feng. T'ao Hung-king.

One of the three prefectures of Ch'ang an, to the N.W. of the capital. See *infra*, 265, San fu.

45.—傅高 Fu kao. *Pie lu.*

Not ascertained.

46.—福 州 Fu chou. Su Sung.

T'ang and Sung: *chou*. Now: Fu kien, Fu chou fu.

47.—福 禄 縣 Fu lu hien. Han Pao-sheng.

Later Han: *hien*, T'ang: *hien*. Now: Kan su, Su chou fu, Kao t'ai hien.

48.—海 州 Hai chou. Su Sung.

T'ang and Sung: *chou*. Now: Kiang su, Hai chou.

49.—海 濵 Hai pin. *Pie lu* [*Med. plants*, 139].

Hai pin means "sea-shore." The *Pie lu* intends a locality. In the *Li tai*, etc. the name Hia pin appears first in the Liao dynasty: *hien*. Now: Chi li, Yung p'ing fu, Fu ning hien.

50.—海 西 Hai si. *Pie lu*.

Early Han: *hien*. Now: Kiang su, Hai chou.

51.—海 鹽 Hai yen. Ta Ming [10th cent.].

Early Han, T'ang and Sung: *hien*. Now: Che kiang, Kia hing fu, Hai yen hien.

52.—漢 Han. Su Kung and Su Sung.

Han is the name of a river in Shen si and Hu pei, a northern tributary of the Yang tsz'.—Han was also an ancient name for Sz ch'uan. Liu Pang, the founder of the Han dynasty [B.C. 202], was prince of 漢 Han, which principality comprised 蜀 Shu and 巴 Pa (Western and Eastern Sz ch'uan). In the San kuo period [3rd cent.] 蜀漢 Shu Han was one of the three kingdoms and occupied present Sz ch'uan.

53.—漢 州 Han chou. Su Sung.

T'ang and Sung: *chou*. Now: Sz ch'uan, Ch'eng te fu, Han chou.

54.—漢 中 Han chung. *Pie lu* and T'ao Hung-king.

Ts'in and Han: province occupying the southern part of present Shen si (Han chung fu, etc.) and the N.W. of Hu pei.

55.—函谷 Han ku. *Pie lu.*

In Han ku LAO TSZ' wrote his *Tao te king.* See *W.D.,*
163. The name was in use during the Ts'in dynasty. Now:
Ho nan, Shen chou, Ling pao hien. Comp. also *infra,* 359,
Ts'in kuan.

56.—邯鄲 Han tan. *Pie lu* and WU P'U.

Ts'in: *kŭn,* Han: *hien.* Now: Southern Chi li, Kuang
p'ing fu and Cheng te fu.

57.—寒石山 Han shi shan. WU P'U.

Not ascertained.

58.—杭 Hang or 杭州 Hang chou. SU SUNG.

T'ang and Sung: *chou.* Now: Che kiang, Hang chou fu.

59.—濠 Hao. SU SUNG.

T'ang and Sung: *chou.* Now: An hui, Feng yang fu.

60.—黑水 Hei shui. T'AO HUNG-KING.

Name of one of the northern affluents of the Yang tsz'
[*v. supra,* 26] in S. Kan su and N.E. Sz ch'uan. The Hei
shui is mentioned in the *Yŭ kung* or Tribute of Emperor Yŭ.

61.—衡山 Heng shan. *Pie lu* and SU SUNG.

One of the five sacred mountains of China, in Hu nan.
—Also name of a district. Tsin, T'ang and Sung: *hien.*
Now: Hu nan, Heng chou fu, Heng shan hien.

62.—歙州 Hi chou. SU SUNG.

T'ang and Sung: *chou.* Now: An hui, Hui chou fu.

63.—下邳 Hia P'ei. *Pie lu.*

Early Han: *hien.* Now: Kiang su, Sü chou fu, P'ei
chou.

64.—硤州 Hia chou, also written 峽州. SU KUNG and
HAN PAO-SHENG.

T'ang and Sung: *chou.* Now: Hu pei, I ch'ang fu.

65.—咸陽 Hien yang. *Pie lu.*

Hien yang was the residence of TS'IN SHI HUANG-TI
[B.C. 221]. It was situated near ancient Ch'ang an.

T'ang: *hien.* Now: Shen si, Si an fu, Hien yang hien.

66.—奭 州 Hing chou. Su Sung.

T'ang and Sung : *chou.* Now : Shen si, Han chung fu, Lo yang hien.

67.—奭 國 Hing kuo. Su Sung.

Sung : 軍 *kün.* Now : Hu pei, Wu ch'ang fu, Hing kuo chou.

68.—奭 元 府 Hing yüan fu. Su Sung.

T'ang and Sung : *fu.* Now : Shen si, Han chung fu.

69.—熊 耳 Hiung rh. *Pie lu.*

Name of a mountain in Ho nan, S.W. of Shen chou.

Also Sui : *hien.* Now : Ho nan, Ho nan fu, Yung ning hien.

69b.—合 州 Ho chou. Su Kung and Su Sung.

T'ang : *chou.* Now : Sz ch'uan, Ch'ung k'ing fu, Ho chou.

70.—合 浦 Ho p'u. *Nan fang ts'ao mu chuang.*

Han : *kün.* South-west part of Kuang tung province. There is now Ho p'u hien in Lien chou fu.

71.—和 州 Ho chou. Su Sung.

T'ang and Sung : *chou.* Now : An hui, Ho chou.

72.—河 Ho. Su Sung.

Ho, the Yellow River. The provinces near it, as Ho pei, Ho nan, Ho tung and Ho nei [see 78, 76, 80, 77] are likewise termed Ho.

73.—河 州 Ho chou. Su Sung.

T'ang and Sung : *chou.* Now : Kan su, Lan chou fu, Ho chou.

74.—河 中 Ho chung and 河中府 Ho chung fu. Su Kung and Su Sung.

T'ang and Sung : *fu.* Now : Shan si, P'u chou fu.

75.—河 間 Ho kien. *Pie lu.*

Han : *kün.* Now : Chi li, Ho kien fu.

71

76.—河南 Ho nan. *Pie lu.*

Early Han : *kün*. Now : Ho nan fu in Ho nan province.
The name Ho nan as that of a province dates from the
Mongol period.

77.—河內 Ho nei. *Pie lu.*

Han : *kün*, occupied South-east Shan si and North Ho nan.
—Ho nei hien is now a district dependent on Huai k'ing fu.

78.—河北 Ho pei. T'AO HUNG-KING, SU KUNG and SU
SUNG.

Early Han : *hien*. Now : Shan si, Kie chou, Jui ch'eng
hien.

In the T'ang and Sung periods Ho pei was the name
of a province (*tao*) and occupied South Chi li and West Shan
tung.

79.—河西 Ho si. *Pie lu.*

According to the History of the Later Han, Ho si
(west of the Yellow River) comprised the districts 武威
Wu wei, 張掖 Chang i, 酒泉 Tsiu ts'üan, 敦煌 Tun huang
and 金城 Kin ch'eng, *i.e.* present Kan su from Lan chou fu
to An si chou. TOU YUNG [† A.D. 62. *See* MAYERS' *Chin.
R. Man.*, 679] was Viceroy of Ho si.

80.—河東 Ho tung. *Pie lu* and SU SUNG.

Ts'in and Han : *kün*. South-west corner of present Shan
si. The province Ho tung in the T'ang and Sung periods
occupied almost the whole of present Shan si.

81.—河陽 Ho yang. SU SUNG.

Early Han, T'ang and Sung : *hien*. Now : Ho nan,
Huai k'ing fu, Meng hien.

82.—胡戎 Hu Jung. LI SHI-CHEN.

Western Barbarians. North-east Tibet, Kukonor.

83.—湖 Hu : The lakes, *i.e.* the lakes in Central China,
especially the Tung t'ing hu, and the provinces south and
north of it, 湖南 Hu nan and 湖北 Hu pei or 湖廣 Hu

kuang. All these appellations occur in the authors of the Sung period. In this period the present Hu kuang was divided into two provinces—荆湖南 King Hu nan and 荆湖北 King Hu pei. See *infra*, 147.

84.—湖湘 Hu siang. Su Sung.

A name for present Hu nan. See 307, sub Siang.

85.—華州 Hua chou. Su Kung.

T'ang and Sung: *chou*. Now: Shen si, T'ung chou fu, Hua chou.

86.—華山 Hua shan. *Pie lu*.

One of the sacred mountains of China, in Shen si, T'ung chou fu, Hua yin hien.

Also Pei Wei: *kün*. Now: Shen si, T'ung chou fu.

87.—華陰 Hua yin. *Pie lu*.

Early Han: *hien*. Now: Shen si, T'ung chou fu, Hua yin hien.

88.—華原 Hua yüan. Su Kung.

Sui, T'ang and Sung: *hien*. Now: Shen si, Si an fu, Yao chou.

89.—淮 Huai. Su Sung.

Name of a great river in Mid China, between the Yang tsz' and the Yellow River. The name is also applied to the country north and south of it, present An hui and Kiang su. Huai is frequently coupled with Kiang [see 124], 淮江 Huai Kiang or also 江淮 Kiang Huai.

90.—淮南 Huai nan. *Pie lu* and Su Sung.

The Huai nan of the Early Han occupied the middle part of present Kiang su between the Yang tsz' and the Huai river. It was the 廣陵郡 Kuang ling *kün* of which the celebrated LIU AN or HUAI NAN WANG [† B.C. 122] was the feudal prince. [*See* MAYERS' *Chin. R. Man.*, 412.]

Tsin: Huai nan kün. Now: An hui, Feng yang fu.

T'ang and Sung: Huai nan, name of a province occupying the southern part of present An hui and Kiang su, *i.e.* the land between the Yang tsz' and the Huai.

91.—淮陽 Huai yang. *Pie lu.*

Early Han : *kuo.*   Now : Ho nan, Ch'en chou fu.

Nan Sung : *kün.*   Now : Kiang su, Huai an fu.

92.—淮源 Huai yüan. *Pie lu.*

In the *Li tai,* etc. this name is first mentioned in the Sui period : *hien.*   Now : Ho nan, Ju ning fu, Sin yang chou.

93.—懷州 Huai chou. Su Sung.

T'ang and Sung : *chou.*   Now : Ho nan, Huai k'ing fu.

94.—懷慶 Huai k'ing. Su Sung [*Med. plants,* 101].

According to the *Li tai,* etc. this name appears first in the Yüan period : *lu* (prefecture).   Now : Ho nan, Huai k'ing fu.

95.—槐里 Huai li. *Pie lu.*

Name of an ancient capital of China. " Bamboo annals " [Legge's *Shu king,* Proleg., 152]:—King E [B.C. 934-909] removed to Huai li.

In the Later Han Huai li was the chief city in the prefecture of 右扶鳳 Yu Fu feng.   Now : Shen si, Si an fu, Hing p'ing hien.

96.—黃澤 Huang tse. Su Sung [*Med. plants,* 51].

Not ascertained.

97.—徽州 Hui chou. Wan Ki [16th cent.].

Now : An hui, Hui chou fu.

98.—會稽 Hui ki. *Pie lu* and T'ao Hung-king.

Hui ki was originally the name of the capital of the ancient kingdom of Yüe [*v. infra,* 418].

Ts'in : province, Eastern Che kiang and Southern Kiang su.

Han and Tsin : *kün.*   Now : Che kiang, Shao hing fu.

99.—弘農 Hung nung. *Pie lu.*

Early Han: *kün*. In the North-west part of Ho nan, Ho nan fu and Shen chou, and Shen si, Hua chou.

100.—霍山 Huo shan. *Pie lu.*

According to T'AO HUNG-KING [*Med. plants*, 266] this is the same as the Heng shan mountain in Hu nan. See 61.

101.—易州 I chou. HAN PAO-SHENG.

T'ang and Sung: *chou*. Now: Chi li, I chou.

102.—益州 I chou. *Pie lu*, WU P'U and T'AO HUNG-KING.

In the Han period I chou, name of a province, occupying present Sz ch'uan, a part of Kui chou and Yün nan. I chou was then also the name of a *kün* = present Yün nan.—Tsin dynasty: I chou, name of a province = present Sz ch'uan and part of Kui chou.

103.—宜州 I chou. SU KUNG and SU SUNG.

T'ang and Sung: *chou*. Now: Kuang si, K'ing yüan fu.

104.—宜都 I tu. T'AO HUNG-KING.

Tsin: *kün*. Now: Hu pei, King chou fu, I tu hien and I ch'ang fu.

105.—翼州 I chou. SU KUNG.

T'ang: *chou*. Now: Sz ch'uan, Lung an fu.

106.—沂州 I chou. SU SUNG.

T'ang and Sung: *chou*. Now: Shan tung, I chou fu.

107.—義陽縣 I yang hien. T'AO HUNG-KING.

Tsin: *hien*. Now: Ho nan, Nan yang fu, T'ung po hien.

108.—伊陽縣 I yang hien. K'OU TSUNG-SHI.

T'ang: *hien*. Now: Ho nan, Ho nan fu, Sung hien. The present I yang hien lies east of Sung hien.

109.—饒州 Jao chou. K'OU TSUNG-SHI.

T'ang and Sung: *chou*. Now: Kiang si, Jao chou fu.

110.—汝南 Ju nan. *Pie lu.*

Early Han : *kün.* Now : Ho nan, Ju ning fu.

111.—潤州 Jun chou. Su Kung and Su Sung.

T'ang and Wu tai : *chou.* Now : Kiang su, Chen kiang fu, Tan t'u hien.

112.—戎州 Jung chou. Su Sung.

T'ang and Sung : *chou.* Now : Sz ch'uan, Sü chou fu.

113.—開州 K'ai chou. Su Kung.

T'ang : *chou.* Now : Sz ch'uan, K'uí chou fu, K'ai hien.

114.—甘松 Kan sung. T'ao Hung-king.

Nan Ts'i : *kün.* Now : Sz ch'uan [unknown in what part of it].

115.—藁城 Kao ch'eng. *Pie lu.*

Early Han : *hien.* Now : Chi li, Cheng ting fu, Kao ch'eng hien.

116.—高麗 Kao li. T'ao Hung-king.

Northern part of present Corea. Mentioned in the Chinese annals since the 5th century.

117.—高昃 Kao liang or 高凉 Kao liang. *Pie lu.*

Early Han : *hien.* Now : Kuang tung, Ch'ao king fu, Yang kiang hien. Biot [56] identifies Kao liang with Kao chou fu, which lies west of Yang kiang hien.

118.—高山 Kao shan. *Pie lu.*

Early Han : *hien.* In present Kiang su [unknown in what part of it].

119.—冀州 Ki chou. T'ao Hung-king and Su Sung.

Han : province = Northern Chi li.—T'ang and Sung : *chou.* Now : Ki chou in Chi li.

120.—岐州 K'i chou. Su Kung.

Sui : *chou.* Now : Shen si, Feng siang fu [*see* Biot, 23].

121.—蘄 州 K'i chou. Li Shi-chen.

Now: Hu pei, Huang chou fu, K'i chou.

122.—嘉 州 Kia chou. Su Sung.

T'ang and Sung: *chou*. Now: Sz ch'uan, Kia ting fu.

123.—絳 州 Kiang chou. Su Sung.

T'ang and Sung: *chou*. Now: Shan si, Kiang chou.

124.—江 Kiang. Su Kung and Su Sung.

Kiang is the Chinese name of the great river in Mid China, which Europeans are accustomed to term Yang tsz' kiang. It denotes also the provinces of Mid China, situated south of the Yang tsz'.

Southern Sung and Nan Ts'i: 江 Kiang or 江 州 Kiang chou, name of a province, occupying modern Kiang si and a part of Fu kien and Hu pei. The *Pie lu*, Kuo P'o and T'ao Hung-king use the name 江 南 Kiang nan (South of the Kiang) to designate the regions south of the Kiang. The term 江 東 Kiang tung (East of the Kiang) in the San kuo period [3rd cent.] referred to the eastern part of the same regions, *i.e.* Southern An hui, Kiang su, Che kiang.

In the T'ang period 江 南 Kiang nan was a vast province (*tao*) occupying present Hu nan, Kiang si, Southern An hui and Kiang su, Che kiang, Fu kien. In A.D. 734 Kiang nan was divided into two provinces—江 南 西 道 Kiang nan si (western) *tao* and 江 南 東 道 Kiang nan tung (eastern) *tao*. The latter occupied present Fu kien, Che kiang and the southern part of Kiang su.—The same names subsisted during the Sung dynasty, only the provinces were much smaller. 江 南 西 路 Kiang nan si lu corresponded to present Kiang si and 江 南 東 路 Kiang nan tung lu to the southern half or An hui.

The terms 江 西 Kiang si and 江 右 Kiang yu (right hand), frequently used by the authors of the T'ang and Sung, have the same meaning as 江南西 Kiang nan si,—and 江東 Kiang tung or 江左 Kiang tso are used for 江南東 Kiang nan tung. [126]

Now 江南 Kiang nan means the provinces of An hui and Kiang su.

125.—江州 Kiang chou.　Su Sung.

T'ang and Sung : *chou.*　Now : Kiang si, Kiu kiang fu.

126.—江夏 Kiang hia.　*Pie lu.*

Early Han : *kün.*　Now : Hu pei, Te an fu.

127.—江林 Kiang lin mountains.　*Pie lu.*

According to T'ao Hung-king [*Med. plants*, 141] same as Kiang yang mountains [see 130].

128.—江陵府 Kiang ling fu.　Su Sung.

T'ang and Sung : *fu.*　Now : Hu pei, King chou fu.

129.—江寧 Kiang ning.　T'ao Hung-king, Ch'en Ts'ang-k'i and Su Sung.

Tsin : *hien,* Sui : *hien,* Sung : *ju.*　Now : Kiang su, Kiang ning fu (Nan king).

130.—江陽 Kiang yang.　T'ao Hung-king.

Early Han : *hien,* Tsin : *kün.* Now : Sz ch'uan, Lu chou.

Also Nan Sung : *kün.* Now : Sz ch'uan, Mei chou, P'eng shan hien.

131.—羌 K'iang.　*Pie lu* and T'ao Hung-king.

North-east Tibet, Kukonor.　See 300, Si K'iang.

132.—交 Kiao or 交州 Kiao chou.　*Pie lu,* T'ao Hung-king and Su Kung.

---

[126] Williams [*Dict.,* 862] is wrong in translating 江右 and 江左 by south and north sides or right and left banks of the Yang tsz' river. Right and left (or west and east) are here to be understood as referring to Kiang nan or the country south of the Kiang and in the same sense as in 山右 Shan yu, the province of Shan si, and 山左 Shan tso, the province of Shan tung.　See W.D. 1002.

Han: province = present province of Kuang tung and part of Kuang si.

San kuo, Tsin and Nan Ts'i: province = South-west part of Kuang tung.

133.—交 趾 Kiao chi. *Pie lu* and T'AO HUNG-KING.
Ancient name for Cochinchina. Han period.

134.—喬 山 K'iao shan. *Pie lu.*
A mountain in Shen si. See *Med. plants*, 86.

135.—解 州 Kie chou. SU SUNG.
Early Han: *hien.* Wu tai and Sung: *chou.* Now: Shan si, Kie chou.

136.—劍 南 Kien nan. SU KUNG.
T'ang: name of a province occupying the greater part of present Sz ch'uan.

137.—建 康 Kien k'ang. LI SHI-CHEN.
Capital of the Tsin dynasty [A.D. 317]. Now: Nan king in Kiang su.

138.—建 寧 Kien ning. LI SHI-CHEN.
Yüan: *lu*, Ming: *fu.* Now: Kien ning fu, in Fu kien.

139.—建 平 Kien p'ing. T'AO HUNG-KING.
Tsin: *kün.* Now: Hu pei, Shi nan fu and a part of K'ui chou fu in Sz ch'uan.

In the Han dynasty Kien p'ing was in Ho nan, now Kui te fu.

Nan Ts'i: Kien p'ing in Yün nan.

140.—犍 爲 Kien wei. *Pie lu.*
Early Han: *kün.* Part of present Sz ch'uan: Kia ting fu, Sü chou fu, etc., and north-east corner of Yün nan.

141.—黔 K'ien. SU SUNG.
T'ang and Sung: *chou.* Northern part of present province Kui chou [Sz' nan fu, etc.].

72

142.—黔 中 K'ien chung.   HAN PAO-SHENG.

Ts'in: province = W. part of present Hu nan and E.
Kui chou.

T'ang: province.   About the same extent.

143.—金 州 Kin chou.   SU KUNG and SU SUNG.

T'ang and Sung: *chou.*   Now: Shen si, Hing an fu.

144.—欽 州 K'in chou.   SU SUNG and LI SHI-CHEN.

Sung: *chou.*   Now: Kuang tung, Lien chou fu, Ling
shan hien.   Present K'in chou lies south of Ling shan hien.

145.—荆 K'ing, an ancient feudal state, same as 楚 Ch'u
[*see above*, 24].   The *Shi king* writes 荆楚 King Ch'u, as
does also the *Pie lu.*   It corresponds to the present provinces
Hu nan and Hu pei and to King chou in the Tribute of Yü.

. 146.—荆 州 King chou.   *Pie lu*, SU KUNG and SU SUNG.

King chou was one of the ancient nine provinces, as
enumerated in the Tribute of Yü, Hu nan, Hu pei and a part
of Ho nan.

King chou was a province, of about the same extent,
during the Han, Tsin and Sui.

Since the Wu tai period King chou name of a prefecture.
Now: King chou fu in Hu pei.

147.—荆 湖 King hu.   SU SUNG.

Name of two provinces in Central China during the
Sung dynasty, *viz.* :—

     荆 湖 北 King hu pei (northern) = North Hu nan
       and greater part of Hu pei.

     荆 湖 南 King hu nan (southern) = South Hu nan.

148.—荆 南 King nan.   SU SUNG.

Probably King hu nan [*see* 147].

149.—荆 山 King shan.   *Pie lu.*

Not found in *Li tai*, etc.   BIOT [81] says :—King shan,

an old city founded in the Liang dynasty in present An hui, Feng yang fu, north of Huai yüan hien.

Su Kung [*Med. plants*, 152] mentions 荊山縣 King shan hien, a district in Siang chou in Hu pei [see 305]. Not in the *Li tai*, etc.

150.—京兆 King chao. *Pie lu.*

One of the three prefectures which comprised and surrounded the imperial city Ch'ang an [Shen si, Si an fu] in the time of the Early Han. See 265, San fu.

151.—京口 King k'ou.

According to the *Kuang yü ki*, 京口鎭 Kin k'ou chen was an ancient name for Chen kiang fu in Kiang su.

152.—京洛 King Lo. K'ou Tsung-shi.

A name for Lo yang, the ancient capital of China, in Ho nan. See 201.

153.—涇州 King chou. Su Kung.

T'ang and Sung: *chou*. Now: Kan su, King chou.

154.—九眞 Kiu chen. *Pie lu.*

Han: *kün*. In the northern part of Cochinchina.

155.—九疑 Kiu i. *Pie lu.*

Name of a mountain. Kiang si, Yüan chou fu.

156.—九具 Kiu kü. *Pie lu.*

Not ascertained.

157.—九原 Kiu yüan. Li Shi-chen.

Ts'in: *kün*. In the north-west corner of the Ordos.

Han: *hien*. North of the Ordos in 五原郡 Wu yüan kün.

158.—關 Kuan, 關中 Kuan chung. Su Kung and Su Sung.

Kuan = a pass or barrier. Here the celebrated defile 潼關 T'ung kuan in Shen si, near the elbow of the Yellow River, is meant. Kuan chung means "within the pass." This

term was also used in the Ts'in dynasty to designate
Shen si and Kan su. Kuan is frequently used for Kuan
chung. 關西 Kuan si (west of the barrier) has the same
meaning. 關內 Kuan nei (within the barrier) was the name
of a province during the T'ang dynasty occupying the greater
part of present Shen si and Eastern Kan su.

K'ou Tsung-shi uses the term 關陝 Kuan Shen for
Shen si.

159.—觀州 Kuan chou. Su Sung.

Sung. *chou.* Now: Kuang si, K'ing yüan fu, Nan tan
chou.

According to Biot [251] Kuan chou, in the Sui dynasty,
was in Chi li, present Tung kuan hien in Ho kien fu.

160.—廣 Kuang or 廣州 Kuang chou. T'ao Hung-king,
Su Sung and Li Shi-chen.

Tsin: Kuang chou, name of a province comprising the
greater part of present Kuang tung and Kuang si.

Since the T'ang dynasty the name Kuang chou is applied
to present Kuang chou fu or Canton.

Nowadays 廣 Kuang means the provinces of Kuang
tung and Kuang si, also 兩廣 Liang (two) Kuang.

161.—廣漢 Kuang Han. *Pie lu.*

Early Han: *kün.* In present Sz ch'uan, occupying
Ch'eng tu fu, Pao ning fu, Lung an fu, T'ung ch'uan fu and
Mien chou.

162.—廣南 Kuang nan. Su Sung.

Sung: province. Present Kuang tung and part of
Kuang si.

163.—光州 Kuang chou. Su Sung.

T'ang, Sung, Yüan and Ming: *chou.* Now: Ho nan,
Kuang chou.

164.—桂 州 Kui chou.　T‘ao Hung-king and Su Kung.
Liang and T‘ang : *chou*.　Now : Kuang si, Kui lin fu.

165.—桂 林 Kui lin.　*Pie lu* and Ch‘en Ts‘ang-k‘i.
Ts‘in : province.　North-east part of modern Kuang si,
of which Kui lin fu is now the capital.

166.—桂 嶺 Kui ling.　Ch‘en Ts‘ang-k‘i [*Med. plants*,
303.]
Mountain chain in Kuang si.　On Klaproth's map it is
marked east of P‘ing lo fu.

167.—桂 陽 Kui yang.　*Pie lu* and T‘ao Hung-king.
Early Han : *kün*.　South-east part of Hu nan and a
part of Kuang tung province.
Tsin : *hien*, dependent on Shi hing kün [see 289].
Now : Hu nan, Kui yang chou.

168.—媯 州 Kui chou.　Han Pao-sheng.
T‘ang and Sung : *chou*.　Now : Chi li, Süan hua fu, Huai
lai hien.

169.—歸 州 Kui chou.　Su Sung.
T‘ang and Sung : *chou*.　Now : Hu pei, I ch‘ang fu, Kui
chou.

170.—夔 州 K‘ui chou.　Su Sung.
T‘ang and Sung : *chou*.　Now : Sz ch‘uan, K‘ui chou fu.

171.—崑 崙 K‘un lun.　Su Kung.
Name of a celebrated mountain chain north of Tibet.
By the same name in ancient times [T‘ang period] the island
Pu lu Condor was designated.　See my memoir on the
Arabs [14].

172.—均 州 Kün chou.　Su Sung.
T‘ang and Sung : *chou*.　Now : Hu pei, Siang yang fu,
Kün chou.

173.—虢 州 Kuo chou.　Su Kung.
T‘ang : *chou*.　Now : Ho nan, Shen chou, Ling pao hien.

174.—蘭陵 Lan ling. T‘ao Hung-king.

Han and Tsin : *hien*.   Now : Shan tung, Yen chou fu, I hien.

Also Nan Sung : *hien*.   Now : Kiang su, Ch‘ang chou fu.

175.—藍田 Lan t‘ien. *Pie lu* and Su Kung.

Early Han : *hien*, T‘ang : *hien*.   Now : Shen si, Si an fu, Lan t‘ien hien.

176.—郎陵 Lang ling. *Pie lu*.

Early Han : *hien*.   Now : Ho nan, Ju ning fu, K‘io shan hien.

177.—狼山 Lang shan. T‘ao Hung-king.

Early Han : *hien*.   In the Tsin period it depended on I tu kün [see 104].   Now : Hu pei, I ch‘ang fu, Ch‘ang yang hien.

178.—琅邪 Lang ye. *Pie lu*.   Han authors.

Ts‘in : province.   South-east part of Shan tung on the sea-shore.   Han, Tsin and Sui : *kün*.   Now : Shan tung, I chou fu.

179.—牟山 Lao shan. Ch‘en Tsz‘-ming [13th cent.].

Nan Ts‘i : *hien*.   Now : Hu pei, Te an fu.

180.—劳山 Lao shan. *K‘ai pao Pen ts‘ao*.

There is a mountain of this name in Shan tung, Lai chou fu.

181.—雷池 Lei ch‘i. *Pie lu*.

Not ascertained.   Probably identical with the next.

182.—雷澤 Lei tse. *Pie lu*.

Probably the marsh 雷夏 Lei hia in Shan tung, Ts‘ao chou fu, P‘u chou, which is mentioned in the Tribute of Yü.

In the *Li tai*, etc. we find Lei tse.   Sui : *hien*.   Now : Shan tung, Ts‘ao chou fu, P‘u chou.

183.—酈縣 Li hien.　T‘AO HUNG-KING.

The *Li tai*, etc. mentions a district of this name in the
T‘ang period.

Now : Ho nan, Nan yang fu, Nei hiang hien.

184.—利州 Li chou.　*K‘ai pao Pen ts‘ao.*

T‘ang and Sung : *chou*.　Now : Sz ch‘uan, Pao ning fu,
Kuang yüan hien.　Sung dynasty :　Li chou, name of province
(*lu*).　It occupied North-east Sz ch‘uan and South Shen si.

185.—澧州 Li chou.　SU KUNG and SU SUNG.

T‘ang and Sung : *chou*.　Now : Hu nan, Li chou.

186.—歷陽 Li yang.　T‘AO HUNG-KING.

Later Han : *hien*.　Now : An hui, Ho chou.

187.—梁 Liang, 梁州 Liang chou.　*Pie lu*, WU P‘U and
SU KUNG.

Liang chou was one of the nine provinces of ancient
China mentioned in the Tribute of Yü : present Sz ch‘uan
and parts of Hu pei and Shen si.　In the Ch‘un ts‘iu period
Liang was a small feudal state in present Shen si.

Han, Tsin and Sui : Liang = *kuo*, subsequently *kün*.
Now :　Ho nan, Kui te fu.　San kuo period, Tsin, Nan Sung
and Sui : Liang chou, name of a province occupying North-
east Sz ch‘uan, parts of Shen si and Hu pei.

Pei Wei : Liang chou, a prefecture.　Now : K‘ai feng
fu in Ho nan.

188.—梁漢 Liang Han.　MA CHI [*Med. plants*, 228].

Probably Sz ch‘uan is meant.　*See* Liang and Han.

189.—凉州 Liang chou.　T‘AO HUNG-KING and SU KUNG.

San kuo period : name of a province in the Wei
kingdom.　Southern Kan su.—T‘ang and Sung : prefecture,
chou.　Now : Kan su, Liang chou fu.

190.—遼州 Liao chou.　HAN PAO-SHENG and SU SUNG.

T‘ang and Sung : *chou*.　Now : Shan si, Liao chou.

191.—遼東 Liao tung. *Pie lu.*

Ts'in and Han : name for the country east of the Liao river, Southern Manchuria.

192.—臨海 Lin hai. T'AO HUNG-KING.

Tsin : *kün.* Now : Che kiang, T'ai chou fu, Lin hai hien.

193.—臨朐 Lin k'ü. *Pie lu.*

From the Early Han down to the present dynasty : *hien.* Now : Shan tung, Ts'ing chou fu, Lin k'ü hien.

194.—臨淄 Lin tsz'. *Pie lu* and SU KUNG.

Since the Early Han : *hien.* Now : Shan tung, Ts'ing chou fu, Lin tsz' hien.

195.—靈州 Ling chou. HAN PAO-SHENG.

Since the T'ang period : *chou.* Now : Kan su, Ning hia fu, Ling chou.

196.—零陵 Ling ling. *Pie lu* and MA CHI.

Early Han : *kün,* occupying a part of Hu nan, *viz.* Ch'ang sha fu, Heng chou fu, Pao k'ing fu, Yung chou fu and of Kuang si : Kui lin fu. T'ang : *chou.* Now : Hu nan, Yung chou fu.

197.—嶺南 Ling nan. HAN PAO-SHENG and SU SUNG.

Ling nan means "south of the mountain range," *i.e.* the Mei ling range, etc. which separate the southern provinces (Kuang tung and Kuang si) from Mid China. These regions are sometimes also termed 嶺表 Ling piao, which has a similar meaning.

T'ang dynasty : Ling nan, name of a province, *tao,* occupying present Kuang tung and Kuang si.

198.—柳城 Liu ch'eng. SU SUNG.

Sung and Yüan : *hien.* Now : Kuang si, Liu chou fu, Liu ch'eng hien.

Also Han and T'ang : *hien.* Now : Chi li, Yung p'ing fu.

199.—柳州 Liu chou. CH'EN TS'ANG-K'I and HAN PAO-SHENG.

T'ang and Sung: *chou*. Now: Kuang si, Liu chou fu.

200.—六安 Liu an. *Pie lu.*

Early Han: *kuo*. Now: An hui, Liu an chou.

201.—洛陽 Lo yang or simply 洛 Lo. SU SUNG.

The ancient (eastern) capital of the Han dynasty, near present Ho nan fu. During the Pei Wei dynasty it was a prefecture, 洛州 Lo chou.

202.—魯 Lu. *Pie lu* and SU SUNG.

Name of an ancient feudal state in South-west Shan tung, the native country of CONFUCIUS.

203.—魯山 Lu shan. *Pie lu.*

Name of a mountain in Ho nan, Chang te fu.
T'ang: *hien*. Now: Ho nan, Ju chou, Lu shan hien.

204.—潞州 Lu chou. SU KUNG and SU SUNG.

T'ang, Wu tai and Kin: *chou*. Now: Lu an fu in South-east Shan si.

205.—鹿臺 Lu t'ai. SU KUNG.

Not ascertained.

206.—廬州 Lü chou. SU SUNG.

T'ang, Sung and Yüan: *chou*. Now: An hui, Lü chou fu.

207.—廬江 Lü kiang. *Pie lu.*

Early Han: *kün*. Now: An hui, Lü chou fu, Lü kiang hien.

208.—廬陵 Lü ling. T'AO HUNG-KING.

Tsin: *kün*. Now: Kiang si, Ki an fu, Lü ling hien.

209.—廬山 Lü shan. SU SUNG.

Mountain in Kiang si near Kiu kiang fu.

210.—龍州 Lung chou. Su Kung and Su Sung.

T'ang and Sung : *chou*. Now : Kuang si, T'ai p'ing fu, Lung chou t'ing. Also T'ang and Wu tai : *chou*. Now : Sz ch'uan, Lung an fu.

211.—龍門 Lung men. *Pie lu*.

According to T'ao Hung-king [*Med. plants*, 110] north of Ch'ang an or Si an fu in Shen si.

212.—龍門山 Lung men shan. Li Shi-chen [*Med. plants*, 152]. A mountain in North China.

213.—龍山 Lung shan. Su Kung.

T'ang : *hien*, which was in Chi li near present Peking.

214.—龍洞 Lung tung. T'ao Hung-king.

Not ascertained.

215.—隴州 Lung chou. Su Sung.

T'ang and Sung : *chou*. Now : Shen si, Feng siang fu, Lung chou.

216.—隴西 Lung si. *Pie lu*, T'ao Hung-king and Su Kung.

隴 Lung was the name of a mountain in West Shen si, in Feng siang fu, Lung chou. Lung si, or West of Lung, was the name of a province in the Ts'in dynasty. It occupied the middle part of present Kan su, the prefectures of Kung ch'ang fu and Lan chou fu.—Han and Tsin : *kün*, T'ang : *hien*. There is now Lung si hien in Kung ch'ang fu.

217.—蠻洞 Man tung. Su Sung.

An ancient name for Nan tan chou in K'ing yüan fu, Kuang si. *See* Biot, 135.

218.—茅山 Mao shan. T'ao Hung-king and Su Sung.

A celebrated mountain in Kiang su, south-east of Kü yung hien and Nan king.

219.—眉州 Mei chou. Su Kung, Su Sung and Li Shi-chen.

T'ang and Sung : *chou*. Now : Sz ch'uan, Mei chou.

220.—孟 州 Meng chou. Su Sung.

T‘ang and Sung: *chou.* Now: Ho nan, Huai k‘ing fu, Meng hien.

221.—綿 州 Mien chou. Su Kung.

T‘ang: *chou.* Now: Sz ch‘uan, Mien chou.

222.—閩 Min. Su Sung and Li Shi-chen.

Ancient name of the province of Fu kien. In the Ts‘in period it was 閩 中 Min chung.

223.—岷 州 Min chou. K‘ou Tsung-shi.

Since the Sung period: *chou.* Now: Kan su, Kung ch‘ang fu, Min chou.

224.—明 州 Ming chou. Su Sung.

T‘ang and Sung: *chou.* Now: Che kiang, Ning po fu.

225.—茂 陵 Mou ling. Li Shi-chen.

Early Han: *hien.* Now: Shen si, Si an fu, Hing p‘ing hien.

226.—南 鄭 Nan cheng. *Pie lu* and T‘ao Hung-king.

Early Han: *hien.* Now: Shen si, Han chung fu, Nan cheng hien.

227.—南 方 Nan fang. Su Sung.

Nan fang means "Southern Region." By this term South China is generally understood, but sometimes also foreign southern countries.

228.—南 海 Nan hai. *Pie lu* and Li Sün.

Ts‘in: province, Han: *kün.* Now: Kuang tung, Kuang chou fu (Canton). Nan hai properly means "Southern Sea," and sometimes we have to understand by this term the Indian Archipelago. [See *Med. plants,* 58.]

229.—南 康 Nan k‘ang. T‘ao Hung-king.

Tsin, Nan Sung and Sui: *kün.* Now: Kiang si, Kan chou fu.

Tsin and down to the Ming period: *hien.* Now: Kiang si, Nan an fu, Nan k'ang hien.

Sung: *kün* (軍), Yüan: *lu,* Ming: *fu.* Now: Kiang si, Nan k'ang fu.

230.—南山 Nan shan. SU KUNG.

T'ang: *hien.* Now: Kuang si, K'ing yüan fu, Hin ch'eng hien.

It seems, however [*Med. plants,* 109] that SU KUNG by Nan shan (southern mountains) means a mountain chain in South Shen si, probably the Chung nan shan [*supra,* 28].

231.—南陽 Nan yang. *Pie lu* and T'AO HUNG-KING.

Early Han: *kün.* Now: Ho nan, Nan yang fu.

232.—南夔 Nan yao. *Pie lu.*

According to T'AO HUNG-KING [*Med. plants,* 32] a locality in North-east Tibet.

233.—南越 Nan Yüe. WU P'U.

An ancient name for South China first appearing in the Ts'in period. Comp. MARQUIS D'HERVEY DE ST. DENYS' *Ethn. d. peuples étrang. Méridionaux* [p. 307].

234.—寧州 Ning chou. T'AO HUNG-KING, SU KUNG and HAN PAO-SHENG.

Tsin: Ning chou = province, present Yün nan and part of Kui chou. Nan Sung: province, present Yün nan and S. Sz ch'uan.

T'ang and Sung: *chou* (prefecture). Now: Kan su, K'ing yang fu, Ning chou. In the Sung period there was also a Ning chou in Sz ch'uan.

235.—巴 Pa and 巴郡 Pa kün. *Pie lu.*

Ts'in: *kün,* province. Eastern part of Sz ch'uan. Comp. also 26, Ch'uan.

236.—巴西 Pa si (western). *Pie lu* and SU KUNG.

Tsin: *kün,* occupied Pao ning fu, Shun king fu and Mien chou in Sz ch'uan.

237.—巴東 Pa tung (eastern). T'AO HUNG-KING.

Tsin: *kün*, occupied K'ui chou fu in Eastern Sz ch'uan and part of Hu pei. There is now in West Hu pei the district Pa tung hien depending on I ch'ang fu.

238.—白山 Pai shan. T'AO HUNG-KING [*Med. plants*, 12]. A mountain near Nan king.

239.—白水 Pai shui. *Pie lu.*

One of the four great rivers of Sz ch'uan. See 26.

Pai shui, Early Han: *hien.* Now: Sz ch'uan, Pao ning fu, Chao hua hien. There is now a district Pai shui in T'ung chou fu, Shen si. According to BIOT [156] this name dates from the Ts'in dynasty.

240.—板橋 Pan k'iao. T'AO HUNG-KING.

From T'AO HUNG-KING [*Med. plants*, 100] it would appear that this locality was near Nan king.

In the Sung period there was a fort Pan k'iao in Sz ch'uan, in present Lu chou fu.

241.—般陽 Pan yang. *Pie lu.*

Early Han: *hien.* Now: Shan tung, Tsi nan fu, Tsz' ch'uan hien.

242.—抱罕 Pao han. T'AO HUNG-KING.

Early Han: *hien.* Now: Kan su, Lan chou fu, Ho chou.

243.—北郡 Pei kün. T'AO HUNG-KING.

Not ascertained.

244.—北部 Pei pu. WU P'U and T'AO HUNG-KING.

Nan Ts'i: *kün.* Now: Mou chou in Sz ch'uan.

Pei Wei: *hien.* Now: Kie chou in Kan su.

245.—北地 Pei ti. SU SUNG.

Pei ti means "northern country, North China." But in the Ts'in and Han dynasties there was a *kün* Pei ti which occupied K'ing yang fu, P'ing liang fu, Ning hia fu in Kan su and a part of Shen si.

246.—北都 Pei tu.   T‘AO HUNG-KING.

In the T‘ang period Pei tu (northern capital) was a name for T‘ai yüan fu in Shan si.

247.—彭城 P‘eng ch‘eng.   T‘AO HUNG-KING.

Early Han and T‘ang: *hien.*   Now: Kiang su, Sü chou fu.

248.—汴 Pien or 汴京 Pien king, also 汴梁 Pien liang. SU SUNG.

T‘ang and Wu tai: 汴州 Pien chou.   During the Sung period it had the above names and was the capital of the Sung.   Now: K‘ai feng fu in Ho nan.

249.—汴西 Pien si or 汴京西 Pien king si (west of the capital).   SU SUNG and K‘OU TS‘UNG-SHI.

Sung: name of a province (*lu*) — Ho nan and parts of Shen si and Hu pei.

250.—汴東 Pien tung or 汴京東 Pien king tung (east of the capital).   SU SUNG.

Sung: name of a province (*lu*) == present Shan tung.

251.—瀕 or 濱 Pin.   SU SUNG.

Wu tai, Sung: *chou.*   Now: Shan tung, Wu ting fu, Pin chou.

252.—賓州 Pin chou.   SU SUNG.

T‘ang: *chou.*   Now: Kuang si, Sz‘ en fu, Pin chou.

253.—并州 Ping chou.   T‘AO HUNG-KING, HAN PAO-SHENG and SU SUNG.

Ping chou was one of the nine provinces in the Chou dynasty, occupying North Shan si.   Same during the Han and Tsin dynasties.

Wu tai: Ping chou, a prefecture, not ascertained, probably in Chi li.

254.—平昌 P‘ing ch‘ang.   Authors of the Han and Tsin dynasties.

Name applied during the Han, Nan Sung and Pei Wei to various districts in Shan tung, An hui, Chi li, Ho nan and Shan si.

255.—平 州 P'ing chou. HAN PAO-SHENG.

T'ang, Wu tai, Sung and Kin: *chou*. Now: Chi li, Yung p'ing fu, Lu lung hien.

256.—平 壽 P'ing shou. *Pie lu*.

Early Han: *hien*. Now: Shan tung, Lai chou fu, Wei hien.

257.—平 陽 P'ing yang in 晉 Tsin. *Pie lu*.

Early Han: *hien*. Now: Shan si, P'ing yang fu.

258.—平 原 P'ing yüan. *Pie lu*.

Early Han: *kün*. Now: Shan tung, Tsi nan fu, P'ing yüan hien.

259.—亳 Po. SU SUNG.

T'ang and Sung: *chou*. Now: An hui, Ying chou fu, Po chou.

260.—博 平 Po p'ing. SU KUNG and SU SUNG.

Early Han: *hien*, T'ang: *hien*, Sung: *hien*. Now: Shan tung, Tung ch'ang fu, Po p'ing hien.

261.—百 濟 Po tsi. T'AO HUNG-KING.

An ancient kingdom in the south-west of Corea. First mentioned in the History of the Later Han.

262.—渤 海 Pu hai. *Pie lu*.

Early Han: *kün*. Now: Chi li, T'ien tsin fu, Ts'ang chou.

263.—濮 陽 縣 Pu yang hien. CHANG YÜ-HI.

From the Early Han down to the Yüan period: *hien*. Now: Chi li, Ta ming fu, K'ai chou.

264.—蒲 州 P'u chou. SU KUNG.

Since Wu tai: *chou*. Now: P'u chou fu in Shan si.

265.—三 輔 San fu. FAN TSZ' KI JAN.

The three prefectures surrounding and comprising the imperial city Ch'ang an of the Han dynasty :—

       1.—京 兆 King chao [see 150].

       2.—馮 翊 Feng i [see 40].

       3.—扶 鳳 Fu feng [see 44].

266.—沙州 Sha chou. Li Sün and Su Kung.

T'ang: *chou.* Now: Kan su, An si chou, Tun huang hien.

267.—沙苑 Sha yüan. *Pie lu* and Su Sung.

In the *Li tai,* etc. a place Sha yüan first appears in the Sung dynasty. It was in Shen si in T'ung chou fu.

268.—山南 Shan nan. *Pie lu, K'ai pao Pen ts'ao* and Li Shi-chen.

Not found in the *Li tai,* etc. In the T'ang period 山南道 Shan nan tao, name of a province south of the Ts'ing ling mountains, *i.e.* Southern Shen si and part of Ho nan.

269.—山東 Shan tung. Su Sung and K'ou Tsung-shi.

Not ascertained for the Sung period. I do not think that the present province of Shan tung is meant. As applied to these regions, this name first appears in the geography of the Kin, in the 12th century.

270.—山陽 Shan yang. *Pie lu.*

Early Han: *kün.* Now: Shan tung, Tsi ning chou, Kin hiang hien.

Also Early Han: *hien.* Now: Ho nan, Huai k'ing fu, Siu wu hien.

There is now a district Shan yang hien in Shen si. Shang chou. This name dates from the Ming period. The same name, now applied to a district in Huai an fu, in Kiang su, can be traced to the Nan Sung dynasty.

271.—山陰 Shan yin. *Pie lu.*

Early Han: *hien.* Now: Che kiang, Shao hing fu, Shan yin hien.

The present district Shan yin hien in Ta t'ung fu, Shan si, dates from the Kin period.

272.—上谷 Shang ku. *Pie lu.*

Ts'in, Han, Tsin and Sui: *kün.* Now: North-west part of Chi li, west of Peking.

273.—上郡 Shang kün. *Pie lu* and T'AO HUNG-KING.

Ts'in, Han and Sui : *kün.* North-east part of Shen si, Yen an fu, Yü lin fu.

274.—上洛 Shang lo. *Pie lu.*

Early Han : *hien.* Now : Shen si, Shang chou.

275.—上黨 Shang tang. *Pie lu.*

Ts'in, Han, Tsin and Sui : *kün.* South-east part of Shan si and Chang te fu in Ho nan.

276.—上蔡縣 Shang ts'ai hien. *Pie lu* and SU SUNG.

Since Early Han : *hien.* Now : Ho nan, Ju ning fu, Shang ts'ai hien.

277.—上虞 Shang yü. *Pie lu.*

Early Han : *hien.* Now : Che kiang, Shao hing fu, Shang yü hien.

But T'AO HUNG-KING [*Med. plants,* 317] thinks that the Shang yü of the *Pie lu* is a locality near the Yellow River.

278.—商州 Shang chou. SU KUNG and SU SUNG.

T'ang and Sung : *chou.* Now : Shen si, Shang chou.

279.—韶州 Shao chou. SU KUNG and SU SUNG.

T'ang and Sung : *chou.* Now : Kuang tung, Shao chou fu.

280.—邵陵 Shao ling. T'AO HUNG-KING.

Ts'in and Nan Sung : *kün.* Now : Hu nan, Pao k'ing fu.

281.—少室 Shao shi. *Pie lu.*

Not found in the *Li tai,* etc. According to the *K.D.* Shao shi is one of the peaks of the celebrated Sung kao shan mountain. [See 317.] But the *Pie lu* seems to keep the two names apart. [See *Med. plants,* 133.]

282.— 申州 Shen chou. HAN PAO-SHENG.

T'ang and Wu tai : *chou.* Now : Ho nan, Ju ning fu, Sin yang chou.

74

The feudal state 申 Shen of the Ch'un ts'iu period was present Nan yang fu in Ho nan.

283.—陝 州 Shen chou.　Su Kung and Su Sung.

Early Han: *hien*, T'ang and Sung: *chou*. Now: Ho nan, Shen chou.

284.—陝 西 Shen si or simply 陝 Shen.　Su Sung.

Shen si, name of a province during the Sung dynasty. It comprised present Shen si and Kan su.　The name Kan su as that of a province appears first in the Yüan dynasty.

285.—石 城 Shi ch'eng.　*Pie lu.*

Early Han down to Nan Ts'i: *hien*.　Now: An hui, Ch'i chou fu.

The name Shi ch'eng was applied in various times to many other districts in China, in Shen king, Shan si, Ho nan, Shen si and Sz ch'uan.

The present Shi ch'eng hien, in Kiang si, Ning tu fu, dates from the Sung period, as does also the district of the same name in Kuang tung, Kao chou fu.

286.—石 州 Shi chou.　Ch'en Ts'ang-k'i and Su Sung.

T'ang, Sung and Kin: *chou*.　Now: Shan si, Fen chou fu, Yung ning chou.

287.—石 山 Shi shan.　*Pie lu.*

Later Han: *hou kuo* (feudal state) in Lang ye kün [see 178] in South-east Shan tung.

288.—施 州 Shi chou.　Su Sung.

T'ang and Sung: *chou*.　Now: Hu pei, Shi nan fu.

289.—始 興 Shi hing.　T'ao Hung-king.

Tsin: *kün*.　Now: Kuang tung, Shao chou fu.

Nan Sung: *hien*.　Now: Kuang tung, Nan hiung chou, Shi hing hien.

Also Nan Sung: *hien*.　Now: Sz ch'uan, Sui ting fu.

290.—壽 州 Shou chou. Su Sung.

T'ang and Wu tai : *chou.* Now : An hui, Feng yang fu, Shou chou.

291.—壽 春 Shou ch'un. Su Sung.
Early Han : *hien.*

Tsin, Sui and T'ang : *hien,* Sung : *fu,* Yüan : *hien.* Now : An hui, Feng yang fu, Shou chou.

292.—蜀 Shu. *Pie lu* and T'ao Hung-king.

Ts'in : province. Western part of present Sz ch'uan, Ch'eng te fu, etc. In the San kuo period the kingdom of Shu comprised nearly the whole of present Sz ch'uan.

Su Sung uses the term 蜀 川 Shu ch'uan or 川 蜀 Ch'uan Shu for Sz ch'uan. [*V. supra,* 26.]

蜀 州 Shu chou [Su Sung] in the T'ang and Sung period was present Ch'eng te fu, the capital of Sz ch'uan.

293.—蜀 漢 Shu Han. *Pie lu,* T'ao Hung-king and Su Sung.

In the San kuo period Shu Han was one of the three kingdoms and corresponds to present Sz ch'uan.

294.—舒 州 Shu chou. Su Sung.

T'ang and Wu tai : *chou.* Now : An hui, An k'ing fu, Ts'ien shan hien.

295.—西 城 Si ch'eng. Su Kung.
Sui : *kün,* T'ang : *hien.* Now : Shen si, Hing an fu.

296.—西 川 Si ch'uan. T'ao Hung-king and Su Sung.

In the period of the Later Han Si ch'uan comprised the four prefectures 天 水 T'ien shui, 安 定 An ting, 北 地 Pei ti and 龍 西 Lung si, where Wei Hiao [† 33 A.D.] maintained for some years an independent sovereignty. See Mayers' *Chin. Read. Man.,* 835. Si ch'uan occupied the north-east part of Kan su.

Tsin : *hien.* Now : Kan su, P'ing liang fu, Ku yŭan chou.

According to BIOT [p. 215] Si ch'uan in the T'ang period was a name for Ch'eng tu fu in Sz ch'uan.

297.—西方 Si fang. SU KUNG [*Med. plants*, 133].

Si fang means "Western Regions." It is not clear whether Western China or Tibet is meant.

298.—西海 Si hai. *Pie lu.*

Si hai means "Western Sea." Name applied in various times to various localities. The famous Chinese general CHANG K'IEN, who in the 2nd cent. B.C. first visited the countries of Western Asia, calls the Mediterranean Sea "Si hai."

Later Han, Si hai : *hien.* Now : Shan tung, I chou fu, Ji chao hien.

Tsin : *kün.* Now : Kan su, prefecture of Kan chou fu.—Sui : *kün.* Near the Ts'ing hai or Kukonor lake.—In the T'ang period there was a district Si hai near present Turfan.

299.—西湖 Si hu. SU KUNG [*Med. plants*, 54].

Si hu = Western Lake. Probably the lake of this name near Hang chou fu in Che kiang is meant.

300.—西羌 Si (western) K'iang or 羌 K'iang. T'AO HUNG-KING, SU KUNG and LI SHI-CHEN.

Ancient name [Han period] for the Tangut tribes living in North-east Tibet and Kukonor.

301.—西嶺 Si ling. *Pie lu.*

Si ling, or Western mountain range, was, according to T'AO HUNG-KING [*Med. plants*, 47], near ancient Ch'ang an (Si an fu, Shen si).

302.—西陽 Si yang. T'AO HUNG-KING.

Tsin : *kün.* Now : Hu pei, Huang chou fu.

303.—錫山 Si shan. Shui king chou [5th cent.].

Si shan mountains in Wei hing [Shen si, Hing an fu. *V. infra,* 384]. See also *Med. plants,* 79.

304.—習 州 Si chou. Su Sung [*Med. plants*, 352].

Not found either in the *Li tai*, etc. or Biot.

305.—襄 Siang, **襄 州** Siang chou. T'ao Hung-king, Su Kung, Han Pao-sheng and Su Sung.

Early Han : Siang hien, in present Ho nan province.

Nan Ts'i : Siang hien. Now : Shen si, Han chung fu.

T'ang and Wu tai : Siang chou. Now : Hu pei, Siang yang fu.

306.—襄陽 Siang yang. T'ao Hung-king and Su Kung.

Early Han : *hien*, Tsin, Nan Sung and Sui : *kün*, T'ang : *hien*, Sung : *fu*.

Now : Hu pei, Siang yang fu.

307.—湘 州 Siang chou. T'ao Hung-king.

Nan Sung : Siang chou, name of a province. Present Hu nan.

308.—象 州 Siang chou. Ch'en Ts'ang-k'i.

T'ang : *chou*. Now : Kuang si, Liu chou fu, Siang chou.

309.—斜谷 Sie ku. *Pie lu* and Han Pao-sheng [*Med. plants*, 347].

According to T'ao Hung-king [*Med. plants*, 47] Sie ku was situated near Ch'ang an (Si an fu, Shen si).

310.—新安 Sin an. T'ao Hung-king.

Tsin : *kün*. Now : Che kiang, Yen chou fu, Shun an hien.

The name Sin an was applied in different times to a great number of different districts in Ho nan, Hu pei, Sz ch'uan, Yün nan, Kuang tung, Shan si, Shan tung, An hui, Chi li and Kui chou. There are still three districts of this name extant, *viz.* in Chi li, Ho nan and Kuang tung.

311.—新羅 Sin lo. Li Sün.

Name of a kingdom in Southern Corea mentioned in the Chinese annals since the 4th century.

312.—新野 Sin ye.   T'AO HUNG-KING.

Early Han and the subsequent dynasties, down to the Sui: *hien.*

Now: Ho nan, Nan yang fu, Sin ye hien.

313.—蕭 州 Su chou.   HAN PAO-SHENG.

T'ang: *chou.*   Now: Kan su, Su chou fu.

314.—隨 州 Sui chou.   SU SUNG.

T'ang, Wu tai and Sung: *chou.*   Now: Hu pei, Te an fu, Sui chou.

315.—宣城 Süan ch'eng.   T'AO HUNG-KING and SU SUNG.

Early Han: *hien,* Tsin and Sui: *kün.*

In the T'ang and Sung periods it was 宣 州 Süan chou.

Now: An hui, Ning kuo fu, Süan ch'eng hien.

316.—宋 Sung.   SU SUNG.

Name of an ancient feudal state in the Chou dynasty. Eastern part of Ho nan and north-western part of An hui.

Later Han: *kuo,* subsequent dynasties down to Sung: *hien.*   Now: An hui, Ying chou fu, T'ai ho hien.

The prefecture 宋 州 Sung chou in the T'ang, Wu tai, and Sung dynasties is present Kui te fu in Ho nan. Another Sung chou during the same periods was in Sz ch'uan, present Lu chou.

317.—嵩 高 Sung kao, 嵩 山 Sung shan.   *Pie lu* and MA CHI.

Sung kao, the name of one of the sacred mountains of China, in Ho nan, north of present Teng feng hien, Ho nan fu. This district in the time of the Han was called Sung kao.

318.—松 州 Sung chou.   SU KUNG.

T'ang: *chou.*   Now: Sz ch'uan, Lung an fu, Sung p'an t'ing.

319.—泗 州 Sz' chou.   SU SUNG.

T'ang and Sung: *chou.*   Now: An hui, Sz' chou.

320.—大安 Ta an. *Pie lu.*

According to the *Li tai*, etc., first mentioned in the Pei Wei period = *kün*.

Now: Shan si, P'ing ting chou, Shou yang hien.

321.—代 Tai or 代郡 Tai kün. *Pie lu.*

Ts'in and Han: *kün*. Northern part of Shan si, Ta t'ung fu.

代州 Tai chou. T'AO HUNG-KING.

T'ang, Sung and Kin: *Tai chou*. Now: Shan si, Tai chou.

322.—泰山 T'ai shan or 太山 T'ai shan. *Pie lu*, WU P'U and SU SUNG.

Name of one of the sacred mountains of China, in Shan tung, T'ai an fu.

Present T'ai an fu was, in the time of the Han, T'ai shan *kün*.

323.—太行山 T'ai hang shan. SU KUNG and SU SUNG.

Name of the range of mountains stretching from north to south and separating Shan si from Chi li and Ho nan.

324.—太吳 T'ai Wu. *Pie lu.*

According to T'AO HUNG-KING [*Med. plants*, 61] same as Wu [Kiang su. See *infra*, 389].

325.—太原 T'ai yüan. SU SUNG.

Early Han, Tsin and Sui: *kün*, T'ang and Sung: *fu*. Now: Shan si, T'ai yüan fu.

326.—台州 T'ai chou. SU SUNG.

T'ang and Sung: *chou*. Now: Che kiang, T'ai chou fu.

327.—丹州 Tan chou. SU SUNG.

T'ang, Wu tai and Kin: *chou*. Now: Shen si, Yen an fu, I ch'uan hien.

328.—丹陽 Tan yang. T'AO HUNG-KING.

Early Han, Tsin, Nan Ts'i and Sui: *kün*. Present Ning kuo fu and T'ai p'ing fu in An hui.

T'ang, Sung and Yüan : *hien.* Now : Kiang su, Chen kiang fu, Tan yang hien.

329.—檀 州 T'an chou. HAN PAO-SHENG.

T'ang and Sung : *chou.* Now : Chi li, Shun t'ien fu, Mi yün hien.

330.—宕 昌 Tang ch'ang. T'AO HUNG-KING.

Sui : *kün.* Now : Kan su, Kung ch'ang fu, Min chou.

T'ang : *hien.* Now : Kuang si, Wu chou fu, Yung hien.

331.—宕 州 Tang chou. SU KUNG.

Tang chou, name applied in the T'ang period to two different prefectures, *viz.* one in present Min chou, Kung ch'ang fu, Kan su, the other in present Mou chou in Sz ch'uan.

332.—當 州 Tang chou. SU KUNG.

T'ang and Sung : *chou.* Now : Sz ch'uan, Lung an fu, Sung p'an ting.

333.—當 陽 縣 Tang yang hien. SU KUNG.

Han and T'ang : *hien.* Now : Hu pei, King men chou, Tang yang hien.

334.—碭 山 Tang shan. *Pie lu.*

Ts'in : *kün.* It was situated where now the provinces Kiang su, Ho nan and Shan tung meet.

Later Han : *hien.* Now : Kiang su, Sü chou fu, Tang shan hien.

335.—湯 陰 T'ang yin. SU SUNG and LI SHI-CHEN.

Since Sui : *hien.* Now : Ho nan, Chang te fu, T'ang yin hien.

336.—洮 陽 T'ao yang. T'AO HUNG-KING.

Early Han : *hien.* Now : Kuang si, Kui ling fu, Ts'üan chou.

Sui : *hien.* Now : Kan su, Kung ch'ang fu, T'ao chou ting.

337.—鄧州 Teng chou. Su Kung.

Since T'ang : *chou*. Now : Ho nan, Nan yang fu, Teng chou.

338.—滇 Tien or 滇南 Tien nan, 滇中 Tien chung. Li Shi-chen.

Ancient names for present Yün nan, dating from the Han period.

339.—天水 T'ien shui. Fan tsz' ki jan.

Early Han, Tsin, Pei Wei and Sui : *kün*. Now : Kan su, Kung ch'ang fu, Ts'in chou.

340.—天台 T'ien t'ai. Ch'en Ts'ang-k'i, Su Sung and Li Shi-chen.

Since Liang : *hien* [Biot, 231]. Now : Che kiang, T'ai chou fu, T'ien t'ai hien.—In the same district is the celebrated T'ien t'ai mountain, the earliest seat of Buddhism in China.

341.—鼎州 Ting chou. Su Sung.

Biot, 202 :—Ting chou in the Sung period = present Chang te fu in Hu nan.

342.—蔡州 Ts'ai chou. T'ao Hung-king and Su Kung.

T'ang, Sung and Kin : *chou*. Now : Ho nan, Ju ning fu.—Here was situated, in the Chou dynasty, the feudal state 蔡 Ts'ai.

343.—滄州 Ts'ang chou. Han Pao-sheng.

T'ang, Sung and Kin : *chou*. Now : Chi li, T'ien tsin fu, Ts'ang chou.

344.—曹州 Ts'ao chou. Han Pao-sheng and Su Sung.

T'ang, Wu tai and Kin : *chou*. Now : Shan tung, Ts'ao chou fu.

345.—澤州 Tse chou. Su Sung.

Since T'ang : *chou*. Now : Shan si, Tse chou fu.

346.—濟陽 Tsi yang. Su Kung.

Name applied, since the Han dynasty, to various places (*kün, hien*) in Ho nan, Kiang su, An hui and Shan tung.

75

The name of the present Tsi yang hien in Tsi nan fu,
Shan tung, dates from the Kin dynasty.

347.—濟 陰 Tsi yin. *Pie lu.*

Early Han : *kün.* Now : Shan tung, Ts'ao chou fu.

348.—齊 Ts'i. SU SUNG.

An important ancient feudal state in the Chou dynasty,
occupying North Shan tung and South Chi li.

Ts'in and Han down to Sui : Ts'i kün, in North-west
Shan tung.

齊 州 Ts'i chou. SU KUNG and SU SUNG.

T'ang and Wu tai : *chou.* Now : Shan tung, Tsi nan fu.

349.—齊 胸 Ts'i k'ü. *Pie lu.*

Not ascertained. Probably 臨 胸 Lin kü in 齊 郡 Ts'i
kün [Early Han], which now is Ts'ing chou fu in Shan tung
[see 193].

350.—齊 山 Ts'i shan. *Pie lu.*

A mountain of this name is in An hui, Ch'i chou fu.

351.—蔣 山 Tsiang shan. T'AO HUNG-KING.

Not ascertained.

352.—錢 塘 Ts'ien t'ang. T'AO HUNG-KING.

Name of a river in Che kiang which flows into
the sea near Hang chou fu. This prefecture comprises
the district Ts'ien t'ang, which name dates from the Ts'in
dynasty.

353.—晉 Tsin, 晉 地 Tsin ti (the country of Tsin). *Pie
lu* and LI SHI-CHEN.

Tsin, an ancient feudal state in the Chou dynasty. It
comprised the southern half of Shan si and the north-west
of Ho nan along the Yellow River.

晉 州 Tsin chou. SU SUNG.

T'ang and Sung : *chou.* Now : Shan si, P'ing yang fu.

354.—晉安 Tsin an. T'AO HUNG-KING.

Tsin : *kün*. Now : Fu kien, Fu chou fu.

355.—晉康 Tsin k'ang. T'AO HUNG-KING.

Nan Sung: *kün*. Now : Kuang tung, Chao k'ing fu, Te k'ing chou.

Nan Ts'i : Tsin k'ang, in Sz ch'uan, in Ch'eng tu fu, Ch'ung k'ing chou.

356.—晉山 Tsin shan. *Pie lu.*

Not ascertained.

357.—晉陽 Tsin yang. *Pie lu.*

Ts'in and down to T'ang : *hien*. Now : Shau si, T'ai yüan fu.

358.—秦 Ts'in, 秦地 Ts'in ti (country of Ts'in). HAN PAO-SHENG and LI SHI-CHEN.

Name of an ancient feudal state during the Chou dynasty, comprising the eastern part of present Kan su and the middle part of Shen si.

秦州 Ts'in chou. SU KUNG, SU SUNG and LI SHI-CHEN.

T'ang, Wu tai and Sung : *chou*. Now : Kan su, Ts'in chou.

359.—秦關 Ts'in kuan (the barrier of Ts'in).

According to T'AO HUNG-KING [*Med. plants*, 104] same as Han ku [*supra*, 55] in North-west Ho nan.

360.—秦山 Ts'in shan. *Pie lu.*

Not ascertained.

361.—秦亭 Ts'in t'ing. *Pie lu.*

According to SU KUNG [*Med. plants*, 132] this locality was between Ts'in chou and Ch'eng chou (Kie chou) in Kan su.

362.—沁州 Ts'in chou. CH'EN TS'ANG-K'I and HAN PAO-SHENG.

Since T'ang : *chou*. Now : Shan si, Ts'in chou.

363.—青州 Ts'ing chou or 菁 Ts'ing. T'AO HUNG-KING and SU KUNG.

One of the nine ancient provinces of China in the Tribute of Yü. It comprised the greater part of present Shan tung.

Han, Tsin and Sui: province occupying the northern part of Shan tung.

Since T'ang: 青州 Ts'ing chou, a prefecture = present Ts'ing chou fu in Shan tung.

364.—青衣 Ts'ing i. *Pie lu.*

Early Han: *hien.* Now: Sz ch'uan, Ya chou fu.

365.—鄒縣 Tsou hien. *Pie lu.*

A district in the ancient feudal state of Lu [see 202]. Now: Shan tung, Yen chou fu, Tsou hien.

366.—梓州 Tsz' chou. SU SUNG.

T'ang and Wu tai: *chou.* Now: Sz ch'uan, T'ung ch'uan fu.

367.—淄州 Tsz' chou. SU KUNG.

T'ang, Wu tai, Sung and Kin: *chou.* Now: Shan tung, Tsi nan fu, Tsz' ch'uan hien.

368.—秭歸 Tsz' kui. *Pie lu.*

Early Han down to T'ang: *hien.* Now: Hu pei, I chang fu, Kui chou.

369.—都鄉 Tu hiang. *Pie lu.*

Early Han: *kuo* in 常山郡 Ch'ang shan kün [see 8]. Now: Chi li, Cheng ting fu.

370.—都梁 Tu liang. LI TANG-CHI.

Early Han: *huo kuo* (small feudal state). Later Han down to Nan Ts'i: *hien.* Now: Hu nan, Pao k'ing fu, Wu kang chou.

371.—燉煌 Tun huang. *Pie lu.*

Early Han and down to Sui: *kün,* T'ang: *hien.* Now: Kan su, An si chou, Tun huang hien.

372.—東海 Tung hai. *Pie lu.*

Han and Tsin : *kün*. Now : Shan tung, I chou fu, T'an ch'eng hien.

Tung hai means "Eastern Sea," and the name is sometimes used in this sense. Comp. *Med. plants*, 147, 200, 201.

373.—東萊 Tung lai. *Pie lu.*

From Han down to Sui : *kün*. Now : Shan tung, Lai chou fu.

374.—東門 Tung men. T'AO HUNG-KING.

Not ascertained. The name means "Eastern Gate."

375.—東山 Tung shan. T'AO HUNG-KING.

Tung shan (Eastern mountains) name of several mountains in various provinces.

376.—東陽 Tung yang. T'AO HUNG-KING.

Tsin and Sui : *kün*. Now : Che kiang, Kin hua fu, Tung yang hien.

Early Han : *hou kuo*. Now : Shan tung, Tung ch'ang fu.

Early Han and Tsin : *hien*. Now : An hui, Sz' chou.

Later Han : *hien*. Now : in Kiang su. Not ascertained.

377.—東野 Tung ye. *Pie lu.*

Not found either in the *Li tai*, etc. or BIOT. Probably in Che kiang. See *Med. plants*, 162.

378.—同州 T'ung chou. SU KUNG and SU SUNG.

From T'ang down to Ming : *chou*. Now : Shen si, T'ung chou fu.

379.—桐柏 T'ung po. *Pie lu.*

The T'ung po mountain is mentioned in the Tribute of Yü. T'AO HUNG-KING [*Med. plants*, 20] says it is situated in I yang hien [*supra*, 107] which is now the district of T'ung po hien in Nan yang fu, Ho nan. The name of T'ung po hien dates only from the time of the Sui.

380.—望 楚 山 Wang ch'u shan. Su Kung.

Mountain in Hu pei, Siang yang fu.

381.—衞 州 Wei chou. Su Sung.

T'ang, Sung and Kin: *chou*. Now: Ho nan, Wei hui
fu.

382.—渭 城 Wei ch'eng. T'ao Hung-king.

Wei ch'eng in the Han dynasty was the same as Hien
yang [*see* 65], north-west of Si an fu, Shen si.

383.—渭 州 Wei chou. Han Pao-sheng.

T'ang: *chou*. Now: Kan su, Kung ch'ang fu.

T'ang, Wu tai and Sung: *chou*. Now: Kan su, P'ing
liang fu.

384.—魏 與 Wei hing. *Pie lu, Shui king chu* and Li Shi-
chen.

Tsin, Nan Sung and Nan Ts'i: *kün*. Now: Shen si,
Hing an fu.

385.—溫 州 Wen chou. Su Kung and Su Sung.

From T'ang down to Ming: *chou*. Now: Che kiang,
Wen chou fu.

386.—文 州 Wen chou. Su Sung.

T'ang and Sung: *chou*. Now: Kan su, Kie chou, Wen
hien.

Sung: *chou*. Now: Kuang si, K'ing yüan fu, Tung
lan chou.

387.—汝 州 Wen chou. Li Shi-chen.

Tsin: *hien*. Two districts of this name, in South Manchuria and in An hui [Biot, 267]: districts in Sz ch'uan in
the T'ang dynasty.

388.—汝 山 Wen shan. *Pie lu* and T'ao Hung-king.

Tsin: *kün* in the province of 孟 州 I chou. Now:
Mou chou in Northern Sz ch'uan.

389.—吳 Wu, 吳 國 Wu kuo (kingdom) and 吳 地 Wu ti
(country of Wu). Frequently mentioned by authors of
various periods,

The ancient kingdom of Wu, mentioned in the Ch'un ts'iu period, occupied present Kiang su and a part of An hui and Che kiang. The capital was near the present Su chou fu, which is still called 吳縣 Wu hien. This district has been so named since the Earlier Han.

The kingdom Wu of the San kuo period [3rd cent.] occupied Southern Kiang su, South An hui, Hu pei and Hu nan, Kiang su, Che kiang and a part of Fu kien.

After the Ch'un ts'iu period the kingdom of Wu was conquered by the kingdom of Yüe [see 418, present Che kiang]. The two names are therefore frequently coupled, 吳越 Wu Yüe. In the Wu tai period Wu Yüe was the name of a province comprising Che kiang and a part of Kiang su.

390.—吳興 Wu hing. T'AO HUNG-KING.

San kuo, Tsin, Nan Sung and Nan Ts'i: kün. Now: Che kiang, Hu chou fu.

391.—吳會 Wu hui. SU KUNG.

Not ascertained. Probably Wu and Hui ki [see 98] in Che kiang are meant.

392.—武昌 Wu ch'ang. T'AO HUNG-KING.

Ts'in, Nan Sung and Nan Ts'i: kün, T'ang and Sung: hien, Yüan: lu, Ming: fu. Now: Hu pei, Wu ch'ang fu.

393.—武功 Wu kung. Pie lu.

According to T'AO HUNG-KING [Med. plants, 47] this locality was near Ch'ang an [Si an fu, Shen si].

Earlier Han: hien. Now: Shen si, Feng siang fu, Mei hien.

Later Han: hien. Now: Shen si, K'ien chou, Wu kung hien.

394.—武陵 Wu ling. Pie lu.

Early Han: kün. Now: Hu nan, Ch'ang te fu, Wu ling hien.

395.—武都 Wu tu. FAN TSZ' KI JAN and LI SHI-CHEN.

Han: *kün*. Now: Kie chou fu, Ch'eng hien, in Kan su.

396.—巫陽 Wu yang. *Pie lu.*

According to T'AO HUNG-KING [*Med. plants*, 26] Wu yang was in 建平 Kien p'ing [*see* 139]. Wu yang is not found either in the *Li tai*, etc. or in BIOT, but there was a district 巫山 Wu shan in ancient Kien p'ing, which still bears the same name and is now comprised in K'ui chou fu, Sz ch'uan.

397.—五原 Wu yüan. *Pie lu.*

Early and Later Han: *kün*. It was situated north of the present Ordos in the country of the Oirats.

398.—雅州 Ya chou. LI SHI-CHEN.

T'ang and Sung: *chou*. Now: Sz ch'uan, Ya chou fu.

399.—陽山 Yang shan. *Pie lu.*

Name of several mountains in North China.

400.—揭州 Yang chou or simply 揭. *Pie lu*, HAN PAO-SHENG and LI SHI-CHEN.

Yang chou was one of the nine ancient provinces of China as enumerated in the Tribute of Yü. It occupied present An hui, Kiang su, Che kiang and a part of Kiang si.

The province Yang chou in the Han dynasty was the same, but included also the whole of Kiang si and Fu kien. It was the same in the Tsin dynasty. The province Yang chou during the Sui was still larger, for it comprised also Kuang tung and Kuang si.

In the T'ang period there was a district Yang hien. Since the Wu tai period: *chou*. Now: Yang chou fu in Kiang su.

401.—耀州 Yao chou. SU SUNG.

Since Wu tai: *chou*. Now: Shen si, Si an fu, Yao chou.

402.—猺獞人 Yao chuang jen. LI SHI-CHEN.

Savages in Southern China. See *W.D.*, 114, 1076.

403.—延州 Yen chou. SU KUNG and SU SUNG.

T'ang and Wu tai: *chou*. Now: Shen si, Yen an fu.

404.—兗州 Yen chou. T'AO HUNG-KING, SU KUNG and SU SUNG.

One of the nine ancient provinces of China, enumerated in the Tribute of Yü, comprising North-west Shan tung and East Chi li.

The province Yen chou in the Han, Tsin and Sui dynasties was of about the same extent.

Since the T'ang, Yen chou a prefecture, now Shan tung, Yen chou fu.

405.—雁門 Yen men. *Pie lu.*

Ts'in: *kün*. In Mid Shan si, between Ta t'ung fu and T'ai yüan fu.

Eastern Han, Tsin and Sui: *kün*. Now: Shan si, Tai chou.

406.—嚴道 Yen tao. *Pie lu.*

From the Early Han down to Yüan: *hien*. Now: Sz ch'uan, Ya chou fu.

407.—銀州 Yin chou. SU SUNG.

T'ang, Wu tai and Sung: *chou*. Now: Shen si, Sui te chou, Mi chi hien.

408.—潁川 Ying ch'uan. *Pie lu.*

Ts'in: province comprising a part of Ho nan: Hü chou, K'ai feng fu, Yü chou, Ju ning fu, Ying chou fu.

Later Han down to Sui: *kün*, situated in the same regions.

The T'ang established the prefecture 潁州 Ying chou, the present Ying chou fu, in North-west An hui.

409.—郢州 Ying chou. T'AO HUNG-KING.

Nan Sung and Nan Ts'i: Ying chou, province corresponding in its extent nearly to present Hu pei.

T'ang: *chou.* Now: Hu pei, An lu fu.

Pei Wei: Ying chou. Now: Ho nan, Ju ning fu.

410.—岳州 Yo chou. Su Sung.

Since T'ang: *chou.* Now: Hu nan, Yo chou fu.

411.—幽 Yu. Su Kung and Han Pao-sheng.

Legge's *Shu king,* 21:—Emperor Yao commanded his brother Ho to reside in the northern regions, in what was called the sombre capital 幽都 Yu tu.—The *Rh ya* says:— 幽州 Yu chou is the same as 燕 Yen. It stretches from the river 易水 I shui (the river on which I chou in North Chi li is situated) to the land of the 北狄 Pei ti (northern barbarians). Yen was an ancient feudal state in the north of China, mentioned in the *Shi king* and in the *Ch'un ts'iu.*

Han period: Yu chou, province, northern part of Chi li. In the San kuo period Yu chou was a province of the kingdom of Wei, in extent as above.

T'ang and Wu tai: Yu chou, prefecture, modern Peking.

412.—鬱州 Yü chou. T'ao Hung-king.

Not mentioned in the *Li tai,* etc. According to Biot [291]:—T'ang: *chou.* Now: Yü lin chou in Kuang si.

413.—豫州 Yü chou. *Pie lu.*

One of the nine ancient provinces of China mentioned in the Tribute of Yü. It corresponds to the present province of Ho nan.

During the Han and Tsin dynasties the province of Yü chou occupied only the eastern part of Ho nan, and parts of An hui and Shan tung belonged to it.—In the Sui dynasty the province of Yü chou comprised almost the whole of present Ho nan, south of the Yellow River, and a small part of An hui.

414. —原州 Yüan chou. Su Kung and Han Pao-sheng.

T'ang, Wu tai and Kin: *chou.* Now: Kan su, P'ing liang fu, Ku yüan chou,—and King chou, Chen yüan hien.

415.—莧句 Yüan kü. *Pie lu* and Su Sung.

T'ao Hung-king [*Med. plants*, 16] says that Yüan ki
is the same as the prefecture 清陰郡 Tsi yin kün in the
province of 兗州 Yen chou [see 347, 404].

According to the *Li tai*, etc. Yüan kü was a *hien* from
the Han down to the T'ang. Now : Shan tung, Ts'ao chou
fu, Ho tse hien.

416.—元山 Yüan shan. *Pie lu.*
Not ascertained.

417.—遠安縣 Yüan an hien. Su Kung.
Since Sui : *hien.* Now : Hu pei, King men chou, Yüan
an hien.

418.—越 Yüe. *Po wu chi,* Su Kung and Su Sung.

Name of an ancient kingdom of the Ch'un ts'iu period,
in present Che kiang. Its capital was 會稽 Hui ki [see 98].
Yüe is frequently mentioned together with Wu [see 389].

越州 Yüe chou, a prefecture noticed by Su Sung, was
etablished by the T'ang. Now : Che kiang, Shao hing fu.

The kingdom of Yüe in Che kiang is sometimes also
called 東越 Tung (Eastern) Yüe, to distinguish it from
南越 Nan (Southern) Yüe. The regions called Nan Yüe
(Southern China) were first conquered by Emperor Ts'in
Shi Huang-ti [B.C. 246-209]. See *Shi ki,* chap. 113.

419.—粵 Yüe. Su Sung.

Ancient term for Southern China, Kuang tung, Kuang
si and Fu kien. The *K.D.* says it is the same as 越 Yüe
[see 418].

The 粵地 Yüe ti (country of the Yüe) is mentioned
in the Chinese annals referring to the Chan kuo period
[B.C. 481-221]. In the History of the Early Han [chap.
45] there is an account of the 南粵王 Nan Yüe wang
(king of Southern Yüe) and the 閩粵王 Min Yüe wang
(king of Fu kien).

東 粤 Tung (Eastern) Yüe and 西 粤 Si (Western) Yüe are terms still used to designate the provinces of Kuang tung and Kuang si.

420.—越 山 Yüe shan. *Pie lu.*

Not ascertained.

421.—郓 州 Yün chou. *K'ai pao Pen ts'ao.*

T'ang and Wu tai: *chou.* Now: Shan tung, T'ai an fu, Tung p'ing chou.

422.—雲 中 Yün chung. *Pie lu.*

Ts'in and Han: *kün.* North-eastern part of the Ordos and North-east Shen si.

423.—雲 夢 Yün meng. Lü Shi *Ch'un ts'iu* [3rd cent. B.C].

The marshes of Yün meng are mentioned in the Tribute of Yü, in the province of King chou (Hu kuang). Legge's *Shu king*, p. 115.

424.—雍 州 Yung chou. *Pie lu,* Su Kung and Han Pao-sheng.

Name of one of the nine ancient provinces of China in the Tribute of Yü, situated, as the ancient account says, between the Ho (Yellow River) and the Hei shui [one of the northern affluents of the Yang tsz' kiang, see *supra*, 26. Its sources are in South Kan su]. The Yung chou of the Tribute of Yü corresponds to Northern Shen si and Eastern Kan su.

In the San kuo period Yung chou was a province of the kingdom of Wei and comprised Shen si north of the Wei River and Eastern Kan su. Yung chou was also a province in the Tsin and Sui dynasties. In the latter period it extended farther to the west into Kan su.

Yung chou in the Wu tai period was a prefecture corresponding to present Si an fu in Shen si.

425.—永 州 Yung chou. Authors of the Sung dynasty.

Since T'ang: *chou.* Now: Hu nan, Yung chou fu.

426.—永昌 Yung ch'ang. *Pie lu.*

Early Han and Tsin: *kūn.* Now: Yün nan, Yung ch'ang fu.

427.—荥陽 Yung yang. T'AO HUNG-KING.

Early Han: *hien*, Tsin: *kūn.* Now: Ho nan, K'ai feng fu, Yung yang hien.

428.—容州 Yung chou. Authors of the T'ang.

T'ang: *chou.* Now: Kuang si, Wu chou fu, Yung hien.

429.—邕州 Yung chou. Authors of the T'ang.

T'ang: *chou.* Now: Kuang si, Nan ning fu.

430.—融州 Yung chou. SU KUNG and SU SUNG.

T'ang: *chou.* Now: Kuang si, Liu chou fu, Yung hien.

ALPHABETICAL INDEX

OF

CHINESE NAMES OF PLANTS.

艾 *ai* 72

安石榴 *an shi liu* ... 280

安息香 *An si hiang* ... 313

菴藺 *an lü* 70

柴胡 *ch'ai hu* 29

餳糖香 *chan t'ang hiang* 314

占斯 *chan sz'* 356

章柳根 *chang liu ken* ... 131

昌陽 *ch'ang yang* ... 194

菖蒲 *ch'ang p'u* 194

常山 *ch'ang shan* ... 141

藷魁 *che k'ui* 181

車前 *ch'e ts'ien* 115

陳皮 *ch'en p'i* 281

沈香 *ch'en hiang* ... 307

芝 *chi* 266

枳 *chi* 334

陟釐 *chi li* 206

脂麻 *chi ma* 216

知母 *chi mu* 9

梔子 *chi tsz'* 335

赤莧 *ch'i hien* 256

赤雹 *ch'i pao* 173

赤參 *ch'i shen* 20

赤尤 *ch'i shu* 12

赤小豆 *ch'i siao tou* ... 231

赤箭 *ch'i tsien* 11

竹 *chu* 357

豬苓 *chu ling* 352

楮 *ch'u* 333

苧麻 *ch'u ma* 88

川穀 *Ch'uan ku* ... 228

川芎 *Ch'uan kung* ... 47

川斷 *Ch'uan tuan* ... 84

川烏頭 *Ch'uan wu t'ou* { 143 / 146

茺蔚 *ch'ung wei* 78

繁縷 *fan lü* 253

防風 *fang feng* 31

防己 *fang ki* 183

防葵 *fang k'ui* 133

榧實 *fei shi* 286

飛廉 *fei lien* 87

肥皂莢 *fei tsao kia* ... 325

風蘭 *feng lan* ... 62, 202

楓 *feng* 312, 352

浮萍 *fou p'ing* 198

伏神 *fu shen* ... ..., 350

茯苓 *fu ling* 350	槐 *huai* 322
腐婢 *fu pi* 232	淮木 *huai mu* 358
覆盆 *fu p'en* ... 165, 166	藋菌 *huan kün* 268
附子 *fu tsz'* ... 143, 146	玄及 *hüan ki* 164
海藻 *hai tsao* 200	玄參 *hüan shen* ... 18, 20
海菜 *hai ts'ai* 200	縣鉤子 *hüan kou tsz'* ... 165
孩兒茶 *hai rh ch'a* ... 185	黃環 *huang huan* ... 175
漢防已 *Han fang ki* ... 183	黃耆 *huang k'i* 2
旱芹 *han k'in* 250	黃芪 *huang k'i* 2
旱蓮草 *han lien ts'ao* ... 120	黃芩 *huang k'in* ... 27
寒莓 *han mei* 165	黃荊 *huang king* ... 348
黑丑 *hei ch'ou* 168	黃葵 *huang k'ui* ... 105
黑參 *hei shen* 18	黃連 *huang lien* ... 26
夏枯草 *hia ku ts'ao* ... 80	黃麻 *huang ma* 217
薤 *hiai* 242	黃蘗 (栢) *huang po* ... 315
香附子 *hiang fu tsz'* ... 59	黃草 *huang ts'ao* ... 128
香蒿 *hiang hao* 74	黃精 *huang tsing* ... 7
香薷 *hiang ju* 63	蕙草 *hui ts'ao* 60
香蒲 *hiang p'u* 196	薰陸香 *hün lu hiang* ... 312
莧 *hien* 256	薰草 *hün ts'ao* 60
杏 *hing* 271	紅姑娘 *hung ku niang*... 106
杏葉沙參 *hing ye sha shen* } 5	紅豆蔻 *hung tou k'ou* ... 57
	葒草 *hung ts'ao* ... 125
合歡 *ho huan* 324	火麻 *huo ma* 217
鶴虱 *ho shi* 93	醫草 *i ts'ao* 72
厚朴 *hou p'o* 316	益智子 *i chi tsz'* ... 285
胡麻 *hu ma* 216	益母 *i mu* 78
胡面莽 *hu mien mang* ... 100	薏苡仁 *i i jen* 228
葫 *hu* 244	飴餹 *i t'ang* 236
虎杖 *hu chang* 126	蘘荷 *jang ho* 96
虎掌 *hu chang* ... 143, 148	藝花 *jao hua* 157
琥珀 *hu p'o* 351	荏 *jen* 67
花椒 *hua tsiao* 288	人參 *jen shen* 3, 20
花王 *hua wang* 53	忍冬 *jen tung* 191

肉豆蔻 *jou tou k'ou* ... 58
肉蓯蓉 *jou ts'ung yung* 10
肉桂 *jou kui* 303
乳香 *ju hiang* 312
蕤核 *jui ho* 338
甘蔗 *kan che* 294
甘菊 *kan kü* ... 69
甘遂 *kan sui* 138
甘草 *kan ts'ao* 1
甘蕉 *kan tsiao* 95
乾薑 *kan kiang* 249
高良薑 *kao liang kiang* 57
藁本 *kao pen* 50
秔 or 秔 *keng* 222
雞腸 *ki ch'ang* ... 253, 254
雞舌香 *ki she hiang* ... 308
雞屎藤 *ki shi t'eng* ... 118
雞頭 *ki t'ou* 297
及己 *ki ki* 42
芰實 *ki shi* 296
蘄艾 *k'i ai* 72
假蘇 *kia su* 65
江蘺 *kiang li* 48
茳芒決明 *kiang mang küe ming* } 110
薑 *kiang* 248, 249
薑芥 *kiang kie* 65
羌活 *K'iang huo* ... 32
茭筍 *kiao sun* 197
芥 *kie* 246
桔梗 *kie keng* 6
芡實 *k'ien shi* 297
牽牛子 *k'ien niu tsz'* ... 168
金釵石斛 *kin ch'ai shi hu* 202

金鈴子 *kin ling tsz'* ... 321
金錢花 *kin ts'ien hua* ... 81
金銀花 *kin yin hua* ... 191
錦葵 *kin k'ui* 105
芹 *k'in* 250
荊芥 *king kie* 65
荊條 *king t'iao* 349
景天 *king t'ien* 205
韭 *kiu* 240
灸草 *kiu ts'ao* 72
樛子 *kiu tsz'* 290
葛 *ko* 174
鈎藤 *kou t'eng* 185
鈎吻 *kou wen* 162
狗脊 *kou tsi* 13
枸骨 *kou ku* 342
枸杞 *kou k'i* 345
菰 *ku* 197
鼓子花 *ku tsz' hua* ... 169
苦瓠 *k'u hu* 264
苦苣 *k'u kü* 257
苦楝子 *k'u lien tsz'* ... 321
苦蕒 *k'u mai* 257
苦參 *k'u shen* 34
苦蘵 *k'u shi* ... 34, 106
苦菜 *k'u ts'ai* 257
苦芺 *k'u yao* 85
橘 *kü* 281
菊 *kü* 69
欅 *kü* 327
巨勝 *kü sheng* 216
瞿麥 *k'ü mai* 112
蒟蒻 *k'ü jo* 148
屈草 *k'ü ts'ao* 214

瓜蒂 *kua ti* 292

栝樓 *kua lou* 172

貫粱 *kuan chung* ... 14

鸛虱 *kuan shi* 257

欵冬花 *k'uan tung hua* 109

卷栢 *kŭan po* 211

卷耳 *kŭan rh* 92

拳參 *k'ŭan shen* 21

决明 *kŭe ming* 110

鬼臼 *kui kiu* 152

鬼目 *kui mu* 187

鬼箭 *kui tsien* 343

鬼督郵 *kui tu yu* 11, 43

鬼油麻 *kui yu ma* ... 86

桂 *kui* 303

桂心 *kui sin* 303

葵 *k'ui* 105

昆布 *kun pu* 201

箘桂 *kŭn kui* ... 303, 304

穬麥 *kung mai* 220

芎藭 *kung k'iung* ... 47

藍 *lan* 123

蘭 *lan* 62

蘭草 *lan ts'ao* 61

狼跋子 *lang po tsz'* ... 175

狼毒 *lang tu* 132

狼牙 *lang ya* 134

蕳菪 *lang tang* 139

蠡豆 *lao tou* 260

雷丸 *lei huan* 353

冷飯團 *leng fan t'uan* ... 179

李 *li* 270

梨 *li* 276

栗 *li* 274

蒺藘 *li lu* 142

體腸 *li ch'ang* 120

蠡實 *li shi* 90

粱 *liang* 225

良薑 *liang kiang* ... 55

蓼 *liao* 124

蓼藍 *liao lan* 123

列當 *lie tang* 10

楝 *lien* 321

連翹 *lien k'iao* 120

蓮藕 *lien ou* 295

淩 *ling* 296

凌霄花 *ling siao hua* ... 170

陵苕 *ling t'iao* 170

零陵香 *ling ling hiang* 60

柳 *liu* 328

劉寄奴 *liu ki niu* ... 86

落帚子 *lo chou tsz'* ... 111

落葵 *lo k'ui* 258

絡石 *lo shi* 189

漏盧 *lou lu* 86

蘆 *lu* 94

鹿藿 *lu huo* 260

鹿豆 *lu tou* 260

陸英 *lu ying* 121

䕡茹 *lü ju* 135

欒華 *luan hua* 326

龍常草 *lung chang ts'ao* 99

龍腦薄荷 *lung nao po ho* 68

龍鬚 *lung sü* 98

龍膽 *lung tan* 39

龍牙草 *lung ya ts'ao* ... 116

龍眼 *lung yen* 285

麻黃 *ma huang* 97

馬蘭 *ma lan* 90
馬藺 *ma lin* 90
馬蓼 *ma liao* 124
馬鞭草 *ma pien ts'ao* 116
馬勃 *ma pu* 213
馬先蒿 *ma sien hao* ... 76
馬蹄決明 *ma t'i küe ming* } 110
馬蹄香 *ma t'i hiang* ... 41
馬兜鈴 *ma tou ling* ... 54
馬尾藻 *ma wei tsao* ... 200
薓斛 *mai hu* 202
薓門冬 *mai men tung* ... 104
蔓荊 *man king* 349
蔓椒 *man tsiao* 290
蔓菁 *man tsing* 247
莽草 *mang ts'ao* ... 158
茅針 *mao chen* 37
茅根 *mao ken* 37
貓兒眼睛草 *mao rh yen tsing ts'ao* } 137
莓 *mei* 165, 166
梅 *mei* 272
美人蕉 *mei jen tsiao* ... 56
美草 *mei ts'ao* ... 56
蘪蕪 *mi wu* 48
牡蒿 *mou hao* 77
牡荊 *mou king* 348
牡桂 *mou kui* 303
牡蒙 *mou meng* ... 21, 22
牡丹 *mou tan* 53
木防己 *mu fang ki* ... 183
木香 *mu hiang* 54
木瓜 *mu kua* 277

木藍 *mu lan* 123
木蘭 *mu lan* 305
木蓮花 *mu lien hua* ... 305
木耳 *mu rh* 267
木通 *mu t'ung* 184
苜蓿 *mu su* 255
奈 *nai* 278
南星 *nan sing* 148
楠 *nan* 310
鬧羊花 *nao yang hua* ... 155
牛李 *niu li* 341
牛蒡 *niu p'ang* ... 91
牛扁 *niu pien* 161
牛舌菜 *niu she ts'ai* ... 193
牛膝 *niu si* 101
女貞 *nü cheng* 342
女青 *nü ts'ing* 118
女菀 *nü yüan* 103
鵝腸 *ó ch'ang* 253
鵝抱蛋 *ó pao tun* 180
菝葜 *pa k'ia* 179
八寶兒 *pa pao rh* ... 205
巴戟天 *pa ki t'ien* ... 15
巴豆 *pa tou* 331
白菖 *pai ch'ang*... 194, 195
白芷 *pai chi* 51
白丑 *pai ch'ou* 168
白附子 *pai fu tsz'* ... 147
白蒿 *pai hao* 75
白棘 *pai ki* 337
白及 *pai ki* 25
白蘝 *pai lien* 180
白茅 *pai mao* 37
白蔘 *pai shen* 4

白朮 *pai shu*	12	
白鮮 *pai sien*	35	
白蘇 *pai su*	67	
白菜 *pai ts'ai*	245	
白豆蔲 *pai tou k'ou* ...	58	
白頭翁 *pai t'ou weng* ...	24	
白蒺蔾 *pai tsi li* ...	129	
白前 *pai ts'ien*	45	
白兔藿 *pai t'u huo* ...	186	
白微 *pai wei*	44	
白藥 *pai yao*	6	
白英 *pai ying*	187	
敗醬 *pai tsiang* ...	108	
排風子 *p'ai feng tsz'* ...	187	
半夏 *pan hia*	150	
斑杖 *pan chang* ...	126	
蔍 *pao*	165, 166	
貝母 *pei mu*	36	
萆薢 *pei hiai* ...	178	
蓬藟 *p'eng lei* ...	165	
彼子 *pi tsz'* ...	286	
篳管菜 *pi kuan ts'ai* ...	7	
蘩蔞 *pi ts'i*	298	
枇杷 *p'i p'a*	282	
別羈 *pie ki*	215	
萹蓄 *pien ch'u* ...	127	
藊豆 *pien tou*	233	
檳榔 *pin lang* ...	287	
蘋 *p'in*	198	
栢 *po*	300	
櫱木 *po mu* ...	315	
百合 *po ho* ...	263	
百部 *po pu* ...	177	
蒲黃 *p'u huang*	196	

葡萄 *p'u t'ao*	293	
三稜草 *san leng ts'ao* ...	59	
桑 *sang*	332	
桑上寄生 *sang shang ki sheng* }	354	
掃帚菜 *sao chou ts'ai* ...	111	
沙參 *sha shen* 4, 20		
山茱萸 *shan chu yü* ...	339	
山薊 *shan ki*	12	
山薑 *shan kiang* 55, 56		
山石榴 *shan shi liu* ...	155	
山藥 *shan yao*	262	
山櫻桃 *shan ying t'ao* ...	284	
杉 *shan (sha)* ...	302	
商陸 *shang lu*	131	
芍藥 *shao yo*	52	
蛇牀 *she ch'uang* ...	49	
蛇含 *she han* ...	117	
蛇銜 *she hien* ...	117	
蛇莓 *she mei* ...	167	
射干 *she kan* ...	153	
愼火 *shen huo* ...	205	
生薑 *sheng kiung* ...	248	
升麻 *sheng ma* ...	33	
薯 *shi*	71	
柿 *shi*	279	
蒒草 *shi ts'ao*	179	
豕首 *shi shou* ... 90, 98		
石長生 *shi ch'ang sheng*	204	
石菖蒲 *shi ch'ang p'u* ...	194	
石竹 *shi chu* ...	112	
石髮 *shi fa* ... 206, 210		
石斛 *shi hu*	202	
石龍 *shi lung*	125	

石龍蒭 *shi lung ch'u* ... 98	辛夷 *sin i* 306
石龍芮 *shi lung jui* ... 160	莎草 *so ts'ao* 59
石南 *shi nan* 347	蒴藋 *so t'iao* 122
石松 *shi sung* 212	銷陽 *so yang* 10
石韋 *shi wei* 203	蘇 *su* 67
溲疏 *shou shu* 346	蘇合香 *su ho hiang* ... 313
黍 *shu* 224	粟 *su* 226
秫 *shu* 227	續骨木 *su ku mu* ... 122
尤 *shu* 12	續斷 *su tuan* 84
鼠李 *shu li* 341	徐長卿 *sü ch'ang k'ing* 43
鼠尾草 *shu wei ts'ao* ... 119	蒜 *suan* 243
薯蕷 *shu yü* 262	酸棗 *suan tsao* 336
蜀葵 *Shu k'ui* 105	酸醤 *suan tsiang* ... 106
蜀漆 *Shu ts'i* 141	旋葍 *süan fu* 169
蜀椒 *Shu tsiao* 289	旋覆花 *süan fu hua* ... 81
蜀羊泉 *Shu yang ts'üan* 107	旋花 *süan hua* 169
水菖蒲 *shui ch'ang p'u* {194 / 195}	松 *sung* 301
	蓯 *sung* 245
水衣 *shui i* 206	菘藍 *sung lan* 123
水蘄 *shui k'in* 250	松蘿 *sung lo* 355
水蓼 *shui liao* ... 124, 125	四葉菜 *sz ye ts'ai* ... 198
水綿 *shui mien* 206	大黃 *ta huang* 130
水萍 *shui p'ing* 198	大薊 *ta ki* 83
水蘇 *shui su* 68	大戟 *ta ki* 136
水苔 *shui t'ai* 206	大力子 *ta li tsz'* .. 91
蓴 *shun* 199	大蓼 *ta liao* 125
蓁藄 *si ming* 252	大麻 *ta ma* 217
桌耳 *si rh* 92	大麥 *ta mai* 219
細辛 *si sin* 40	大萍 *ta p'ing* 198
小薊 *siao ki* 83	大豆 *ta tou* 229
小麥 *siao mai* 218	大豆豉 *ta tou shi* ... 234
小蘗 *siao po* 315	大薺 *ta tsi* 252
小草 *siao ts'ao* 16	大青 *ta ts'ing* 89
小青 *siao ts'ing* 89	丹桂 *tan kui* 303

丹 皮 *tan p'i* 53
丹 沙 草 *tan sha ts'ao* ... 204
丹 參 *tan shen* 20
檀 香 *t'an hiang* 309
檀 桓 *tan huan* 315
黨 參 *Tang shen* 3, 4
當 歸 *tang kui* 46
潟 潟 青 *t'ang t'ang ts'ing* 211
稻 *tao* 221
桃 *t'ao* 273
燈 籠 草 *teng lung ts'ao* 106
地 膚 *ti fu* 111
地 血 *ti hüe* ... 23, 182
地 黃 *ti huang* 100
地 筋 *ti kin* ... 38, 100
地 骨 皮 *ti ku p'i* ... 345
地 栗 *ti li* 298
地 耳 *ti rh* 269
地 髓 *ti sui* 100
地 錢 草 *ti ts'ien ts'ao* ... 66
地 榆 *ti yü* 19
釣 樟 *tiao chang* 311
吊 蘭 *tiao lan* ... 62, 202
鐵 色 草 *t'ie se ts'ao* ... 80
鐵 掃 帚 *t'ie sao chou* ... 90
瀳 *tien* 123
甜 瓜 *t'ien kua* 292
天 竺 桂 *T'ien chu kui* ... 304
天 雄 *t'ien hiung* 143, 144, 146
天 花 粉 *t'ien hua fen* ... 172
天 蓼 *t'ien liao* ... 124, 125
天 麻 *t'ien ma* 11
天 門 冬 *t'ien men tung* ... 176
天 名 精 *t'ien ming tsing* 93

天 南 星 *t'ien nan sing* ... 148
天 泡 草 *t'ien p'ao ts'ao* ... 106
天 豆 *t'ien tou* 140
丁 香 *ting hiang* 308
葶 藶 *t'ing li* 114
橐 吾 *t'o wu* 109
豆 姑 娘 *tou ku niang* ... 106
豆 蔻 *tou k'ou* 58
頭 髮 菜 *t'ou fa ts'ai* ... 200
蒼 耳 *ts'ang rh* 92
蒼 朮 *ts'ang shu* 12
璪 *tsao* 275
蚤 休 *tsao hiu* 151
皂 莢 *tsao kia* 325
草 蒿 *ts'ao hao* 74
草 決 明 *ts'ao küe ming* ... 82
草 果 *ts'ao kuo* 58
草 豆 蔻 *ts'ao tou k'ou* ... 58
草 蓯 蓉 *ts'ao ts'ung yung* 10
草 子 兒 *ts'ao tsz' rh* ... 228
草 烏 頭 *ts'ao wu t'ou* 143, 146
澤 蘭 草 *tse lan ts'ao* ... 62
澤 瀉 *tse sie* 192
澤 漆 *tse ts'i* 137
側 子 *tse tsz'* ... 143-146
稷 *tsi* 223
薺 *tsi* 251
薺 苨 *tsi ni* 5
薺 薴 *tsi ning* 68
蒺 藜 *tsi li* 129
積 雪 草 *tsi süe ts'ao* ... 66
蕺 *ts'i* 259
漆 *ts'i* 318
漆 姑 *ts'i ku* 107

漆頭 *ts'i t'ou* ...	...	135
醬 *tsiang* ...	...	237
墻蘼 *ts'iang mi*	...	171
薔薇 *ts'iang wei*	...	171
接骨 *tsie ku* ...		84, 122
藉姑 *tsie ku* ...		299
茜草 *ts'ien ts'ao* ...	...	182
前胡 *ts'ien hu* ...	...	30
千年艾 *ts'ien nien ai*	...	72
千年柏 *ts'ien nien po*	...	212
千歲虆 *ts'ien sui lei*	...	190
錢蒲 *ts'ien p'u* ...	...	194
藎草 *tsin ts'ao* ...	...	128
秦芃 *Ts'in kiao* ...	...	28
秦皮 *Ts'in p'i* ...	...	323
秦椒 *Ts'in tsiao*	...	288
并中苦 *tsing chung t'ai*		207
青蒿 *ts'ing hao* ...	...	74
青木香 *ts'ing mu hiang*		54
青皮 *ts'ing p'i* ...	...	281
青葙 *ts'ing siang*	...	82
爵牀 *tsio chuang*	...	64
雀髀斛 *tsio pi hu*	...	202
雀瓢 *tsio p'iao* ...	...	118
鵲豆 *ts'io tou* ...	...	233
楸 *ts'iu* ...	...	319
秋葵 *ts'iu k'ui* ...	...	105
昨葉荷草 *tso ye ho ts'ao*		209
醋 *ts'u* ...	...	238
蔥 *ts'ung* ...	...	241
蓯蓉 *ts'ung yung*	...	10
梓 *tsz'* ...	...	319
紫背龍牙 *tsz' pei lung ya*		117
紫葬 *tsz' p'ing* ...	...	198
紫參 *tsz' shen* ...	...	21
紫蘇 *tsz' su* ...	...	67
紫草 *tsz' ts'ao* ...	...	23
紫葳 *tsz' wei* ...	...	170
紫菀 *tsz' yüan* ...	...	102
刺虆 *tsz' po* ...	...	315
刺楸 *tsz' ts'iu* ...	...	319
茈胡 *ts'z' hu* ...	...	29
慈姑 *ts'z' ku* ...	...	299
毒魚 *tu yü* ...	...	156
獨帚 *tu chou* ...	...	111
獨活 *tu huo* ...	...	32
獨脚蓮 *tu kio lien*	...	152
杜仲 *tu chung* ...	...	317
杜衡 *tu heng* ...		41, 55
杜若 *tu jo* ...	...	55
杜鵑花 *tu küun hua*	...	155
杜蘭 *tu lan* ...	...	202
土茯苓 *t'u fu ling*	...	179
土瓜 *t'u kua* ...	...	173
土當歸 *t'u tang kui*	...	46
土青木香 *t'u ts'ing mu hiang* }		54
菟葵 *t'u k'ui* ...	...	105
菟絲子 *t'u sz' tsz'*	...	163
冬瓜 *tung kua* ...	...	265
冬葵子 *tung k'ui tsz'*	...	105
冬青 *tung ts'ing*	...	342
桐 *t'ung* ...	...	320
通草 *t'ung ts'ao*	...	184
瓦松 *wa sung* ...	...	209
萬年松 *wan nien sung*	...	212
望江南 *wan kiang nan*		110
王瓜 *wang kua* ...	...	173

王不留行 *wang pu liu hing*		{	106 113
王孫 *wang sun*	...	...	22
葦 *wei*	... ...	...	94
衛矛 *wei mou*	...	...	343
萎蕤 *wei jui*	...	...	8
薇銜 *wei hien*	...	...	79
惡實 *wu shi*	...	...	91
屋遊 *wu yu*	...	...	209
吳茱黃 *Wu chu yü*	...	...	291
蕪荑 *wu i*	...	...	330
蕪姑 *wu ku*	...	...	330
蕪菁 *wu tsing*	...	...	247
五加 *wu kia*	...	...	344
五味子 *wu wei tsz'*	...	...	164
烏喙 *wu hui*	...	143,	146
烏韭 *wu kiu*	...	...	210
烏頭 *wu t'ou*	...	143,	146
烏草 *wu ts'ao*	...	...	119
烏芋 *wu yü*	...	...	298
牙皂 *ya tsao*	...	...	325
楊櫨 *yang lu*	...	...	346
羊躑躅 *yang chi chu*	...	...	155
羊負來 *yang fu lai*	...	...	92
羊韭 *yang kiu*	...	...	104
羊桃 *yang t'ao*	...	...	188
羊泉 *yang ts'üan*	...	...	107
羊蹄 *yang t'i*	...	...	193
野苣 *ye hien*	...	...	256
野槐 *ye huai*	...	...	34
野橘 *ye kü*	...	...	69
糵米 *ye mi*	...	...	235
鴈來紅 *yen lai hung*	...	...	256
胭脂豆 *yen chi tou*	...	...	258
茵蔯蒿 *yin ch'en hao*	...	...	73
茵芋 *yin yü*	...	...	159
陰行草 *yin hing ts'ao*	...	...	86
淫羊藿 *yin yang huo*	...	...	17
櫻桃 *ying t'ao*	...	...	283
蘡實 *ying shi*	...	...	171
映山紅 *ying shan hung*			155
柚 *yu*	... ...	...	281
由跋 *yu po*	...	...	149
禹餘糧 *Yü yü liang*		104,	179
芋 *yü,* 芋頭 *yü t'ou*		...	261
榆 *yü*	...	...	329
郁李 *yü li*	...	...	340
玉竹 *yü chu*	...	...	8
玉蘭 *yü lun*	...	...	306
玉柏 *yü po*	...	...	212
芫花 *yüan hua*	...	...	156
元參 *yüan shen*	...	...	18
遠志 *yüan chi*	...	...	16
蔦尾 *yüan wei*	...	...	154
垣衣 *yüan i*	...	...	208
月桂 *yüe kui*	...	...	304
雲實 *yün shi*	...	...	140

ALPHABETICAL INDEX

OF

GENUS NAMES OF PLANTS.

Abrus	...	...	... 231	Amarantus	 256
Abutilon	...	...	... 105	Amber	 351
Acacia	...	...	... 324	Amomum	... 57, 58, 96
Acanthopanax	...	319, 344	Amyris	 313
Aceranthus	...	... 17	Anchusa	 23
Achillea	...	...	... 71	Anemarrhena	 9
Achyranthes	...	... 101	Anemone	 24
Aconitum	143-146, 161	Anemonopsis	 33
Acorus	...	...	194, 195	Anethum	 250
Adenophora	...	.. 4, 5	Angelica	... 30, 32, 47, 48, 51
Adiantum	...	... 204	Apium	... ... 46, 250
Æginetia	...	... 10	Aplotaxis	 54
Ægle	...	...	... 334	Apricot	 271
Aërides	...	... 202	Aralia	... 32, 46, 184, 344
Agaricus	...	... 266	Arctium	 91
Agrimonia	...	... 116	Ardisia	 89
Ajuga	...	...	... 80	Areca	 287
Akebia	...	... 184	Arisæma	 148-150
Aletris	...	...	... 9	Aristolochia	 54
Alga	...	...	... 200	Artemisia	... 70-77
Alisma	...	... 192	Arum	... ... 148, 150
Allium	...	... 240-244	Arundo	 94
Alpinia	...	... 56, 57	Asarum	... ... 40, 41
Aloëxylon	...	... 307	Asclepias	 186

Asparagus 176	Carpesium 93
Aster 90, 102, 103	Cassia 110
Astilbe 33	Castanea 274
Astragalus 2	Catalpa 319
Atractylis 12	Catechu 185
Aucklandia 54	Celastrus 141
Azalea 155	Celery 250
	Celosia 80, 82
Balanophora 10	Ceraja 202
Bambusa 357	Ceramium 206
Basella 258	Cerastium 92
Benincasa 265	Cerasus 283, 284
Benzoin 313	Chenopodium 111
Berberis 27, 315, 338	Chloranthus 42
Bignonia 170	Chrysanthemum . 69
Bletia 25	Cibotium... 13
Boehmeria 88	Cichorium 257
Boymia 291	Cicuta 50
Brasenia 199	Cimicifuga 33
Brassica 245, 247	Cinnamomum ... 303, 304
Broussonetia 333	Cirsium 84
Bryonia 172	Citrus 281, 334
Bupleurum 29	Clematis 184
	Cloves 308
Cæsalpinia 140	Cnicus 83, 85
Calamintha 68	Cnidium 49
Calystegia 169	Cocculus 183
Campsis 170	Coix 228
Canna 56	Colocasia... 261
Cannabis 217	Codonopsis 3
Capsella 251	Conophallus 148
Cardamom ... 58, 285	Convallaria 8
Carduus 83, 84, 87	Convolvulus 169
Carex 59, 179	Coptis 26

78

Coreopsis	... ...	... 31
Cornus ...	...	... 339
Costus ...	...	... 54
Cotyledon	...	... 209
Croomia ...	...	... 162
Croton ...	...	... 331
Cunninghamia ...	...	... 302
Cuscuta ...	...	... 163
Cutch ...	...	... 185
Cyathula	...	... 101
Cydonia ...	...	... 277
Cynomorium	...	... 10
Cyperus ...	...	... 59
Daphne ...	...	... 156
Dendrobium	...	... 202
Deutzia ...	...	... 346
Dianthus	...	... 112
Dianthera	...	... 89
Diervilla ...	...	... 346
Dichroa ...	...	... 141
Dicliptera	...	... 89
Dictamnus	...	... 35
Dioscorea	...	178, 262
Diospyros	...	... 279
Diphylleia	...	... 152
Dipsacus...	...	83, 84
Dolichos ...	...	... 233
Dorstenia	...	... 51
Draba ...	...	... 114
Dumasia ...	...	... 260
Eclipta ...	...	... 120
Elettaria ...	...	... 58
Eleutherococcus ...	...	... 344
Elsholtzia	...	... 63
Ephedra ...	...	... 97
Epidendron	...	... 202
Epimedium	...	... 17
Equisetum	...	... 97
Eriobotrya	...	109, 282
Eritrichium	...	... 254
Eupatorium	...	24, 62
Euphorbia	135–138, 205	
Euryale ...	...	... 297
Euxolus ...	...	... 256
Evodia ...	...	291, 315
Evonymus	...	317, 343
Exidia ...	...	... 267
Farfugium	...	... 109
Fatsia ...	...	... 184
Forsythia	...	... 120
Fragaria ...	...	... 167
Frankincense	...	... 312
Fraxinus...	...	... 323
Fritillaria	...	... 36
Galanga ...	...	... 57
Galium ...	...	... 7
Gambir ...	...	... 185
Gardenia...	...	... 335
Gastrodia	...	... 11
Gelsemium	...	... 162
Gentiana	...	39, 118
Geranium	...	... 161
Geum ...	...	... 117
Ginger ...	...	248, 249

Ginseng	...	3
Glechoma	...	66
Gleditschia	...	325
Glycyrrhiza	...	1
Gymnocladus	...	325
Hedysarum	...	2, 19
Herpestis	...	15
Heterotropa	...	41
Hibiscus	...	105
Hirneola	...	267
Hordeum	...	219, 220
Houttuynia	...	259
Hydrangea	...	141
Hydrocharis	...	198
Hydrocotyle	...	66
Hydropyrum	...	197
Hyoscyamus	...	139, 155
Hypericum	...	120
Ilex	...	342
Illicium	...	158
Imperata	...	37
Incarvillea	...	76
Indigofera	...	123
Inula	...	54, 81
Ipomœa	...	168
Iris	...	90, 154
Isatis	...	123
Ixeris	...	216
Jatropha	...	147
Justicia	...	26, 89

Kadsura	...	164
Kochia	...	111
Kœlreuteria	...	326
Lablab	...	233
Lactuca	...	257
Lagenaria	...	264
Laminaria	...	200, 201
Lamium	...	84
Lappa	...	91
Lemna	...	198
Leontice	...	26
Leonurus	...	78
Lespedeza	...	90
Levisticum	...	46
Libanotis	...	31
Ligularia	...	109
Ligusticum	...	46
Ligustrum	...	342
Lilium	...	263
Limnanthemum	...	199
Limodorum	...	202
Lindera	...	311
Liquidambar	...	312, 313
Lithospermum	...	23
Litsæa	...	304
Lomaria	...	14
Lonicera	...	191
Lophanthus	...	80
Loranthus	...	354, 355
Luisia	...	202
Lycium	...	345
Lycoperdon	...	213

Lycopodium ...	211, 212	
Lysimachia ...	... 141	
Macroclinidium ...	... 43	
Magnolia...	305, 306, 316	
Malouetia ...	... 189	
Malt	... 235	
Malva	... 105	
Mandragora ...	... 132	
Manihot	... 147	
Marsilea	... 198	
Medicago ...	... 255	
Mélandrium ...	... 113	
Melanthium ...	... 176	
Melia	... 321	
Melon	... 292	
Menispermum ...	... 183	
Menyanthes ...	... 199	
Morus	... 332	
Mosla	... 64	
Moss	... 288	
Moxa	... 72	
Mulgedium ...	... 216	
Musa	... 95	
Mushroom ...	268, 269	
Mylitta	... 353	
Myristica ...	... 58	
Nauclea	... 185	
Nelumbium ...	... 295	
Nepeta	65, 66	
Nephelium ...	... 285	
Nephrodium ...	... 14	

Nerium	... 189	
Nigella	... 216	
Niphobolus ...	... 203	
Nitraria	... 345	
Nothosmyrnium ...	... 50	
Nutmeg	... 58	
Ocimum	... 60	
Œcœoclades ...	... 202	
Œnanthe ...	... 250	
Olea	... 342	
Olibanum ...	... 312	
Onoclea	13, 14	
Ophioglossum ...	... 203	
Ophiopogon ...	... 104	
Origanum ...	31, 65	
Orixa	... 141	
Orobanchacea ...	... 10	
Orontium ...	... 195	
Oryza	221, 222	
Pachyma ...	... 350	
Pachyrhizus ...	... 174	
Pæderia	... 118	
Pæonia	... 52	
Paliurus	... 337	
Panax	3, 344	
Panicum	223, 224	
Pardanthus ...	... 153	
Paris	22, 151	
Passerina ...	138, 157	
Patrinia	... 108	
Paulownia ...	... 320	

Peach 273	Premna 232
Pear 276	Prosopis 325
Perilla 67	Prunella 80
Peristrophe 89	Prunus ... { 270, 272, 283, 284, 338, 340
Persea 310	
Petasites 109	Pterocarpus 315
Peucedanum ... 31, 32, 133	Pterocarya 327
Phalaris 128	Pueraria 174
Pharbitis 168	Pulsatilla 24
Phaseolus 231	Punica 280
Phelipæa 10	Pupalia 101
Phellodendron 315	Putchuk 54
Philadelphus 346	Pycnostelma 43
Phragmites 94	Pyrethrum 69
Physalis 106	Pyrus 276
Phytolacca 131	
Pinellia 150	
Pinus 301	Ranunculus 160
Pistia 198	Rape 247
Plantago 115	Rehmannia 100
Platycodon 6	Rhamnus ... 336, 341
Plectranthus 103	Rheum 130
Pleione 36	Rhodea 142
Podophyllum 152	Rhododendron ... 155, 347
Polemonium 113	Rhus 162, 318
Pollia 55	Rhynchosia 260
Polygala ... 15, 16	Rhynchospermum ... 189
Polygonatum 7, 8	Robinia 2, 7, 34
Polygonum { 21, 89, 101, 123-127	Rosa 54, 171
	Rose-maloes 313
Polypara 259	Roxburghia 177
Polypodium ... 13, 203	Rubia 182
Potentilla ... 117, 134	Rubus 165, 166
Poterium 19	Rumex 193

Saccharum	... ...	294
Sagina ...	... ...	107
Sagittaria	... ...	299
Salix ...	... ...	328
Salvia ...	... 20,	65, 119
Salvinia ...	... ...	198
Sambucus	... 121,	122
Sanguisorba	... ...	19
Santalum	... ...	309
Saponaria	... ...	113
Sargassum	... ...	200
Schizandra	... ...	164
Scirpus ...	...	59, 298
Scopolia ...	... ...	139
Scrophularia	... ...	18
Scutellaria	... ..	27
Sedum ...	... ...	205
Selaginella	... ...	211
Senecio ...	... 79,	86, 109
Septas ...	... ...	15
Sesamum	... ...	216
Setaria ...	...	225–227
Silene ...	... ...	113
Siler ...	... ...	31
Sinapis ...	... ...	246
Siphonostegia	...	70, 86
Sisymbrium	... ...	114
Skimmia ...	... ...	159
Smilax ...	... ...	179
Soja ...	...	229, 230
Solanum ...	...	107, 187
Solena ...	... ...	172
Solidago ...	... ...	86
Sonchus ...	... ...	257

Sophora ...	...	2, 34, 322
Soy ...	...	... 234
Stellaria ...	...	- ... 253
Stellera ...	...	... 157
Stemona ...	...	... 177
Stenocoelium	...	... 31
Stephania	...	... 183
Tanacetum	...	... 72
Tecoma ...	...	... 170
Terra japonica	...	... 185
Thalictrum	...	33, 36
Thladiantha	...	... 173
Thlaspi ...	...	... 252
Thuja ...	...	... 300
Torreya ...	...	... 286
Trapa ...	...	... 296
Tribulus ...	...	... 129
Trichomanes	...	... 210
Trichosanthes	...	172, 173
Tricertandra	...	... 42
Trillidium	...	... 151
Trillium ...	...	... 151
Triticum ...	...	... 218
Tussilago	...	... 109
Typha ...	...	... 196
Typhonium	...	... 152
Ulmus ...		327, 329, 330
Umbilicus	...	... 209
Uncaria ...	...	... 185
Urtica ...	...	... 11
Uvularia ...	...	... 36

Veratrum 142	Wickstrœmia 138, 156, 182
Verbena 116	Woodwardia 14
Vicia 129	
Villarsia 199	
Vincetoxicum ... 44, 45, 118	Xanthium 92
Vinegar 238	
Viscum 354, 355	
Vitex 348, 349	Zala 198
Vitis ... 180, 190, 293	Zanthoxylon 288-291, 315
	Zingiber 96, 248, 249
Wahlenbergia 4, 5	Zizyphus... 275, 336, 337